Soil Microbiology, Ecology, and Biochemistry

Soil Microbiology, Ecology, and Biochemistry

Fourth edition

Edited by

Eldor A. Paul
Natural Resource Ecology Laboratory and
Department of Soil and Crop Sciences
Colorado State University
Fort Collins, CO 80523
USA

AMSTERDAM • BOSTON • HEIDELBERG • LONDON
NEW YORK • OXFORD • PARIS • SAN DIEGO
SAN FRANCISCO • SINGAPORE • SYDNEY • TOKYO
Academic Press is an imprint of Elsevier

Academic Press is an imprint of Elsevier
32 Jamestown Road, London NW1 7BY, UK
525 B Street, Suite 1800, San Diego, CA 92101-4495, USA
225 Wyman Street, Waltham, MA 02451, USA
The Boulevard, Langford Lane, Kidlington, Oxford OX5 1GB, UK

Fourth edition **2015**

Library of Congress Cataloging-in-Publication Data
Soil microbiology, ecology, and biochemistry / editor, Eldor A. Paul. – Fourth edition.
      pages cm
   Includes bibliographical references and index.
   ISBN 978-0-12-415955-6
1. Soil microbiology. 2. Soil biochemistry. I. Paul, Eldor Alvin, editor.
   QR111.P335 2015
   579'.1757–dc23

                                                                        2014025057

British Library Cataloguing in Publication Data
A catalogue record for this book is available from the British Library

For information on all Academic Press publications
visit our web site at store.elsevier.com

ISBN: 978-0-12-415955-6

Contents

7. **Physiological and Biochemical Methods for Studying Soil Biota and Their Functions**   **187**

Ellen Kandeler

11. Plant-Soil Biota Interactions   311

R. Balestrini, E. Lumini, R. Borriello and V. Bianciotto

12. Carbon Cycling: The Dynamics and Formation of Organic Matter 339

William Horwath

13. Methods for Studying Soil Organic Matter: Nature, Dynamics, Spatial Accessibility, and Interactions with Minerals   383

Claire Chenu, Cornelia Rumpel and Johannes Lehmann

Contributors

Numbers in parentheses indicate the pages on which the authors' contributions begin.

E. Carol Adair (501), Rubenstein School of Environment and Natural Resources, University of Vermont, Burlington, Vermont, USA

Vanessa L. Bailey (535), Pacific Northwest Laboratory, Richland, WA, USA

R. Balestrini (311), Istituto per la Protezione Sostenibile delle Piante, UOS Torino, Viale Mattioli, 10125 Torino, Italy

V. Bianciotto (311), Istituto per la Protezione Sostenibile delle Piante, UOS Torino, Viale Mattioli, 10125 Torino, Italy

Christopher B. Blackwood (273), Department of Biological Sciences, Kent State University, Kent, OH, USA

R. Borriello (311), Istituto per la Protezione Sostenibile delle Piante, UOS Torino, Viale Mattioli, 10125 Torino, Italy

Peter J. Bottomley (443), Department of Crop and Soil Science, Oregon State University, Corvallis, OR, USA

Claire Chenu (379), Agro Paris Tech, UMR Bioemco, Thiverval Grignon, France

David C. Coleman (111), Odum School of Ecology, University of Georgia, Athens, GA, USA

Harold P. Collins (535), USDA-Agriculture Research Service, Temple, TX, USA

Alex R. Crump (535), Department of Crop and Soil Sciences, Washington State University, Pullman, WA, USA

Stephen J. Del Grosso (501), Natural Resource Ecology Laboratory, Colorado State University, Fort Collins, CO, USA; USDA Agricultural Research Service and Natural Resource Ecology Laboratory, Colorado State University, Fort Collins, CO, USA

Serita D. Frey (223), Department of Natural Resources & the Environment, University of New Hampshire, Durham, NH, USA

Emmanuel Frossard (467), Institute of Agricultural Sciences, ETH Zurich, Zurich, Switzerland

P.M. Groffman (417), Institute of Ecosystem Studies, Millbrook, NY, USA

R.J. Heck (15), School of Environmental Sciences, University of Guelph, Ontario, Canada

William Horwath (339), Department of Land, Air, and Water Resources, University of CA, Davis, CA, USA

Ellen Kandeler (187), Institute of Soil Science and Land Evaluation, Soil Biology, University of Hohenheim, Stuttgart, Baden-Württemberg, Germany

Michael A. Kertesz (467), Department of Environmental Sciences, The University of Sydney, Sydney, Australia

Ken Killham (41), James Hutton Institute, Invergowrie, Dundee, Scotland, UK

Johannes Lehmann (379), Department of Crop and Soil Sciences, Cornell University, Ithaca, NY, USA

E. Lumini (311), Istituto per la Protezione Sostenibile delle Piante, UOS Torino, Viale Mattioli, 10125 Torino, Italy

Susan M. Lutz (501), Natural Resource Ecology Laboratory, Colorado State University, Fort Collins, CO, USA

William B. McGill (245), Department of Earth and Environmental Science, University of Pennsylvania, Philadelphia, PA, USA

Sherri J. Morris (273), Biology Department, Bradley University, Peoria, IL, USA

David D. Myrold (443), Department of Crop and Soil Science, Oregon State University, Corvallis, OR, USA

William J. Parton (501), Natural Resource Ecology Laboratory, Colorado State University, Fort Collins, CO, USA

Eldor A. Paul (1), Natural Resource Ecology Laboratory and Department of Soil and Crop Sciences, Colorado State University, Fort Collins, CO, USA

Alain F. Plante (245, 501), Department of Earth and Environmental Science, University of Pennsylvania, Philadelphia, PA, USA

Jim I. Prosser (41), Institute of Biological and Environmental Sciences, University of Aberdeen, Aberdeen, Scotland, UK

G.P. Robertson (417), Department of Plant, Soil, and Microbial Sciences, Michigan State University, East Lansing, MI, USA

Cornelia Rumpel (379), CNRS, UMR Bioemco, Thiverval Grignon, France

Robert L. Sinsabaugh (77), Department of Biology, University of New Mexico, Albuquerque, NM, USA

Jeffrey L. Smith (535), USDA-Agriculture Research Service, Pullman, WA, USA (deceased)

Maddie M. Stone (245), Department of Earth and Environmental Science, University of Pennsylvania, Philadelphia, PA, USA

D. Lee Taylor (77), Department of Biology, University of New Mexico, Albuquerque, NM, USA

Janice E. Thies (151), Department of Crop and Soil Sciences, Cornell University, Ithaca, NY, USA

R.P. Voroney (15), School of Environmental Sciences, University of Guelph, Ontario, Canada

Diana H. Wall (111), Department of Biology and Natural Resource Ecology Laboratory, Colorado State University, Fort Collins, CO, USA

Preface

Soil microbiology, although strictly defined as the study of soil organisms that can be seen under the microscope, has historically been studied and taught in a much broader context. Soil biota would be an appropriate term to cover the great diversity of organisms, such as archaea, bacteria, fungi, and soil fauna, yet the more traditional term, microbiology, will continue to be used in this edition's title. The interaction of organisms with each other and their environment is studied in soil ecology. The physiological processes that contribute to ecosystem functioning, nutrient cycling, and biogeochemical processes, involving soil enzymes and the formation and dynamics of soil organic matter (SOM), are best described as biochemistry. The few years since the advent of the twenty-first century have seen greater changes in our field than during any previous period, including the exciting time at the beginning of the twentieth century when the identification of organisms associated with the recognized nutrient cycles was called the "Golden Age of Soil Microbiology."

Nucleic acid methodologies, after 30 years of promise, have matured to the stage where questions and hypothesis can now be developed. This is especially applicable to the 99% of soil biota known from microscopic and genomic studies to be nonculturable. Techniques have improved and costs have decreased to such an extent that individual laboratories across the globe can now address important soil microbiology, ecology, and biochemistry questions to substantially advance our field. Numerous types of new "omics research" are growing beyond their descriptive phase to yield hypotheses about function and should increase the ability of our field to better serve humankind and to conduct socially relevant science. New information on community diversity and process genes, obtained by collaboration among soil microbiological–biochemical and gene-processing laboratories, is bringing exciting insights. This text reflects the modern information age. The individual chapters provide numerous electronic links to available databases and websites, such as those for common analysis of 10,000 soil samples, using modern techniques. The use of new methodology for better characterizing SOM also provides great promise. The ability to perform 100 respiration or enzyme measurements simultaneously allows us to obtain the amount of data necessary to characterize the vast number of ecosystems, ranging from tropical rainforests to Arctic tundra, across our planet.

Greatly improved instrumental analyses and data handling are allowing us to ask questions such as: (1) What is the true role of humics in soil? (2) Why does lignin appear to be important in litter-decomposition dynamics, but not important in soil formation? (3) What is the relative role of chemical resistance versus matrix stabilization of SOM? (4) Do differences in the temperature sensitivity of different types of SOM affect their response to global change? (5) What is the role of microbial products in soil formation? (6) What is the role of SOM dynamics in food sustainability and global change? (7) How much does community diversity change with soil type, vegetation, and abiotic conditions? (8) Does the size of the soil biotic biomass and its composition affect ecosystem functioning? (9) What is the role of food webs? (10) Can we use improved models that include microbial characteristics and enzyme kinetics to better describe nature's processes?

Although current biogeochemical models that only implicitly recognize the role of biota have served us well, the possibility of including microbial biomass, fungal-to-bacterial ratios, gene abundance, and enzyme activities into meaningful, biogeochemical models across multiple scales is exciting. Soil microbiologists have long recognized the importance of scale. Eldor Paul can remember his PhD supervisor stating that microbiologists could always blame their problems, or interpret their data, on the basis of microsites. Now more than ever, we recognize the need to relate and interpret enzyme reactions and clay-protection studies at the nanometer scale, microbial activities at the micron scale, and plant-process-pedon interactions at the meter scale. Our data must be appropriate to model and ask questions at the kilometer and megameter scales required to interpret questions that range from agricultural sustainability to global climate change. The need to measure processes at meaningful time scales and interpret them over scales ranging from seconds to millennia is equally important.

The authors of this volume are very cognizant of the multiple teaching, research, and societal aspects and responsibilities of our field. This is ever more so as our readership rapidly expands in both numbers and interests. Our readers range from biology-ecology-biogeochemistry-soil science students to members of online courses, such as range management. Engineers, foresters, biogeochemists, agronomists, science teachers, extension workers, and researchers across the globe will critically look to our volume for insights applicable to their fields. When this edition was envisioned, it was assumed that approximately one-third of the volume would be new material. It has been updated to an even greater extent. The ability to add definitions, additional diagrams, and explanations in the supplemental material, available online and in the electronic version, will hopefully increase the usefulness of this volume. It is our expectation that the readers of both the hard cover and electronic editions will ensure that soil microbiology, ecology, and biochemistry will build on its great past and exciting present to produce a great future based on basic knowledge and unifying concepts.

We greatly appreciate the superb, editorial work and patience of Laurie Richards. We also note with sadness and regret the recent passing of our friend, coworker, and fellow author, Jeff Smith.

Eldor Paul, Paul Voroney, Richard Heck, Ken Killham, Jim Prosser, Lee Taylor, Bob Sinsabaugh, Dave Coleman, Diana Wall, Janice Thies, Ellen Kandeler, Serita Frey, Alain Plante, Maddie Stone, Bill McGill, Sherri Morris, Chris Blackwood, Raffaella Balestrini, Erica Lumini, Roberto Borriello, Valeria Bianciotto, Will Horwath, Claire Chenu, Cornelia Rumpel, Johann Lehman, Phil Robertson, Peter Groffman, Peter Bottomley, David Myrold, Michael Kertesz, Emmanuel Frossard, Bill Parton, Steve del Grosso, Carol Adair, Susan Lutz, Hal Collins, Alex Crump, and Vanessa Bailey.

Chapter 1

Soil Microbiology, Ecology, and Biochemistry: An Exciting Present and Great Future Built on Basic Knowledge and Unifying Concepts

Eldor A. Paul

Natural Resource Ecology Laboratory and Department of Soil and Crop Sciences, Colorado State University, Fort Collins, CO, USA

Chapter Contents

I SCOPE AND CHALLENGES

The study of soil biota, their interactions, and biochemistry, the subject matter of this book, must strive for excellence in both its research and impacts as it gains ever-increasing importance in science, education, and applications. Textbooks have a fundamental role to play in synthesizing information in a readable manner and making it available to a global audience. Knowledge in this field is expanding at an exponential rate at a time when global society faces a multitude of challenges to help maintain environmental sustainability. Many great opportunities exist. Advances in molecular techniques and analytical instrumentation are revolutionizing our knowledge about microbial community structure and are facilitating the integration of this knowledge with concepts concerning the composition and formation of soil organic matter (SOM), its interactions with the soil matrix, and its role in ecosystem functioning. Responses to climate

Soil Microbiology, Ecology, and Biochemistry. http://dx.doi.org/10.1016/B978-0-12-415955-6.00001-3

change and the possibility of increased natural disasters are at the forefront of our need to supply information on possible impacts. Questions about the role of soil biota and their processes, relative to food security for an increasing global population that needs to improve its diet at a time of economic globalization, will also need to be answered with sound science. Invasive species, water and air pollution, and plant diseases will probably be exacerbated by climate change and by intensification of management for food production and biofuels at the same time that we strive to protect our natural environments. Soil microbiology, ecology, and biochemistry will increasingly be called on to help provide the basic information required for biologically sustainable ecosystem services at a reasonable cost (Cheeke et al., 2013).

The history of our science is important to interpret today's knowledge and challenges. Perusing a few of the volumes of older literature can provide scientific insights and demonstrate approaches to problem solving that are very applicable today. The *Textbook of Agricultural Bacteriology* by Löhnis and Fred (1923) is an English translation and revision of the early 1913 German text. This book highlights the 1890 to 1910 "Golden Age of Microbiology" when representative microorganisms responsible for the major biogeochemical cycles were discovered. Although entitled "Agricultural Bacteriology," fungi and fauna are also discussed. Waksman and Starkey (1931) recognized the role of decomposition in the carbon (C) cycle. They calculated that the atmosphere over each acre of land at 0.03% CO_2 represented 5.84 tons of C when at that time a good yield of sugar cane consumed 20 tons of C. They also estimated that atmospheric CO_2 had a global turnover time of 35 years. The general content and format of that text were followed in subsequent volumes, including the current one.

Principles of Soil Microbiology (Waksman, 1932), a more extensive volume, recognized the rapid growth in the knowledge of mycorrhyzal fungi, the solid foundation developed in the study of decomposition, the relationships between plant growth and microbial activity, and the interdependence between the activities of microorganisms and chemical transformations in soil. The availability of direct, microscopic counts showed that only 1-5% of the microscopic microbial count could be cultured and that the bacteria and fungi occupied only a tiny fraction of the soil volume. Waksman's, (1952) volume contains a good history of the field with pictures of our founding parents and grandparents. It did something I hope this volume also does by pointing out some of the more promising lines of advancement in our field and in suggesting some likely paths for future study.

Alexander (1961) recognized the interplay of microbiology, soil science, and biochemistry, and his book contains an initial section on microbial ecology and ecological interactions that summarizes community composition as understood before the molecular age. His chapters on the microbiology of plant component and pesticide decomposition, written before the major impact of isotope tracer research, are still well worth a trip to the library. Swift et al. (1979), in their volume, *Decomposition in Terrestrial Ecosystems*, highlighted the importance of faunal-microbial interactions, as did *Fundamentals of Soil Ecology* (Coleman et al., 2004). Binkley's (2006) chapter, "Soils in Ecology and Ecology in Soils," highlights the integration of soils, plants, and animals and discusses interactions of soil science and ecology in the twenty-first century. Berthelin et al. (2006) wrote an interesting history of soil biology, and the writings of Feller et al. (2003) and Feller (1997) review the role of humus in soils. Earlier editions of this text have also provided a history of this field. The individual chapters of this volume provide further background information that highlights the rapid advances that will allow soil microbiology, ecology, and biochemistry to successfully advance into the future based on a solid past and exciting present. They also strive to follow the example of Waksman (1952) in suggesting some likely paths for the future.

II THE CONTROLS AND UNIFYING PRINCIPLES IN OUR FIELD

No field of knowledge stands alone, and modern science must provide impacts and applications for society, as well as for teaching and research. It is important that readers gain an integrated knowledge of studies on (1) all soil biota, which for historical purposes we still often lump under the term "microbiology"; (2) the relationship between organisms and their physical surroundings, which is referred to as their "ecology"; and (3) the physiology of organisms, enzymes and their relationships to SOM, nutrient cycling, and biogeochemistry, which we call "biochemistry." It is most important to have an understanding of how the different subjects are integrated. A number of unifying concepts can assist in such integration. Figure 1.1 shows some of the areas that define our field and also the multiple biotic and abiotic controls. The interactions would have to be shown in a third dimension and are best discussed with the use of models that adequately incorporate these concepts (Chapter 17). The discussion of biochemistry and physiology (Chapter 9) and that on the application of concepts in ecology (Chapter 10) help provide readers from a variety of backgrounds with some of the required information we need in integrating our diverse field of studies.

This volume will be used for teaching and research in biogeochemistry, microbiology, soil science, ecology, and biology classes. Applications in food, biofuel, and fiber production include forestry, agriculture, and range sciences (Chapter 18). Engineers and industry consultants are applying soil biological information to many studies, including pollution control. Today's societal questions include soil biotic responses to global change. We must be able to supply information on how to mitigate some of the negative effects of CO_2, CH_4, and N_2O as greenhouse gases. The role of tundra soils and peats in the global-C cycle must also be considered as the earth warms and changes in its precipitation patterns. The finding that frozen soils contain an amount of C equivalent to the rest of the terrestrial soil C supply is mind-boggling. However, much of this occurs in deep deposits, cryoperturbed sites, and peat deposits, often found in

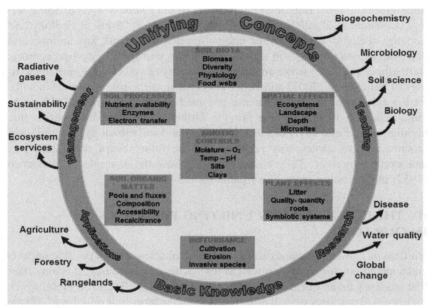

FIG 1.1   Controls and unifying concepts in soil microbiology, ecology, and biochemistry.

estuaries, where the water content may control its decomposition more than global warming. Increased human populations, along with a common desire across the globe for better living standards, will place demands on food and fiber production. Soils are the major resource for these efforts. Efficient utilization of nutrients involves not only C and N, but also P, S, Fe, and the minor elements (Chapter 16). Soil organisms will be called on to help maintain soil tilth, water penetration, resistance to erosion, and other ecosystem services, including water quality, invasive species, and disease prevention. The following section will highlight spatial distribution and community composition of organisms and their substrate, SOM composition, matrix interactions, microbial products, and spatial scaling. More detailed information and references are found in the succeeding individual chapters.

III   THE SPECIAL ROLE OF ACCESSIBILITY AND SPATIAL SCALING OF BIOTA AND SOIL ORGANIC MATTER

The size, accessibility, and spatial distribution of the soil biota, enzymes, microbial products, and organic matter are recognized as being of prime importance in the reactions and processes discussed in this volume. Löhnis and Fred (1923) recognized that the very small size of microorganisms resulted in large surface areas and allowed very large numbers to persist in soil, water, air, and food. Waksman and Starkey (1931) showed pictures that related microorganisms

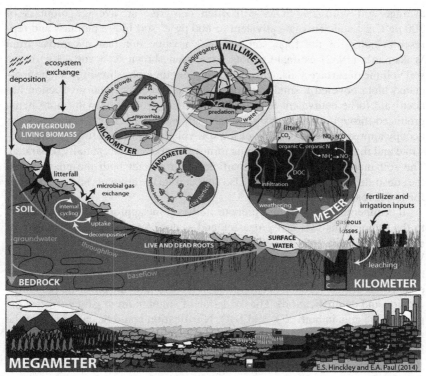

FIG 1.2   Spatial relationships from the nanometer to the kilometer and megameter must be understood to interpret soil microbiology, ecology, and biochemistry. *Drawn and copyrighted by E. Hinckley and E. Paul.*

to the soil's physical structure, recognized the colloidal nature of SOM, and calculated that 100 million bacteria in a gram of soil would occupy only 1/10,000th of the total soil volume.

There is a widespread recognition of the need to interpret processes at multiple scales, including molecular (nm), microorganism (μm), soil fauna, aggregates and roots (mm), whole plant and soil pedon (m), field and landscape (km), and especially now to global levels (megamater), as shown in Fig. 1.2. The biochemical processes occur at the nm scale with individual bonds and atoms occurring at the even smaller Angstrom scale (not shown). Typically, enzymes are 3-4 nm in size, as are the micropores of minerals, such as allophone. Clay particles with lengths of 2 μm, but with an edge of 1 nm, have a surface area approaching 1000 m^2g^{-1}. We often assume that the soil organisms are associated with the clay fractions by attachment to the particles, but particles can often be attached to the microbiota, especially where sesquioxides are involved. Soil colloids, which are operationally defined, usually have a size of 1-1000 nm. Nanoparticles are defined as being 1-100 nm. The microorganisms in an average soil with 2% SOM will often comprise an average biomass of 300 μg C g^{-1} soil. This is equivalent to 600 μg g^{1} soil biotic biomass and represents 0.06% of the total soil volume, a value not that dissimilar from Waksman's (1932) estimate. Roots have been shown to occupy 1% of the soil volume of surface soils, with early microphotographs showing that the associated biota covered a small portion of that surface. Enzyme production has been said to be equivalent to 1-4% of microbial production with more being produced through biomass lysis (Sinsabaugh and Shah, 2012). Thus, enzymes occupy approximately 0.0025% of the soil volume demonstrating that the substrate and the enzymes mediating its turnover can easily be spatially separated. The soil matrix, especially clay particles, can protect both the enzyme and the substrate from attack, but may also concentrate the amounts and interactions for faster reactions.

It is imperative that we interpret our studies at the appropriate scales relative to the interactions between processes and scales. An appropriate example of spatial scale challenges can be found for nitrogen (N) reactions in the biosphere. Where earlier editions stressed the need for increased N for agricultural production, present overuse of Angstrom-based, industrial N for fertilization at the plant (m) scale has resulted in two microbial reactions at the μm scale (i.e., nitrification and denitrification). The oxidation of ammonium to gain energy at the μm scale results in both NO_3 and N_2O. Nitrification could be called one of the worst diseases attributable to disturbance and agriculture. Soil microorganisms in the absence of oxygen at the μm scale use denitrification to oxidize other substrates producing some N_2O. Although these processes help close the N cycle, they also are one of the great ecological challenges at the megameter scale.

Visualization of the soil can now be achieved at the molecular (nm) level (Chapter 7). The μm to mm scale, found in thin sections, can also be useful. Figure 1.3 shows the great differences in habitat and spatial accessibility between the parent material (C horizon) devoid of SOM and the surface horizon of a grassland Mollisol. It shows a cross section of a root as well as plant-residue particles that tend to concentrate substrates and organisms. The SOM, although accounting for only 3% of the volume of this soil, appears to be uniformly distributed due to its nm size and high surface area. This is very important in exchange reactions, aeration, and water-holding capacity due to its colloidal characteristics. The fungi and fauna will be concentrated on the plant particles and the outsides of aggregates (Chapter 8), whereas the aggregate interiors have more bacteria. From this figure, it also is easy to see why later chapters will show that much of the biological activity occurs in the upper soil surfaces.

The illustration "Mechanisms of soil organic matter stabilization: evolution in understanding" by J. D. Jastrow (Fig. 1.4), adapted from Jastrow and Miller (1998), shows (in green) three major categories of mechanisms responsible for stabilizing soil organic matter—biochemical recalcitrance, chemical stabilization, and physical protection (Christensen, 1996; Sollins et al., 1996) and updates these to reflect evolving insights into the factors controlling

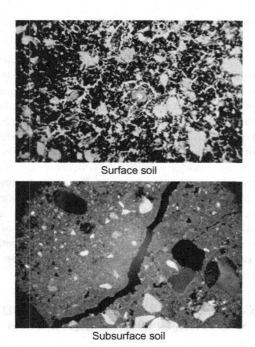

FIG 1.3 Thin section of subsurface and surface soil showing the general distribution of soil organic matter, aggregates, and plant particle concentration in the surface relative to the unaggregated, low SOM subsurface.

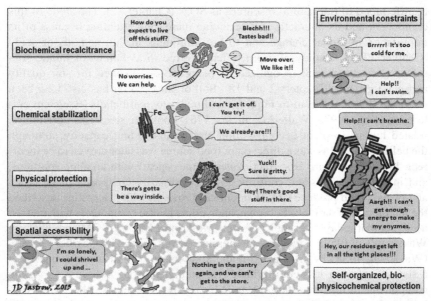

FIG. 1.4 Mechanisms of soil organic matter stabilization: evolution in understanding. *Courtesy of J. D. Jastrow, 2013.*

stabilization. Although biochemical recalcitrance may affect the rate of decomposition or transformation, recent research indicates that all types of organic materials are decomposable under the right conditions (illustrated in yellow). Thus, the intrinsic chemical characteristics of SOM are no longer believed to be a key determinant of its long-term persistence in most soils. The important role of environmental constraints (particularly at high latitudes and for wetlands) is highlighted in blue. In addition to the physical protection afforded by occlusion of organic matter in aggregates, decomposition can be limited due to spatial accessibility (shown in purple), that is, sometimes decomposer populations are simply not co-located with their substrates or are prevented from reaching them due to physical constraints, such as the distributions of water and air within the pore network. New understanding of the feedbacks between soil biota and soil structure suggests several integrated biophysicochemical stabilization mechanisms (shown in magenta) can occur via the dynamics of this self-organized system.

IV SOIL ORGANIC MATTER AS A CONTROL AND INFORMATIONAL STOREHOUSE OF BIOTIC FUNCTIONS

Löhnis and Fred (1923), who produced the first "cycle of matter" diagram, stated that all organic residues must be mineralized; otherwise the earth would long since have been littered with corpses. That proportion of primary production and animal products, stabilized against immediate mineralization, forms the life blood of our natural resources as SOM (Chapter 12). The study of this component and its interactions is receiving increasing attention because of its importance in global change and to breakthroughs in the methods of study (Chapter 13). The study of SOM, often referred to as humus or humic substances, has long been recognized as being of significance for soil quality and plant growth (Chapters 2 and 18). Soil organic matter has also long been recognized as significant to religious and mythological beliefs in ancient cultures (Feller, 1997). Wallerius, in his 1761 book on scientific agriculture, related humus to plant decomposition and water-holding capacity, although the belief that humus was a direct plant food source was later proven to be incorrect. In 1766, Archard first fractionated humus (peat) with alkaline solutions, and in 1804, de Saussure confirmed it was primarily composed of C, H, O, and N and could produce CO_2 (Kononova, 1961). In 1806, Berzelius described the interaction of dark black, brown, and yellow humic compounds with metals, a subject that today still requires further work in describing SOM stabilization. Waksman's (1952) chapter on humus summarized his earlier review (Waksman, 1938) by stating, "Humus represents a natural organic system in a state of more or less equilibrium." He predicted today's discussion on the components of this important substrate by saying, "Frequently, the narrow definition does not differentiate between humus and humic acid, another ill-defined term occasionally used for the alkali soluble or alcohol soluble humus constituents" (p. 125). The Russian author Kononova (1961) reviewed the extensive humic studies and argued that Waksman's comments were inaccurate. However, today there are new questions regarding humic substance synthesis by polycondensation reactions of microbially transformed plant products. Many now believe that humus is a complex mixture of microbial and plant polymers and that their degradation products are associated in super structures, stabilized by hydrogen and hydrophobic bonds.

Earlier versions of this text stated that the materials stabilized by clays are the older, complex, humified components. Our present research is questioning this concept. It indicates that microbial products are associated with the clays and thus protected. It also shows that all soil fractions contain some proportion of young and old materials and that they all participate, to varying degrees, in the stabilization of SOM. The particulate fraction (>53 μm), although primarily consisting of plant residues and associated microbial products, can also have older charcoal and organic-micelle protected residues. The older, mineral-associated fractions have a large concentration of materials that can be thousands of years old, but because clays also react with the recently produced microbial products, they also contain some young components. The wish to isolate defined fractions with very specific mean residence times will never be completely possible. All fractions constitute a part of a dynamic soil system and will have varying degrees of both old and young materials depending on the fraction isolated. Modern methodology now allows us to use a number of approaches that can give a great deal of information regardless of whether the SOM is defined on a biochemical, functional, or physical kinetic basis or as humic compounds (Supplemental Fig. 1.1; see online supplemental material at http://booksite.elsevier.com/9780124159556).

An SOM-spatial distribution complex in soils that deserves more attention involves extracellular polysaccharides. Extracellular polymeric slime has long been recognized in soils, especially in aggregation. It is also known to be involved in biofilms and on root surfaces (Fig. 1.2). The assemblage of micro-colonies within a biofilm can involve protozoa, nematodes, fungi, and a broad range of prokaryotes. Biota embedded in biofilms often have altered phenotypic expressions and complementary functions (Dick, 2013). They are also considered to have cell-to-cell signaling and nutrient and energy transfer functions. There is a need to understand their role in interpreting community diversity and in soil enzyme functioning. Other complex interactions will no doubt be found as we delve deeper into the interesting field we call soil microbiology, ecology, and biochemistry.

V   BIOTIC DIVERSITY AND MICROBIAL PRODUCTS

The significant breakthroughs in molecular biology that have occurred in the last 30 years, in conjunction with the recent advances in automated-gene sequencing (Chapter 6), will allow us to characterize a greater proportion of the soil biota as well as determine their functional genes. There have been recent advances in both the organisms and processes in the N cycle (Chapters 14 and 15). Advances in automated analysis of physiological reactions, enzyme assays, and soil respiration (Chapter 7) are giving us the opportunity to relate information on community diversity to microbial ecology (Chapter 10), physiology and biochemistry (Chapter 9), and global distribution (Chapter 8). These techniques have identified the archea and recognized many nonculturable soil inhabitants. It is important that we ask how this new information can be related to the controls shown in Fig. 1.1. We must ensure that the best available modern methodology is used to ask questions about the processes involved. Biodiversity by itself is an important question, with soil being the greatest repository of genes in nature (Chapters 3, 4, 5). As we ask questions about community composition, the role of genetic redundancy becomes important. The various chapters in this volume will show that the generalized processes such as decomposition, which can be carried out by numerous organisms, have significant genetic redundancy. Food web interactions (Chapter 5) can be complex and may require more information on specific organisms and genes as well as on interactions (Moore and de Ruiter, 2012). Specialized biogeochemical processes, such as those in nitrification (Chapter 14) and N fixation (Chapter 15), are carried out by restricted populations that can become limiting (Bardgett, 2005). Arbuscular and ectomycorrhizal fungi (Chapter 11) are associated with many plants, but not all associations will provide benefits and can even be parasitic in nature. Some of these fungi still cannot be grown in pure culture. Genomic analysis is vital for understanding the complex biotic interactions involved (Chapters 3, 4, and 5). The same is true for physiological processes such as symbiotic relationships (Chapters 11 and 15) and nutrient transformations (Chapters 16 and 18), in which the most competitive colonizers may not be the most efficient ones.

VI UNIFYING CONCEPTS

There is now significant research showing that unifying concepts can be developed for the controls in soil microbiology, ecology, and biochemistry (Fierer et al., 2009). Individual controls (Fig. 1.1) may dominate at a specific microsite; however, at the plant-pedon level and above (Fig. 1.2), many soils have a large number of similarities. Figure 1.5 summarizes some of the major biotic-abiotic components and controls that occur in soils. The plants affect the biotic community structure and, eventually, the amount of SOM, due to both the quality and quantity of their above- and belowground inputs and whether or not they produce a litter layer. The two bacteria, pictured in Fig. 1.5, were originally shown as colored inserts in Löhnis and Fred (1923). Decomposition involves many organisms providing genetic redundancy, but the enzymes involved in lignin degradation are somewhat specialized and have a slight home field advantage, where the organism in the litter below a certain type of plant will decompose that litter at a slightly faster rate. The lignin to N ratio is a good

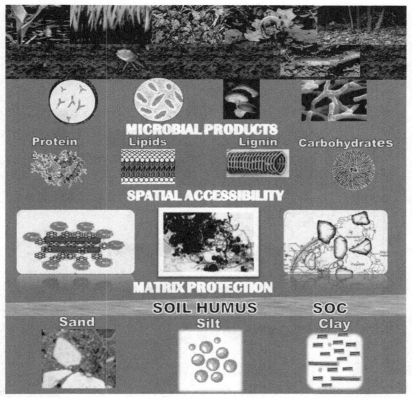

FIG. 1.5 Representation of the effects of plant litter quality and quantity on litter accumulation, microbial community structure, spatial complexity, humus composition, and soil matrix interactions involved in the control of the interactions in soil microbiology, ecology, and biochemistry.

general indicator of litter decomposition, but not of humus formation. In addition, lignin does not appear to persist in soils. Cotrufo et al. (2013) proposed the Microbial Efficiency-Matrix Stabilization (MEMS) framework (Fig. 1.6) to study and integrate litter decomposition and SOM formation. This was based on the current understanding of the importance of microbial C use efficiency (CUE) and C and N allocation in controlling the proportion of plant-derived C and N that is incorporated into SOM. The MEMS diagram also stresses the importance of soil-matrix interactions in controlling SOM stabilization. The factors directly controlling the proportion of plant-derived C and N retained in SOM pools versus mineralized, in the short-term, include microbial allocation of C and N to growth, enzyme production, and microbial products that interact with the soil matrix.

Results from NMR, pyrolysis molecular-beam mass spectrometry, XANES, and mid-infrared analyses show that humus from many different soils has a

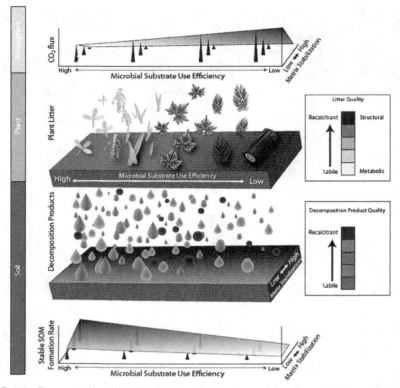

FIG. 1.6   Representation of the effects of plant litter quality on CO_2 efflux and soil organic matter (SOM) stabilization in the microbial efficiency-matrix stabilization (MEMS) framework. Above- and belowground plant litters undergo microbial processing, which determines the quantity and chemical nature of decomposition products. More dissolved organic matter, carbohydrates, and peptides are formed from high-quality (e.g., fine roots and herbaceous) litter than from low-quality (e.g., needle and wood) litter, which loses most of the C as CO_2. The fate of the decomposition products depends on their interactions with the soil matrix. More stable SOM accumulates in soils with a high soil matrix stabilization, including expandable and nonexpandable phyllosilicates: Fe-, Al-, Mn-oxides, polyvalent cation, or high allophane content. *Cotrufo et al. (2013).*

somewhat similar basic complex of functional groups related to plant and microbial products. These darkly colored materials can be protected by self-aggregation, especially for micelles where hydrophobic lipids of both plant and microbial origin provide water repellency. The stabilization of substrate by clays, sesquioxides, and microaggregate formation results in the 1000-year-old SOM products so often found in soils. The protection of SOM by silts and clays results in SOM levels from the 0.6-8% soil organic carbon (SOC) usually found in mineral soils (Fig. 1.5). Protection by calcium and sesquioxide has long been known to exist, but has not been studied enough to allow quantification for model parameterization.

The advances in biotic, community structure analysis by techniques, such as pyrosequencing, should also provide enough information on biotic community composition to allow its incorporation into further understanding and model development. The interactive controls and the importance of enzymes and matrix stabilization of amino compounds tend to produce humus components low in C:N ratio and are somewhat similar in all soils allowing us to further seek unifying concepts. An example of the development of understanding through multiple approaches and modeling that will be covered in this volume is given in Supplemental Fig. 1.2 (see online supplemental at http://booksite.elsevier.com/9780124159556). The SOM at a long-term experimental site in Michigan was shown to have a kinetically determined, active fraction representing 5% of its C with a rapid turnover rate largely dependent on recent inputs. The slow pool, determined with tracers and long-term incubation due to its significant size and moderate turnover, is the most important pool in nutrient cycling and biogeochemistry. The large, resistant pool supplies long-term stability, but is still sensitive to recent management.

Symbiotic systems are an important part of nearly all plant soil relationships. Aboveground, beneath-ground photosynthate distribution, and root production are major factors in controlling many of the biota and organisms discussed in this volume. The high energy-requiring, symbiotic-N fixers require more photosynthate than the mycorrhizal fungi, but the mycorrhizal fungi contribute significantly to SOM formation. Due to an increased understanding of the size, activity, and community composition of soil biota, we can now assign some numbers to their roles in C cycling at this site. The bacteria are responsible for more of the C respired from the maize soybean rotation. However, the fungi dominate respiration in the poplar plantation illustrating that although similar controls exist, specific populations and abiotic controls affect individual ecosystem processes. The fauna play their major role as ecosystem engineers. The nanometer, micrometer, and kilometer effects and interactions shown in Fig. 1.2 all relate to the understanding and modeling (Chapter 17) of the global (megameter) aspects of the C cycle reflected in the rapidly rising atmospheric CO_2 contents. The controls and interactions must be considered as we ensure that the exciting present information on soil microbiology, ecology, and biochemistry outlined in this volume builds toward a great future created from a basic knowledge and unifying concepts.

REFERENCES

Alexander, M., 1961. Introduction to Soil Microbiology. Wiley, New York.

Bardgett, R., 2005. The Biology of Soil. Oxford University Press, Oxford, UK.

Berthelin, J., Babel, U., Toutain, F., 2006. History of soil biology. In: Warkenten, B. (Ed.), Footprints in the Soil. Elsevier, Amsterdam, pp. 279–306.

Binkley, D., 2006. Soils in ecology and ecology in soils. In: Warkenten, B. (Ed.), Footprints in the Soil. Elsevier, Amsterdam, pp. 259–278.

Cheeke, T.C., Coleman, D.C., Wall, D.H., 2013. Microbial Ecology in Sustainable Agroecosystems. CRC Press, Boca Raton, FL.

Christensen, B.T., 1996. Carbon in primary and secondary organomineral complexes. In: Carter, M.R., Stewart, B.A. (Eds.), Structure and Organic Matter Storage in Agricultural Soils. CRC Press, Inc., Boca Raton, pp. 97–165.

Coleman, D.C., Crossley Jr., D.A., Hendrix, P.E., 2004. Fundamental of Soil Ecology. Elsevier-Academic Press, New York.

Cotrufo, F., Wallenstein, M., Boot, C., Denef, K., Paul, E.A., 2013. The molecular efficiency-matrix stabilization (MEMS) framework integrates plant litter decomposition with soil organic matter stabilization. Do labile plant inputs form stable soil organic matter? Glob. Chang. Biol. http://dx.doi.org/10.1111/gcb.12113.

Dick, R.P., 2013. Manipulation of beneficial organisms in crop rhizoshperes. In: Cheeke, T.C., Coleman, D.C., Wall, D.H. (Eds.), Microbial Ecology in Sustainable Agroecosystems. CRC Press, Boca Raton, pp. 23–48.

Fierer, N., Grandy, S., Six, J., Paul, E.A., 2009. Searching for unifying principles in soil ecology. Soil Biol. Biochem. 41, 2249–2256.

Feller, C.L., 1997. The concept of humus in the past three centuries. In: Yaaalon, D.H., Berkowits, S. (Eds.), History of Soil Science, Vol. 29. Catena Verlag, GmbH Reiskirchen, Germany, pp. 15–46.

Feller, C., Brown, G.G., Blanchart, E., Deleporte, P., Chernyanskii, S.S., 2003. Charles Darwin, earthworms and the natural sciences: various lessons from past to future. Agric. Ecosyst. Environ. 99, 29–49.

Jastrow, J.D., Miller, R.M., 1998. Soil aggregate stabilization and carbon sequestration: feedbacks through organomineral associations. In: Lal, R., Kimble, J.M., Follett, R.F., Stewart, B.A. (Eds.), Soil Processes and the Carbon Cycle. CRC Press LLC, Boca Raton, pp. 207–223.

Kononova, M.M., 1961. Soil Organic Matter: Its Nature, Its Role in Soil Formation and in Soil Fertility. Pergammon, Oxford.

Löhnis, F., Fred, E.B., 1923. Textbook of Agricultural Bacteriology. McGraw Hill, New York.

Moore, J.C., de Ruiter, P.C., 2012. Energetic Food Webs: An Analysis of Real and Model Ecosystems. Oxford University Press, Oxford, UK.

Sinsabaugh, R.L., Shah, J.H., 2012. Ecoenzymatic stoichiometry and ecological theory. Annu. Rev. Ecol. Evol. Syst. 43, 313–343.

Sollins, P., Homann, P., Caldwell, B.A., 1996. Stabilization and destabilization of soil organic matter: mechanisms and controls. Geoderma 74, 65–105.

Swift, M.J., Heal, J., Anderson, J.M., 1979. Decomposition in Terrestrial Ecosystems. University of California Press, Berkeley.

Waksman, S., 1932. Principles of Soil Microbiology. Williams & Wilkins, Baltimore.

Waksman, S.A., 1938. Humus: Origin, Chemical Composition and Importance in Nature. Williams & Wilkins, Baltimore.

Waksman, S.A., 1952. Soil Microbiology. John Wiley, New York.

Waksman, S.A., Starkey, R.L., 1931. The Soil and the Microbe. John Wiley, New York.

Chapter 2

The Soil Habitat

R.P. Voroney and R.J. Heck
School of Environmental Sciences, University of Guelph, Ontario, Canada

Chapter Contents

I INTRODUCTION

Soil, the naturally occurring unconsolidated mineral and organic material at the earth's surface, provides an essential natural resource for living organisms. It is a central component of the earth's critical zone and deserves special status due to its role in regulating the earth's environment, thus affecting the sustainability of life on the planet. The soil environment is the most complex habitat on earth. This complexity governs soil biodiversity, as soil is estimated to contain one-third of all living organisms and regulates the activity of the organisms responsible for ecosystem functioning and evolution. The concept that the earth's physicochemical properties are tightly coupled to the activity of the living organisms it supports was proposed in the early 1970s by James Lovelock and Lynn Margulis as the Gaia hypothesis. They theorized that the earth behaves as a superorganism, with an intrinsic ability to control its own climate and chemistry and thus maintain an environment favorable for life. Soil has the intrinsic ability to both support terrestrial life and provide a habitat for the interdependent existence and evolution of organisms living within it. In 2011,

Soil Microbiology, Ecology, and Biochemistry. http://dx.doi.org/10.1016/B978-0-12-415955-6.00002-5

scientists embarked on a Global Soil Biodiversity Initiative to assess soil life in all biomes across the globe for the essential ecosystem services that soils provide (i.e., plant biomass production, decomposition, and nutrient cycling) and to identify where soil quality is endangered due to human activities. The ultimate goal of the initiative was to guide environmental policy for sustainable land management (Soil Stories, The Whole Story, http://youtube.com/watch?v? Ego6LI-IjbY; see online supplemental material at: http://booksite.elsevier. com/9780124159556).

Soils (pedosphere) develop at the interface where organisms (biosphere) interact with rocks and minerals (lithosphere), water (hydrosphere), and air (atmosphere), with climate regulating the intensity of these interactions. In terrestrial ecosystems, the soil affects energy, water and nutrient storage and exchange, and ecosystem productivity. Scientists study soil because of the fundamental need to understand the dynamics of geochemical–biochemical– biophysical interactions at the earth's surface, especially in light of recent and ongoing changes in global climate and the impact of human activity. Geochemical fluxes between the hydrosphere, atmosphere, and lithosphere take place over the time span of hundreds to millions of years. Within the pedosphere, biologically induced fluxes between the lithosphere, atmosphere, and biosphere take place over a much shorter time frame, hours and days to months, which complicates the study of soils.

The soil habitat is defined as the totality of living organisms inhabiting soil, including plants, animals, and microorganisms, and their abiotic environment. The exact nature of the habitat in which the community of organisms lives is determined by a complex interplay of geology, climate, and plant vegetation. The interactions of rock and parent material with temperature, rainfall, elevation, latitude, exposure to sun and wind, and numerous other factors, over broad geographical regions, environmental conditions, and plant communities, have evolved into the current terrestrial biomes with their associated soils (Fig. 2.1).

Because soils provide such a tremendous range of habitats, they support an enormous biomass, with an estimated 2.6×10^{29} for prokaryotic cells alone, and harbor much of the earth's genetic diversity. A single gram of soil contains kilometers of fungal hyphae, more than 10^9 bacterial and archaeal cells and organisms belonging to tens of thousands of different species. Zones of good aeration may be only millimeters away from areas that are poorly aerated. Areas near the soil surface may be enriched with decaying organic matter and other accessible nutrients, whereas the subsoil may be nutrient poor. The variance of temperature and water content of surface soils is much greater than that of subsoils. The soil solution in some pores may be acidic, yet in others more basic, or may vary in salinity depending on soil mineralogy, location within the landscape, and biological activity. The microenvironment of the surfaces of soil particles, where nutrients are concentrated, is very different from that of the soil solution.

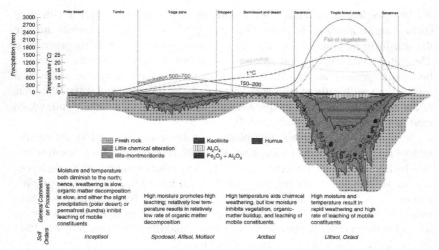

FIG. 2.1 Environmental factors affecting the distribution of terrestrial biomes and formation of soils along a transect from the equator to the North Polar region *(from Birkeland, 1999)*.

II SOIL GENESIS AND FORMATION OF THE SOIL HABITAT

Soils derived from weathered rocks and minerals are referred to as mineral soils. When plant residues are submerged in water for prolonged periods, biological decay is slowed. Accumulations of organic matter at various stages of decomposition become organic soils and include peat land, muck land, or bogs and fens. Soil can also be formed in coastal tidal marshes or inland water areas supporting plant growth where areas are periodically submerged.

Mineral soils are formed by the physical and chemical weathering of the rocks and minerals brought to the earth's surface by geological processes. They extend from the earth's terrestrial surface into the underlying, relatively unweathered, parent material. The parent material of mineral soils can be the residual material weathered from solid rock masses or the loose, unconsolidated materials that often have been transported from one location and deposited at another by such processes as sedimentation, erosion, and glaciation. The disintegration of rocks into smaller mineral particles is a physical-chemical process brought about by cycles of heating and cooling, freezing and thawing, and also by abrasion from wind, water, and ice masses. Chemical and biochemical weathering processes are enhanced by the presence of water, oxygen, and the organic compounds resulting from biological activity. These reactions convert primary minerals, such as feldspars and micas, to secondary minerals, such as silicate clays and oxides of aluminum, iron, and silica. Soluble constituent elements in inorganic forms provide nutrients to support the growth of various organisms and plants.

The physical and chemical weathering of rocks into fine particles with large surface areas, along with the accompanying release of plant nutrients, initiate the soil-forming process (Fig. 2.2) by providing a habitat for living organisms. The initial colonizers of soil parent material are usually organisms capable of both photosynthesis and N_2 fixation. Intimate root-bacterial/fungal/actinomycetal associations with early plants assist with supplying nutrients and water. Products of the biological decay of organic residues accumulate in the surface soil, which then forms soil organic matter (SOM).

Soil organisms, together with plants, constitute one of the five interactive factors responsible for soil formation. By 1880, Russian and Danish soil scientists had developed the concept of soils as independent natural bodies, each possessing unique properties resulting from parent material, climate, topography, and living matter, interacting over time. The approach for describing soil genesis in the landscape and as a unique biochemical product of organisms participating in the genesis of their own habitat was quantified by Hans Jenny in 1941 in his classic equation of soil forming factors:

$$\text{Soil} = f[\text{Parent material}, \text{climate}, \text{living organisms}, \text{topography}, \text{time}] \quad (2.1)$$

It took 10,000-30,000 years to form soils in the glaciated areas; alternatively, soil formation can take hundreds of thousands of years, depending on deposition and erosion processes. (The Five Factors of Soil Formation: http://www.youtube.com/watch?v=bTzslvAD1Es; see online supplemental material at: http://booksite.elsevier.com/9780124159556).

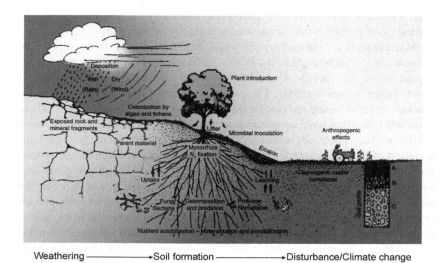

Weathering ————————→Soil formation ————————→Disturbance/Climate change
      time   time

FIG. 2.2 Soils are formed as a function of the five interacting factors of climate, parent material, topography, organisms, and time. Human activities have altered soil formation and promoted soil degradation through cultivation and cropping *(from Paul and Clark, 1996).*

Humans have had a negative effect on soil formation due to some of the agricultural processes they use. The clearing of native vegetation, tillage/cultivation of the surface soil, and cropping for agriculture are known to degrade soils due to promoting erosion and enhancing losses of SOM. Yet alternatively, humans have also contributed to the improvement of soil conditions by installing drainage and irrigation systems and adding amendments of nutrients and organic matter for remediation of mine sites and other areas of exposed parent material. Hans Jenny Memorial Lecture in Soil Science - The Genius of Soil. (http://www.youtube.com/watch?v=y3q0mg54Li4; see online supplemental material at: http://booksite.elsevier.com/9780124159556).

A   Soil Profile

During formation, soils develop horizontal layers, or horizons, that appear different from one another (Fig. 2.3). The horizons within a soil profile vary in thickness depending on the intensity of the soil-forming factors, although their boundaries are not always easy to distinguish. Uppermost layers of mineral soils are the most altered, whereas the deeper layers are most similar to the original parent material. Alterations of the solum, the parent material most affected during soil formation, involve (1) decay of organic matter from plant residues and roots and accumulating as dark-colored humus (the organic matter-enriched horizons nearest the soil surface are called A horizons); (2) eluviation by water of soluble and colloidal inorganic and organic constituents from surface soils to varying depths in the profile; and (3) accumulation of inorganic and organic precipitates in subsurface layers. These underlying, enriched layers are referred to as B horizons. The C horizons are the least weathered of the mineral soil profile. Organic soils are commonly saturated with water and mainly consist of mosses, sedges, or other hydrophytic vegetation, with the upper material being referred to as the O layer. In upland forested areas where drainage is better, folic-derived organic materials accumulate as an L-F-H layer.

The vadose zone is the underlying, unsaturated, parent material extending downward from the soil surface to where it reaches the water table and the soil becomes saturated. Below the solum, this zone contains relatively unweathered parent material, low in organic matter and nutrients, and intermittently deficient in O_2. The thickness of the vadose zone can fluctuate considerably during the season, depending on soil texture, soil water content, and height of the soil water table. When the water table is near the surface, for example as in wetlands, it may be narrow or nonexistent. But in arid or semiarid areas, where soils are well drained, the vadose zone can extend for several meters.

III   PHYSICAL ASPECTS OF SOIL

Dimensions of the features, commonly encountered when considering the soil habitat, range from a few meters (the pedon, the fundamental unit of soil within the landscape), down through a few millimeters (soil aggregates and the fine earth fraction), to a few micrometers (living microorganisms and clay minerals) and nanometers (fragments of microbial cell walls).

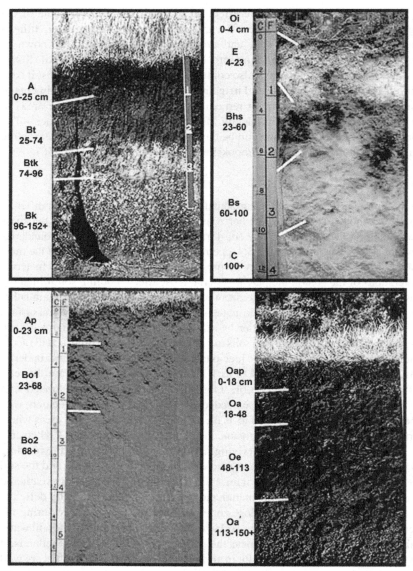

FIG. 2.3 Profiles of common mineral and organic soils: Mollisol (top left), Spodosol (top right), Oxisol (bottom left), and Histosol (bottom right).

A Soil Texture

The larger mineral particles include stones, gravels, sands, and coarse silts that are generally derived from ground-up rock and mineral fragments. Although particles > 2 mm in diameter may affect the physical attributes of a soil, it is the fine earth fraction, those individual mineral particles ≤ 2 mm in diameter, that describe soil texture. The fine earth fraction of soil particles ranges in size over four orders of magnitude: from 2.0 mm to smaller than 0.002 mm in diameter. Sand-sized particles are individually large enough (2.0 to 0.05 mm) to be seen by the naked eye and feel gritty when rubbed between the fingers in a moist state. Somewhat smaller, silt-sized particles (0.05 to 0.002 mm) are microscopic and feel smooth and slippery even when wet. Clay-sized particles are the smallest of the mineral particles ($<$0.002 mm), seen only with the aid of a microscope, and when wet, they form a sticky mass. The proportions of sand, silt, and clay are referred to as soil texture, and terms such as *sandy loam*, *silty clay*, and *clay loam* are textural classes used to identify the soil's texture. When investigating a field site, considerable insight to the behavior and properties of the soil can be inferred from its texture (e.g., soil water characteristics, nutrient retention, susceptibility to compaction); thus, it is often one of the first properties to be measured (How to determine soil texture by feel; http://www.youtube.com/watch?v¼GWZwbVJCNec; see online supplemental material at: http://booksite.elsevier.com/9780124159556).

The surface of mineral soils contains an accumulation of living biomass and dead and decomposing organic material. The SOM typically accounts for 1-10% of the total soil mass, and most is intimately associated with the mineral fraction, especially the surfaces of clays, making it is difficult to isolate from the soil for study. The larger, recognizable remains of plant, animal, and soil organisms that can be separated from soils by hand picking, sieving techniques, and floatation are referred to as particulate organic matter or light fraction organic matter. These tissues undergo continuous microbial decay and turnover and, over periods of years to decades, accumulate in soils as brown to black-colored, chemically complex colloids.

B Aggregation of Soil Mineral Particles

Typically the individual mineral particles in surface soils are coated and become glued together with organic matter and by inorganic cements, forming spatial clusters within the soil structural matrix known as aggregates or peds. The arrangement of aggregates is referred to as soil structure and, together with soil texture, is of particular importance in regulating biotic activity because of its influence on water content and aeration.

SOM and clay minerals play particularly important roles in aggregate formation due to their large surface area and negative electrostatic charge (i.e., are colloidal in nature). Loamy or clayey soils are usually strongly aggregated, whereas sandy and silty soils are weakly aggregated.

Tisdall and Oades (1982) presented a conceptual model of the hierarchical nature of soil aggregation, which described the linkages between the architecture of the soil habitat and the role of microbial activity in its genesis (Fig. 2.4). Three assemblages of aggregates are generally recognized with

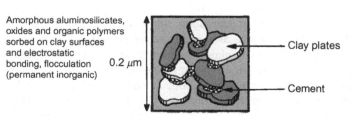

FIG. 2.4   Model of the aggregated, hierarchical nature of the soil system and major binding agents *(from Tisdall and Oades, 1982).*

diameter classes of: 0.002-0.020 mm, 0.020-0.250 mm, and > 0.250 mm and are referred to as microaggregates, mesoaggregates, and macroaggregates, respectively.

Microaggregates are formed by flocculation of fine silt and clay particles, amorphous minerals (composed of oxides and hydroxides of aluminum, silicon, iron, and manganese, and silicates of aluminum and iron), and nonhumic and humic substances, which are largely dominated by electrostatic and van der Waals forces. Polyvalent cations, such as Al^{3+}, Fe^{3+}, Ca^{2+}, and Mg^{2+} adsorbing onto their surfaces and reacting with exposed functional groups, promote these flocculation reactions. Sticky polysaccharides and proteins, derived from plant and animal tissues, microbial cells and exudates from roots, hyphae, and bacteria further enhance these stabilization reactions. In particular, wherever intense biological activity associated with organic matter decomposition occurs, the extensive exopolysaccharides microorganisms produce glue minerals and organic particles that form microaggregates. The core of mesoaggregates is usually the residual debris left from the decay of plant and microbial tissues. Bits of this decaying particulate organic matter, and their colonizing microbial biofilms are encrusted with fine mineral particles, which act as nuclei for the formation of larger aggregates. In the rhizosphere, hyphae of arbuscular mycorrhizal fungi (AMF) contribute to aggregation.

Macroaggregates are formed where a network of living plant roots and root hairs, fungal hyphae, and fibrous organic matter physically enmesh clusters of micro- and mesoaggregates for a period sufficient for them to be chemically linked. Casts deposited by earthworms and fecal pellets left by microarthropods, mites, and colembola also contribute to aggregate formation. This hierarchical organization of aggregate formation (i.e., large aggregates being composed of smaller aggregates), which in turn are composed of even smaller aggregates, is characteristic of most undisturbed surface soils.

Micro- and mesoaggregates tend to be especially resistant to mechanical breakdown, for example, from the impact of rainfall, or from slaking-rapid rewetting of dry soil, or from freezing and thawing. The restricted size of the pores contained within these aggregates can limit accessibility of the associated colloidal humus to microbial decay and restrict interactions of soil organisms, thereby protecting microorganisms from predation by fauna. Macroaggregates usually remain intact as long as the soil is not disturbed, for example, by earthworm and other faunal activity, by the impact of intense rainfall, or by tillage machinery. The pore spaces contained within macroaggregates, referred to as intraaggregate pore space, are important for providing soil aeration and water retention for plant growth. The pore space surrounding macroaggregates, collectively referred to as the interaggregate pore space, is where plant roots and larger fragments of plant residues are found. Macroaggregation is important for controlling biotic activity and SOM turnover in surface soils because the interaggregate pore space allows for the exchange of oxygen and other gases with the atmosphere (Ball, 2013). It may also regulate the accessibility of particulate organic matter to decay by soil organisms. By determining the nature and pore-size distribution within soil, macroaggregation can give fine-textured, clayey- and loamy-textured soils the beneficial pore space characteristics for aeration, water infiltration, and drainage of sandy soils. (Method for evaluation of aggregation in the field shown in Fig. 2.7: http://www.sac.ac.uk/vess; see online supplemental material at: http://booksite.elsevier.com/9780124159556); Guimarães et al., 2011.

IV   SOIL HABITAT AND SCALE OF OBSERVATION

A   Scale of Soil Habitat

The habitat provided by the soil is characterized by heterogeneity, measured across scales from nanometers to kilometers (Chapter 1), differing in chemical, physical, and biological characteristics in both space and time. At various levels within this continuum of scales, different soil properties used to characterize the habitat can assume greater or lesser importance, depending on the function or attribute that is under consideration. For higher organisms, such as animals that range over wide territories, the habitat may be on the scale of a landscape and beyond. For studies of climate change effects on soil respiration, the distribution of hydrologic features within a watershed affecting soil warming may be appropriate. At the other extreme, evaluations of more specific processes that impact individual microbial species' functioning (e.g., denitrifier activity, and oxygen and substrate availability) may be possible only at the microhabitat scale. The habitat of a particular soil organism includes the physical and chemical attributes of its location and also the biological component of the habitat that influences the growth, activities, interactions, and survival of other organisms associated with this space (i.e., other microorganisms, fauna, plants, and animals). The spatial attributes across all scales must be considered when describing soil organism activity. Soil habitat spatial heterogeneity is an important contributor to the coexistence of species in soil microbial communities, thereby enhancing overall soil biodiversity by promoting the persistence of individual populations. It is central to the explanation of the high species-richness of the soil population.

Studies have confirmed that soil organisms are usually not randomly distributed, but exhibit predictable spatial patterns over wide spatial scales (Fig. 2.5). Spatial patterns of soil biota also affect the spatial patterns of activity and the processes they carry out. As an example, inorganic N production from ammonification of SOM can accumulate if the microbial processes of mineralization and immobilization are physically separated in soil space. This also occurs where plant residues are left on the soil surface compared to when they are incorporated into the soil.

Although the main factors influencing the gross behavior of soil organisms are known, the role of spatial distribution has not been studied in detail. There are few methods currently available that enable the study of microbial activity

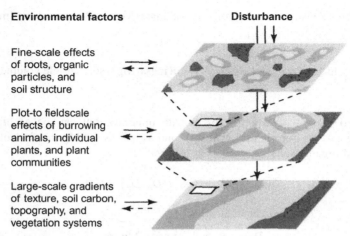

FIG. 2.5 Determinants of spatial heterogeneity of soil organisms. Spatial heterogeneity in soil organism distributions occurs on nested scales and is shaped by a spatial hierarchy of environmental factors, intrinsic population processes, and disturbance. Disturbance operates at all spatial scales and can be a key driver of spatial heterogeneity, for example, through biomass reduction of dominant organisms or alteration of the physical structure of the soil substrate. Feedback between spatial patterns of soil biotic activity and heterogeneity of environmental factors adds further complexity (dotted arrows) *(from Ettema and Wardle, 2002).*

in situ at the level of the soil microhabitat. It is common practice for soil scientists, after collecting samples in the field from a soil profile, to pass the soil through a 2-mm sieve to remove the plant debris and macrofauna and then homogenize the samples before analysis.

B Pore Space

On a volume basis, mineral soils are about 35-55% pore space, whereas organic soils are 80-90% pore space. Total soil pore space can vary widely for a variety of reasons, including soil mineralogy, bulk density, organic matter content, and disturbance. Pore space can range from as low as 25% for compacted subsoils in the lower vadose zone to more than 60% in well-aggregated clay-textured surface soils. Even though sand-textured soils have a higher mean pore size, they tend to have less total pore space than do clay soils.

Soil pore space is defined as the percentage of the total soil volume occupied by soil pores:

$$\%\text{pore space} = [\text{pore volume}/\text{soil volume}] \times 100 \qquad (2.2)$$

Direct measurements of soil pore volume are difficult to perform, but estimations can be obtained from data on soil bulk density and soil particle density, using the following formulas:

Soil bulk density : $D_b (Mg\ m^{-3}) = \text{soil mass}\,(Mg)/\text{soil bulk volume}\,(m^3)$

$$(2.3)$$

Soil particle density : $D_p (Mg\ m^{-3}) = \text{soil mass}\,(Mg)/\text{soil particle volume}\,(m^3)$

$$(2.4)$$

(assumed to be 2.65 Mg m^{-3} for silicate minerals, but can be as high as 3.25 Mg m^{-3} for iron-rich tropical soils and as low as 1.3 Mg m^{-3} for volcanic soils and organic soils)

$$\text{\%pore space} = 100 - \left[(D_b/D_p) \times 100\right] \qquad (2.5)$$

where:

$$D_b = \text{soil bulk density, Mg m}^{-3} \qquad (2.6)$$

$$D_p = \text{soil particle density, Mg m}^{-3} \qquad (2.7)$$

Although total pore space is important, the size and interconnection of the pores are the key in determining the habitability of the soil. Total pore space is usually divided into two size classes, macropores and micropores, largely based on their ability to retain water left after drainage under the influence of gravity (see soil water content). Macropores are ≥ 10 μm in diameter and allow rapid diffusion of air and rapid water infiltration and drainage. They can occur as the spaces between individual sand and coarse silt grains in coarse-textured soils and in the interaggregate pore space of well-aggregated loam- and clay-textured soils. Macropores can be created by roots, earthworms, and other soil organisms forming a special type of pore termed biopore. Biopores are typically lined with organic matter and clay and are ideal habitats for soil biota. They provide continuous channels often extending from the soil surface throughout the soil for lengths of one or more meters.

Soil pores < 10 μm in diameter are referred to as micropores and are important for water retention for plants and for providing an aqueous habitat for microorganisms. Water flow and gaseous diffusion in micropores are slow, but help to provide a stable, local environment for soil biota. Although the larger micropores, together with the smaller macropores, can accommodate plant root hairs and microorganisms, pores ≤ 5 μm in diameter are not habitable by most microorganisms. They may even restrict diffusion of exoenzymes and nutrients, thereby inhibiting the uptake of otherwise potential substrates. Although surface soils are typically about 50% pore space on a volume basis, only a quarter to a half of this pore space may be habitable by soil microorganisms due to restricted pore size (Table 2.1).

TABLE 2.1 Relationship of soil water potential and equivalent pore diameter to soil pore space (% of total soil porosity) across a textural gradient

General pore categories	Soil water potential (kPa)	Equivalent pore diameter (μm)	Sandy loam	Loam	Clay loam
Macropores (aeration)	0	All pores filled (saturated)	100	100	100
	-1	≤ 300	5	3	2
	-10	≤ 30	24	14	12
Micropores (capillary)	-33	≤ 10	8	5	3
	-100	≤ 3	3	14	7
	-1000	≤ 0.3	2	6	17
	-10,000	≤ 0.03	2	7	12
		Total pore space (%)	44	49	53

The work of Kubiena in the 1930s contributed significantly to our understanding of the nature of soil solid and pore space at the microscopic scale. Much of this early research was based on the examination of thin sections (25 μm thick) of intact blocks of soil. Adaptation of advancements in data acquisition and computer-assisted analysis of digital imagery during the past quarter century have led to the quantitative spatial analysis of soil components. Recent developments in microcomputerized X-ray tomography (CT scanning) allow for the study of the properties of the soil's intact three-dimensional structure. These systems have high resolution capabilities (10-30 μm), which allow differentiation of solids for quantifying the distribution of organic and mineral materials. The technology is also able to readily distinguish air-filled and water-filled pore space. However, it is still not possible to distinguish microorganisms from soil particles with this technology. A CT image of a soil core in three orthogonal planes is shown in Fig. 2.6. Highly attenuating features, such as iron oxide nodules, appear bright in the imagery, whereas features with low attenuation capability, such as pore space, appear dark. Though distinguishing microorganisms from soil particles with this technology is still limited, newer technologies, such as secondary ion mass spectrometry (SIMS), have the potential to do so (www.sciencemag.org/cgi/content/full/304/5677/1634/DC1; Movies S1 and S2; see online supplemental material at http://booksite.elsevier.com/9780124159556).

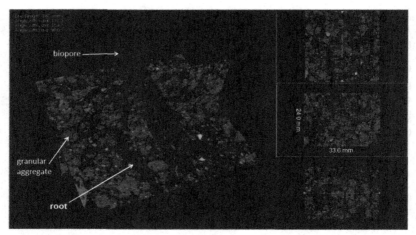

FIG. 2.6 X-ray CT image of forest mull Ah horizon from Nepean, Ontario, Canada. The three images on the right correspond to the orthogonal planes in the main image. Voxel size of imagery is 40 μm; dimensions of the full image are 33.6 mm (width) × 33.6 mm (length) × 24.0 mm (height). By convention, air-filled pore space is dark, and solid materials are lighter in tone. Diameter of the measured biopore, an earthworm channel, is ~3.7 mm.

V   SOIL SOLUTION CHEMISTRY

An understanding of the chemistry of the soil solution, providing an environment for soil organisms, needs to take into account the nature and quantity of its major components: water, dissolved organic matter and inorganic constituents, and O_2 and CO_2. The biogeochemistry of the soil solution is mainly determined by acid-base and redox reactions (Chapter 9). The thermodynamic activities of protons and electrons in soil solution define the chemical environment that controls biotic activity. Conceptually, both can be considered as flowing from regions of high concentration to regions of low concentration, with soil microbial activity having a profound effect on regulating this flow.

A   Soil pH

Protons supplied to the soil from atmospheric and organic sources react with bases contained in aluminosilicates, carbonates, and other mineralogical and humic constituents. In a humid climate with excess precipitation, and given sufficient time, basic cations (Na^+, K^+, Ca^{2+}, Mg^{2+}) will be exchanged from mineral and organic constituents by H^+ and be leached from the surface soil. The presence of calcite and clay minerals, such as smectites, which are saturated with basic cations, retards the rate of acidification. Continued hydrolysis results in the formation of the secondary minerals (kaolinite, gibbsite, and goethite) and a soil solution buffered between pH 3.5 and 5. Semi-arid and arid conditions lead to an opposite trend: a soil solution buffered at an alkaline pH. Soil pH influences a number of factors affecting microbial activity, such as solubility and ionization of inorganic and organic soil solution constituents, which in turn affect soil enzyme activity. There are large numbers of both organic and inorganic acids found in soils, although the majority of these acids are relatively weak.

Measurements of soil solution pH provide important data for predicting potential microbial reactions and enzyme activity in soil. Though easily measured in a soil paste with a pH electrode, interpretation of its effects on microbial processes can be difficult. This is largely because concentrations of cations sorbed to the surfaces of negatively charged soil colloids are 10-100 times higher than those of the soil solution. Thus, if enzymes are sorbed to colloid surfaces, their apparent pH optimum would be 1-2 pH units higher than if they were not sorbed. An example of this is soil urease activity, which has an apparent pH optimum of 8.5-9.0 in soil that is about 2 pH units higher than optimal urease activity measured in solution.

B Soil Redox

The most reduced material in the biosphere is the organic matter contained in living biomass. Organic matter in soils ranges from total dominance, as in peatlands, to the minor amounts found in young soils or at depth in the vadose zone. The metabolic activity of soil organisms produces electrons during the oxidation of organic matter, which then must be transferred to an electron acceptor, the largest of which is O_2 contained in freely drained, aerobic soils. The O_2 contained in soil air or present in soil solution can be consumed within hours depending on the activity of soil organisms and is replenished by O_2 diffusion. If O_2 consumption rates by soil organisms are high, due to an abundant supply of readily decomposable organic C, or if O_2 diffusion into the soil is impeded because of waterlogging or restricted pore sizes, due to clay texture or soil compaction, soil solution O_2 concentrations continue to decrease. When all available dissolved O_2 has been used, the solution changes from aerobic (oxic) to anaerobic (anoxic). Microbial activity will then be controlled by the movement of electrons to alternative electron acceptors.

Development of anaerobic conditions results in a shift in the activity of the soil microbial populations. The activity of aerobic and facultative organisms, which dominate well-drained oxic soils, decreases while the activity of obligate anaerobic organisms and fermentative organisms increases. This switch in electron acceptors promotes the reduction of several important elements in soil, including N, Mn, Fe, and S in a process known as anaerobic respiration and of CO_2 by methanogenisis.

Redox potential (E_H) measurements provide an indication of the soil aeration status. They are a measure of electron availability occurring as a result of electron transfer between oxidized (chemical species that have lost electrons) and reduced (chemical species that have gained electrons) chemical species. The measurements are often used to predict the most probable products of biological reactions. For example, N_2O can be produced from nitrification under aerobic conditions and from denitrification under moderately reducing conditions where the reduction intensity is not strong enough to completely reduce nitrate to N_2 gas.

The magnitude of E_H depends on $E°$, and also on the relative activities of the oxidant and the reductant. These various quantities are related by the Nernst equation,

$$E_H(V) = E° - (0.0591/n) \log (\text{reduct})/(\text{oxid}) + (0.0591\, m/n)\text{pH} \qquad (2.8)$$

The E_H is the electrode potential of the standard hydrogen electrode, $E°$ is the standard half-cell potential, n is the number of electrons transferred, m is the number of protons exchanged, *reduct* is the activity of the reduced species, and *oxid* is the activity of the oxidized species.

The major redox reactions occurring in soils and the electrode potentials for these transformations are shown in Table 2.2. Typically, dissolved O_2 and NO_3^- serve as electron acceptors at $E_H \sim 350$ to $400\, mV$ and above. When their concentrations in soil solution become low, Mn, Fe, and SO_4^{2-} serve as alternate electron acceptors. Their reduction processes occur over a wider range compared with O_2 and NO_3^- reduction and methane production. Mn and Fe serve as electron acceptors at $\sim 350\, mV$ for Mn and at $250\, mV$ to $\sim 100\, mV$ for Fe. When Fe^{3+} in Fe oxide is reduced, the oxide dissolves and Fe^{2+} goes into solution. Sulfate reduction occurs at an E_H as high as $350\, mV$ and to as low as $\sim 100\, mV$. Methane production begins when E_H is close to $\sim 100\, mV$.

Although the activity of electrons can be described by pE, E_H has the advantage of being a standard measurement for investigations of soil redox potential both in the laboratory and the field. Soil E_H can be obtained relatively easily from measurements of the pore water using a platinum (Pt)

TABLE 2.2 Important Redox Pairs and the Approximate E_H Values at the Occurrence of Transitions at the Reference Soil pH of 7.0

	Oxidized Form	Reduced Form	Approximate E_H at Transformations (mV)
Oxygen	O_2	H_2O	+600 to +400
Nitrogen	NO_3^-	N_2O, N_2, NH_4^+	250
Manganese	Mn^{4+}	Mn^{2+}	225
Iron	Fe^{3+}	Fe^{2+}	+100 to -100
Sulfur	SO_4^{2-}	S^{2-}	-100 to -200
Carbon	CO_2	CH_4	Less than -200

electrode. However, the values for soil E_H can be difficult to interpret as the Pt electrode measurement does not reflect changes in all the chemical species involved in redox reactions and also responds to changes in pH. Often two or more redox reactions occur simultaneously, and thus measured E_H usually reflects a mixed potential.

Platinum-electrode E_H measurements are still useful and can be interpreted as a semiquantitative assessment of a soil's redox status. In studies of paddy soils, for rice production, E_H measurements can be used to monitor progressive development of reducing conditions and can distinguish oxic and anoxic conditions. These fields provide a unique environment for studying the relationship between soil E_H and greenhouse gas emissions because of controlled irrigation and drainage practices (Fig. 2.7). During the flood season, the paddy soils are a major source of CH_4 and an important source of N_2O when they are drained. Strategies designed to mitigate CH_4 emissions from submerged rice fields can adversely affect greenhouse warming potential by stimulating higher N_2O emissions. The different E_H conditions required for N_2O and CH_4 formations and the trade-off pattern of their emissions as found in rice fields makes it a challenge to abate the production of one gas without enhancing the production of the other. An E_H greater than -150 mV, but less than +180 mV, offers the minimum global warming potential contribution from these rice soils.

C Soil Aeration

Molecular diffusion dominates the transport of gases in the soil. Diffusion through the air-filled pores maintains the gaseous exchange between the atmosphere and the soil, and diffusion through water films of varying thickness maintains the exchange of gases with soil organisms. Diffusivity through both pathways can be described by Fick's law:

$$J = D \, dc/dx, \qquad (2.9)$$

where J is the rate of gas diffusion (g cm^{-2} sec^{-1}), D is the diffusion coefficient for soil air and for water (cm^2 sec^{-1}), c is the gas concentration (g cm^{-3}), x is the distance (cm), and dc/dx is the concentration gradient. The gaseous diffusion coefficient in soil air is much smaller than that in the atmosphere. The limited fraction of total pore volume is occupied by continuous air-filled pores, pore tortuosity, and soil particles. Water reduces the cross-sectional area and increases the mean path length available for diffusion. For soil air, this is referred to as the effective diffusion coefficient, D_e, and is a function of the air-filled porosity. Likewise, tortuosity due to particulate material present in soil solution reduces rates of gaseous diffusion. As shown in Table 2.3, diffusion of gases in water is ~1/10,000 of that in air. Gaseous diffusion through a 10-μm water film would take the same time as diffusion through a 10-cm air-filled pore.

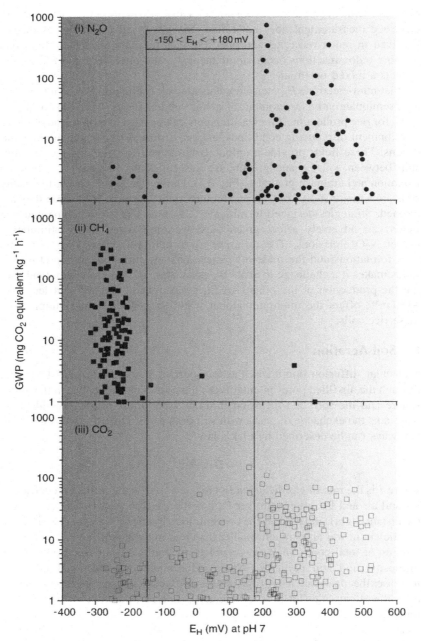

FIG. 2.7   Global warming potential (GWP) contribution of N_2O, CH_4, and CO_2 as a function of soil E_H. All eight soils showed the same pattern of (i) N_2O, (ii) CH_4, and (iii) CO_2 dynamics with soil E_H change from high to low. The figure is plotted in a logarithmic scale to cover a wide range of values. Global warming potential contributions below 1 mg CO_2 equivalent $kg^{-1} h^{-1}$ were considered insignificant and not illustrated in the figure for clarity. *(from Kewei and Patrick, 2004).*

TABLE 2.3 Temperature Effects on Gaseous Diffusion in (A) Air, in (B) Water, and (C) Gas Solubility in Water

Temperature (°C)	N_2	O_2	CO_2	N_2O
A. Gaseous Diffusion Coefficients in Air (cm^2/sec)				
0	0.148	0.178	0.139	0.179
10	0.157	0.189	0.150	0..190
20	0.170	0.205	0.161	0.206
30	0.180	0.217	0.172	0.218
B. Gaseous diffusion Coefficients in Water ($\times 10^{-4}$ cm^2/sec)				
0	0.091	0.110	0.088	0.111
10	0.130	0.157	0.125	0.158
20	0.175	0.210	0.167	0.211
30	0..228	0.275	0.219	0.276
C. Solubility Coefficients (volume of dissolved gas relative to volume of water, cm^3/cm^3)				
0	0.0235	0.0489	1.713	1.30
10	0.0186	0.0380	1.194	1.01
20	0.0154	0.0310	0.878	0.71
30	0.0134	0.0261	0.665	0.42

VI ENVIRONMENTAL FACTORS, TEMPERATURE, AND MOISTURE INTERACTIONS

A Soil Temperature

The physical, chemical, and biological processes that occur in soil are influenced by temperature in vastly different ways and are the most important environmental factors that must be considered. Whereas rates of molecular diffusion always increase with increasing temperature, solubility of gases in soil solution does not and can even decrease.

The relation between a chemical reaction rate and temperature was first proposed by Arrhenius:

$$k = A\,e^{-Ea/RT} \tag{2.10}$$

The constant A is called the frequency factor and is related to the frequency of molecular collisions; Ea is the activation energy required to initiate the reaction; R is the universal gas constant and has a value of 8.314 J mol^{-1} T^{-1}; e is the base of the natural logarithm; T is the absolute temperature (in °K); and k is the specific reaction rate constant (time^{-1}).

Conversion of Eq. (2.10) to natural logarithmic form gives

$$\ln k = (-Ea/RT) + \ln A \tag{2.11}$$

By determining values for k over a moderate range of soil temperatures, a plot of ln k versus 1/T results in a linear relationship, providing the activation energy is constant over the temperature interval; Ea is obtained from the slope of the line and A from the Y intercept.

A similar equation can be used to describe temperature effects on enzyme activity:

$$k_{(cat)} = \kappa(k_B T/h)e^{-\Delta G^{\#}/RT} \tag{2.12}$$

where $k_{(cat)}$ is the reaction rate (time^{-1}), κ is the transmission coefficient, k_B is the Bolzmann constant, h is the Planck constant, $\Delta G^{\#}$ is the activation energy, R is the universal gas constant, and T is the absolute temperature (in °K). The transmission coefficient varies significantly with the viscosity of the soil solution, which increases by a factor of almost 2 over a temperature drop from 20°C to near 0°C. In both instances, a change in temperature results in an exponential change in the reaction rate, the magnitude of which is a function of the activation energy. Note also that there is an inverse relationship between reaction rate and activation energy.

The relationship between temperature and biologically mediated processes is complicated as individual species differ in their optimal temperature response, different microbial communities are active as temperatures change, and microorganisms are able to adapt by altering their physiology and cellular mechanisms, membrane fluidity and permeability, and structural flexibility of enzymes and proteins.

The relative sensitivity of soil microbial activity to temperature can be expressed as a Q_{10} function, which is the proportional change in activity associated with a 10°C temperature change

$$Q_{10} = (k_2/k_1)^{[(10/(T_2-T_1)]} \tag{2.13}$$

where k_2 and k_1 are the rate constants for a microbial process under study at temperatures differing by 10°C. It is generally accepted that a Q_{10} of ~2 can be used to describe the temperature sensitivity of soil biochemical processes, such as soil respiration, over the mesophilic temperature range (20-45°C); that is, microbial activity at 30°C is twofold higher than it is at 20°C. At temperatures beyond 45°C, the microbial community composition shifts from mesophilic to thermophilic, and microorganisms adapt by increasing concentrations of saturated fatty acids in their cytoplasmic membrane and by production of heat-stable proteins.

Microorganisms that have an upper growth temperature limit of $\leq 20°C$, commonly referred to as psychrophiles, are capable of growth at low temperatures by adjusting upward both the osmotic concentration of their cytoplasmic constituents to permit cell interiors to remain unfrozen and the proportion of unsaturated fatty acids in their cytoplasmic membrane. A common adaptive feature of psychrophiles to low temperatures is that their enzymes have much lower activation energies and much higher (up to 10-fold) specific activities than do those of mesophiles, resulting in a reaction rate, $k_{(cat)}$, that is largely independent of temperature. Although microbial activity slows at lower temperatures, rates are significantly higher and more sensitive to temperature changes than those predicted from studies over the mesophilic temperature range. As an example, researchers studying decomposition of SOM, soil respiration, and N mineralization have reported values for Q_{10} increasing to near 8-10 with soil temperatures approaching 0°C (Kirschbaum, 2013). Díaz-Raviña et al. (1994) reported Q_{10} values between 3.7 and 6.7 for the 0-10°C interval for thymidine incorporation and between 5.0 and 13.9 for acetate incorporation for a soil bacterial community.

Low temperatures are common over vast areas of the earth and include soils in temperate, polar, and alpine regions where mean annual temperatures are < 5°C. Soils in these environments contain a great diversity of cold-adapted microorganisms, able to thrive even at subzero temperatures and to survive repeated freeze/thaw events. A bacterium isolated from a permafrost soil in northern Canada can grow at -15°C and remain metabolically active at temperatures down to -25°C. Microbial activity during cold periods when plants are dormant or soils are barren can play a significant role in overwinter losses of soil nutrients, in particular for N with overwinter leaching and by denitrification during freeze/thaw.

Very few soils maintain a uniform temperature in their upper layers. Variations may be either seasonal or diurnal. Because of the high specific heat of water, wet soils are less subject to large diurnal temperature fluctuations than are dry soils. Among factors affecting the rate of soil warming, the intensity and reflectance of solar irradiation are critical. The soil's aspects (south- versus north-facing slopes), steepness of slope, degree of shading, and surface cover (vegetation, litter, mulches) determine effective solar irradiation. Given the importance of soil temperature in controlling soil processes, models of energy movement into the surface soil profile have been developed. They are based on physical laws of soil heat transport and thermal diffusivity and include empirical parameters related to the temporal (seasonal) and sinusoidal variations in the diurnal pattern of near-surface air temperatures. The amplitude of the diurnal soil temperature variation is greatly dampened with profile depth.

B Soil Water

Soil water affects the moisture available to organisms as well as soil aeration status, the nature and amount of soluble materials, the osmotic pressure, and the pH of the soil solution. Water acts physically as an agent of transport by mass flow and as a medium through which reactants diffuse to and from sites of reaction. Chemically, water acts as a solvent and as a reactant in important chemical and biological reactions. Of special significance in the soil system, and to microbial cells in particular, is the fact that water adsorbs strongly to itself and to surfaces of soil particles by hydrogen bonding and dipole interactions. Soil water content can be measured on a mass or volume basis. Gravimetric soil water content is the mass of water in the soil, measured as the mass loss in a soil dried at 105°C (oven-dry weight) and is expressed per unit mass of oven-dried soil. Volumetric soil water content is the volume of water per unit volume of soil. Soil water is also described in terms of its potential free energy, based on the concept of matric, osmotic, and gravitational forces affecting water potential. Soil water potential is expressed in units of pascals (Pa), or more commonly, kilopascals (kPa), with pure water having a potential of 0. Matric forces are attributed to the adhesive or adsorption forces of water attraction to surfaces of mineral and organic particles and to cohesive forces or attraction to itself. These forces reduce the free energy status of the water. Solutes dissolved in soil solution also contribute to a reduction in the free energy of water and give rise to an osmotic potential that also is negative. Combined, the matric and osmotic forces are responsible for the retention of water in soils. They act against gravitational forces that tend to draw water downward and out of the soil.

When the gravitational forces that drain water downward are exactly counterbalanced by the matric and osmotic forces that hold onto the water, the soil is said to be at field capacity, or at its water-holding capacity. This will occur after irrigation, after a heavy rainfall, or after spring thaw, which leave the soil saturated with a soil water potential = 0 kPa. Gravitational forces begin to immediately drain away water in excess of what can be retained by matric + osmotic forces, leaving the soil after one to two days at field capacity. By definition, the field capacity for loam and clay loam soils is a soil water potential of -33 kPa, and for sandy soils it is -10 kPa. (How soil properties affect soil water storage and movement: http://www.youtube.com/watch?v=jWwtDKT6NAw; see online supplemental material at: http://booksite.elsevier.com/9780124159556).

Water retention, or soil water content at a given soil water potential, is a function of the size of pores present in the soil, or pore size distribution (Table 2.1). Soils of different texture have very different water contents, even though they have the same water potential. Surface tension is an important property of water influencing its behavior in soil pores. In addition, due to water's strong cohesive forces, it has a high surface tension. Based on matric forces and properties of surface tension, the maximum diameter of pores filled with water at a given soil water potential can be estimated using the Young-Laplace equation:

$$\frac{\text{Maximum pore diameter } (\mu m)}{\text{Retaining water}} = \frac{300}{\text{Soil water potential (kPa)}}$$

(2.14)

Those soil pores ≥ 10 μm in diameter drain under the influence of gravitation forces, given that the soil water potential at field capacity is -33 kPa.

Soil water characteristics are difficult to assess in laboratory and field studies. A program developed by Solomon and Rauls can be used to estimate soil water characteristics based on soil texture and organic matter content, which are commonly measured physical and chemical soil properties (http://hydrolab. arsusda.gov/soilwater/Index.htm; see online supplemental material at: http:// booksite.elsevier.com/9780124159556). Soil water potential determines the energy that an organism must expend to obtain water from the soil solution. Generally, aerobic microbial activity in soil is considered optimal over soil water potentials ranging from about -50 to -150 kPa, which is 30-50% of the soil's total pore space, depending on its texture and bulk density (Table 2.1). Aerobic activity decreases as soil becomes wetter and eventually saturated, due to restricted O_2 diffusion. When greater than 60% of the pore space is water-filled, activity of microorganisms able to use alternative electron acceptors increases (e.g., that of anaerobic denitrifiers).

As the soil dries and water potential decreases, water films on soil particles become thinner and more disconnected, restricting substrate and nutrient diffusion, and increasing the concentration of salts in the soil solution. Although many plants grown for agricultural purposes wilt permanently when the soil water potential reaches -1500 kPa, rates of soil microbial activity are less affected as the relative humidity within the soil remains high. Respiration rates, for example, can still be ∼90% of maximum at this low soil water potential. Studies both in laboratory incubations and in the field have reported that the logarithm of soil matric potential is a good predictor of the effect of water content on soil respiration.

Although some microorganisms are able to adapt to low soil water potentials by accumulating osmolytes (amino acids and polyols) or changing the properties of their outer membrane, rapid changes in soil water potential associated with drying/rewetting cause microbes to undergo osmotic shock and induce cell lysis. A flush of activity by the remaining microbes, known as the Birch effect, results from mineralization of the labile cell constituents released.

Different communities of organisms are active over the range of water potentials commonly found in soils. Protozoa are active at water potentials near field capacity in water films ≥ 5 μm thick, whereas microorganisms can be active at lower water potentials, due to their size and association with the surfaces of soil particles. Even though fungi are generally considered to be more tolerant of lower soil water potentials than are bacteria, presumably because soil bacteria are relatively immobile and rely more on diffusion processes for nutrition, the opposite has also been shown to be true due to differences in community structure. Table 2.4 shows differences in the ability of various soil organisms to tolerate water stress. Sulfur and ammonium oxidizers, typified by *Nitrosomonas* and *Thiobacillus* species, respectively, are less tolerant of water stress than are the ammonifiers and typified by

TABLE 2.4 Ability of Different Organisms to Tolerate Water Stress

Soil Water Potential (kPa)	Water Activity (A_w)	Organism
-0.5	> 0.99	Protozoa
-1,500	0.99	*Rhizobium, Nitrosomonas*
-4,000	0.97	*Bacillus*
-10,000	0.93	*Clostridium, Fusarium, Eurotium*
-25,000	0.83	*Micrococcus, Penicillium, Aspergillus*
-65,000	0.62	*Xeromyces, Chrysosporium, Monascus*

Clostridium and *Penicillium*. Ammonium may accumulate in droughty soils at the water potentials where ammonifiers are still active but restrict nitrification.

Soil moisture and temperature are the critical factors affected by climate regulating soil biological activity. This control is affected by changes in the underlying rates of enzyme-catalyzed reactions and sizes of the substrate organic and inorganic pools. Where water is nonlimiting, biological activity may depend primarily on temperature. Standard Arrhenius theory can be used to predict these temperature effects. However, as soils dry, moisture is a greater determining factor of biological processes than is temperature. Likely these two environmental influences do not affect microbial activity in linear fashion, but display complex, nonlinear interactions that reflect the individual responses of the various microorganisms and their associated enzyme systems.

The interactions of temperature, moisture, and organisms are exemplified by the current concerns about climate change on soil biology. A hundred years ago, Swedish scientist Svante Arrhenius asked the important question, "Is the mean temperature of the ground in any way influenced by the presence of the heat-absorbing gases in the atmosphere?" He went on to become the first person to investigate the effect that doubling atmospheric CO_2 would have on global climate. The question was debated throughout the early part of the 20th century and is a main concern of Earth systems scientists today. The earth's surface is warming for several reasons, one of which is increased emissions of greenhouse gases from soils due to past and current agricultural management practices, deforestation, combustion of fossil fuels, and industrial pollution. Global temperatures have increased by $\sim$0.5°C over the past 100 years and are expected to increase by 1°C to 6°C by 2100. Although this represents only a few degrees temperature change, global warming will dramatically increase microbial decay rates of the organic matter stored in the boreal forests and tundra regions, estimated to contain $\sim$30% of the global soil C (Kirschbaum, 1995). Global warming is already melting glaciers and ice sheets at accelerated rates, and the permafrost thaw rate has more than tripled over the past half century.

The critical concern is that SOM decomposition is stimulated to a greater extent than is net plant productivity, the C input to SOM. Theory also suggests that the more resistant organic matter compounds, with high activation energies, would become more decomposable at higher temperatures (Davidson and Janssens, 2006). Degradation of cellulose, hemicellulose, and other components of SOM by extracellular enzymes is the rate-limiting step in CO_2 emissions. However, feedback mechanisms characteristic of all biogeochemical cycles may dampen effects of temperature changes. Soils are complex, with interactions such as change in rainfall patterns that affect plant productivity, soil water storage, and nutrient cycling. Many of the environmental constraints affect decomposition reactions by altering organic matter (substrate) concentrations at the site at which all decomposition occurs, that of the enzyme reaction site. We must also consider decomposition rates at the enzyme affinity level; Michaelis-Menten models of enzyme kinetics are covered in Chapter 16 and energy yields in Chapter 9. The kinetic and thermodynamic properties of extracellular enzymes and their responses to environmental factors are now being considered in models of the effects of global warming on carbon cycling. Changes in microbial community structure (Chapter 8) will also have profound influences. The goal of this chapter is to provide an environmental boundary of the soil habitat and a description of its fundamental physical and chemical properties. With this as a foundation, later chapters in this volume explore in detail information about organisms, their biochemistry, and interactions.

REFERENCES

Ball, B.C., 2013. Soil structure and greenhouse gas emissions: a synthesis of 20 years of experimentation. Eur. J. Soil Sci. 64, 357–373.

Birkeland, P.W., 1999. Soils and Geomorphology, third ed. Oxford University Press, Oxford.

Davidson, E.A., Janssens, I.A., 2006. Temperature sensitivity of soil carbon decomposition and feedbacks to climate change. Nature 440, 165–173.

Díaz-Raviña, M., Frostegård, Å., Bååth, E., 1994. Thymidine, leucine and acetate incorporation into soil bacterial assemblages at different temperatures. FEMS Microbiol. Ecol. 14, 221–232.

Ettema, C.H., Wardle, D.A., 2002. Spatial soil ecology. Trends Ecol. Evol. 17, 177–183.

Guimarães, R.M.L., Ball, B.C., Tormena, C.A., 2011. Improvements in the visual evaluation of soil structure. Soil Use Manage. 27, 395–403. download method for evaluation of aggregation in the field, http://www.sac.ac.uk/vess.

Kewei, Y., Patrick Jr., W.H., 2004. Redox window with minimum global warming potential contribution from rice soils. Soil Sci. Soc. Am. J. 68, 2086–2091.

Kirschbaum, M.U.F., 2013. Seasonal variations in the availability of labile substrate confound the temperature dependence of organic matter decomposition. Soil Biol. Biochem. 57, 568–576.

Kirschbaum, M.U.F., 1995. The temperature dependence of soil organic-matter decomposition, and the effect of global warming on soil C storage. Soil Biol. Biochem. 27, 753–760.

Paul, E.A., Clark, F.E., 1996. Soil Microbiology and Biochemistry, second ed. Academic Press, San Diego, CA, USA.

Tisdall, J.M., Oades, J.M., 1982. Organic matter and water stable aggregates in soils. J. Soil Sci. 33, 141–163.

Chapter 3

The Bacteria and Archaea

Ken Killham[1] and Jim I. Prosser[2]

[1]*James Hutton Institute, Invergowrie, Dundee, Scotland, UK*
[2]*Institute of Biological and Environmental Sciences, University of Aberdeen, Aberdeen, Scotland, UK*

Chapter Contents

I INTRODUCTION

Bacteria and archaea have been distinguished by the term *prokaryotes* from *eukaryotes*, which comprise algae, protozoa, fungi, plants, and animals. Prokaryotes were originally defined as those organisms whose nucleus, respiratory, and photosynthetic machinery were not separated from cytoplasm by membranes; where nuclear division occurred by fission rather than mitosis; and whose cell walls contained mucopeptide. In fact, the mitochondria and chloroplasts of eukaryotic cells originated as endosymbiotic prokaryotic cells. Ribosomes in prokaryotes are smaller (70S) than those of eukaryotes (80S), and no eukaryote is able to fix atmospheric N_2 or produce methane. There are many exceptions to this early definition, and subsequent attempts to distinguish prokaryotes and eukaryotes in terms of morphological and physiological characteristics and boundaries between the two domains are blurred. Although bacteria and archaea may resemble each other in their microscopic size, in many other ways they are as different from each other as they are from eukaryotes.

Despite the apparent relative simplicity of prokaryotes as a group, they have substantially greater phylogenetic and functional diversity than eukaryotes and possess the most efficient dispersal and survival mechanisms of all organisms. They also vastly outnumber eukaryotes. Global organic C in prokaryotes is equivalent to that in plants; soil bacteria and archaea contain 10-fold more N. As a consequence, prokaryotes play an essential role in all global biogeochemical cycles and other important ecosystem functions, including the creation, maintenance, and functioning of soil. This chapter describes the phylogeny and physiological characteristics of soil bacteria and archaea in the context of soil ecosystems and shows how major recent discoveries are changing our view of their ecology. There is considerable debate regarding use of the term *prokaryote*, rather than specifying bacteria and archaea (Doolittle and Zhaxybayeva, 2013), but the term will be used in this chapter for brevity, while acknowledging the considerable differences between bacteria and archaea.

II PHYLOGENY

A Cultivated Organisms

Traditional classification of prokaryotes was based on a large number of phenotypic characteristics (e.g., morphology, motility, biochemical characteristics, antibiotic sensitivity). In the 1980s, the potential to establish evolutionary relationships through differences in gene sequence and the development of early sequencing techniques led to phylogenetic studies based on sequences of 16S rRNA and 18S rRNA genes, which are present in all prokaryote and eukaryote cells, respectively. These genes possess regions of highly conserved sequence, facilitating alignment, and variable and hypervariable regions, which enable discrimination between different organisms. Quantification of genetic distance, by comparing sequence differences between organisms, allows estimation of evolutionary distance. These studies showed that the archaea and the bacteria are, in terms of phylogenetic or evolutionary distance, as distinct from each other as they are from eukaryotes (Fig. 3.1; Woese et al., 1990). Rather than two domains of life, prokaryotes and eukaryotes, there are three: archaea, bacteria, and eukarya. Figure 3.1 also illustrates the remarkably high diversity of bacteria, archaea, and eukaryotic microorganisms compared to that of plants and animals.

Evidence suggests that archaea are, in many ways, more similar to eukaryotes than they are to bacteria. In fact, there is now a return to a two-domain view of life, but the two domains are archaea and bacteria, with eukaryotes branching

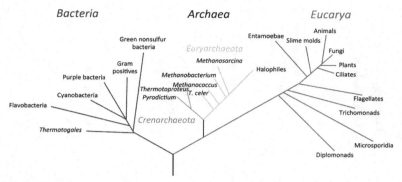

FIG. 3.1 The universal tree of life constructed by sequence analysis of single subunit rRNA genes. *(Redrawn from Wheelis et al., 1992.)*

from within the archaea. The major distinction in life is therefore between archaea and bacteria, and not between prokaryotes and eukaryotes (Williams et al., 2012). These developments influence the way that we consider the term *prokaryote* and, indeed, whether it should be used at all (Doolittle and Zhaxybayeva, 2013). Molecular phylogeny also led to reanalysis and reappraisal of classification within the bacteria and archaea. These major divisions are presented in Tables 3.1 and 3.2.

B Uncultivated Organisms

Cultivation-based studies indicated that soil prokaryotes were highly diverse, but there was increasing evidence that only a small fraction of cultivated organisms could grow on laboratory media (Torsvik et al., 1996). This was confirmed by use of the polymerase chain reaction (PCR) to amplify 16S rRNA genes directly from DNA extracted from soil, thereby avoiding the requirement of laboratory cultivation (Pace et al., 1986). Analysis of these sequences demonstrated the presence and abundance of novel, high-level taxonomic groups with very few or no cultivated representatives (Fig. 3.2). There was tremendous 16S rRNA gene-sequence diversity within these uncultivated groups and previously unsuspected diversity within representatives of cultivated soil prokaryotes that had been studied extensively in the laboratory for many years.

The impact of these molecular studies is illustrated by phylogenetic analysis of bacterial 16S rRNA gene sequences from cultivated organisms and sequences amplified directly from environmental DNA performed just a few years after introduction of these molecular techniques (Hugenholtz et al., 1998). The analysis identified 40 high-level groups within the bacterial domain. Almost 50% of these groups contained sequences from bacteria that had never been seen in culture. In addition, several of the remaining groups contained very few sequences from cultivated organisms and were dominated by environmental sequences. In just a few years, this approach had nearly doubled the number

TABLE 3.1 Characteristics of Bacterial Phylogenetic Groups with Cultivated Representatives

	Environmental Origin	Metabolism	Other Characteristics	Examples
Aquificales	Extreme environments (hot, sulfur pools, thermal vents)	Microaerophilic; chemolithotrophic; can oxidize hydrogen and reduce sulfur		*Aquifex aeolicus*
Thermodesulfobacterium	Thermal vent	Sulfate reducers; autotrophic or organotrophic; anaerobic	Motile rods; Gram-negative cell wall	*Thermodesulfobacterium hydrogeniphilum*
Thermotogales	Hot vents and springs; moderate pH and salinity	Sulfur reducers; organotrophic; some produce hydrogen	Prominent cell envelope	*Thermotoga maritima*
Coprothermobacter	Anaerobic digestors, cattle manure	Heterotrophic, methanogenic, sulfate reduction	Rod-shaped cells	*Coprothermobacter platensis*
Dictyoglomus	Hot environments	Chemoorganotrophic	Degrade xylan	*Dictyoglomus thermophilum*
Green non-sulfur bacteria and relatives	Wide range, but few cultured	Anoxygenic photosynthesis (*Chloroflexus*); organotrophic (*Thermomicrobium*)		*Chloroflexus, Herpetosiphon, Thermomicrobium roseum*
Actinobacteria (high G + C Gram-positives, including actinomycetes)	Soil, some are pathogens	Aerobes, Heterotrophic—major role in decomposition	Gram-positive, includes mycelial forms	

Group	Environment	Metabolism	Characteristics	Examples
Planctomycetes	Soil and water	Obligate aerobes	Flagellated swarmer cells; budding bacteria, ovoid, holdfast, cell wall lacks murein	Planctomyces, Pasteuria, Isocystis pallida
Chlamydia	Intracellular parasites	Heterotrophic	No peptidoglycan	Chlamydia psittaci, Trachomatis
Verrucomicrobia	Freshwater and soil. Few cultured			
Nitrospira	Soil and aquatic environments	Autotrophic nitrite oxidizers, facultative heterotrophs	Spiral shaped	Nitrospira
Acidobacterium	Wide range of environments, including soil	Acidophilic or anaerobic (very few cultured)		
Synergistes	Anaerobic environments (termite guts, soil, anaerobic digestors)	Anaerobic		
Flexistipes	Animals		Spiral shaped	
Cyanobacteria	Aquatic, but found in soil	Oxygenic, photosynthetic; some fix N_2	Gliding; unicellular, colonial, or filamentous	Aphanocapsa, Oscillatoria, Nostoc, Synechococcus, Gleobacter, Prochloron
Firmicutes (low G+C Gram-positive)	Soil, water, some are pathogens	Aerobic or anaerobic (rarely photosynthetic)	Cocci or rods; includes endospore formers	Clostridium Peptococcus, Bacillus, Mycoplasma
Fibrobacter				
Green sulfur bacteria	Anaerobic and sulfur-containing muds, freshwater and marine	Photosynthetic; anaerobic; autotrophic (S oxidation) or heterotrophic	Non-motile	Chlorobium, Chloroherpeton

Continued

TABLE 3.1 Characteristics of Bacterial Phylogenetic Groups with Cultivated Representatives—Cont'd

	Environmental Origin	Metabolism	Other Characteristics	Examples
Bacteroides-Cytophaga-Flexibacter group	Wide variety, including soil, dung, decaying organic matter	Aerobic, microaerophilic or facultatively anaerobic, organotrophs, some strict anaerobes (*Bacteroides*)	Gliding (*Cytophaga*), Gram-negative, rods, some pleomorphic, some helical, unbranched filaments	*Flavobacterium, Sphingobacterium, Cytophaga, Saprospira Bacteroides, Prevotella, Porphyromonas*
Thermus/Deinococcus	High temperature environments, nuclear waste		Coccoid, rods; radioresistant or thermophilic; thick cell wall	*Deinococcus radiodurans, Thermus aquaticus*
Spirochetes and relatives (spirochetes and leptospiras)	Wide range	Chemoheterotrophic	Motile with flagellum; long, helical, coils; Gram-negative	*Spirochaeta, Treponema, Borrelia, Leptospira, Leptonema*
Fusobacteria	Pathogens	Anaerobic	Gram-negative cell wall	
Proteobacteria	"Classical" Gram-negative bacteria	Heterotrophs; chemolithotrophs; chemophototrophs; anaerobic (most) or aerobic; some photosynthetic; some fix N_2	Often motile (flagella or gliding); Gram-negative cell wall structure	*Rhizobacterium, Agrobacterium, Rickettsia, Nitrobacter, Pseudomonas, Nitrosomonas, Thiobacillus, Alcaligenes, Spirillum, Nitrosospira, Legionella,* (some) *Beggiatoa, Desulfovibrio, Myxobacteria, Bdellovibrio*

Some groups are represented by many cultivated organisms. In general, these are found in a wide range of habitats and exhibit a diversity of physiological and morphological characteristics. Other groups are represented by very few cultivated strains. The range of environments and physiologies listed here are likely to expand as additional environments are explored and isolation techniques improve.

TABLE 3.2 Characteristics of Archaeal Phylogenetic Groups

	Environmental Origin	Metabolism	Other Characteristics	Examples
Euryarchaeota				
Extreme halophiles	Salt lakes	Heterotrophic; aerobic; non-photosynthetic phosphorylation	Require salt for growth, cell walls and enzymes stabilized by Na$^+$	*Halobacterium, Natronobacterium*
Methanogens	Swamps, marshes, marine sediments, guts, sewage treatment	Anaerobic; generate methane; fix CO_2; electrons from hydrogen		*Methanobacterium, Methanospirillum, Methanococcus*
Sulfur-metabolizing thermophiles	Hydrothermal vents	Anaerobic sulfur oxidizers, thermophiles, heterotrophs, methanogens, sulfate reducers	Polar flagella	*Thermococcus Pyrococcus Archaeoglobus*
Crenarchaeota				
Hypothermophiles	Hot sulfur rich environments (hot springs, thermal vents)	Oxidize elemental sulfur, aerobic or anaerobic	Some lack cell wall, e.g., *Thermoplasma*; group includes *Thermus aquaticus*	*Thermoplasma Sulfolobus, Acidothermus Pyrodictium occultum*
Thaumarchaeota				
Ammonia oxidizers	Soil, marine	Autotrophic/mixotrophic; oxidize ammonia to nitrite		*Nitrososphaera viennensis, Nitrosotalea devanaterra*
Korarchaeota/Xenarchaeota	Hot springs	None cultivated		
Nanoarchaeota	Hot vents	Symbiont of archaea	Coccoid; <400 nm in diam	*Nanoarchaeum equitans*

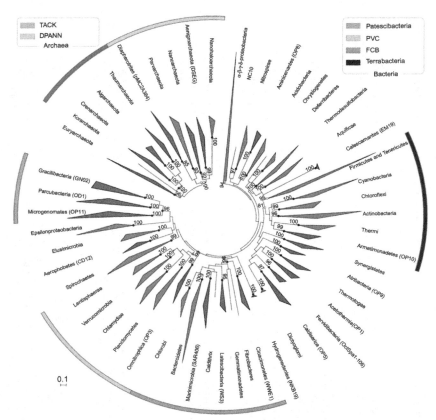

FIG. 3.2 Phylogenetic analysis of ≤ 38 genes from genomes of cultivated and uncultivated (in red) archaea and bacteria. Superphyla are indicated by different colors. *(From Rinke et al., 2013, with permission.)*

of known bacterial high-level taxonomic groups and revolutionized our view of diversity within cultivated organisms.

Soil bacterial and archaeal communities are now routinely characterized by analysis of 16S rRNA genes due to avoidance of the requirement for laboratory cultivation, although this approach does have some limitations. Comparison of environmental sequences with those in ever-expanding databases enables putative identification of organisms and comparison of communities in different environments. Bacteria that were previously considered to be "typical" of soil (bacilli within the phylum Firmicutes; pseudomonads, e.g., *Pseudomonas aeruginosa*, *P. fluorescens*, *P. putida*; actinobacteria, including the former actinomycetes) are often rare, whereas many novel, "yet-to-be-cultured" organisms are ubiquitous and dominant (Rappe and Giovannoni, 2003). Similarly, archaea were considered to be extremophiles, adapted to conditions atypical of most soils (high temperature, salt concentration, acidity, or anaerobicity), but one group, the crenarchaea, typically represent 1-2% of temperate soil prokaryote communities (Buckley and Schmidt, 2003).

The development of modern sequencing methods and associated technologies has increased the depth and detail with which we can characterize natural communities and has decreased reliance on cultivation for phylogenetic studies. It is now possible to isolate a single bacterial or archaeal cell, to amplify and sequence its DNA, and to reconstruct its genome without cultivation. This approach (single-cell genomics) enables investigation of the metabolic potential of uncultivated organisms, and information from genomes already obtained by this process has provided a much more detailed and robust phylogeny than that obtained using a single gene, such as the 16S rRNA gene (Fig. 3.2; Rinke et al., 2013). This is a rapidly developing area of research, and as a result, regularly updated interactive web sites are now available to facilitate viewing of current phylogenies (e.g., http://tolweb.org/tree/, http://itol.embl.de/).

C Phylogeny and Function

Molecular characterization of microbial communities would be of enormous value if the presence of a particular sequence or organism provided information on its function in the soil. In some cases, phylogenetic groupings are informative. For example, bacilli form resistant spores and rhizobia fix N_2. However, individual phylogenetic groups can display considerable physiological versatility, and many functional characteristics are distributed among varied and evolutionarily distant groups. In addition, we know little of the physiological characteristics of novel groups and subgroups.

The scale of this problem is illustrated by analysis of cultivated prokaryotes. Traditional phenetic classification of prokaryotes led to the ordering and naming of hierarchical groups as for plants and animals, with the species as the basic unit of classification. For higher organisms, the species is defined through the Biological Species Concept as the ability of members within a species to interbreed and not breed with members of other species (Cohan, 2002). Prokaryotes do not have sexual reproduction mechanisms, preventing application of this species concept. This has led to many attempts to define species in other ways and the absence of a reliable, universal species definition. Prokaryotes can transfer genes by horizontal (or lateral) gene transfer, HGT, which bypasses standard evolutionary processes. The most obvious example is the spread of plasmid-borne antibiotic resistance. HGT influences community structure and activity enormously, further reducing the relevance of "species" in ecological studies.

Sequencing of complete genomes of cultivated prokaryotes has increased, rather than decreased, the importance of these issues. Comparative genomics has shown HGT to be much greater than previously thought, such that evolutionary relationships must be represented as complex webs or networks, rather than the dichotomously branching trees illustrated in Figs. 3.1 and 3.2. In addition, genomes of different members of the same species contain a "core" of genes, common to all, but comprising only a minority of the species "pangenome" (Koonin, 2012). For example, the core genome of seven *Escherichia coli* strains contained 2800 genes, from a pangenome of approximately 6000 genes (Chen et al., 2006). Most genes, including those responsible for metabolism and substrate uptake, form a "shell" and each will be present in only a proportion of species members. The remaining genes (10-30%) form a "cloud," each present in a small minority of members. The functions of some of these genes are unknown, but many encode functions important for bacterial or archaeal ecology (e.g., response to environmental stress). There is therefore considerable diversity, even within a single species, no reliable "unit of diversity," much weaker links between phylogeny and function, and other methodological and conceptual implications for our understanding of evolution and diversity.

The continuing decrease in sequencing costs facilitates molecular characterization of soil prokaryotes, and complete coverage of 16S rRNA gene sequences is being approached in soils. Weak links between function and phylogeny and the lack of availability of cultivated representatives of soil organisms deny us knowledge of associated physiological characteristics and potential; thus we can only speculate on the role of most soil prokaryotes. There is a need to improve methods for enrichment, isolation, and cultivation of these organisms and/or to develop further molecular, or at least cultivation-independent, approaches to establish their ecosystem function *in situ*. As a consequence, soil bacterial and archaeal community ecology is dominated by descriptive studies and correlations between community composition and soil characteristics rather than studies that increase the understanding of the mechanisms driving community composition and the significance of prokaryote diversity for soil ecosystem function.

III GENERAL FEATURES OF PROKARYOTES

The majority of prokaryotes are smaller than eukaryotes, and cell size *per se* has significant influences on their ecology, methods of study, and perceptions of importance. Prokaryotic cells are in the order of portions of microns to several microns in length or diameter, although there are notable exceptions (Schulz and Jorgensen, 2001). Microscopic size makes observational studies difficult and leads to the study of populations or communities, rather than individuals, and to estimation of characteristics (e.g., cell concentrations) on the basis of properties of samples.

Small size is associated with high surface area: volume ratio, which explains, in part, the ability of prokaryotes to sequester nutrients at extremely low concentrations. Cells are in intimate contact with their physical and chemical environment. Although homeostatic mechanisms exist for maintenance of internal solute concentrations and pH, prokaryotes respond much more rapidly to, and are influenced more by, changes in environmental conditions than complex eukaryotic cells. This, in turn, demands greater consideration of micro-environments or microhabitats and the physicochemical characteristics of the environment immediately surrounding the cell. The 1-10 μm scale will be of greater significance for growth and activity of unicellular prokaryotes than bulk soil properties. This introduces technical challenges, but also conceptual issues. Cell size will influence the distribution and movement of organisms. For example, prokaryotes are able to penetrate and colonize small soil pores, potentially protecting them from predation. Spatial heterogeneity, coupled with limited transport of soluble nutrients and cells through soil, has enormous implications for biological interactions, including competition, with consequent impact on evolutionary processes and diversity.

IV CELL STRUCTURE

A Unicellular Growth Forms

Cell shape depends on an interaction between generation of internal turgor pressure and variation in the elasticity of cell wall components during cell growth. These interactions can be used to explain the differences between rod-shaped and coccoid cells and some of the more detailed aspects of cell shape and morphology. In addition, there is increasing evidence for the involvement of an actin-like skeleton controlling cell growth and shape in bacterial cells (Carballido-Lopez and Errington, 2003; Daniel and Errington, 2003). Unicellular bacteria exhibit a wide range of cell forms, including spiral, vibrioids, pleomorphic cells, stalked cells, bacteroids, and even square bacteria. Archaea also exhibit a range of unusual growth forms, although the mechanisms involved are unclear.

The shape, form, and size of prokaryotic cells are important characteristics when considering the ecology of bacteria in the soil. Nutrient uptake will be determined, to some extent, by the surface area: volume ratio for a cell. This factor has been shown to contribute to the ability of spiral-shaped bacteria (e.g., *Spirillum*) to outcompete rod-shaped pseudomonads under conditions of substrate limitation. Shape and size may also be important in susceptibility to predation, with evidence that protozoa and flagellates choose prey partly on the basis of these factors.

Bacterial shape and size are not fixed, and for *E. coli*, cell volume can vary 27-fold depending on growth rate. In addition, many bacteria take on different forms depending on environmental conditions. A striking example is the N_2-fixing *Rhizobium* that forms nodules on the roots of leguminous crops. Initial contact with the plant root is through attraction to the root of flagellated, motile, rod-shaped bacteria. After infection, flagella are shed and bacteria form swarmer cells. Rapid division of these cells leads to formation of nodules that contain masses of *Rhizobium* cells, the majority of which are misshapen "bacteroids," with bulging cell walls and unusual morphologies. Another example is *Arthrobacter*, commonly isolated from the soil, which grows as rod-shaped, blue-pigmented cells at high growth rates and as coccoid purple cells under nutrient limiting conditions.

These effects are accentuated in starving cells, with the transition from growth to starvation frequently associated with a significant decrease in cell size, changes in cell characteristics (Fig. 3.3), rapid turnover of cell material, a decrease in ribosome number, and expression of a suite of starvation genes, including those involved in the "stringent response" (Boutte et al., 2013; Kjelleberg, 1993). Many of these genes encode for high affinity, nutrient uptake systems with broad substrate specificity. They enable cells to sequester a wider range of substrates at low concentrations, giving them an advantage over organisms that, under conditions of nutrient excess, have much greater maximum specific growth rates. Starvation forms are resistant to environmental stress, and although vegetative cells, they can have greatly increased survival capability. Direct observation of prokaryotic cells in soil indicates that many are much smaller than typical laboratory-grown organisms, with significant proportions of cells passing through 0.4-μm-pore-size filters. There is evidence that these cells have a low ability to grow in laboratory culture.

A limited number of microbial groups, notably bacilli and clostridia, produce internal spores, termed endospores (Fig. 3.4). These structures are highly resistant to extremes of temperature, radiation, pressure, and other forms of environmental stress. They are the most resistant biological structures known, and although the environmental extremes to which they are resistant are rarely

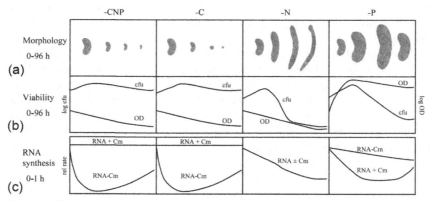

FIG. 3.3 Changes in morphology (A), viability (B), and RNA synthesis (C) (in the presence and absence of chloramphenicol) in *Vibrio* during starvation for combined and individual starvation for C, N, and S. The different forms of starvation lead to significant changes in cell size and/or cell morphology and have different effects on changes in RNA content and in biomass and cell concentrations. Chloramphenicol was added to inhibit protein synthesis. *(Redrawn from Östling et al., 1993.)*

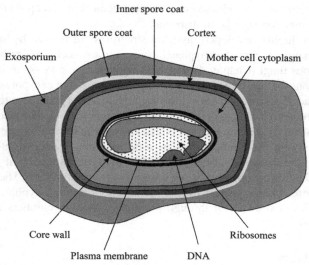

FIG. 3.4   Major characteristics of a bacterial spore.

encountered in soil, they are responsible for the persistence and survival of these organisms over many years and even centuries.

B   Filamentous and Mycelial Growth

Even though prokaryotes are typically considered unicellular, a number of groups exhibit filamentous growth. For example, some bacteria (e.g., strepto-cocci) and cyanobacteria (e.g., *Nostoc*) grow filamentously, not as continuous hyphae, but as chains of cells. In some cases, this is due to incomplete separation of dividing cells, but in others, it provides some of the advantages of true mycelial organisms, with regard to compartmentalization and differentiation. In N_2-fixing cyanobacteria, anaerobic conditions required for N_2-fixation can be localized in some cells, whereas others carry out oxygenic photosynthesis. Chemical communication between cells allows for a two-way flow of nutrients and signaling processes that lead to regular distribution of N_2-fixing cells along filaments.

The actinobacteria exhibit the greatest variety of growth forms, ranging from single-celled rods and cocci to mycelial structures (Prosser and Tough, 1991). For example, arthrobacters and some rhodococci grow as unicells and do not form true mycelia, but develop filamentous cells, which may subsequently fragment. Others, such as *Nocardia*, are dimorphic and form true, branching hyphal structures during early growth, which then fragment as con-ditions become less favorable. The hyphal form facilitates colonization of soil particles and, potentially, movement across barren regions to new nutrient sources. Fragments may subsequently develop centers of mycelial growth, and their major function may be in dispersal.

The most highly developed mycelial structures are formed by streptomycetes, which grow as branched hyphae forming a true mycelium, similar to those of filamentous fungi (Chapter 4). Hyphal fragmentation may occur under certain conditions, but the major means of dispersal is through exospores borne on aerial hyphae. Exospores do not exhibit the high levels of resistance to environmental extremes of endospores, but are very resistant to desiccation. This enables their survival in the soil, but is also an important factor in their effective dispersal through the atmosphere. On solid medium, spores develop at the center of actinobacterial colonies, fuelled by lytic products of substrate mycelium that have stopped growing through nutrient limitation. These organisms therefore achieve growth over short distances through hyphal extension and branching and dispersal through production of high numbers of single-celled spores.

C Cell Walls

Prokaryotic cells are surrounded by a rigid cell wall, protecting the cell from osmotic lysis (Figs. 3.5 and 3.6). The cell wall is important in diagnosis and phenotypic classification of bacteria, providing a major division between Gr^+ and Gr^- bacteria. The cell wall of Gr^+ cells consists of a single layer of

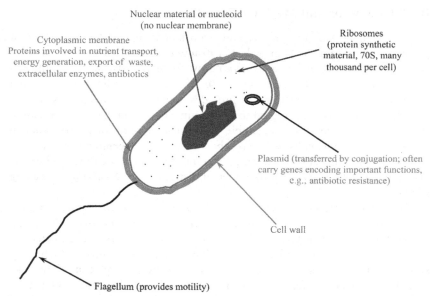

FIG. 3.5 Illustration of the major characteristics of prokaryote cell structure. See Fig. 3.6 for detailed structure of the bacterial cell wall.

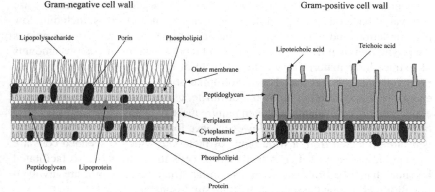

FIG. 3.6 Detailed structure of gram-negative and gram-positive bacterial cell walls.

peptidoglycan, surrounding the cytoplasmic membrane. Peptidoglycan is a polymer consisting of a backbone of alternative N-acetylglucosamine and N-acetylmuramic acid residues connected to cross-linked peptide chains of four amino acids. Gram-positive cell walls also usually contain teichoic acids, polymers of glycerol or ribitol, linked by phosphate groups, and containing amino acids and sugars. In the more complex Gr^+ cell wall, the peptidoglycan layer is much thinner and is surrounded by an outer membrane enclosing a periplasmic space, which contains enzymes involved in nutrient acquisition, electron transport, and protection from toxins.

Archaeal cell walls differ from those of bacteria and provide a major exception to the original definition of prokaryotes. They lack an outer membrane, and peptidoglycan is replaced by a range of compounds, including pseudomurein (similar to peptidoglycan) or other polysaccharides. Many archaeal cell walls contain paracrystalline protein or glycoprotein, the S-layer, which is also found in some bacteria. In addition, some prokaryotes do not possess a cell wall (e.g., intracellular mycoplasmas and the archaeon Thermoplasma, which have strengthened membranes).

The cell walls of many bacteria are encased within extracellular material (Fig. 3.5), ranging from apparently rigid and distinct capsules of specific thickness to more diffuse (chemically and physically) extracellular polymeric substances. Many roles have been assigned to this material, including protection from predation, adhesion to solid surfaces, and biofilm formation. In the free-living, N_2-fixing bacterium *Azotobacter* (a member of the Pseudomonadales), limited O_2 diffusion through extracellular material creates anaerobic regions required for N_2 fixation. Biofilm formation is particularly important, with suggestions that the majority of the soil microbial community is attached to particulate matter (clay minerals, soil organic matter [SOM], plant roots, and animals). Adsorption of nutrients to particulate material provides concentrations necessary for microbial growth. Surface attachment can increase survival of bacteria and protect them from environmental stress, including low pH, starvation, and inhibition by antibiotics and heavy metals. An example is the production by nitrifying bacteria in soil model systems of copious amounts of extracellular material that effectively form a blanket over colonies. This occurs despite the fact that these autotrophic organisms barely gain sufficient energy from oxidation of ammonia or nitrite, use much of this energy to generate reducing equivalents, and then require more reducing equivalents because of the requirement to fix CO_2. However, once formed, biofilms of these organisms are protected from a wide range of factors to which suspended cells are susceptible (Prosser, 2011). Attachment of cells to surfaces is also facilitated by short, hair-like fimbriae, whereas similar structures, sex pili, are involved in cell-to-cell contact associated with plasmid transfer.

D Internal Structure

Prokaryotes are defined as cells that lack internal, membrane-bound organelles, but exhibit diverse internal structures and primitive differentiation (Fig. 3.5). In bacteria, the cytoplasm is enclosed by a membrane consisting of ester-linked, straight-chained fatty acids. The lipids comprising the membrane form a bilayer with non-polar hydrophobic ends associating with each other and externalized polar hydrophilic ends. Cytoplasmic membranes in archaea contain only a single lipid layer consisting of ether-linked, branched aliphatic acids. The cytoplasmic membrane provides both a permeability barrier and an important link between the cell and its environment. It contains many proteins required for the import of nutrients, export of waste products, and production of extracellular enzymes for breakdown of high-molecular-weight compounds. It is also involved in energy generation in respiring cells through oxidative phosphorylation and can provide protection through toxin or antibiotic degrading enzymes.

Bacterial and archaeal chromosomes are generally present as a single, circular, double-stranded DNA molecule forming a nucleoid, but some (e.g., streptomycetes) have linear chromosomes. The molecular biology of archaea is, in many ways, more similar to that of eukaryotes than bacteria. In bacteria, DNA is condensed within the cell by the enzyme DNA gyrase. The same mechanism operates in many archaea, but in others, DNA is folded around clusters of histones, as in eukaryotes. Some archaea have several origins of replication, like eukaryotes, whereas bacteria have just one. The mechanisms for gene transcription of archaea are also closer to those of eukaryotes than bacteria.

Additional extrachromosomal genetic material is often present as one or many small DNA molecules, termed plasmids. These can be transmitted vertically (from generation to generation) and horizontally between different strains. Other mechanisms of genetic exchange in prokaryotes are transformation, which involves direct uptake of DNA and incorporation of genes into the host chromosome, and transduction, in which genes are transferred by bacteriophage. HGT is also mediated by gene transfer agents (GTA), which are phage-like elements that package and transfer random pieces of chromosomal DNA (Lang et al., 2012). GTAs and transduction highlight the potential importance of phage and phage-like particles in prokaryote ecology. Many genes in prokaryote genomes appear to be of viral origin, and the "virome" is thought to be of major significance for gene transfer in marine environments. Very little is known of the soil virome. Spatial heterogeneity will influence the extent and location of HGT activity, but HGT may represent an important driver of phylogenetic and physiological diversity in soil communities. It also represents an important difference between prokaryotes and eukaryotes, in that genetic exchange between the former is unidirectional.

Transcription and translation are closely coupled in bacteria and archaea, unlike eukaryotes, where transcription occurs within a membrane-bound nucleus. Proteins are synthesized within the cytoplasm by thousands of ribosomes often forming structures termed polysomes and attached to mRNA. Storage products can also accumulate in cells, forming microscopically visible inclusion bodies, including storage compounds, such as poly-β-hydroxybutyrate (a glucose polymer), polyphosphate, and glycogen. The sulfur oxidizer *Beggiatoa* stores elemental sulfur, which can be seen as yellow granules within the cell. Another example is magnetosomes, which are magnetite particles (Fe_3O_4) that enable magnetotactic responses.

Some intracellular structures are associated with specific metabolic processes. Autotrophic bacteria possess carboxysomes, which are particulate bodies involved in fixation of CO_2. Photosynthetic bacteria possess complex intracellular membrane structures, which are the site of energy-trapping photosynthetic processes. They show relatively close evolutionary relationships to other functional groups with similar complex membrane structures involved in other reactions, including ammonia, methane, and iron oxidation. Membranes can occur in distinctive patterns (e.g., by forming a layer within the cytoplasmic membrane or in an equatorial plane), and are diagnostic for some groups.

Prokaryotes are distinguished from eukaryotes by the lack of a unit membrane-bound nucleus and other organelles, but again, there are exceptions. The Planctomycete, Verrucomicrobia, and Chlamydia (PVC) superphylum is a collection of species from the Planctomycetes, Verrucomicrobia, Chlamydiae, and Lentisphaerae phyla and the Candidate Poribacteria, Candidate phylum OP3, and Candidate division WWE2. These organisms are found in a range of environments and possess a variety of functions, but fall within the same 16S rRNA gene group. Many members of this group possess internal membranes that appear similar to eukaryotic endomembranes, but evidence suggests that they have arisen through convergent evolution (McInerney et al., 2011). Some planctomycetes are involved in anaerobic ammonia oxidation (the anammox process), in which ammonia is oxidized using nitrite as an electron acceptor to form N_2. These organisms possess unit-membrane-bound anammoxosomes, which are believed to contain enzymes involved in the anammox process and in protecting the cell from the toxic intermediate hydrazine.

E   Motility

Many prokaryotes are motile, with the most obvious mechanism being "swimming" by rotation of one (polar), two, or many (peritrichous) flagella, or archaella in archaea. The bacterial flagellum consists of a long, helically shaped protein (flagellin) anchored in the cytoplasmic membrane and extending through the cell wall. Rotation of the flagellum is powered by proton motive force generated across the cell membrane and leads to movement of the cell. Archaella consist of archaellins and have different structure. Movement in the soil environment is important for unicellular organisms searching for new sources of nutrients. Flagellar motility is unlikely to be useful for transport over significant distances, as rates of movement are not great. Nevertheless, flagellar motility can be linked to chemotaxis (movement toward a chemical attractant), which appears to be important for communication between cells and plants in the soil environment (e.g., in movement of rhizobia toward roots prior to nodule formation). Movement over long distances is more likely to be achieved through suspension of cells in water channels and bulk flow of water or carriage on roots or soil animals. Lack of motility is probably important in the formation of microniches occupied by microcolonies of organisms, reducing competitive interactions, reducing selective sweeps, and increasing soil bacterial diversity. Other mechanisms exist for bacterial motility. For example, spirochetes possess an axial filament that enables the cell to move by flexing and spinning. Other organisms, including cyanobacteria, *Cytophaga* (a member of the Bacteroidaceae), and mycobacteria (a member of the Actinobacteria) can move by gliding over surfaces (Harshey, 2003; McBride, 2001).

V   METABOLISM AND PHYSIOLOGY

A   The Diversity of Prokaryotic Metabolism

The availability of molecular and genomic techniques has alerted us to the tremendous diversity of prokaryotes in the soil environment (Delmont et al., 2011). This diversity is critical to the maintenance of soil health and quality, driving the many soil functions that underpin ecosystem productivity (Garbeva et al., 2004). Although phylogenetic diversity is impressive, metabolic diversity provides the key to understanding their importance.

The prokaryotes carry out all the basic types of metabolism that occur in eukaryotes, but they also operate several other types of energy-generating metabolism. The diversity of prokaryotic metabolism is expressed by a great variation in modes of energy generation and metabolism. It is this diversity that underpins soil function and enables prokaryotes to occupy the widest range of soil habitats (Lengeler et al., 2009). For example, eukaryotic soil microorganisms produce energy through aerobic respiration (e.g., saprotrophic fungi, protozoa, and other members of the soil micro- and mesofauna) or oxygenic photosynthesis (e.g., algae, plants). Many prokaryotes generate energy by these processes, but there are additional mechanisms of energy production and associated metabolism, unique to prokaryotes, which are summarized in Table 3.3.

The application of molecular techniques has demonstrated very high diversity within soil microbial communities and the existence of novel groups at all phylogenetic levels. It would be surprising if this phylogenetic diversity was not reflected in physiological diversity. Physiology is best studied in pure, laboratory cultures, yet the majority of the newly discovered organisms have not been cultivated. However, molecular approaches can provide clues to physiology and in subsequent assessment of relevance to ecosystem function. This is exemplified by evidence from a metagenomic study that indicated that archaea could oxidize ammonia. This study identified 16S rRNA genes of non-thermophilic

TABLE 3.3 Uniquely Prokaryotic Forms of Energy Production and Related Metabolism

Metabolic System	Basis of Metabolism
Anaerobic respiration	Respiration using alternative electron acceptors to oxygen, such as nitrate and sulfate
Prokaryotic fermentation	Unique prokaryotic pathways, such as the Embden-Meyerhof pathway where glucose is converted into pyruvate
Lithotrophy	Use of inorganic sources of energy, such as ammonium, iron, and sulfur
Photoheterotrophy	Photosynthesis during which organic compounds are used as a carbon source
Anoxygenic photosynthesis	Photosynthetic phosphorylation in absence of oxygen
Archaean methanogenesis and light driven, non-photosynthetic photophosphorylation	Unique archaean metabolisms, using H_2 as an energy source in methane production and conversion of light energy into chemical energy
Alternative autotrophic CO_2 fixation	Fixation of carbon dioxide through the acetyl CoA pathway and the reverse TCA cycle in bacteria and several other pathways in archaea

crenarchaea on the same chromosomal DNA as genes homologous to those encoding bacterial ammonia monooxygenase, which catalyzes the first step in ammonia oxidation (Treusch et al., 2005). Archaea that contain these genes are now placed within a separate phylogenetic group, the thaumarchaea. Coincidentally, an autotrophic archaeal ammonia oxidizer was isolated at the same time (Könneke et al., 2005), demonstrating that the process occurred in thaumarchaea, but subsequent assessment of the importance of thaumarchaea, and of bacteria, for soil ammonia oxidation required the use of a range of molecular techniques and experimental systems (Prosser and Nicol, 2012). Further developments in molecular techniques, in particular single-cell genomics, are likely to greatly assist in determining the physiological diversity and potential of uncultivated soil prokaryotes.

B   Carbon and Energy Sources

The phylogenetic classification of prokaryotes described earlier in this chapter provides an indication of evolutionary relationships that will have arisen, in part, through differences in physiological characteristics. As a consequence, molecular phylogeny often agrees, to some extent, with traditional classification based on combinations of physiologies. Microbial physiological response has been linked to ecosystem function (Schimel et al., 2007), and physiological classification is also valuable when considering the role of prokaryotes in soil processes. Many physiological processes have evolved at different times in different organisms, with distinct physiological characteristics being found in many genera and species. This is illustrated in Tables 3.1 and 3.2. For example, approximately 50% of phylogenetic groups of bacteria contain some members capable of denitrification.

Physiological classification of soil microorganisms is of great value to those interested in ecosystem functions, as it provides one of the underpinning bases for consideration of ecological roles (Fig. 3.7). The first order of such a classification relates to the energy source of the microorganisms; phototrophs use light, and chemotrophs use chemical energy. The second order usually relates to the C source. Autotrophs or lithotrophs use CO_2, whereas organotrophs or heterotrophs use organic compounds. Cyanobacteria and green sulfur bacteria are examples of photoautotrophs, whereas the purple non-sulfur bacteria are photoheterotrophs. S- and Fe-oxidizing thiobacilli and nitrifiers, which oxidize reduced forms of N, are examples of chemoautotrophs, whereas *Pseudomonas* and *Rhizobium* are examples of chemoheterotrophs. The phylogenetic affiliations of these two genera based on 16S and 23S rRNA gene sequence data can be found in Anzai et al. (2000) and Yarza et al. (2010).

Although the standard physiological classification based on energy/C source described earlier enables links to be made to potential ecosystem function, there are exceptions to these rules. For example, heterotrophic, bacterial nitrifiers have a range of complex physiologies that cannot always be readily classified

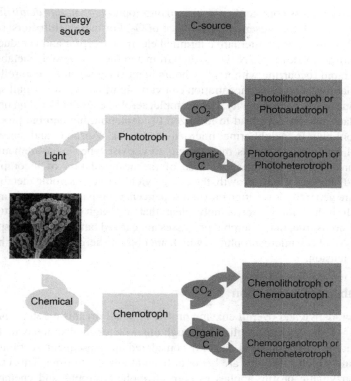

FIG. 3.7 Physiological classification of soil bacteria in terms of utilization of C and energy sources.

according to Fig. 3.7, not least because some appear to change their physiological strategy with time and with the availability of energy sources (Behrendt et al., 2010).

C   Oxygen Requirements

The requirement for the presence or absence of molecular oxygen provides a useful basis for further classification of soil bacteria and a useful indicator of ecological niche. Obligate aerobes, such as *Rhizobium* (a chemoheterotroph), *Thiobacillus/Acidithiobacillus* (chemoautotrophs), and many other soil bacteria, require oxygen, which acts as a terminal electron acceptor. The phylogenetic characteristics of *Thiobacillus/Acidithiobacillus* can be found in Kelly and Wood (2000). Obligate anaerobes, such as *Clostridium pasteurianum*, require the absence of molecular oxygen, and instead of using inorganic electron acceptors, they remove hydrogen atoms from organic compounds and dissipate them through reaction with products of carbohydrate breakdown. Facultative anaerobes, such as the denitrifying chemoheterotroph *Pseudomonas aeruginosa* or, less commonly, the chemoautotroph *Thiobacillus denitrificans*, can grow in both the presence and absence of O_2. For these denitrifiers, nitrate substitutes for O_2 as an alternative terminal electron acceptor and is reduced to either nitrous oxide or free N. The switch from aerobic to anaerobic metabolism is quite rapid (occurring within a few hours in most cases) and is controlled by O_2 availability. Oxygen concentration can vary significantly over small spatial scales due to soil heterogeneity. For example, aerobic degradation of an organic particulate material will lead to localized O_2 depletion, but aerobic processes may continue in neighboring microenvironments. Aerobic and anaerobic metabolisms and organisms may appear to coexist when measurements are made on bulk soil samples. Because of the accumulation of incompletely oxidized products during growth, the energy yields from anaerobic metabolism in soil are generally low (often only a few per cent) compared to aerobic metabolism. It is becoming increasingly clear that although the classification in Fig. 3.7 holds true, many aerobic processes are carried out in soil using sparing supplies of O_2 by microaerophiles, which are obligate aerobes that grow best at low O_2 tension.

D Substrate Utilization

The degradative and overall enzyme profile of a soil organism determines the range of substrates it can utilize, although other factors (substrate availability, competition, environmental factors) considered in subsequent sections also determine which substrates are being utilized at any given time. This concept of an enzymatic profile applies to both chemoheterotrophs and chemoautotrophs. Table 3.4 highlights some of the enzymes controlling both intracellular and extracellular utilization of substrates. The size of many of the substrates

TABLE 3.4 Selected Examples of Enzymes and Associated Bacteria Involved in Organic Substrate Utilization

Degradative Enzymes Used for Organic Substrate Utilization	Distribution of Enzymes/Examples of Soil Bacteria with Ecologically Significant Activity of These Enzymes
Cellulase (cellulose → glucose subunits)	Species of *Bacillus*, *Cellulomonas*, and *Pseudomonas*
Glucose oxidase (glucose → CO_2)	Ubiquitous enzyme among soil bacteria
Protease (protein → amino acids)	Widespread among soil prokaryotes, but species of *Pseudomonas* and *Flavobacterium* are strongly proteolytic

TABLE 3.4 Selected Examples of Enzymes and Associated Bacteria Involved in Organic Substrate Utilization—Cont'd

Degradative Enzymes Used for Organic Substrate Utilization	Distribution of Enzymes/Examples of Soil Bacteria with Ecologically Significant Activity of These Enzymes
Deaminase/amino transferase; amino acid decarboxylase (removal of amino and carboxyl groups to liberate NH_3 and CO_2 from amino acids)	More common enzymes than proteases, although major differences in rates between amino acids
Urease (urea → ammonia + carbon dioxide)	About 50% of heterotrophic soil bacteria are ureolytic
Amylase and glucosidase (starch → glucose)	Species of *Bacillus, Pseudomonas*, and *Chromobacterium*
Ligninase (lignin → aromatic subunits)	Although lignin degradation is primarily the domain of the white rot fungi, species of *Arthrobacter, Flavobacterium,* and *Pseudomonas* are sometimes involved
Pectinase (pectin → galacturonic acid subunits)	Species of *Arthrobacter, Pseudomonas*, and *Bacillus* (some species possess all of the pectinase enzymes— polygalacturonase, pectate lyase, pectin lyase, and pectin esterase). Many plant pathogens possess pectinase to assist in plant host penetration
Phosphatase (phosphate esters → phosphate)	About 30% of heterotrophic soil bacteria possess phosphatase enzymes
Sulfatase (sulfate esters → sulfate)	Much fewer possess sulfatase
Invertase (sucrose → fructose + glucose)	Particularly active in saprotrophic soil bacteria such as species of *Acinetobacter* and *Bacillus*
Chitinase (chitin → amino sugar subunits)	The actinomycetes *Streptomyces* and *Nocardia*
Amino acid decarboxylase	Both aromatic and non-aromatic amino acid decarboxylases are found in a wide range of soil bacteria synthesizing amino acids

(lignin, cellulose) ensures that most depolymerization is extracellular, with only the final utilization of the simple building blocks being intracellular. The table is by necessity a simplification as, in a number of cases, there is a suite of enzymes rather than a single enzyme operating on the substrate in question. The enzymes listed in Table 3.4 can be classified as oxidoreductases (glucose oxidase, ammonia monooxygenase, hydroxylamine oxidoreductase, nitrite oxidoreductase, methane monooxygenase), transferases (amino transferases), hydrolases (protease, urease, amylase, ligninase, pectinase, phosphatase, sulfatase, invertase, chitinase), and lyases (amino acid decarboxylase). Some of the enzymes highlighted in Table 3.5 are constitutive, whereas others are inducible. More comprehensive accounts of soil enzymes can be found in recent reviews (Burns and Dick, 2002; Burns et al., 2013).

E Autochthony and Zymogeny

Identifying a soil bacterium as a chemoheterotroph, according to the preceding classification scheme for metabolism, provides little information on how competitive that bacterium will be under particular conditions of substrate(s) supply. The great soil microbiologist Winogradsky addressed this issue through reference to the comparative kinetics of growth, which relate substrate concentration and specific growth rate. In Fig. 3.8, which illustrates some simple and contrasting growth curves, population Z will outcompete populations Y and Z at high substrate concentrations. At low substrate concentrations, population X will outcompete populations X and Z. There is no substrate concentration at which population Y is the best competitor. Soil bacteria that exhibit the growth kinetics of population Z, with a relatively high maximum specific growth rate (μ_{max}) and substrate affinity (K_s), will be more competitive at high substrate

TABLE 3.5 Selected Examples of Enzymes and Associated Soil Bacteria Involved in Autotrophic Oxidation of Substrates

Ammonia monooxygenase (ammonia → hydroxylamine)	*Nitrosomonas, Nitrosospira, Nitrosococcus, Nitrososphaera viennensis, Nitrosotalea devanaterra*
Hydroxylamine oxidoreductase (hydroxylamine → nitrite)	*Nitrosomonas, Nitrosospira, Nitrosococcus*
Nitrite oxidoreductase (nitrite → nitrate)	*Nitrobacter, Nitrospira*
Methane monooxygenase	Methanotrophs
	Methylotrophs such as *Methylomonas* and *Methylococcus*

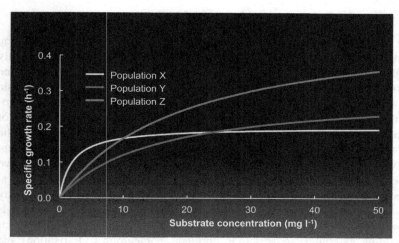

FIG. 3.8 Comparative Monod kinetics demonstrating the influence of substrate concentration on the specific growth rate of organisms with different maximum specific growth rates and substrate affinities. (Populations X, Y, and Z have maximum specific growth rates of 0.2, 0.3, and 0.5 h^{-1} and substrate saturation constants of 2, 15, and 20 mg l^{-1}, respectively.)

concentrations and are termed *zymogenous*. Not surprisingly, soil micro-environments, such as the rhizosphere, are dominated by zymogenous bacteria, such as fluorescent pseudomonads, which grow rapidly on simple C-substrates (primarily glucose).

Soil bacteria that exhibit the growth kinetics of population Z with low μ_{max} but relatively low K_s (i.e., a high substrate affinity) are termed *autochthonous*. The bacterial populations found in some of the less accessible soil microenvironments (the smaller pores inside soil aggregates), where substrate C-flow is rarely more than a trickle, are generally dominated by autochthonous bacteria (Killham et al., 1993). The spatial compartmentalization of soil, with regard to microbial populations, highlights the importance of substrate and nutrient availability (sometimes referred to as bioaccessibility) as a driver of both the diversity and function of the bacterial population (Killham et al., 1993).

In reality, there are degrees of autochthony and zymogeny, as indicated by the growth kinetics of the populations in Fig. 3.8. There is a continuum of growth kinetics that ensures that the most competitive soil bacteria will change with substrate concentration. Successions will therefore often occur in environments, such as the rhizosphere where C-flow changes, although the picture is further complicated by the proliferation of substrates with varying recalcitrance, enzyme specificity, and availability. It must also be remembered that bacteria grow rapidly and that those with high μ_{max} will rapidly deplete substrates, making conditions more favorable for autochthonous organisms. In addition, a high substrate concentration may indicate that conditions are unfavorable for metabolism of that substrate, rather than favorable for those that can metabolize it.

F   Oligotrophy, Copiotrophy, and the r-K Continuum

The physiologically based autochthony–zymogeny classification pioneered by Winogradsky is often considered analogous to the r-K continuum commonly used in plant and animal ecology, based on logistic models, and to the scheme of oligotrophy-copiotrophy. The terms r and K refer to maximum intrinsic growth rate and carrying capacity, respectively. The K-strategists and oligotrophic bacteria are considered to be adapted to growth under conditions of C/nutrient starvation (*oligocarbotrophy* specifies C-starvation, whereas the terms *oligonitrotrophy*, or *oligophosphotrophy* specify other types of nutrient starvation). Copiotrophs are adapted to nutrient excess. Although the two schemes are used in similar ways, the contrast in physiological versus logistical approaches represents a significant conceptual difference, and carrying capacity (K) and substrate affinity (K_s) are not analogous. Furthermore, oligotrophy is considered to include unusual forms of C and nutrient scavenging, such as exploiting gaseous C-sources (e.g., through anapleurotic CO_2 fixation) and volatile organic acids.

G   Facultative Activities

Although formal functional/physiological classification of soil bacteria is a useful foundation, it does not embrace the flexibility exhibited by many organisms. For example, we know that many soil bacteria that grow readily on "standard" organic substrates can adapt and continue to metabolize under the harsh conditions of C-starvation often found in soil. The physiological strategies and mechanisms utilized by the "facultative oligotrophs" are only partially characterized and include scavenging of gaseous forms of C, which diffuse through the soil pore network. The ever-changing conditions in soil also demand significant physiological flexibility.

Perhaps the best-known example of facultativeness relates to O_2 requirements. Facultative anaerobes metabolize most efficiently as aerobes. Indeed, the denitrifying pseudomonads that often colonize the rhizosphere as aerobic chemoheterotrophs may seldom experience low O_2 concentrations that induce the nitrate-reductase enzyme system associated with denitrification. Many microaerophiles also often operate under well-aerated conditions, but as facultative microaerophiles, they are physiologically adapted to life at the low O_2 concentrations sometimes experienced in wet soils, particularly in microsites such as water-saturated aggregates beyond a critical radius, which prevents adequate diffusive resupply after removal by respirational (roots, animals, microorganisms) demand (Greenwood, 1975). This concept of critical aggregate radius provides the potential for coupled aerobic (nitrification) and anaerobic (denitrification) activities in soil aggregates (Kremen et al., 2005).

Another aspect of flexibility is the capacity to utilize more than one substrate, often simultaneously (for example, through cometabolism), which may be the norm rather than the exception. This is reflected in the diversity of catabolic enzymes, both intracellular and extracellular. Among the most versatile degraders in the soil are species of the genus *Pseudomonas* (Timmis, 2002). They can degrade the most complex aromatic structures through to the simplest sugars (Powlowsky and Shingler, 1994). *Pseudomonas cepacia*, for example, can utilize more than one hundred different C substrates (Parke and Gurian-Sherman, 2001).

Flexibility is also evident from the way the soil microbial population adapts rapidly to the degradation of new substrates through the contamination of soil with organoxenobiotics that have been synthesized *de novo* by industry. The development of populations with appropriate catabolic genes is one of the greatest phenomena exhibited by the soil microbial community and is discussed further in Section VI.

One of the most fascinating examples of facultative physiology among the prokaryotes comes about when bacteria grow and reach population densities at which functional genes are activated. Some processes can only occur when bacteria act in unison, which does not occur in single cells or at low population densities. These functional genes are activated by a critical concentration of a molecule produced by the bacteria called an autoinducer, which is detected by specialized membrane-borne receptors. This process is known as quorum-sensing. Many processes of ecological importance are activated by this type of cell density-dependent signaling, including production of extracellular enzymes, biofilm formation, antibiotic resistance, and other secondary metabolic processes. These influence metabolic processes, but are also of potential significance in microbial interactions. Cell density-dependent signaling has been recognized in Gr⁻ bacteria for some time (Bassler, 1999), where it is mediated by production and detection of autoinducing signal molecules (*N*-acyl homoserine lactones or AHLs). Its relevance to soil has remained mysterious until recently. This has been largely because of the lack of suitable techniques and the challenges presented by working with soil. Molecular approaches, which use whole cell biosensors engineered to fluoresce in the presence of the AHL signal molecules, are beginning to reveal active, bacterial cell signaling in soil (Burmølle et al., 2003). Even though this technique has only been applied to resolve questions at the bulk soil level, it has the potential for *in situ* resolution of the functional significance of cell signaling between soil bacteria. Equivalent quorum-sensing mechanisms also exist in Gr⁺ bacteria. In *Bacillus subtilis*, the signaling compounds, termed pheromones, are peptides. Groups of strains (pherotypes) producing the same compound can communicate with each other, but not with different pherotypes. This leads to cooperation and competition within and between different pherotypes that potentially influence diversity and community composition of bacilli in the soil (Štefanič and Mandič-Mulec, 2009).

VI BIODEGRADATION CAPACITY

A Cellulose

Plant residues provide the major source of SOM, and their biodegradation is critical to ecosystem productivity. Because plants typically contain up to 60% cellulose (Kögel-Knabner, 2002), the decomposition of cellulose is a key activity of soil bacteria and is vital to the energy flow through soils and the cycling of N, P, and S, where immobilization generally accompanies cellulose decomposition. The decomposition of cellulose is a relatively specialized depolymerization exercise (involving a restricted number of saprophytes) followed by hydrolysis to glucose, which is rapidly utilized as an energy source by most heterotrophic soil microorganisms. The cellulose polymer (molecular weight approximately 1 million) occurs in plant residues in a semicrystalline state and consists of glucose units joined by β-1,4 linkages, with chains held together by hydrogen bonding (Fig. 3.9).

The cellulase enzyme complex, which catalyzes cellulose decomposition, occurs in a large number of cellulolytic bacteria (species of *Bacillus*, *Pseudomonas*, *Streptomyces*, and *Clostridium*; Coughlan and Mayer, 1992) and fungi and operates via a two-stage process. The first involves "conditioning," through recrystallization of cellulose. In the second stage, the molecule is depolymerized extracellularly into shorter and shorter units, eventually forming double to single sugar units by the enzyme cellobiase. These depolymerases attack both the ends and the "body" of the polymer (exo- and endo-depolymerases, respectively). Although the half-lives and turnover times of cellulose and hemicellulose in soil are in the order of days and weeks, the utilization of glucose after cellulose depolymerization is extremely rapid (in the order of hours to a day; Killham, 1994). The recent review of Wang et al. (2013) provides a good overview of the structural and biochemical basis for cellulose biodegradation.

The term *hemicellulose* describes various sugar (hexoses and pentoses) and uronic acid polymers that, like cellulose, are decomposed by a relatively specialized depolymerization process, followed by a much more rapid assimilation and oxidation of the simple monomer. Pectin, a polymer of galacturonic acid subunits, provides a good example of this process, with specialist pectinolytic bacteria, such as species of *Arthrobacter* and *Streptomyces*, which produce the

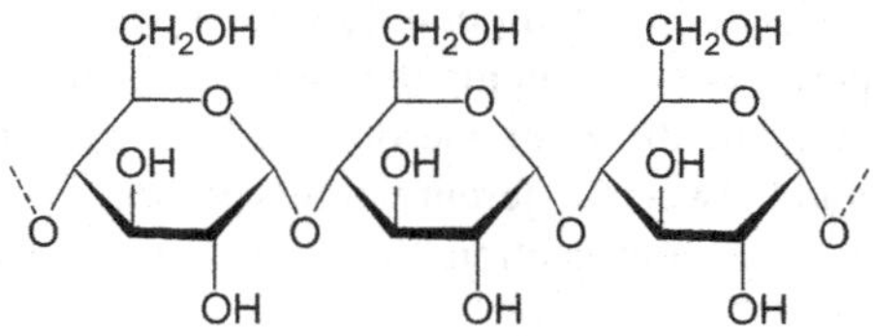

FIG. 3.9 The chemical structure of cellulose, which consists of β-1,4-linked glucose molecules.

extracellular pectin depolymerases (exo- and endo-). Subsequently, a much wider range of heterotrophic soil microorganisms uses galacturonic acid oxidase to exploit the energy bound in the subunit itself (Killham, 1994).

B Pollutants

With continually expanding industrialization and a global dependence on fossil fuel hydrocarbons as well as agrochemicals, there are few environments that are not in some way affected by a spectrum of organic pollutants. These organic pollutants include the aliphatic hydrocarbons (alkanes from oil spills and petrochemical industry activities), alicyclic hydrocarbons (terpenoid plant products), aromatic hydrocarbons (single aromatics such as the petrochemical solvents benzene and toluene, and polyaromatics, such as pyrene), the chlorinated hydrocarbons (chlorinated aliphatics, such as chloroform), chlorinated aromatics (chlorobenzenes and chlorinated polyaromatics, such as PCBs, DDT, and dioxins), and N-containing aromatics (TNT). Because of the increasing awareness of the adverse effects of these organoxenobiotic pollutants, the soil microorganisms and associated genes (largely carried on plasmids, which are extrachromosomal pieces of DNA) involved in their degradation have become sources of great interest to soil microbiologists focusing on bioremediation of contaminated environments (Crawford and Crawford, 1996).

Some of the organic pollutants entering the soil environment are toxic to bacteria and are recalcitrant to mineralization. The reasons for the latter are numerous and include key physicochemical characteristics such as the way in which the pollutant partitions into the soil matrix—only relevant if related to metabolic processes. For example, the more hydrophobic polyaromatic hydrocarbons (PAHs) and polychlorinated biphenyls (PCBs) tend to bind strongly to the soil solid phase and limit microbial/enzymatic access (Dick et al., 2005). Another reason for recalcitrance is the large number of enzyme-regulated steps and the number of well-regulated genes required for mineralization (Kanaly and Harayama, 2000). Furthermore, because many of the organic pollutants under consideration are synthetic, there are often no corresponding genes existing in soil bacteria that can immediately specify the degradation process.

The highly chlorinated organic pollutants have been studied extensively in terms of their degradation and the evolution of their degradative genes. Although degradation is generally slow, particularly for the more chlorinated compounds, gene clusters have evolved for their degradation. Structural and regulatory genes have high sequence homology, even though they have evolved as plasmid-borne genes in different soil bacterial genera from across the world (Daubaras and Chakrabarty, 1992). This evolution of gene-regulated bacterial degradative pathways (usually involving 4-8 genes in each case) enables complete mineralization of a range of chlorinated pollutants, including chlorobenzenes used in a number of industrial processes and the chlorophenoxyacetic acids associated with some of the hormonal herbicides used in agricultural weed control. The degradation of many of the chlorinated organic pollutants is somewhat complicated by the fact that bacterial removal of chlorine from the molecule (dehalogenation) occurs most readily under anaerobic conditions. Bioremediation of contaminated environments that contain both chlorinated and non-chlorinated hydrocarbons can be a complicated process, which involves management of both anaerobic and aerobic conditions. Although the anaerobic dehalogenation of PCBs is relatively well understood, obtaining reliable and useful bacterial isolates is fraught with difficulties (May et al., 1992).

The soil bacteria and associated genes involved in the degradation of PAHs have, like the PCBs, been extensively studied and exploited in both bioremediation and the generation of bacterial biosensors for detection of these complex organic pollutants. For the bacteria that aerobically degrade PAHs (naphthalene, acenaphthylene, fluorene, anthracene, phenanthrene, pyrene, benz(a)anthracene, benzo(a)pyrene), attack often involves oxidation of the

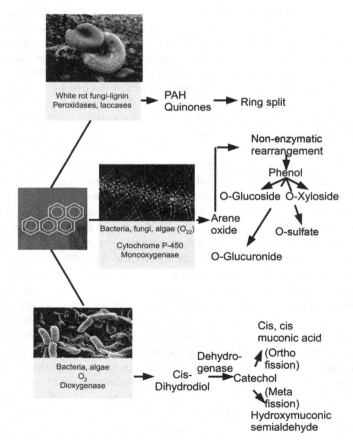

FIG. 3.10 Schematic of soil microbial metabolism of polyaromatic hydrocarbons.

rings by dioxygenases to form cis-dihydrodiols (Fig. 3.10). The bacteria involved in these oxidative reactions include species of *Mycobacterium* (Kelley et al., 1993) and *Pseudomonas* (Selifonov et al., 1993). The dihydrodiols are transformed further to diphenols that are then cleaved by other dioxygenases. In many bacteria, this precedes conversion to salicylate and catechol (Sutherland et al., 1995). The latter is a rather toxic, relatively mobile intermediate, and any bioremediation of PAHs must consider and manage issues such as this both because of possible environmental hazards and because the catechol may inhibit the bacteria being harnessed for the bioremediation itself!

VII DIFFERENTIATION, SECONDARY METABOLISM, AND ANTIBIOTIC PRODUCTION

Differentiation is a change in bacterial activities from those associated with vegetative growth. This phenomenon is particularly associated with cell starvation, where intracellular components are often transported and resynthesized into new compounds through secondary (non-growth linked) metabolism. The compounds formed can include antibiotics, pigments, and even agents such as melanin that are protective against enzymatic attack. Antibiotics are often powerful inhibitors of growth and metabolism of other microbial groups, with varying degrees of specificity. Streptomycin, for example, produced by certain species of *Streptomyces*, strongly inhibits a wide range of both Gr^+ and Gr^- bacteria. Cycloheximide, on the other hand, inhibits only eukaryotes and is used as a fungal inhibitor in the bacterial plate count.

Antibiotic production by soil bacteria involving secondary metabolism has been harnessed for decades for a wide range of medical applications. Although antibiotic production has long been linked with chemical defense, the factors determining antibiotic production in soil suggest that antibiosis only occurs when supply of available C is high (Thomashow and Weller, 1991). These conditions are likely to be met in the rhizosphere, the zone around seeds or "spermosphere," and relatively fresh plant or animal residues (Whipps, 2001).

Although a considerable number of antibiotic-producing bacteria has been identified, evidence that antibiosis is a significant chemical defense strategy in the rhizosphere tends to be indirect. For example, strains of fluorescent pseudomonads can demonstrate high specific rates of antibiotic phenazine production, which is strongly inhibitory against *Gaeumannomyces graminis* var *tritici* (Ggt), the causal agent of take-all in wheat (Freeman and Ward, 2004). When *Tn5* mutants of such pseudomonads have been introduced into the wheat rhizosphere, the removal of antibiotic production has been associated with reduced control of Ggt (Thomashow and Weller, 1991).

The key genes and associated enzymes involved in antibiotic synthesis have been characterized in some cases (e.g., for phenazine; Blankenfeldt et al., 2004) and hence offer new and more powerful approaches to investigating these secondary metabolic, chemical defense strategies of bacteria in soil by probing for transcripts of antibiotic biosynthesis genes and/or by use of stable isotope (^{13}C) probing of the rhizobacterial nucleic acid pool. Questions of both the activation and significance of antibiotic-mediated chemical defense can now be resolved.

The potential rewards of applying new molecular approaches in antibiotic discovery using soil bacteria are exciting, and may be of great significance in combating the problems of ever-increasing antibiotic resistance in health care. Recent sequencing of the genome of a *Rhodococcus* strain bearing a mega-plasmid (acquired through horizontal gene transfer from *Streptomyces*) has revealed an exceptionally large number of genes coding for secondary metabolic products (antibiotics, toxins, and pigments; Kurosawa et al., 2008).

When the *Rhodococcus* is stressed by being forced to grow in the presence of a competing soil streptomycetes (itself an antibiotic producer), it produces its own, novel antibiotic, rhodostreptomycin (an aminoglycoside), which kills the streptomycetes. The final two sections of this chapter have considered two forms of xenobiotics-antibiotics and persistent organic pollutants, such as PCBs and PAHs. Nakatsu et al. (1991) pointed out that these two forms of organoxenobiotic have a key role in presenting soil bacteria with a major selection pressure. The molecular bases for organoxenobiotic resistance and catabolism are generally located in soil bacteria on extrachromosomal plasmid DNA, the maintenance of which requires this selection pressure.

VIII   CONCLUSION

The structure and physiology of prokaryotes have been studied for more than a century, and knowledge of their taxonomic and metabolic diversity and relevancy to their central role and essential activities in soil ecosystems is increasing rapidly. Although this chapter summarizes our current knowledge of the cell structure differences of prokaryotes and eukaryotes, we have also indicated the vast amount of important information that remains to be discovered. The majority of phylogenetic groups and strains that are abundant and important for soil processes have yet to be isolated and characterized. The diversity of the prokaryotes is significantly greater than that of higher organisms, but the lack of clear species definition limits our ability to apply concepts linking diversity to important ecosystem properties, such as stability and resilience. Elucidation of mechanisms that generate and control prokaryote diversity within the soil environment and increase our understanding of the two-way relationships between prokaryotes and the soil, with spatial and temporal heterogeneity, represent enormous and exciting challenges. It is essential that these challenges are met to provide a basis for understanding and potentially predicting the impact of environmental change on prokaryote diversity. Increased understanding of the links between phylogenetic, functional, and physiological diversity is also essential to determine the consequences for changes in prokaryote diversity and community structure on terrestrial biogeochemical cycling processes.

REFERENCES

Anzai, Y., Kim, H., Park, J.Y., Wakabayashi, H., 2000. Phylogenetic affiliation of the pseudomonads based on 16S rRNA sequence. Int. J. Syst. Evol. Microbiol. 50, 1563–1589.

Bassler, B.L., 1999. How bacteria talk to each other: regulation of gene expression by quorum sensing. Curr. Opin. Microbiol. 2, 582–587.

Behrendt, U., Schumann, P., Stieglmeier, M., Pukall, R., Augustin, J., Spröer, C., Schwendner, P., Moissl-Eichinger, C., Ulrich, A., 2010. Characterization of heterotrophic nitrifying bacteria with respiratory ammonification and denitrification activity-description of *Paenibacillus uliginis* sp. nov., an inhabitant of fen peat soil and *Paenibacillus purispatii* sp. nov., isolated from a spacecraft assembly clean room. Syst. Appl. Microbiol. 33, 328–336.

Blankenfeldt, W., Kuzin, A.P., Skarina, T., Korniyenko, Y., Tong, L., Bayer, P., Janning, P., Thomashow, L.S., Mavrod, D.V., 2004. Structure and function of the phenazine biosynthetic protein PhzF from *Pseudomonas fluorescens*. Proc. Natl. Acad. Sci. U. S. A. 101, 16431–16436.

Boutte, C.C., Cara, C., Crosson, S., 2013. Bacterial lifestyle shapes stringent response activation. Trends Microbiol. 21, 174–180.

Buckley, D.H., Schmidt, T.M., 2003. Diversity and dynamics of microbial communities in soils from agro-ecosystems. Environ. Microbiol. 5, 441–452.

Burmølle, M., Hansen, L.H., Sørensen, S.J., 2003. Presence of N-acyl homoserine lactones in soil detected by a whole-cell biosensor and flow cytometry. Microb. Ecol. 45, 226–236.

Burns, R.G., Dick, R.P., 2002. Enzymes in the Environment. Anonymous Marcel Dekker, New York.

Burns, R.G., Marxsen, J., Sinsabaugh, R.L., Stromberger, M.E., Wallenstein, M.D., Weintraub, M.-N., Zoppini, A., 2013. Soil enzymes in a changing environment: current knowledge and future directions. Soil Biol. Biochem. 58, 216–234.

Carballido-Lopez, R., Errington, J., 2003. A dynamic bacterial cytoskeleton. Trends Cell Biol. 13, 577–583.

Chen, W.L., Hung, C.-H., Xu, J., Reigstad, C.S., Magrini, V., Sabo, A., Blasiar, D., Bieri, T., Meyer, R.R., Ozersky, P., Armstrong, J.R., Fulton, R.S., Latreille, J.P., Spieth, J., Hooton, J.M., Mardis, E.R., Hultgren, S.J., Gordon, J.I., 2006. Identification of genes subject to positive selection in uropathogenic strains of *Escherichia coli*: a comparative genomics approach. Proc. Natl. Acad. Sci. U. S. A. 103, 5977–5982.

Cohan, F.M., 2002. What are bacterial species? Annu. Rev. Microbiol. 56, 457–487.

Coughlan, M.P., Mayer, F., 1992. The cellulose-decomposing bacteria and their enzyme systems. In: Balows, A., Trüper, H.G., Dworkin, M., Harder, W., Schleifer, K.H. (Eds.), The Prokaryotes: a handbook on the biology of bacteria, second ed. Springer-Verlag, New York, pp. 460–516.

Crawford, R.L., Crawford, D.L., 1996. Bioremediation: Principles and Applications. Cambridge University Press, Cambridge.

Daniel, R.A., Errington, J., 2003. Control of cell morphogenesis in bacteria: two distinct ways to make a rod-shaped cell. Cell 113, 767–776.

Daubaras, D., Chakrabarty, A.M., 1992. The environment, microbes and bioremediation: microbial activities modulated by the environment. Biodegradation 3, 125–135.

Delmont, T.O., Robe, P., Cecillon, S., Clark, I.M., Constancias, F., Simonet, P., Hirsch, P.R., Vogel, T.M., 2011. Accessing the soil metagenome for studies of microbial diversity. Appl. Environ. Microbiol. 77, 1315–1324.

Dick, K.J., Burauel, P., Jones, K.C., Semple, K.T., 2005. Distribution of aged ^{14}C-PCB and ^{14}C-PAH residues in particle-size and humic fractions of an agricultural soil. Environ. Sci. Technol. 39, 6575–6583.

Doolittle, W.F., Zhaxybayeva, O., 2013. What is a prokaryote? In: Rosenberg, E., DeLong, E.F., Lory, S., Stackebrandt, E., Thompson, F. (Eds.), The Prokaryotes—Prokaryotic Biology and Symbiotic Associations. Springer-Verlag, Berlin, Heidelberg, pp. 21–37.

Freeman, J., Ward, E., 2004. *Gaeumannomyces graminis*, the take-all fungus and its relatives. Mol. Plant Pathol. 5, 235–252.

Garbeva, P., van Veen, J.A., van Elsas, J.D., 2004. Microbial diversity in soil: selection of microbial populations by plant and soil type and implications for disease suppressiveness. Annu. Rev. Phytopathol. 42, 243–270.

Greenwood, D.J., 1975. Soil physical conditions and crop production. MAFF Bulletin 29, 261–272.

Harshey, R.M., 2003. Bacterial motility on a surface: many ways to a common goal. Annu. Rev. Microbiol. 57, 249–273.

Hugenholtz, P., Goebel, B.M., Pace, N.R., 1998. Impact of culture-independent studies on the emerging phylogenetic view of bacterial diversity. J. Bacteriol. 180, 4765–4774.

Kanaly, R.A., Harayama, S., 2000. Biodegradation of high-molecular-weight polycyclic aromatic hydrocarbons by bacteria. J. Bacteriol. 182, 2059–2067.

Kelley, I., Freeman, J.P., Evans, F.E., Cerniglia, C.E., 1993. Identification of metabolites from the degradation of fluoranthene by *Mycobacterium* sp strain Pyr-1. Appl. Environ. Microbiol. 59, 800–806.

Kelly, D.P., Wood, A.P., 2000. Reclassification of some species of *Thiobacillus* to the newly designated genera *Acidithiobacillus* gen. nov., *Halothiobacillus* gen. nov., and *Thermithiobacillus* gen. nov. Int. J. Syst. Evol. Microbiol. 50, 511–516.

Killham, K., 1994. Soil Ecology. Cambridge University Press, Cambridge.

Killham, K., Amato, M., Ladd, J.N., 1993. Effect of substrate location in soil and soil pore-water regime on carbon turnover. Soil Biol. Biochem. 25, 57–62.

Kjelleberg, S., 1993. Starvation in Bacteria. Plenum Press, New York.

Kögel-Knabner, I., 2002. The macromolecular organic composition of plant and microbial residues as inputs to soil organic matter. Soil Biol. Biochem. 34, 139–162.

Könneke, M., Bernhard, A.E., De La Torre, J.R., Walker, C.B., Waterbury, J.B., Stahl, D.A., 2005. Isolation of an autotrophic ammonia-oxidizing marine archaeon. Nature 437, 543–546.

Koonin, E.V., 2012. The Logic of Chance. The Nature and Origin of Biological Evolution. FT Press Science, New Jersey.

Kremen, A., Bear, J., Shavit, U., Shaviv, A., 2005. Model demonstrating the potential for coupled nitrification denitrification in soil aggregates. Environ. Sci. Technol. 39, 4180–4188.

Kurosawa, K., Ghiviriga, I., Sambandan, T.G., Lessard, P.A., Barbara, J.E., Rha, C., Sinskey, A.J., 2008. Rhodostreptomycins, antibiotics biosynthesized following horizontal gene transfer from *Streptomyces padanus* to *Rhodococcus fascians*. J. Am. Chem. Soc. 130, 1126–1127.

Lang, A.S., Zhaxybayeva, O., Beatty, J.T., 2012. Gene transfer agents: phage-like elements of genetic exchange. Nat. Rev. Microbiol. 10, 472–482.

Lengeler, J.W., Drews, G., Schlegel, H.G. (Eds.), 2009. Habitats of Prokaryotes. In: Biology of the Prokaryotes. Blackwell Science Ltd, Oxford.

May, H.D., Boyle, A.W., Price, W.A., Blake, C.K., 1992. Subculturing of a polychlorinated biphenyl-dechlorinating anaerobic enrichment on solid media. Appl. Environ. Microbiol. 58, 4051–4054.

McBride, M.J., 2001. Bacterial gliding motility: multiple mechanisms for cell movement over surfaces. Annu. Rev. Microbiol. 55, 49–75.

McInerney, J.O., Martin, W.F., Koonin, E.V., Allen, J.F., Galperin, M.Y., Lane, N., Archibald, J.M., Embley, T.M., 2011. Planctomycetes and eukaryotes: a case of analogy not homology. Bioessays 33, 810–817.

Nakatsu, C., Ng, J., Singh, R., Straus, N., Wyndham, C., 1991. Chlorobenzoate catabolic transposon Tn5271 is a composite class-I element with flanking class-II insertion sequences. Proc. Natl. Acad. Sci. U. S. A. 88, 8312–8316.

Östling, J., Holmquist, L., Flärdh, K., Svenblad, B., Jouper-Jaan, Å., Kjelleberg, S., 1993. Starvation and recovery of Vibrio. In: Kjelleberg, S. (Ed.), Starvation in Bacteria. Plenum Press, New York, pp. 103–128.

Pace, N.R., Stahl, D.A., Lane, D.J., Olsen, G.J., 1986. The analysis of microbial populations by ribosomal RNA sequences. Adv. Microb. Ecol. 9, 1–55.

Parke, J.L., Gurian-Sherman, D., 2001. Diversity of the Burkholderia cepacia complex and implications for risk assessment of biological control strains. Annu. Rev. Phytopathol. 39, 225–258.

Powlowsky, J., Shingler, V., 1994. Genetics and biochemistry of phenol degradation by Pseudomonas sp. CF600. Biodegradation 5, 219–236.

Prosser, J.I., 2011. Soil nitrifiers and nitrification. In: Ward, B.B., Klotz, M.G., Arp, D.J. (Eds.), Nitrification. ASM Press, Washington D.C, pp. 347–383.

Prosser, J.I., Nicol, G.W., 2012. Archaeal and bacterial ammonia oxidisers in soil: the quest for niche specialisation. Trends Microbiol. 20, 523–531.

Prosser, J.I., Tough, A.J., 1991. Growth mechanisms and growth kinetics of filamentous microorganisms. Crit. Rev. Biotechnol. 10, 253–274.

Rappe, M.S., Giovannoni, S.J., 2003. The uncultured microbial majority. Annu. Rev. Microbiol. 57, 369–394.

Rinke, C., Schwientek, P., Sczyrba, A., Ivanova, N.N., Anderson, I.J., Cheng, J.-F., Darling, A., Malfatti, S., Swan, B.K., Gies, E.A., Dodsworth, J.A., Hedlund, B.P., Tsiamis, G., Sievert, S.M., Liu, W.-T., Eisen, J.A., Hallam, S.J., Kyrpides, N.C., Stepanauskas, R., Rubin, E.M., Hugenholtz, P., Woyke, T., 2013. Insights into the phylogeny and coding potential of microbial dark matter. Nature 499, 431–437.

Schimel, J., Balser, T.C., Wallenstein, M., 2007. Microbial stress-response physiology and its implications for ecosystem function. Ecology 88, 1386–1394.

Schulz, H.N., Jorgensen, B.B., 2001. Big bacteria. Annu. Rev. Microbiol. 55, 105–137.

Selifonov, S.A., Grifoll, M., Gurst, J.E., Chapman, P.J., 1993. Isolation and characterization of (+)-1,1a-dihydroxy-1-hydrofluoren-9-one formed by angular dioxygenation in the bacterial catabolism of fluorene. Biochem. Biophys. Res. Commun. 193, 67–76.

Štefanič, P., Mandič-Mulec, I., 2009. Social interactions and distribution of Bacillus subtilis pherotypes at microscale. J. Bacteriol. 191, 1756–1764.

Sutherland, J.B., Rafii, F., Kahn, A.A., Cerniglia, C.E., 1995. Mechanisms of polycyclic aromatic hydrocarbon degradation. In: Young, L.L., Cerniglia, C.E. (Eds.), Microbial Transformation and Degradation of Toxic Organic Chemicals. Wiley-Liss, New York, pp. 269–306.

Thomashow, L.S., Weller, D.M., 1991. Role of antibiotics and siderophores in biocontrol of take-all disease. In: Kleister, D.L., Cregan, P.B. (Eds.), The Rhizosphere and Plant Growth. Kluwer Academic Publishers, Dordrecht, pp. 245–251.

Timmis, K.N., 2002. Pseudomonas putida: a cosmopolitan opportunist par excellence. Environ. Microbiol. 4, 779–781.

Torsvik, V., Sorheim, R., Goksoyr, J., 1996. Total bacterial diversity in soil and sediment communities—a review. J. Ind. Microbiol. 17, 170–178.

Treusch, A.H., Leininger, S., Schleper, C., Kietzin, A., Klenk, H.P., Schuster, S.C., 2005. Novel genes for nitrite reductase and amo-related proteins indicate a role of uncultivated mesophilic crenarchaeota in nitrogen cycling. Environ. Microbiol. 7, 1985–1995.

Wang, M., Liu, K., Dai, L., Zhang, J., Fang, X., 2013. The structural and biochemical basis for cellulose biodegradation. J. Chem. Technol. Biotechnol. 88, 491–500.

Wheelis, M.L., Kandler, O., Woese, C.R., 1992. On the nature of global classification. Proc. Natl. Acad. Sci. U. S. A. 89, 2930–2934.

Whipps, J.M., 2001. Microbial interactions and biocontrol in the rhizosphere. J. Exp. Bot. 52, 487–511.

Williams, T.A., Foster, P.G., Nye, T.M.W., Cox, C.J., Embley, T.M., 2012. A congruent phylogenomic signal places eukaryotes within the Archaea. Proc. R. Soc. B 279, 4870–4879.

Woese, C.R., Kandler, O., Wheelis, M.L., 1990. Towards a natural system of organisms—proposal for the domains Archaea, Bacteria, and Eucarya. Proc. Natl. Acad. Sci. U. S. A. 87, 4576–4579.

Yarza, P., Ludwig, W., Euzeby, J., Amann, R., Schleifer, R.H., Glöckner, F.O., Rossello-Mora, R., 2010. Update of the all-species living tree project based on 16S and 23S rRNA sequence analyses. Syst. Appl. Microbiol. 33, 291–299.

Chapter 4

The Soil Fungi: Occurrence, Phylogeny, and Ecology

D. Lee Taylor and Robert L. Sinsabaugh
Department of Biology, University of New Mexico, Albuquerque, NM, USA

Chapter Contents

I INTRODUCTION

There is now considerable circumstantial evidence that true fungi (Kingdom Eumycota) were instrumental in both the colonization of land by the ancestors of terrestrial plants (Simon et al., 1993) and the termination of carbon (C) deposition into geological reserves (i.e., fossil fuels, Floudas et al., 2012). The traits that underlie these major evolutionary and ecological transitions illustrate why fungi play such important roles in soils. Most fungi interact intimately with both living and dead organisms, especially plants. The mycorrhizal symbiosis with plant roots is thought to have permitted aquatic plants to transition into the challenging terrestrial habitat. Mycorrhizal and other interactions with living plants (pathogens, endophytes) may be highly specific or generalized with outcomes

Soil Microbiology, Ecology, and Biochemistry. http://dx.doi.org/10.1016/B978-0-12-415955-6.00004-9
77

that vary among taxa, but influence the structure and function of plant communities. Fungi have profound influences on biogeochemical cycles through their growth habits, which include external digestion of food resources using a powerful arsenal of degradative enzymes and secondary metabolism. It was the innovation and diversification of polyphenolic-degrading enzyme machinery among the white-rot basidiomycete fungi that may have halted the accumulation of undecayed plant materials during the carboniferous (Floudas et al., 2012). The filamentous habit common to the majority of soil-dwelling fungi allows them to bridge gaps between pockets of soil water and nutrients; force their way into substrates such as decaying wood; and redistribute C, minerals, and water through the soil. Filamentous growth may underlie the abilities of some fungi to withstand soil water deficits and cold temperatures that are beyond the tolerance of bacteria and archaea. Fungi constitute large fractions of living and dead soil biomass, particularly in forested habitats. Their growth and production of cell wall materials lead to the creation and stabilization of soil aggregates, which are key elements of soil structure. Rates of turnover of fungal biomass have important consequences for C cycling and long-term sequestration in soil. In this chapter, we summarize recent advances in our understandings of the phylogeny, biodiversity, and ecology of fungi of relevance to their diverse roles in soil environments. Supplementary online files provide definitions of some key terms, a primer on fungal systematics, and an example of the fungal life cycle.

II   PHYLOGENY

A   Definition of Eumycota

The *Fungi* (capitalized italic when used as a formal Linnean taxon) are recognized as a Kingdom (see online Supplemental Material Section I at http://booksite.elsevier.com/9780124159556 for a refresher on the taxonomic hierarchy of life). Although sometimes loosely referred to as *microbes*, fungi are eukaryotes and most are multicellular. A superkingdom of eukaryotes that includes fungi and animals is the *Opisthokonta*. The evolutionary lines that constitute the true fungi, or *Eumycota*, are all descended from a single common ancestor (i.e., constitute a monophyletic clade) within the *Opistokonta* (James et al., 2006; Steenkamp et al., 2006). There is now strong evidence that a small group of protists within the *Opisthokonta*, called the nuclearids, are the closest sister group to the *Eumycota* (Liu et al., 2009). Interestingly, nuclearids are phagotrophs (organisms lacking cell walls that engulf relatively large food items via phagocytosis), suggesting an important transition to osmotrophy (organisms that can only ingest small molecules that can move through pores in the cell wall) early in fungal evolution. Various molecular-clock estimates place the origin of the *Eumycota* from 600 million to > 1 billion years before the present (Berbee and Taylor, 1993; Heckman et al., 2001), but a robust estimate will require more reliable fossil calibration points (Berbee and Taylor, 2010).

A number of shared, derived traits ("synapomorphies" in phylogenetic terms) are characteristic of fungi (James et al., 2006; Stajich et al., 2009). Fungi depend on organic compounds for C, energy, and electrons, taking up these resources via osmotrophy. Although many fungi can fix CO_2 using enzymes of central anabolic cycles (e.g., pyruvate carboxylase) they are considered heterotrophs in a broad sense. Chitin, a polymer of N-acetylglucosamine, is a feature of the cell wall matrix of most fungi, although a few parasitic lineages have life stages that lack cell walls (e.g., *Rozella*), and some groups have little or no chitin in their walls (e.g. ascomycete yeasts). The ancestral state for the true fungi includes a mobile, flagellated meiospore (zoospore) stage; flagella appear to have been lost several times through the evolution of the fungi and today exist only in several early diverging lineages. Most fungi also use the sugar trehalose as an energy store, display apical growth, and have spindle-pole bodies rather than centrioles (with the exception of ancient, flagellated lineages).

The body of a fungal individual may contain one type of nucleus with only a single set of chromosomes—a haploid growth form. Alternatively, the cells of two different haploid individuals may fuse (plasmogamy). In most fungi, this fusion occurs only immediately before nuclear fusion (karyogamy) and meiosis (see Fig. S4.1 in online Supplemental Material at http://booksite.elsevier.com/9780124159556 for an exemplar fungal life cycle). However, in some fungi there is a brief (phylum *Ascomycota*) or prolonged (phylum *Basidiomycota*) stage in which the two different nuclei multiply in a synchronized fashion; this phase of the life cycle is termed *dikaryotic* (Fig. 4.1). In the more ancient fungal groups, the filaments in which the nuclei are housed do not contain cross-walls (septa), whereas in other groups, septa divide hyphal filaments into distinct cells. Although not representing any particular taxon, the features shown in

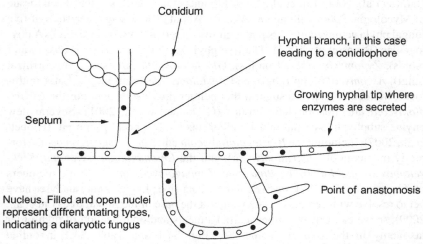

FIG. 4.1 Filamentous fungal growth form. Diagram depicts a mycelium growing from left to right.

Fig. 4.1 are characteristic of the *Dikarya*. Cross walls or septa separate individ-ual cells (numbers of nuclei are usually variable in *Ascomycota*, but are more often fixed in *Basidiomycota*). As a mycelium grows, hyphae branch at regulated intervals in response to external and internal signals. In many fungi, cytoplasm is retracted from older parts of the mycelium, leaving walled-off empty cells. The newly formed, thin and soft hyphal tip extends due to turgor pressure. The growing tip is the area of most active enzyme secretion and nutri-ent uptake.

In the two most recently evolved phyla, the *Ascomycota* and *Basidiomycota*, nuclear division and apportionment to the cells comprising the hyphae is tightly regulated, whereas in groups without septa, nuclei flow freely through the entire mycelium (e.g., in phylum *Glomeromycota*).

Although the evolutionary cohesion of the *Eumycota* is widely accepted, some uncertainties and surprises with respect to membership have attracted attention in the last decade. There is mounting evidence that the *Microsporidia*, a lineage of highly specialized and reduced (unicellular, lacking mitochondria and chitin) animal parasites, fall within the *Eumycota* (Capella-Gutiérrez et al., 2012). The animal parasite *Pneumocystis* is now known to be a divergent fungus, probably falling within the *Ascomycota* (Edman et al., 1988; James et al., 2006). Molecular phylogenies confirm the long-held view that the *Oomy-cota*, which includes the filamentous plant pathogens *Pythium* and *Phytophthora*, belong to the superkingdom Stramenopiles (heterokonts), while the slime molds belong to the superkingdom *Amoebozoa*, both outside the *Opisthokonta*.

B   Present Phylogeny

Molecular systematics of the *Eumycota* are in the midst of radical revision (James et al., 2006; Liu et al., 2009; Floudas et al., 2012; see the special issue of Mycologia, 2006, volume 98, issue 6). About 20 years ago, a system of five fungal phyla achieved universal recognition based, in part, on initial rDNA phy-logenies (Bruns et al., 1992). The five phyla were the *Chytridiomycota* (water molds), *Zygomycota* (bread molds), *Glomeromycota* (arbuscular mycorrhizal fungi), *Ascomycota* (cup fungi), and *Basidiomycota* (club fungi). Later multi-locus, molecular analyses suggest that neither the *Zygomycota* nor the *Chytri-diomycota* are monophyletic groups (O'Donnell et al., 2001). Several new phyla, subphyla, and unranked higher taxa have been proposed (Hibbett et al., 2007), including the *Blastocladiomycota* and *Neocalimastigomycota* (for-merly members of *Chytridiomycota*), and the *Entomophthoromycotina*, *Kick-xellomycotina*, *Mucoromycotina*, and *Zoopagomycotina* (formerly members of *Zygomycota*; Fig. 4.2). Even extremely data-rich phylogenomic analyses have yet to resolve with certainty relationships at the base of the fungal tree (Liu et al., 2009), so we can expect further rearrangements and optimization of high-order taxonomy in the years to come. See Fig. S4.1 and Table S4.1 in online Supplemental Material at http://booksite.elsevier.com/9780124159556 for help

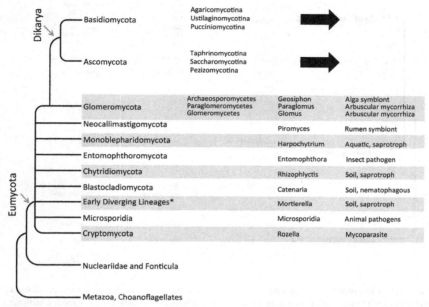

FIG. 4.2 Overview of fungal phylogeny. Currently recognized phyla and subphyla in the Kingdom Fungi. The *Metazoa* are an outgroup within the *Opisthokonta*, whereas the *Nucleariidae* are thought to be the closest relatives of the Fungi. Exemplar taxa and their ecological roles are provided on the right. *The Early Diverging Lineages include the *Kickxellomycotina, Mortierellomycotina, Mucoromycotina, Nephridiophagidae, Olpidiaceae*, and *Zoopagomycotina*. Information for phyla *Basidiomycota* and *Ascomyota* is provided in Figs. 4.3 and 4.4.

in making sense of some confusing terminology in fungal systematics. Good sources for the most up-to-date taxonomic hierarchy for the fungi include Index Fungorum (www.indexfungorum.org/), the Tree of Life (tolweb.org/Fungi/2377), and NCBI (www.ncbi.nlm.nih.gov/taxonomy).

Current understandings of the major evolutionary lines of fungi, along with a few exemplar taxa and their trophic niches, are presented in Figs. 4.2–4.4 (see also Hibbett et al., 2007; Stajich et al., 2009). Major evolutionary groups in soil include the chytrids, or the "water molds" generally viewed as aquatic organisms, many of which degrade pollen or algal biomass. However, as discussed in Section III.A, several orders, including the *Spizellomycetales* and *Rhizophlyctidales*, are found in diverse soils, taking on dominant roles in extreme, high elevations (Freeman et al., 2009). Phylogenetically diverse, filamentous species scattered among the early diverging fungal lineages are widespread in soils. Some are rapidly growing saprotrophs with a preference for labile C compounds; these include *Mortierella* (placed either within *Mortierellomycotina* or *Mucuromycotina*) and the bread molds *Rhizopus* and *Mucor* (*Mucuromycotina*). The phylum *Glomeromycota* encompasses all fungi that form arbuscular mycorrhizae, as well as the enigmatic, algal symbiont

Basidiomycota

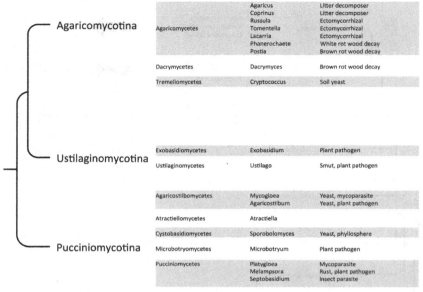

FIG. 4.3 Currently recognized subphyla and classes within the phylum *Basidiomycota*. Several subphyla and classes that contain only a few, rarely encountered species are not shown. Exemplar taxa and their ecological roles are provided on the right.

Ascomycota

FIG. 4.4 Currently recognized subphyla and classes within the phylum *Ascomycota*. Several subphyla and classes that contain only a few, rarely encountered species are not shown. Exemplar taxa and their ecological roles are provided on the right.

Geosiphon (Gehrig et al., 1996). There is some evidence that the *Glomeromycota* is basal to the *Dikarya*, but this is not yet certain. The subkingdom *Dikarya* is comprised of the most recent and derived "crown" phyla *Ascomycota* and *Basidiomycota*. The *Ascomycota* constitute the most species-rich fungal phylum, accounting for roughly 75% of described fungal species; these species fall into three subphyla, as follows. (1) The *Taphrinomycotina*, which has recently gained support as monophyletic (Schoch et al., 2009), encompasses the fission yeasts (*Schizosaccharomyces*); the animal pathogen *Pneumocystis*; the unusual dimorphic (see definition in Section III of online Supplemental Material at http://booksite.elsevier.com/9780124159556) plant pathogen *Taphrina*; the root-associated, sporocarp-forming, filamentous genus *Neolecta*; and the newly erected class of filamentous soil fungi, the *Archaeorhizomyces* (see Section II. C). (2) The *Saccharomycotina* include the budding yeasts, such as *Saccharomyces, Debaromyces, Pichia, Candida*, and others, many of which are found in soils, presumably decomposing labile organic materials both aerobically and anaerobically. (3) The third subphylum is the *Pezizomycotina*, which accounts for the greatest phylogenetic, species, and functional diversity within the *Ascomycota*. The vast majority of lichen-forming fungi fall within this lineage (there are a few basidiomycete lichens), as do a few ectomycorrhizal (EMF) taxa, all dark-septate endophyes (DSE), essentially all ericoid mycorrhizal (ERM) species, and a spectrum of pathogens and saprotrophs.

The diverse phylum *Basidiomycota* is also divided into three well-supported subphyla. The basal subphylum is the *Pucciniomycotina*, which includes all rust fungi, an economically important group of plant pathogens, as well as some yeasts that are common in soil, but relatively little studied, such as *Sporobolomyces* and *Leucosporidium*. The second subphylum, *Ustilaginomycotina*, is also predominantly comprised of plant pathogens, the smuts. Similar to the rusts, the *Ustilagomycotina* includes several yeasts, such as *Malassezia* (a skin pathogen that is also frequently recovered from soils) and *Arcticomyces* (a cold-soil yeast). This subphylum appears to be more closely related to the *Agaricomycotina* than to the rusts (James et al., 2006). The *Agaricomycotina* includes the vast majority of filamentous *Basidiomycota*, including all mushroom-forming taxa. These encompass nearly every conceivable soil niche (except thermophiles and psychrophiles), accounting for all brown rot and white rot fungi as well as most EMF taxa.

C Novel Lineages

Another important impetus for major taxonomic changes has been the discovery of novel lineages from environmental samples. A landmark paper described seasonal changes in soil fungal communities at the Niwot Ridge alpine research site in Colorado (Schadt et al., 2003). A considerable fraction of the community was comprised of taxa that appeared to fall within the *Ascomycota*, but were difficult to assign to any known class. Subsequent studies demonstrated that these fungi occur in a wide range of habitats and geographic regions (Porter et al., 2008). Discovery of an isolate belonging to this lineage allowed the description of a new class of *Ascomycota*, the *Archaeorhizomycetes* (Rosling et al., 2011). Another deeply divergent lineage for which there is only one cultured representative, but evidence for wide occurrence in aquatic and soil habitats, is the *Cryptomycota* (Jones et al., 2011). This early-diverging lineage includes the mycoparasitic genus *Rozella*. The roles of other members of this group, which may comprise a significant fraction of the diversity of the *Eumycota*, remain to be elucidated. Additional deeply diverging lineages, some of which are weakly placed within the euchytrids, have been recovered in marine habitats (Nagano et al., 2010; Richards et al., 2012) and soils (Glass et al., 2013). It is likely that additional deep lineages will be added to the fungal tree in the coming years. Some of these lineages, such as *Archaeorhizomycetes*, likely play important roles in soils.

III OCCURRENCE

A Extremophiles, Distribution Across the Planet

Fungi are present and prominent in all soils. At broad phylogenetic scales, the aphorism "everything is everywhere" does seem to apply to fungi: beyond soils, they are also found in nearly every other habitat on Earth, including deep sea hydrothermal vents and sediments, subglacial sediments, ancient permafrost, sea ice, hot and cold deserts, salterns, and soils of the Dry Valleys of Antarctica (Cantrell et al., 2011). Surprisingly, fungi may be the most abundant eukaryotic members of some deep-sea sediments (Edgcomb et al., 2011). A key question is the extent to which these taxa are active in these hostile environments, as opposed to surviving in highly resistant, dormant spore stages following introduction by wind or by other vectors (Pearce et al., 2009; Bridge and Spooner, 2012). Phylogenetically diverse yeasts dominate the isolates obtained from many of these extreme habitats. For example, species of the dimorphic basidiomycete yeast *Cryptococcus* can be dominant in glacial habitats, permafrost, marine sediments, and unvegetated Antarctic Dry Valley soils (Connell et al., 2006; Bridge and Spooner, 2012; Buzzini et al., 2012). In the marine study, these yeasts were detected by culture-independent DNA and RNA methods, the latter strongly suggesting *in situ* metabolic activity (Edgcomb et al., 2011). Species of *Cryptococcus* are also very abundant in boreal and temperate soils. Ascomycotan yeasts, such as *Pichia*, *Debaromyces*, *Candida*, *Metschnikowia*, and *Aureobasidium pullulans* are also found in extremely cold and/or saline habitats (Gunde-Cimerman et al., 2003; Zalar et al., 2008; Cantrell and Baez-Félix, 2010; Butinar et al., 2011; Cantrell et al., 2011).

Several filamentous *Ascomycota* are also noteworthy extremophiles. Species of *Geomyces* (*Leotiomycetes*; now placed in teleomorph *Pseudogymnoascus*) have been recorded from marine habitats as well as cold soils (Arenz and Blanchette, 2011; Bridge and Spooner, 2012; Richards et al., 2012); one isolate was reported to be metabolically active down to -35°C (Panikov and Sizova, 2007). A cold-soil dwelling member of this genus, *Geomyces destructans*, is the causal agent of white-nose syndrome, an epidemic that threatens numerous bat species in North America (Blehert et al., 2009; Lorch et al., 2011). The so-called meristematic, microcolonial, or black yeasts are distributed among several lineages of *Sordariomycetes*, *Eurotiomycetes*, and *Dothideomycetes* (Onofri et al., 2000; Selbmann et al., 2005; Sterflinger et al., 2012) and commonly occur in extreme habitats (e.g., in canyon walls of the Antarctic Dry Valleys). Some species grow on or within rocks (endolithic) in both hot and cold deserts and at high elevations. Some members of this group display moderate halotolerance, extreme drought tolerance, or can grow at pHs down to 0 (Starkey and Waksman, 1943). Convergent evolution of strong melanization, slow growth, isodiametric meristematic cells (i.e., cells that can reinitiate growth when dislodged from the colony), and other putative stress-related features is seen among the black yeasts (Selbmann et al., 2005; Sterflinger et al., 2012). Lichens are also formed predominantly by filamentous taxa of Ascomycota and are found across the array of extreme hot, cold, dry, and saline environments (discussed earlier), often playing the role of chief primary producer by virtue of photosynthetic activities of their cyanobacterial or algal photobionts (Vitt, 2007). Lichens are also important and widespread in less-extreme terrestrial habitats (Feuerer and Hawksworth, 2007). These same classes of Ascomycota include species with the highest heat-tolerance seen in the *Eukaryota*. Thermophilic and thermotolerant fungi share several key convergent traits. Their spores usually do not germinate below temperatures of 45°C, even though the mycelium can grow at lower temperatures (Maheshwari et al., 2000). These fungi, which belong to several orders within the *Ascomycota* (*Sordariales*, *Eurotiales*, *Onygenales*), one of which is within the early diverging lineages (*Mucorales*), have been recovered from diverse soils in both hot and cold regions. Within the *Ascomycota*, closely related species may be mesophiles and thermophiles, although many members of the *Chaetomiaceae* (*Sordariales*) are thermophilic. The primary niches of thermophiles are concentrated aggregations of moist, well-oxygenated organic material that "self-heat" due to intensive respiration during decomposition (Maheshwari et al., 2000). Humans encourage their growth through various composting methods. Mesophilic taxa initiate decomposition, causing the initial temperature to rise, but then cease to grow as temperatures surpass 40-45°C, at which point thermotolerant and thermophilic taxa germinate, grow, and drive the temperature of the heap toward their maximum growth temperatures of 50-60°C.

In the case of black yeasts, thermophiles, and lichens, there is no question of their activity and adaptation to extremes because they can be observed actively growing under these extreme conditions. This is also true of some of the cold-tolerant taxa, such as *Geomyces*, which can be observed growing in a dense mat across permanently frozen ice lenses in the Fox Permafrost Tunnel near

Fairbanks, Alaska (Waldrop et al., 2008). It seems likely that some of the other extremophiles described, such as yeasts isolated from glacial habitats, are also "indigenous" taxa that are adapted to these extreme environments because they display tolerance to extreme conditions in the laboratory (Onofri et al., 2000; Gunde-Cimerman et al., 2003; Selbmann et al., 2005; Arenz and Blanchette, 2011; Butinar et al., 2011; Buzzini et al., 2012). Some cosmopolitan taxa detected in extreme environments, such as species of *Penicillium* and *Aspergillus*, may be present only as inactive spores.

Although members of the *Dikarya* dominate records for extremophilic fungi, members of the early-diverging fungal lineages have been reported from marine habitats using culture-independent methods (Richards et al., 2012). Culture-independent approaches have also revealed a preponderance of chytrid lineages in soils at globally distributed, high-elevation sites that are above the vegetated zone (Freeman et al., 2009; Gleason et al., 2010). It has been proposed that these fungi may derive nutrients from algae and/or pollen transported by wind from distant locations; their presence in marine habitats might also be due to trophic linkages with algae (Richards et al., 2012).

To summarize, although no fungi match the extreme thermotolerance of *Archaea* and *Bacteria* from hot springs and hydrothermal vents, thermophilic fungi are among the most heat-tolerant eukaryotes. Furthermore, certain species equal or surpass any prokaryotes in cold tolerance and also occupy extreme habitats with respect to salt, desiccation, hydrostatic pressure, and pH. However, the most luxuriant fungal growth can be seen in moist, aerobic soils with large amounts of complex organic C.

B   Biomass, Growth, and Abundance

Fungi generally dominate microbial biomass and activity (i.e., respiration) in soil organic horizons, particularly in forests (Joergensen and Wichern, 2008). Bacterial-to-fungal ratios tend to be lower in acidic, low-nutrient soils with recalcitrant litter and high C-to-N ratios (Fierer et al., 2009), whereas bacteria are increasingly prominent in high N+P, saline, alkaline, and anaerobic (waterlogged) soils (Joergensen and Wichern, 2008).

Fungal biomass varies widely within and across biomes in relation to litter composition, root density, and nutrient availability. Fungi may comprise up to 20% of the mass of decomposing plant litter. In biomes dominated by EMF plants, extraradical mycelium may comprise 30% of the microbial biomass and 80% of the fungal biomass (Högberg and Högberg, 2002). Although fungal abundance and ratios of fungal to bacterial biomass tend to increase as soil pH decreases (Högberg et al., 2007; Joergensen and Wichern, 2008; Rousk et al., 2009), other studies suggest that fungal distributions are more influenced by N and P availability than pH *per se* (Fierer et al., 2009; Lauber et al., 2009). Estimates of fungal biomass turnover in soils are on the order of months (i.e., 130-150 days; Rousk and Baath, 2011), which is comparable to studies of fungal growth rates on submerged litter (e.g., Gulis et al., 2008).

Fungal mycelia generally grow radially as fractal networks in soil, wood, and litter (Fig. 4.1; Bolton and Boddy, 1993). Fungi alter hyphal development in response to environmental conditions to minimize cost-to-benefit ratios in terms of C or nutrient capture versus expenditure on growth. Very fine "feeder hyphae" are elaborated in resource-rich patches, whereas nutrient-poor areas are less densely colonized by hyphae specialized for efficient searching and nutrient transport. The transport hyphae may aggregate into tightly woven bundles called cords, strands, or rhizomorphs depending on their developmental structure. All these aggregations provide larger diameter transport tubes. However, species vary considerably in hyphal growth patterns, the size and structure of transport networks, and the resulting foraging strategies (Boddy, 1999; Agerer, 2001; Donnelly et al., 2004).

IV BIODIVERSITY

A Estimates of Species Richness

The true diversity of the Eumycota is uncertain and controversial. There are roughly 100,000 described taxa (Kirk et al., 2008), which is thought to include many synonyms due to both duplicate descriptions and anamorph-teleomorph pairs (i.e., the tradition in mycology of giving separate names to asexual and sexual forms). New species are being described at a rapid rate, limited primarily by a dearth of fungal taxonomists rather than a lack of fungi in need of description (Blackwell, 2011; Hawksworth, 2012). A variety of approaches has been used to estimate how many species of fungi exist. The approach that has received the most attention has been to census both plants and fungi at focal sites within a region to derive species ratios of fungal to plant taxa. These regional ratios are then multiplied by the estimated numbers of plants on Earth to arrive at an estimate for fungal richness. Hawksworth compiled data from well-studied sites in the United Kingdom and obtained a ratio of six fungi per vascular plant species, giving rise to a widely cited global estimate of 1.5 million fungi (Hawksworth, 1991, 2002). This estimate was criticized as highly exaggerated by May (1991). A recent mathematical approach that used rates of species and higher taxonomic rank accumulation over time, as well as relationships between numbers of species and higher ranks from well-studied groups, to estimate numbers of species in less-well-studied groups yielded an estimate of 660,000 species of fungi (Mora et al., 2011). This is considered too low by many mycologists (Bass and Richards, 2011). Several recent molecular surveys of fungi in environmental samples have applied the fungus-to-plant ratio method and yielded estimates of 5-6 million species (O'Brien et al., 2005; Taylor et al., 2013); estimates based on thorough censuses from single plant species have yielded estimates as high as 10 million (Cannon, 1997). The fungus-to-plant ratio estimates from molecular surveys are more likely to be within the correct order of magnitude, but will require further refinement as we learn more about how fungal specificity varies across biomes (Arnold et al., 2000; Berndt, 2012) and how plant and fungal geographic distributions scale to one another (Schmit et al., 2005; Schmit and Mueller, 2007; Tedersoo et al., 2009a, 2012).

B   Fungal Dispersal and Biogeography

Fungal biogeography and assembly of local communities are determined, in part, by dispersal. Fungi in soil disperse, albeit slowly, through growth of their mycelial networks. They disperse more rapidly and over larger distances through movement of various propagules. A hallmark of fungi is the production of propagules in the forms of meiotic or mitotic single-celled resting structures (zygospores, ascospores, basidiospores, conidia, and so on; see Fig. 4.1 and Sections II and III in the online Supplemental Material at http://booksite. elsevier.com/9780124159556) capable of surviving harsh conditions and dispersing great distances by air. Thick-walled vegetative cells (e.g., monilioid cells) or aggregations of such cells, such as the cannonball-like sclerotia formed by *Cenococcum* and *Thanatephorus*, can also be important propagules. Meiotic basidiospores are very small and are released into the air by fungi that form mushrooms as well as by many plant pathogenic fungi. However, studies of mushrooms have shown that, much like plant seeds, the vast majority of basidiospores fall to the ground within centimeters of the fruiting body (Galante et al., 2011). Studies of EMF fungi colonizing individual pine "islands" in the midst of a "sea" of nonectomycorrhizal plants have shown that distance to forest edge and airborne dispersal capabilities of spores of different fungi have strong effects on potential and actual colonization (Peay et al., 2007, 2010, 2012). On the other hand, many EMF taxa are shared between North America and Europe, suggesting that over longer timescales, rare dispersal events can eventually lead to very large geographic ranges.

A single taxon defined by traditional morphological and anatomical characters may encompass two or more subgroups that are distinct when phylogenetic or biological species concepts are applied; these are cryptic species. Cryptic species nested within traditional taxonomic species often have much narrower geographic distributions. Molecular studies are revealing cryptic species within widespread species complexes in a number of EMF and decomposer fungi (James et al., 1999; Taylor et al., 2006; Geml et al., 2008; Carlsen et al., 2011; Grubisha et al., 2012). These studies also reveal finer host specificity than previously recognized. It seems reasonable to expect that the capacity for aerial dispersal interacts with host and habitat specificity to influence the patterns of successful dispersal that are observed in nature. In contrast to EMF fungi and decomposers, plant pathogens are notorious for very wide dispersal capabilities. However, this perspective may be driven in part by our alarm at the devastation that can ensue when a rare dispersal event (often human-mediated) carries a virulent pathogen to a novel, susceptible host. Native elms and chestnut trees were effectively lost to North America due to the introduction of virulent Dutch elm disease (*Ophiostoma novo-ulmi*), possibly from Asia, and chestnut blight (*Cryphonectria parasitica*) from Japan. However, the survival of these trees for millennia before the arrival of these pathogens again underlines that cross-continental jumps are rare.

V   FUNGAL COMMUNITIES

A   Definition

In mainstream ecology, the term *community* refers to the set of sympatric, metabolically active organisms that interact or can potentially interact. We know little about when, where, and how fungi interact with other organisms in soil, aside from conspicuous manifestations, such as mycorrhizal colonization of plant roots or nematode-trapping fungi. Fungi in soil vary at least four orders of magnitude in size. Single-celled yeasts may be 3-10 μm in diameter. In contrast, a single mycelial individual of the white rot, root pathogen *Armillaria mellea*, spanned 15 ha, with a predicted mass greater than 10,000 kg (Smith et al., 1992). In practice, we often define fungal communities in soil at plot scales similar to those often used in plant ecology. How well this scale agrees with "fungal community" as a theoretical construct remains to be determined. In some areas of mycology, more careful attention has been paid to the spatial definition of *community*. In particular, researchers studying wood, litter, and dung decay have recognized that fungal species must colonize, grow, and reproduce within the confines of a particular substrate, leading to the designation of "unit communities" (Cooke and Rayner, 1984). For example, the fungi that occupy a single, isolated leaf might constitute a unit community. The application of the unit community perspective to soil is difficult due to the complex distribution of resources and lack of distinct spatial boundaries.

Fungal communities in soil can be extremely species rich and patchy at small spatial scales. For example, high throughput clone sequencing of fungi in cores collected approximately one m apart in a boreal forest site revealed in excess of 300 fungal taxa in 0.25 g soil (Fig. 4.5; M. G. Booth & D. L. Taylor, unpublished). Moreover, the dominant taxa in the first core were quite distinct from the dominant taxa in the second core (Fig. 4.6).

B   Abiotic Drivers

A variety of studies demonstrate that species composition within a sample, plot, or site can be influenced by an array of abiotic factors. Communities of EMF fungi from Northern Hemisphere temperate and boreal forests have been studied most extensively. A series of papers report sharp differences in fungal communities as a function of soil horizon (Taylor and Bruns, 1999; Dickie et al., 2002; Lindahl et al., 2006; Taylor et al., 2013). Soil pH, moisture content, and

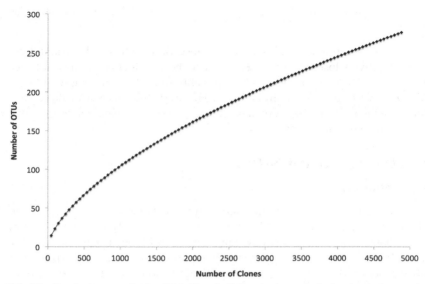

FIG. 4.5 Rarefaction curve for fungi in 0.25 g of soil. Rarefaction analysis showing the increase in number of observed OTUs (Mau Tau) with number of clone sequences resampled in Estimate 7.0 in a single ¼ g soil sample. Over 250 OTUs were recovered when ITS sequences were grouped at 97% identity; the rarefaction curve was not saturated, and the Chao I estimator was much higher, suggesting that many additional species were present. *M. G. Booth and D. L. Taylor, unpublished data.*

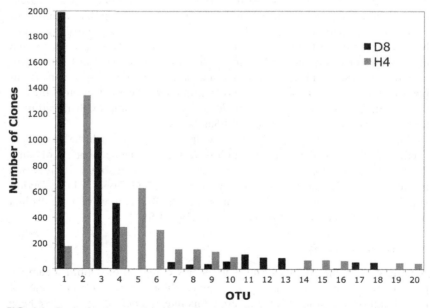

FIG. 4.6 Rank abundances of dominant fungi in adjacent cores. Numbers of clones (Y axis) belonging to dominant OTUs (X axis) documented in fungal ITS clone libraries generated from boreal forest soil samples collected ~1 m apart. Most of the dominant operational taxonomic units (OTUs, roughly equivalent to species) occur preferentially or exclusively in only one of the two samples. *M. G. Booth and D. L. Taylor, unpublished data.*

nutrient levels (particularly N) are also correlated with community composition in soil (Taylor et al., 2000; Toljander et al., 2006; Cox et al., 2010), but are relatively weak drivers compared to soil horizon, except in cases of extreme gradients such as gaseous ammonia pollution from a fertilizer plant (Lilleskov et al., 2002).

Even within guilds, fungal species vary considerably in their niche preferences, presumably due to competition and niche-partitioning over evolutionary time. There is evidence that mycelial exploration type (i.e., hyphal growth and foraging patterns; Agerer, 2001) may be related to strategies for N acquisition from different sources (Hobbie and Agerer, 2010). Taylor et al. (2000) found higher capacities for growth on media containing only organic N sources among fungal isolates from the pristine end of an N-deposition gradient across Europe compared with those from the polluted end of the gradient, where mineral N is more available. Less attention has been paid to how pH or moisture may underlie habitat preferences other than the well-known high levels of enzyme activity and resulting competitive dominance, exhibited by ERM fungi under acidic conditions (Smith and Read, 2008). Several recent papers have suggested that soil K and Ca levels are predictive of the composition of ectomycorrhizal (ECM) communities associated with *Alnus* species (Tedersoo et al., 2009b; Roy et al., 2013). In tandem, there is a great deal of interest in "bioweathering," the contribution of fungi and bacteria to the dissolution of minerals in rock (Taylor et al., 2009; see Chapter 16).

There is increasing evidence that soil fungal communities are influenced by climatic conditions on both geographic and temporal scales. Circumstantial evidence suggests that soil temperatures contribute to differences in fungal communities across latitude. For example, soil fungal communities were more strongly correlated with temperature than with latitude across the five bioclimactic subzones of the Arctic (I. Timling, D. A. Walker, & D. L. Taylor, unpublished). Most studies have undertaken sampling in the summer at higher latitudes or the equivalent at peak growing season (e.g., wet season) in tropical latitudes. Yet much of belowground activity likely occurs at other times. For example, the aforementioned studies of midlatitude, high-elevation alpine sites at Niwot Ridge in Colorado showed that microbial (including fungal) biomass peaked in late winter under the snowpack (Schadt et al., 2003; Lipson and Schmidt, 2004). Given that up to 10 months of the year are snow-covered in boreal, Arctic, and Antarctic regions, activity under snow could strongly influence annual biogeochemical fluxes (Sturm et al., 2005). Even in cold-dominated biomes, fungal communities in soil have been shown to shift predictably across seasons (Schadt et al., 2003; Taylor et al., 2010).

Seasonal climate variation could be considered a form of predictable disturbance. Other, more extreme and less predictable disturbances also perturb the biomass and composition of fungal communities in soil. The disturbance that has received the greatest attention is fire. EMF communities are strongly impacted by fire as a consequence of direct heat injury and consumption of fungal biomass as well as death or injury to host plants, loss of organic matter, and changes in soil chemistry (e.g., transient increases in N availability, higher pH). Studies showing mild effects of fire have usually been conducted in habitats with frequent low-severity burns that do not kill the host trees (Stendell et al., 1999; Dahlberg et al., 2001; Smith et al., 2004). Where hotter, stand-replacing fires occur, impacts on ECM fungi are stronger (Baar et al., 1999; Cairney and Bastias, 2007). In general, fire reduces fungal diversity and preferentially removes those taxa that have strong preferences for the litter layer. When mycorrhizal hosts are killed, vegetation succession is reset and suites of so-called early stage fungi are the dominant colonizers (see next section; Visser, 1995; Treseder et al., 2004; Taylor et al., 2010). The direct effects of fire versus the indirect effects mediated by vegetation changes are difficult to disentangle.

Disruption in the form of agricultural tillage has been shown to impact fungal communities (Helgason et al., 1998). It is likely that smaller-scale disturbances influence fungal communities, but fewer studies are available. In the Arctic, fungal communities in paired vegetated versus cryoturbated microsites ("frostboils") are distinct (I. Timling, D. A. Walker, & D. L. Taylor, unpublished), but again, effects of host plants versus direct effects of cryoturbation on fungi are uncertain.

C   Biotic Drivers

Most fungi consume living or dead plant material for their primary energy source, and a large fraction of fungi display some degree of specialization toward their living or dead plant substrates. Thus, plant community composition plays a dominant role in determining which fungi are present at a site. Historically, only the plant-to-fungus direction of influence has received much recognition. However, there is increasing evidence that these influences are reciprocal: the spectrum of species present, i.e., the plant microbiome, and their relative abundances can have major impacts on plant community composition as well (van der Heijden et al., 1998a, b; Reynolds et al., 2003; Bever et al., 2010). Specific plant pathogens can dramatically reduce or eliminate particular host species, again altering plant community composition. Plant-soil feedbacks (Bever, 1994; Klironomos, 2002), often involving the buildup of host-specific pathogens in soil, can alter coexistence and competitive dynamics within plant communities.

It has been well documented that EMF fungi range from genus, or even species-specific, to quite generalist, associating with both angiosperms and gymnosperms (Molina et al., 1992). Broader associations appear to be the norm in ECM fungi, but host-specialist fungi can be important players, for example in the reciprocally specific associations of alder (Miller et al., 1992). Although the several hundred species of AM fungi found in > 200,000 vascular plant species have been assumed to have low specificity, a more complicated picture is emerging (see Chapter 11). Much like mycorrhizal fungi, decomposer fungi display a range of specialization toward their substrates. The genetic, physiological or ecological bases for such specialization are not known in most cases. Certain white and brown rot wood decay fungi are found on wide arrays of both angiosperm and gymnosperm hosts. Yet many fungi in these guilds favor either angiosperms or gymnosperms and may even prefer families or genera within these lineages.

Competition among fungal species plays an important role in structuring communities. Studies have demonstrated that the arrival order of ECM species can shift competitive dominance in colonization of seedling root systems (Kennedy et al., 2009). Such "priority effects" have also been shown to occur for decomposer communities in wood (Fukami et al., 2010). In this latter example, the order of species arrivals also affected the progress of decay, suggesting that competition and community assembly may have ecosystem consequences. These interactions likely involve indirect exploitative competition for resources as well as various forms of direct interference competition. Further evidence for competition in soil fungi come from statistical analyses of patterns of co-occurrence and avoidance. In general, cooperating/synergistic species should co-occur more often than expected by chance, whereas antagonistic species should co-occur less often than expected by chance; the latter situation is called a "checkerboard" pattern (Stone and Roberts, 1990). In a study of pine ECM communities, avoidance patterns suggestive of competition were more common than were co-occurrence patterns (Koide et al., 2004). Wood decay fungi are well-known for overt signs of competitive interactions, as many use combative strategies, including production of bioactive compounds as well as direct invasion and lysis of opposing hyphae (Cooke and Rayner, 1984).

Fungal communities undergo succession, likely driven by a combination of species interactions and changing resource/environmental conditions, analogous to patterns known from prokaryotes and plants. Patterns of succession are complicated in soil fungal communities due to the wide range of relevant spatial and temporal scales. In vegetated ecosystems, succession of some fungal guilds occurs in tandem with plant succession. These observations led to the classification of "early-stage" and "late-stage" EMF taxa (Deacon et al., 1983). Early-stage species are the first to colonize young tree seedlings in habitats where no mature trees occur (e.g., "old-fields"), whereas late-stage species are typical of mature forests. These changes in fungal communities may be driven by changes in the C-provisioning capacity of trees as they grow (late-stage fungi tend to produce larger mycelial mats and may have larger C demands) or due to changes in the soil environment, particularly the buildup of a well-decomposed organic horizon (late-stage fungi appear to have a greater capacity to degrade complex organic polymers). Studies of fungal communities in soil beyond only ECM taxa also demonstrate strong shifts in composition in concert with plant successional stage (Taylor et al., 2010).

A microhabitat in which fungal succession has been well studied is coarse woody debris. As with EMF communities, wood-decomposer fungi are territorial, usually occupying contiguous patches of substrate to the exclusion of other species. A series of studies have demonstrated that the first colonists of wood are present at low levels in the living tree and grow actively once the dead wood has dried beyond a certain threshold (Chapela and Boddy, 1988). These fungi are primarily soft-rot members of the *Ascomycota*, such as *Xylaria*. These taxa are quickly followed by white and brown rot fungi in the *Basidiomycota*. Observations of fruit body formation over time on downed logs, as well as direct molecular analyses of wood itself, agree that *Basidiomycota*-dominated communities follow predictable series of species appearance and disappearance. Carbohydrate and phenolic composition, as well as C:N:P ratios, change significantly as decomposition progresses, which likely provides a basis for niche-differentiation and successional patterns among decomposers. Yet chemical composition does not entirely explain the successional patterns. For example, certain wood-rot *Basidiomycota* "follower species" seem to occur only when another species of *Basidiomycota* has previously colonized the log. Whether these patterns arise from direct species interactions or from changes in the chemical environment imparted by the primary species (i.e., facilitation in the sense of classical succession theory) is unknown.

Successional patterns in leaf litter are similar, but faster than in wood. The earliest colonizers are often found within the attached, senescent leaves as endophytes, primarily *Sordariomycetes* and *Dothideomycetes* (*Ascomycota*) (Snajdr et al., 2011). R-selected "sugar fungi" that are not present as endophytes, such as *Mortierella*, may also play important roles in the earliest stages of decay, when labile carbohydrates are readily available. Once litter is exposed to the moist forest floor, rhizomorph and cord-forming *Basidiomycota* often aggressively colonize leaves. However, in arid lands, it appears that various drought-tolerant, melanized *Ascomycota* predominate over *Basidiomycota* as both plant symbionts and decomposers (Porras-Alfaro et al., 2011).

VI FUNCTIONS

A Introduction

Fungi mediate nearly every aspect of organic matter production, decomposition, and sequestration, with concomitant roles in the mineralization and cycling of N and P. Filamentous fungi support plant production through mycorrhizal associations that enhance the acquisition of water and nutrients (Chapter 11). Aboveground, ubiquitous fungal endophytes can confer resistance to thermal and drought stress and reduce herbivory (Porras-Alfaro and Bayman, 2011). Fungi also support C and N fixation by algae and cyanobacteria through lichen associations and, in arid and polar regions, the formation of biotic crusts that also mediate soil atmosphere exchanges, water infiltration, and stabilize surface soils against erosion (Pointing and Belnap, 2012). As agents of organic matter decomposition, fungi mediate the creation and protection of soil organic matter (SOM), as well as its mineralization. Surface litter becomes SOM when the residue of humic material and microbial products reaches a threshold composition of 70% acid insoluble material (Berg and McClaugherty, 2003), at which point the cost of further degradation is not energetically favorable without additional inputs of more labile material (Moorhead et al., 2013). However, surface litter decomposition may not be the principal pathway of SOM formation in many ecosystems. Recent papers provide evidence that the growth and turnover of roots and their associated fungi is responsible for most soil C sequestration in forest soils (Wilson et al., 2009; Clemmensen et al., 2013).

Filamentous fungi promote macroaggregate formation by binding soil particles with hypha and producing cell wall materials that act as adhesives (Willis et al., 2013). Aggregate formation promotes soil C sequestration by providing physical protection from decomposers and their degradative enzymes (Wilson et al., 2009).

B Nutrient Cycling

The most commonly measured characteristics of soil fungi are their biomass, elemental stoichiometry, growth rates and efficiencies, and activities of their extracellular enzymes (Sinsabaugh and Follstad Shah, 2011; Sinsabaugh et al., 2013; Wallander et al., 2013). These parameters define the size of fungal C and nutrient pools and rates of fungal-mediated biogeochemical reactions. Fungi must generate small, molecular mass, growth substrates by enzymatically degrading complex organic matter outside the cell. However, enzymes released to the environment by intent or as a result of cell lysis are beyond the control of that organism (Fig. 4.1). As a result, a substantial fraction of the enzymatic potential of soils is a legacy of enzymes that are spatially and temporally displaced from their origins (Burns et al., 2013). In some cases, particularly for oxidative enzymes, these enzymes catalyze degradation and condensation reactions secondary to their original function (Sinsabaugh, 2010).

1 Enzymes/Decomposition

It is becoming increasingly difficult to generalize about the enzymatic capabilities of higher fungal taxonomic categories. Most hydrolytic and oxidative capabilities are broadly distributed across taxonomic groups. As studies of fungal genomics advance, there is increasing emphasis on the expression and ecological purposes of these capabilities. Fungi are considered the principal degraders of plant cell wall material, especially during the early stages of decomposition when their filamentous growth form and their capacity to secrete a variety of glycosidases and oxidases allows them to ramify through the cellular structure of plant litter. The capacity to at least partially degrade cellulose, especially after it has been decrystallized, is widespread in fungi, including basal lineages.

Ascomycota and *Basidiomycota* have the widest genetic and ecological capacity, which is facilitated by the synergistic expression of a variety of other polysaccharide-degrading enzymes. The ecological capacity of *Glomeromycota* to degrade cellulose and other cell wall polysaccharides appears more limited, but recent studies indicate greater capacity than once thought (Talbot et al., 2008; Kowalchuk, 2012).

Production of laccases and other phenol oxidases are also widely distributed among *Ascomycota*, *Basidiomycota*, and *Glomeromycota* (Baldrian, 2006). In some organisms, mainly saprotrophs, these enzymes function primarily in the degradation of lignin and other secondary compounds in plant cell walls, often indirectly through the production of small reactive oxidants known as redox mediators (Rabinovich et al., 2004). Other saprotrophs oxidatively degrade SOM to obtain chemically protected C, N, and P (Burns et al., 2013). Mycorrhizal fungi appear to use the same strategy to mine N and P (Pritsch and Garbaye, 2011). In many taxa, a large but indeterminate portion of oxidative activity is related to morphogenesis (e.g., melanin production), detoxification, and oxidative stress (Baldrian, 2006; Sinsabaugh, 2010), rather than nutrient acquisition. But once released into the environment through biomass turnover, these activities also contribute to the oxidative potential of soils, which catalyzes nonspecific condensation and degradation reactions that contribute to both the creation and loss of SOM (Burns et al., 2013).

Peroxidases, which have greater oxidative potential than laccases, are produced by some members of the *Ascomycota* and *Basidiomycota*. The most widely distributed enzyme, Mn-peroxidase, acts indirectly on organic compounds by generating diffusable Mn^{+3}. The contribution of Mn peroxidase to soil C dynamics is highlighted by manipulation studies showing that Mn availability can limit the decomposition of plant litter (Trum et al., 2011). Some *Basidiomycota*, principally wood-rotting fungi, produce lignin peroxidase, which directly oxidizes aromatic rings (Rabinovich et al., 2004).

The taxonomic distribution of extracellular enzymatic capacity is the basis for the traditional soft rot, brown rot, white rot ecological classification of decomposer fungi on the basis of the appearance of rotting wood. In enzymatic terms, these classifications refer to organisms that primarily attack cell wall polysaccharides, those that deploy lower redox potential laccases in addition to glycosidases, and those that deploy high redox potential laccase-mediated, or peroxidase systems capable of effectively depolymerizing lignin. Interestingly, recent phylogenomic studies suggest that the common ancestor of the *Agaricomycotina* (basidiomycete class containing all ECM, brown rot and white rot fungi) was capable of white rot. The evolution of brown rot appears to have occurred independently several times, in part through the loss of Mn-peroxidase genes.

ECM fungi were once thought to have little saprotrophic capability compared to wood and litter decay fungi. Some recent genomic data support this perspective (Martin et al., 2008). But there is growing evidence that at least some ECM fungi can attack a wide range of organic polymers (Read and Perez-Moreno, 2003; Talbot et al., 2008; Pritsch and Garbaye, 2011), but these activities are likely under different environmental controls than those to which pure decomposer fungi respond. The key issue is that decomposer fungi are usually C-limited, while ECM fungi produce degradative enzymes to access and mobilize N and P (Talbot et al., 2008). Thus the cost-benefit ratios that drive the adaptive evolution of enzyme activities differ for ECM versus decomposer fungi.

2 Nitrogen

The potential to use chitin as an N source is widespread among fungi, largely because fungal cell walls include chitin and related compounds (Geisseler et al., 2010). In some studies, chitinase activity is used as an indicator of fungal bio-mass and metabolism. However, proteins and their degradation products are the largest source of organic N in soils, and it is likely that most saprotrophic and biotrophic fungi obtain much of their N from degradation of peptides (Hofmockel et al., 2010; Sinsabaugh and Follstad Shah, 2011). In contrast to their role in P acquisition, the role of AM fungi in N acquisition is poorly resolved (Veresoglou et al., 2012). The enzymatic capacity of ECM fungi to mine N from SOM has received more attention (Talbot et al., 2008; Pritsch and Garbaye, 2011). The role of fungi in the N cycle was once considered primarily assimilatory. That is, fungi assimilated inorganic N and N-containing organic molecules to support the production of new fungal biomass and supply plant hosts. Recent studies have shown that dissimilatory denitrification and codenitrification pathways are widespread among *Ascomycota* and are responsible for a large fraction of nitrous oxide efflux, especially in arid soils (Spott et al., 2011; Shoun et al., 2012).

Anthropogenic N deposition has been shown to decrease fungal diversity, biomass, and respiration (Knorr et al., 2005; Treseder, 2008). Increased N deposition is associated with decreases in laccase and peroxidase activities in litter and soil, leading to slower decomposition and increased soil C sequestration (Sinsabaugh et al., 2002; Gallo et al., 2004). Nitrogen deposition also alters the economics of plant-fungal symbioses, which may also contribute to observed decreases in fungal biomass and activity. Another contributing factor may be the two- to three-fold difference in biomass C-to-N ratio of fungi (10-20) and bacteria (5-10), which might promote bacterial growth relative to fungi (Van Der Heijden et al., 2008; Strickland and Rousk, 2010) when N availability is high. A recent review noted a positive relationship between fungal dominance as indicated by qPCR and soil C-to-N ratios across biomes (Fierer et al., 2009).

3 Phosphorus

All soil microorganisms produce extracellular phosphatases. Consequently, phosphatase activities in soil are generally greater than any other measured enzymatic activity, making it difficult to resolve the contributions of individual taxa to P mineralization. For fungi at least, it appears that most phosphatases released for extracellular P acquisition have acidic pH optima, whereas those intended for intracellular reactions have optima at neutral to alkaline pH (Plassard et al., 2011). Inositol phosphates, produced mostly as P storage products by plants, account for half of soil organic P (Plassard et al., 2011; Menezes-Blackburn et al., 2013). The abundance of these compounds has more to do with their recalcitrance to degradation than their rate of production. Phytate (inositol hexaphosphate) is crystalline; specific enzymes (phytases) are needed to hydrolyze the phosphate. *Ascomycota* are considered the best producers of phytase, but some *Glomeromycota* and *Basidiomycota* also produce them (Plassard et al., 2011; Menezes-Blackburn et al., 2013). Like all extracellular phosphatases, enzyme expression is induced by P deficiency. Mineral phosphate is also important to fungi and plants. In alkaline soil, calcium phosphates may be abundant, whereas weathered acid soils may have high concentrations of iron and aluminum phosphates. Some fungi, particularly EMF *Basidiomycota*, solubilize phosphate from mineral sources using low molecular weight organic acids, such as oxalate (Courty et al., 2010; Plassard et al., 2011).

C   Bioremediation

The filamentous growth habit and enzymatic versatility of fungi can also be adapted to treat waste streams and remediate soils contaminated with organic pollutants or toxic metals (Harms et al., 2011; Strong and Claus, 2011). The most effective pollutant remediators belong to the phyla *Ascomycota* and *Basidiomycota*. Most of this capacity is related to the production of a broad spectrum of extracellular laccases and peroxidases with varying redox potentials, pH optima, and substrate specificity that oxidatively modify or degrade aliphatic and aromatic pollutants, including halogenated compounds. In addition, some fungi also produce nitroreductases and reductive dehalogenases that further contribute to the degradation of explosive residues and halogenated contaminants. Intracellularly, many fungi also have cytochrome P450 oxidoreductases that can mitigate the toxicity of a broad range of compounds. The toxicity of metal contaminants can be mitigated by translocation and sequestration in chemically inaccessible complexes. In natural systems, improving the bioremediation capabilities of ECM fungi is of particular interest because the C supply from host plants may support fungal growth into contaminated hotspots and stimulate cometabolic reactions.

VII   FUNGUS-LIKE ORGANISMS AND SOIL FOOD WEBS

Fungi are linked to numerous other organisms in complex soil food webs. The spectrum of organisms that feed on or parasitize fungi in soil is extremely diverse, spanning viruses, bacteria, protists, insects, small mammals, and other fauna. Similarly, the spectrum of living organisms that are subject to attack by fungi is equally diverse. Fungi that are placed into one guild designation often engage in other sorts of interactions, and the extent of these phenomena is poorly known. As with fungal linkages as mycorrhizae and decomposers, other food web linkages range from nonspecific to highly specific. Some of the most widespread trophic linkages of soil fungi appear to be related to N demand imparted by the high C:N ratios of plant materials that comprise the main energy source. A number of wood decay fungi obtain supplemental N by feeding on nematodes (Thorn and Barron, 1984; Barron, 2003). Some of these fungi are also able to prey on bacteria (Barron, 1988). Some of the best-known nematode-trapping fungi actively produce enzymes to degrade cellulose or lignocellulose, leading Barron (2003) to suggest that these fungi are primarily decomposers that have particularly dramatic adaptations for N supplementation. Mycorrhizal fungi also engage in various interactions that may seem unexpected given their guild designation. For example, some EMF fungi can supplement N by attacking living animals. such as collembolans (small wingless insects known as springtails; Klironomos and Hart, 2001). ECM *Tuber* species sometimes cause "brulés" or bare patches of surrounding herbaceous vegetation, perhaps by attacking these understory plants (Plattner and Hall, 1995). There are also numerous mycoparasitic fungi, some of which are closely related to EMF taxa. One example is *Gomphidius*, a member of the *Boletales* (*Basidiomycota*) that parasitizes an EMF genus of *Boletales*, *Suillus* (Olsson et al., 2000).

Fungal hyphae in soil are usually abundantly festooned with both extracellular and intracellular bacteria (Bonfante and Anca, 2009; Hoffman and Arnold, 2010). These bacterial communities are not the same as bacterial communities in bulk soil, and certain isolates have been shown to exert specific influences on the fungi. For example, "mycorrhizal helper bacteria" can strongly promote the formation of ectomycorrhizae by some ECM fungi (Garbaye, 1994; Frey-Klett et al., 2007). Bacteria on hyphal surfaces presumably benefit from fungal exudates just as rhizosphere bacteria benefit from root exudates. It is likely that many complex interactions between fungi, bacteria, and other organisms have important impacts on fungal function in soil, although only a few examples have been well studied to date.

Numerous insects are adapted to the consumption of fungal hyphae, either as mycelium in soil (e.g., mites and collembolans) or in the concentrated tissue of sporocarps (e.g., fungus gnats). Several studies have manipulated the presence of these grazers and have usually found strong impacts on competitive dominance among the fungi (Klironomos et al., 1992). In addition to insects, a diverse assemblage of soil protists, nematodes, earthworms, and small mammals consume fungi, often in a fungal species-specific fashion (Hanski, 1989; Jørgensen et al., 2005).

Although fungi and bacteria drive the bulk of primary litter decomposition in most soils, other small eukaryotes make important direct and indirect contributions to soil nutrient cycling. Several nonfungal protists act as primary decomposers through the release of extracellular enzymes, just as do fungi (Adl and Gupta, 2006). Among these are various slime molds (*Amoebozoa*), which can be abundant in soil, and contribute to degradation of wood, bark, and dung. Similarly, some soil stramenopiles (Oomycetes), particularly in the *Saprolegniales*, participate in deconstruction of leaves, wood, dung, and animal remains.

However, nonfungal eukaryotes probably exert stronger effects through their roles as secondary decomposers, i.e., by consuming and reprocessing materials contained within the biomass of fungal and bacterial primary decomposers. Diverse protists grow in soil as naked amoebae, consuming bacteria, yeasts, and other small resources through either phagocytosis or osmotrophy. Important taxa of secondary saprotrophs include *Amoebozoa* (e.g., *Acanthamoebidae Flabellinea, Tubullinea, Mastigamoebidae, Eumycetozoa), Euglenozoa (Euglenida, Kinetoplastea), Heterobolosea (Acrasidae, Gruberellidae, Vahlkampfiidae), Cercozoa (Cercomonadidae, Silicofilosea), and the Ciliata* (Adl and Gupta, 2006). Species in these groups typically display marked substrate preferences and other fine-tuning to their soil environments (e.g., moisture gradients, seasonality). Naked amoebae are the most abundant protistan grazers in soil. Densities are greatest in the rhizosphere, up to 30 times greater than that of bulk soil (Bonkowski, 2004). Because the C-to-N ratio of protozoans is similar to that of their prey, grazing increases N mineralization rates, thereby promoting plant growth. Grazing intensity varies among taxa in relation to cell wall composition, cell size, and chemical signals. Gram-negative proteobacteria are more susceptible than G^+ bacteria. Grazing of mycorrhizal fungi reduces hyphal lengths and can alter fine root architecture. Last, soil protists also contribute to primary production. Specifically, the Chlorophyta and Charophyceae (green algae) can be abundant in the uppermost layers of soil, where sufficient light penetrates to allow photosynthesis (Hoffmann, 1989). Soil food webs may well be the most complex food webs on Earth, with fungi playing central roles as both specialized consumers and specific prey.

ACKNOWLEDGMENTS

This material is based on work supported by the National Science Foundation through awards EF-0333308 and ARC-0632332 to DLT.

REFERENCES

Adl, M., Gupta, V.S., 2006. Protists in soil ecology and forest nutrient cycling. Can. J. For. Res. 36, 1805–1817.

Agerer, R., 2001. Exploration types of ectomycorrhizae. Mycorrhiza 11, 107–114.

Arenz, B., Blanchette, R., 2011. Distribution and abundance of soil fungi in Antarctica at sites on the Peninsula, Ross Sea Region and McMurdo Dry Valleys. Soil Biol. Biochem. 43, 308–315.

Arnold, A.E., Maynard, Z., Gilbert, G.S., Coley, P.D., Kursar, T.A., 2000. Are tropical fungal endophytes hyperdiverse? Ecol. Lett. 3, 267–274.

Baar, J., Horton, T., Kretzer, A., Bruns, T., 1999. Mycorrhizal colonization of *Pinus muricata* from resistant propagules after a stand-replacing wildfire. New Phytol. 143, 409–418.

Baldrian, P., 2006. Fungal laccases: occurrence and properties. FEMS Microbiol. Rev. 30, 215–242.

Barron, G., 1988. Microcolonies of bacteria as a nutrient source for lignicolous and other fungi. Can. J. Bot. 66, 2505–2510.

Barron, G.L., 2003. Predatory fungi, wood decay, and the carbon cycle. Biodiversity 4, 3–9.

Bass, D., Richards, T.A., 2011. Three reasons to re-evaluate fungal diversity "on Earth and in the ocean". Fungal Biol. Rev. 25, 159–164.

Berbee, M.L., Taylor, J.W., 1993. Dating the evolutionary radiations of the true fungi. Can. J. Bot. 71, 1114–1127.

Berbee, M.L., Taylor, J.W., 2010. Dating the molecular clock in fungi-how close are we? Fungal Biol. Rev. 24, 1–16.

Berg, B., McClaugherty, C., 2003. Plant Litter: Decomposition, Humus Formation, Carbon Sequestration, first ed. Springer Verlag, Berlin.

Berndt, R., 2012. Species richness, taxonomy and peculiarities of the neotropical rust fungi: are they more diverse in the Neotropics? Biodivers. Conserv. 21, 2299–2322.

Bever, J.D., 1994. Feeback between plants and their soil communities in an old field community. Ecology 75, 1965–1977.

Bever, J.D., Dickie, I.A., Facelli, E., Facelli, J.M., Klironomos, J., Moora, M., Rillig, M.C., Stock, W.D., Tibbett, M., Zobel, M., 2010. Rooting theories of plant community ecology in microbial Interactions. Trends Ecol. Evol. 25, 468–478.

Blackwell, M., 2011. The fungi: 1, 2, 3 . . . 5.1 million species? Am. J. Bot. 98, 426–438.

Blehert, D.S., Hicks, A.C., Behr, M., Meteyer, C.U., Berlowski-Zier, B.M., Buckles, E.L., Coleman, J.T., Darling, S.R., Gargas, A., Niver, R., et al., 2009. Bat white-nose syndrome: an emerging fungal pathogen? Science 323, 227.

Boddy, L., 1999. Saprotrophic cord-forming fungi: meeting the challenge of heterogeneous environments. Mycologia 91, 13–32.

Bolton, R.G., Boddy, L., 1993. Characterization of the spatial aspects of foraging mycelial cord systems using fractal geometry. Mycol. Res. 97, 762–768.

Bonfante, P., Anca, I.-A., 2009. Plants, mycorrhizal fungi, and bacteria: a network of interactions. Annu. Rev. Microbiol. 63, 363–383.

Bonkowski, M., 2004. Protozoa and plant growth: the microbial loop in soil revisited. New Phytol. 162, 617–631.

Bridge, P., Spooner, B., 2012. Non-lichenized Antarctic fungi: transient visitors or members of a cryptic ecosystem? Fungal Ecol. 5, 381–394.

Bruns, T.D., Vilgalys, R., Barns, S.M., Gonzalez, D., Hibbett, D.S., Lane, D.J., et al., 1992. Evolutionary relationships within the fungi: analyses of nuclear small subunit rRNA sequences. Mol. Phylogenet. Evol. 1, 231–241.

Burns, R., DeForest, J., Marxsen, J., Sinsabaugh, R., Stromberger, M., Wallenstein, M., Weintraub, M., Zoppini, A., 2013. Soil enzyme research: current knowledge and future directions. Soil Biol. Biochem. 58, 216–234.

Butinar, L., Strmole, T., Gunde-Cimerman, N., 2011. Relative incidence of ascomycetous yeasts in Arctic coastal environments. Microb. Ecol. 61, 832–843.

Buzzini, P., Branda, E., Goretti, M., Turchetti, B., 2012. Psychrophilic yeasts from worldwide glacial habitats: diversity, adaptation strategies and biotechnological potential. FEMS Microbiol. Ecol. 82, 217–241.

Cairney, J., Bastias, B., 2007. Influences of fire on forest soil fungal communities. Can. J. For. Res. 37, 207–215.

Cannon, P.F., 1997. Diversity of the Phyllachoraceae with special reference to the tropics. In: Hyde, K.D. (Ed.), Biodiversity of Tropical Microfungi. University Press, Hong Kong, pp. 255–278.

Cantrell, S.A., Baez-Félix, C., 2010. Fungal molecular diversity of a Puerto Rican subtropical hypersaline microbial mat. Fungal Ecol. 3, 402–405.

Cantrell, S.A., Dianese, J.C., Fell, J., Gunde-Cimerman, N., Zalar, P., 2011. Unusual fungal niches. Mycologia 103, 1161–1174.

Capella-Gutiérrez, S., Marcet-Houben, M., Gabaldón, T., 2012. Phylogenomics supports microsporidia as the earliest diverging clade of sequenced fungi. BMC Biol. 10, 47.

Carlsen, T., Engh, I.B., Decock, C., Rajchenberg, M., Kauserud, H., 2011. Multiple cryptic species with divergent substrate affinities in the *Serpula himantioides* species complex. Fungal Biol. 115, 54–61.

Chapela, I., Boddy, L., 1988. Fungal colonization of attached beech branches. I. Early stages of development of fungal communities. New Phytol. 110, 39–45.

Clemmensen, K.E., Bahr, A., Ovaskainen, O., Dahlberg, A., Ekblad, A., Wallander, H., Stenlid, J., Finlay, R.D., Wardle, D.A., Lindahl, B.D., 2013. Roots and associated fungi drive long-term carbon sequestration in boreal forest. Science 339, 1615–1618.

Connell, L., Redman, R., Craig, S., Rodriguez, R., 2006. Distribution and abundance of fungi in the soils of Taylor Valley, Antarctica. Soil Biol. Biochem. 38, 3083–3094.

Cooke, R.C., Rayner, A.D.M., 1984. Ecology of Saprophytic Fungi. Longman, London.

Courty, P.-E., Buée, M., Diedhiou, A.G., Frey-Klett, P., LeTacon, F., Rineau, F., 2010. The role of ectomycorrhizal communities in forest ecosystem processes: new perspectives and emerging concepts. Soil Biol. Biochem. 42, 679–698.

Cox, F., Barsoum, N., Lilleskov, E.A., Bidartondo, M.I., 2010. Nitrogen availability is a primary determinant of conifer mycorrhizas across complex environmental gradients. Ecol. Lett. 13, 1103–1113.

Dahlberg, A., Schimmel, J., Taylor, A., Johannesson, H., 2001. Post-fire legacy of ectomycorrhizal fungal communities in the Swedish boreal forest in relation to fire severity and logging intensity. Biol. Conserv. 100, 151–161.

Deacon, J., Donaldson, S., Last, F., 1983. Sequences and interactions of mycorrhizal fungi on birch. Plant Soil 71, 257–262.

Dickie, I.A., Xu, B., Koide, R.T., 2002. Vertical niche differentiation of ectomycorrhizal hyphae in soil as shown by T-RFLP analysis. New Phytol. 156, 527–535.

Donnelly, D.P., Boddy, L., Leake, J.R., 2004. Development, persistence and regeneration of foraging ectomycorrhizal mycelial systems in soil microcosms. Mycorrhiza 14, 37–45.

Edgcomb, V.P., Beaudoin, D., Gast, R., Biddle, J.F., Teske, A., 2011. Marine subsurface eukaryotes: the fungal majority. Environ. Microbiol. 13, 172–183.

Edman, J.C., Kovacs, J.A., Masur, H., Santi, D.V., Elwood, H.J., Sogin, M.L., 1988. Ribosomal RNA sequence shows *Pneumocystis carinii* to be a member of the fungi. Nature 334, 519–522.

Feuerer, T., Hawksworth, D.L., 2007. Biodiversity of lichens, including a world-wide analysis of checklist data based on Takhtajan's floristic regions. Biodivers. Conserv. 16, 85–98.

Fierer, N., Strickland, M.S., Liptzin, D., Bradford, M.A., Cleveland, C.C., 2009. Global patterns in belowground communities. Ecol. Lett. 12, 1238–1249.

Floudas, D., Binder, M., Riley, R., Barry, K., Blanchette, R.A., Henrissat, B., Martínez, A.T., Otillar, R., Spatafora, J.W., Yadav, J.S., et al., 2012. The Paleozoic origin of enzymatic lignin decomposition reconstructed from 31 fungal genomes. Science 336, 1715–1719.

Freeman, K., Martin, A., Karki, D., Lynch, R., Mitter, M., Meyer, A., Longcore, J., Simmons, D., Schmidt, S., 2009. Evidence that chytrids dominate fungal communities in high-elevation soils. Proc. Natl. Acad. Sci. U. S. A. 106, 18315–18320.

Frey-Klett, P., Garbaye, J., Tarkka, M., 2007. The mycorrhiza helper bacteria revisited. New Phytol. 176, 22–36.

Fukami, T., Dickie, I.A., Paula, Wilkie J., Paulus, B.C., Park, D., Roberts, A., Buchanan, P.K., Allen, R.B., 2010. Assembly history dictates ecosystem functioning: evidence from wood decomposer communities. Ecol. Lett. 13, 675–684.

Galante, T.E., Horton, T.R., Swaney, D.P., 2011. 95% of basidiospores fall within 1 m of the cap: a field-and modeling-based study. Mycologia 103, 1175–1183.

Gallo, M., Amonette, R., Lauber, C., Sinsabaugh, R., Zak, D., 2004. Microbial community structure and oxidative enzyme activity in nitrogen-amended north temperate forest soils. Microb. Ecol. 48, 218–229.

Garbaye, J., 1994. Tansley review no. 76 helper bacteria: a new dimension to the mycorrhizal symbiosis. New Phytol. 128, 197–210.

Gehrig, H., Schüßler, A., Kluge, M., 1996. *Geosiphon pyriforme*, a fungus forming endocytobiosis with *Nostoc* (Cyanobacteria), is an ancestral member of the glomales: evidence by SSU rRNA analysis. J. Mol. Evol. 43, 71–81.

Geisseler, D., Horwath, W.R., Joergensen, R.G., Ludwig, B., 2010. Pathways of nitrogen utilization by soil microorganisms–a review. Soil Biol. Biochem. 42, 2058–2067.

Geml, J., Tulloss, R.E., Laursen, G.A., Sazanova, N.A., Taylor, D.L., 2008. Evidence for strong inter-and intracontinental phylogeographic structure in *Amanita muscaria*, a wind-dispersed ectomycorrhizal basidiomycete. Mol. Phylogenet. Evol. 48, 694–701.

Glass, D.J., Takebayashi, N., Olson, L.E., Taylor, D.L., 2013. Evaluation of the authenticity of a highly novel environmental sequence from boreal forest soil using ribosomal RNA secondary structure modeling. Mol. Phylogenet. Evol. 67, 234–245.

Gleason, F.H., Schmidt, S.K., Marano, A.V., 2010. Can zoosporic true fungi grow or survive in extreme or stressful environments? Extremophiles 14, 417–425.

Grubisha, L.C., Levsen, N., Olson, M.S., Taylor, D.L., 2012. Intercontinental divergence in the *Populus*-associated ectomycorrhizal fungus, *Tricholoma populinum*. New Phytol. 194, 548–560.

Gulis, V., Suberkropp, K., Rosemond, A.D., 2008. Comparison of fungal activities on wood and leaf litter in unaltered and nutrient-enriched headwater streams. Appl. Environ. Microbiol. 74, 1094–1101.

Gunde-Cimerman, N., Sonjak, S., Zalar, P., Frisvad, J.C., Diderichsen, B., Plemenitas, A., 2003. Extremophilic fungi in arctic ice: a relationship between adaptation to low temperature and water activity. Phys. Chem. Earth Parts A/B/C 28, 1273–1278.

Hanski, I., 1989. Fungivory: fungi, insects and ecology. In: Wilding, N., Collins, N.M., Hammond, P.M., Webber, J.F. (Eds.), Insect-Fungus Interactions. Academic Press, London, pp. 25–68.

Harms, H., Schlosser, D., Wick, L.Y., 2011. Untapped potential: exploiting fungi in bioremediation of hazardous chemicals. Nat. Rev. Microbiol. 9, 177–192.

Hawksworth, D.L., 1991. The fungal dimension of biodiversity: magnitude, significance, and conservation. Mycol. Res. 95, 641–655.

Hawksworth, D.L., 2002. The magnitude of fungal diversity: the 1·5 million species estimate revisited. Mycol. Res. 105, 1422–1432.

Hawksworth, D., 2012. Global species numbers of fungi: are tropical studies and molecular approaches contributing to a more robust estimate? Biodivers. Conserv. 21, 2425–2433.

Heckman, D.S., Geiser, D.M., Eidell, B.R., Stauffer, R.L., Kardos, N.L., Hedges, S.B., 2001. Molecular evidence for the early colonization of land by fungi and plants. Science 293, 1129.

Helgason, T., Daniell, T., Husband, R., Fitter, A., Young, J., 1998. Ploughing up the wood-wide web? Nature 394, 431.

Hibbett, D.S., Binder, M., Bischoff, J.F., Blackwell, M., Cannon, P.F., Eriksson, O.E., Huhndorf, S., James, T., Kirk, P.M., Lucking, R., et al., 2007. A higher-level phylogenetic classification of the fungi. Mycol. Res. 111, 509–547.

Hobbie, E.A., Agerer, R., 2010. Nitrogen isotopes in ectomycorrhizal sporocarps correspond to belowground exploration types. Plant Soil 327, 71–83.

Hoffman, M.T., Arnold, A.E., 2010. Diverse bacteria inhabit living hyphae of phylogenetically diverse fungal endophytes. Appl. Environ. Microbiol. 76, 4063–4075.

Hoffmann, L., 1989. Algae of terrestrial habitats. Bot. Rev. 55, 77–105.

Hofmockel, K.S., Fierer, N., Colman, B.P., Jackson, R.B., 2010. Amino acid abundance and proteolytic potential in North American soils. Oecologia 163, 1069–1078.

Högberg, M.N., Högberg, P., 2002. Extramatrical ectomycorrhizal mycelium contributes one-third of microbial biomass and produces, together with associated roots, half the dissolved organic carbon in a forest soil. New Phytol. 154, 791–795.

Högberg, M.N., Högberg, P., Myrold, D.D., 2007. Is microbial community composition in boreal forest soils determined by pH, C-to-N ratio, the trees, or all three? Oecologia 150, 590–601.

James, T.Y., Porter, D., Hamrick, J.L., Vilgalys, R., 1999. Evidence for limited intercontinental gene flow in the cosmopolitan mushroom, *Schizophyllum commune*. Evolution 53, 1665–1677.

James, T.Y., Kauff, F., Schoch, C.L., Matheny, P.B., Hofstetter, V., Cox, C.J., Celio, G., Gueidan, C., Fraker, E., Miadlikowska, J., et al., 2006. Reconstructing the early evolution of fungi using a six-gene phylogeny. Nature 443, 818–822.

Joergensen, R.G., Wichern, F., 2008. Quantitative assessment of the fungal contribution to microbial tissue in soil. Soil Biol. Biochem. 40, 2977–2991.

Jørgensen, H.B., Johansson, T., Canbäck, B., Hedlund, K., Tunlid, A., 2005. Selective foraging of fungi by collembolans in soil. Biol. Lett. 1, 243–246.

Jones, M.D., Forn, I., Gadelha, C., Egan, M.J., Bass, D., Massana, R., Richards, T.A., 2011. Discovery of novel intermediate forms redefines the fungal tree of life. Nature 474, 200–203.

Kennedy, P.G., Peay, K.G., Bruns, T.D., 2009. Root tip competition among ectomycorrhizal fungi: are priority effects a rule or an exception? Ecology 90, 2098–2107.

Kirk, P., Cannon, P., Minter, D., Stalpers, J., 2008. Ainsworth and Bisby's Dictionary of the Fungi, 10th ed. CABI, Wallingford.

Klironomos, J.N., 2002. Feedback with soil biota contributes to plant rarity and invasiveness in communities. Nature 417, 67–70.

Klironomos, J.N., Hart, M.M., 2001. Food-web dynamics: animal nitrogen swap for plant carbon. Nature 410, 651–652.

Klironomos, J.N., Widden, P., Deslandes, I., 1992. Feeding preferences of the collembolan *Folsomia candida* in relation to microfungal successions on decaying litter. Soil Biol. Biochem. 24, 685–692.

Knorr, M., Frey, S., Curtis, P., 2005. Nitrogen additions and litter decomposition: a meta-analysis. Ecology 86, 3252–3257.

Koide, R.T., Xu, B., Sharda, J., Lekberg, Y., Ostiguy, N., 2004. Evidence of species interactions within an ectomycorrhizal fungal community. New Phytol. 165, 305–316.

Kowalchuk, G.A., 2012. Bad news for soil carbon sequestration? Science 337, 1049–1050.

Lauber, C.L., Hamady, M., Knight, R., Fierer, N., 2009. Pyrosequencing-based assessment of soil pH as a predictor of soil bacterial community structure at the continental scale. Appl. Environ. Microbiol. 75, 5111.

Lilleskov, E., Fahey, T., Horton, T., Lovett, G., 2002. Belowground ectomycorrhizal fungal community change over a nitrogen deposition gradient in Alaska. Ecology 83, 104–115.

Lindahl, B.D., Ihrmark, K., Boberg, J., Trumbore, S.E., Högberg, P., Stenlid, J., Finlay, R.D., 2006. Spatial separation of litter decomposition and mycorrhizal nitrogen uptake in a boreal forest. New Phytol. 173, 611–620.

Lipson, D., Schmidt, S., 2004. Seasonal changes in an alpine soil bacterial community in the Colorado Rocky Mountains. Appl. Environ. Microbiol. 70, 2867–2879.

Liu, Y., Steenkamp, E.T., Brinkmann, H., Forget, L., Philippe, H., Lang, B.F., 2009. Phylogenomic analyses predict sistergroup relationship of nucleariids and fungi and paraphyly of zygomycetes with significant support. BMC Evol. Biol. 9, 272.

Lorch, J.M., Meteyer, C.U., Behr, M.J., Boyles, J.G., Cryan, P.M., Hicks, A.C., Ballmann, A.E., Coleman, J.T.H., Redell, D.N., Reeder, D.A.M., et al., 2011. Experimental infection of bats with *Geomyces destructans* causes white-nose syndrome. Nature 480, 376–378.

Maheshwari, R., Bharadwaj, G., Bhat, M.K., 2000. Thermophilic fungi: their physiology and enzymes. Microbiol. Mol. Biol. Rev. 64, 461–488.

Martin, F., Aerts, A., Ahren, D., Brun, A., Danchin, E., Duchaussoy, F., Gibon, J., Kohler, A., Lindquist, E., Pereda, V., et al., 2008. The genome of Laccaria bicolor provides insights into mycorrhizal symbiosis. Nature 452, 88–92.

May, R.M., 1991. A fondness for fungi. Nature 352, 475–476.

Menezes-Blackburn, D., Jorquera, M.A., Greiner, R., Gianfreda, L., de la Luz Mora, M., 2013. Phytases and phytase-labile organic phosphorus in manures and soils. Crit. Rev. Environ. Sci. Technol. 43, 916–954.

Miller, S.L., Koo, C.D., Molina, R., 1992. Early colonization of red alder and Douglas fir by ectomycorrhizal fungi and *Frankia* in soils from the Oregon coast range. Mycorrhiza 2, 53–61.

Molina, R., Massicotte, H., Trappe, J.M., 1992. Specificity phenomena in mycorrhizal symbioses: community-ecological consequences and practical implications. In: Allen, M.F. (Ed.), Mycorrhizal Functioning: an Integrative Plant-Fungal Process. Chapman & Hall, New York, pp. 357–423.

Moorhead, D.L., Lashermes, G., Sinsabaugh, R.L., Weintraub, M.N., 2013. Calculating co-metabolic costs of lignin decay and their impacts on carbon use efficiency. Soil Biol. Biochem. 66, 17–19.

Mora, C., Tittensor, D.P., Adl, S., Simpson, A.G.B., Worm, B., 2011. How many species are there on Earth and in the ocean? PLoS Biol. 9, e1001127.

Nagano, Y., Nagahama, T., Hatada, Y., Nunoura, T., Takami, H., Miyazaki, J., Takai, K., Horikoshi, K., 2010. Fungal diversity in deep-sea sediments-the presence of novel fungal groups. Fungal Ecol. 3, 316–325.

O'Brien, H., Parrent, J., Jackson, J., Moncalvo, J., Vilgalys, R., 2005. Fungal community analysis by large-scale sequencing of environmental samples. Appl. Environ. Microbiol. 71, 5544–5550.

O'Donnell, K., Lutzoni, F.M., Ward, T.J., Benny, G.L., 2001. Evolutionary relationships among mucoralean fungi (Zygomycota): evidence for family polyphyly on a large scale. Mycologia 93, 286–297.

Olsson, P.A., Münzenberger, B., Mahmood, S., Erland, S., 2000. Molecular and anatomical evidence for a three-way association between *Pinus sylvestris* and the ectomycorrhizal fungi *Suillus bovinus* and *Gomphidius roseus*. Mycol. Res. 104, 1372–1378.

Onofri, S., Fenice, M., Cicalini, A.R., Tosi, S., Magrino, A., Pagano, S., Selbmann, L., Zucconi, L., Vishniac, H.S., Ocampo-Friedmann, R., et al., 2000. Ecology and biology of microfungi from Antarctic rocks and soils. Ital. J. Zool. 67, 163–167.

Panikov, N.S., Sizova, M.V., 2007. Growth kinetics of microorganisms isolated from Alaskan soil and permafrost in solid media frozen down to -35°C. FEMS Microbiol. Ecol. 59, 500–512.

Pearce, D.A., Bridge, P.D., Hughes, K.A., Sattler, B., Psenner, R., Russell, N.J., 2009. Microorganisms in the atmosphere over Antarctica. FEMS Microbiol. Ecol. 69, 143–157.

Peay, K.G., Bruns, T.D., Kennedy, P.G., Bergemann, S.E., Garbelotto, M., 2007. A strong species-area relationship for eukaryotic soil microbes: island size matters for ectomycorrhizal fungi. Ecol. Lett. 10, 470–480.

Peay, K.G., Garbelotto, M., Bruns, T.D., 2010. Evidence of dispersal limitation in soil microorganisms: isolation reduces species richness on mycorrhizal tree islands. Ecology 91, 3631–3640.

Peay, K.G., Schubert, M.G., Nguyen, N.H., Bruns, T.D., 2012. Measuring ectomycorrhizal fungal dispersal: macroecological patterns driven by microscopic propagules. Mol. Ecol. 21, 4122–4136.

Plassard, C., Louche, J., Ali, M.A., Duchemin, M., Legname, E., Cloutier-Hurteau, B., 2011. Diversity in phosphorus mobilisation and uptake in ectomycorrhizal fungi. Ann. For. Sci. 68, 33–43.

Plattner, I., Hall, I., 1995. Parasitism of non-host plants by the mycorrhizal fungus *Tuber melanosporum*. Mycol. Res. 99, 1367–1370.

Pointing, S.B., Belnap, J., 2012. Microbial colonization and controls in dryland systems. Nat. Rev. Microbiol. 10, 551–562.

Porras-Alfaro, A., Bayman, P., 2011. Hidden fungi, emergent properties: endophytes and microbiomes. Phytopathology 49, 291.

Porras-Alfaro, A., Herrera, J., Natvig, D.O., Lipinski, K., Sinsabaugh, R.L., 2011. Diversity and distribution of soil fungal communities in a semiarid grassland. Mycologia 103, 10–21.

Porter, T., Schadt, C., Rizvi, L., Martin, A., Schmidt, S., Scott-Denton, L., Vilgalys, R., Moncalvo, J., 2008. Widespread occurrence and phylogenetic placement of a soil clone group adds a prominent new branch to the fungal tree of life. Mol. Phylogenet. Evol. 46, 635–644.

Pritsch, K., Garbaye, J., 2011. Enzyme secretion by ECM fungi and exploitation of mineral nutrients from soil organic matter. Ann. For. Sci. 68, 25–32.

Rabinovich, M., Bolobova, A., Vasil'chenko, L., 2004. Fungal decomposition of natural aromatic structures and xenobiotics: a review. Appl. Biochem. Microbiol. 40, 1–17.

Read, D., Perez-Moreno, J., 2003. Mycorrhizas and nutrient cycling in ecosystems-a journey towards relevance? New Phytol. 157, 475–492.

Reynolds, H.L., Packer, A., Bever, J.D., Clay, K., 2003. Grassroots ecology: plant-microbe-soil interactions as drivers of plant community structure and dynamics. Ecology 84, 2281–2291.

Richards, T.A., Jones, M.D., Leonard, G., Bass, D., 2012. Marine fungi: their ecology and molecular diversity. Ann. Rev. Mar. Sci. 4, 495–522.

Rosling, A., Cox, F., Cruz-Martinez, K., Ihrmark, K., Grelet, G.A., Lindahl, B.D., Menkis, A., James, T.Y., 2011. Archaeorhizomycetes: unearthing an ancient class of ubiquitous soil fungi. Science 333, 876–879.

Rousk, J., Baath, E., 2011. Growth of saprotrophic fungi and bacteria in soil. FEMS Microbiol. Ecol. 78, 17–30.

Rousk, J., Brookes, P.C., Bååth, E., 2009. Contrasting soil pH effects on fungal and bacterial growth suggest functional redundancy in carbon mineralization. Appl. Environ. Microbiol. 75, 1589–1596.

Roy, M., Rochet, J., Manzi, S., Jargeat, P., Gryta, H., Moreau, P.-A., Gardes, M., 2013. What determines *Alnus*-associated ectomycorrhizal community diversity and specificity? A comparison of host and habitat effects at a regional scale. New Phytol. 198, 1228–1238.

Schadt, C.W., Martin, A.P., Lipson, D.A., Schmidt, S.K., 2003. Seasonal dynamics of previously unknown fungal lineages in tundra soils. Science 301, 1359–1361.

Schmit, J.P., Mueller, G.M., 2007. An estimate of the lower limit of global fungal diversity. Biodivers. Conserv. 16, 99–111.

Schmit, J.P., Mueller, G.M., Leacock, P.R., Mata, J.L., Wu, Q., Huang, Y., 2005. Assessment of tree species richness as a surrogate for macrofungal species richness. Biol. Conserv. 121, 99–110.

Schoch, C.L., Sung, G.-H., López-Giráldez, F., Townsend, J.P., Miadlikowska, J., Hofstetter, V., Robbertse, B., Matheny, P.B., Kauff, F., Wang, Z., et al., 2009. The Ascomycota tree of life: a phylum-wide phylogeny clarifies the origin and evolution of fundamental reproductive and ecological traits. Syst. Biol. 58, 224–239.

Selbmann, L., De Hoog, G.S., Mazzaglia, A., Friedmann, E.I., Onofri, S., 2005. Fungi at the edge of life: cryptoendolithic black fungi from Antarctic desert. Stud. Mycol. 51, 1–32.

Shoun, H., Fushinobu, S., Jiang, L., Kim, S.-W., Wakagi, T., 2012. Fungal denitrification and nitric oxide reductase cytochrome P450nor. Philos. Trans. R. Soc. B Biol. Sci. 367, 1186–1194.

Simon, L., Bousquet, J., Levesque, R.C., Lalonde, M., 1993. Origin and diversification of endomy-corrhizal fungi and coincidence with vascular land plants. Nature 363, 67–69.

Sinsabaugh, R.L., 2010. Phenol oxidase, peroxidase and organic matter dynamics of soil. Soil Biol. Biochem. 42, 391–404.

Sinsabaugh, R.L., Follstad Shah, J.J., 2011. Ecoenzymatic stoichiometry of recalcitrant organic matter decomposition: the growth rate hypothesis in reverse. Biogeochemistry 102, 31–43.

Sinsabaugh, R., Carreiro, M., Repert, D., 2002. Allocation of extracellular enzymatic activity in relation to litter composition, N deposition, and mass loss. Biogeochemistry 60, 1–24.

Sinsabaugh, R.L., Manzoni, S., Moorhead, D.L., Richter, A., 2013. Carbon use efficiency of microbial communities: stoichiometry, methodology and modelling. Ecol. Lett. 16, 930–939.

Smith, S.E., Read, D., 2008. Mycorrhizal Symbiosis, third ed. Academic Press, Amsterdam.

Smith, M.L., Bruhn, J.N., Anderson, J.B., 1992. The fungus *Armillaria bulbosa* is among the largest and oldest living organisms. Nature 356, 428–431.

Smith, J.E., McKay, D., Niwa, C.G., Thies, W.G., Brenner, G., Spatafora, J.W., 2004. Short-term effects of seasonal prescribed burning on the ectomycorrhizal fungal community and fine root biomass in ponderosa pine stands in the Blue Mountains of Oregon. Can. J. For. Res. 34, 2477–2491.

Snajdr, J., Cajthaml, T., Valásková, V., Merhautová, V., Petránková, M., Spetz, P., Leppänen, K., Baldrian, P., 2011. Transformation of *Quercus petraea* litter: successive changes in litter chemistry are reflected in differential enzyme activity and changes in the microbial community composition. FEMS Microbiol. Ecol. 75, 291–303.

Spott, O., Russow, R., Stange, C.F., 2011. Formation of hybrid N_2O and hybrid N_2 due to codenitrification: first review of a barely considered process of microbially mediated N-nitrosation. Soil Biol. Biochem. 43, 1995–2011.

Stajich, J.E., Berbee, M.L., Blackwell, M., Hibbett, D.S., James, T.Y., Spatafora, J.W., Taylor, J.W., 2009. Primer–the fungi. Current Biol. 19, R840.

Starkey, R.L., Waksman, S.A., 1943. Fungi tolerant to extreme acidity and high concentrations of copper sulfate. J. Bacteriol. 45, 509–519.

Steenkamp, E.T., Wright, J., Baldauf, S.L., 2006. The protistan origins of animals and fungi. Mol. Biol. Evol. 23, 93–106.

Stendell, E., Horton, T., Bruns, T., 1999. Early effects of prescribed fire on the structure of the ectomycorrhizal fungus community in a Sierra Nevada ponderosa pine forest. Mycol. Res. 103, 1353–1359.

Sterflinger, K., Tesei, D., Zakharova, K., 2012. Fungi in hot and cold deserts with particular reference to microcolonial fungi. Fungal Ecol. 5, 453–462.

Stone, L., Roberts, A., 1990. The checkerboard score and species distributions. Oecologia 85, 74–79.

Strickland, M.S., Rousk, J., 2010. Considering fungal: bacterial dominance in soils-methods, controls, and ecosystem implications. Soil Biol. Biochem. 42, 1385–1395.

Strong, P., Claus, H., 2011. Laccase: a review of its past and its future in bioremediation. Crit. Rev. Environ. Sci. Technol. 41, 373–434.

Sturm, M., Schimel, J., Michaelson, G., Welker, J.M., Oberbauer, S.F., Liston, G.E., Fahnestock, J., Romanovsky, V.E., 2005. Winter biological processes could help convert Arctic tundra to shrubland. Bioscience 55, 17–26.

Talbot, J., Allison, S., Treseder, K., 2008. Decomposers in disguise: mycorrhizal fungi as regulators of soil C dynamics in ecosystems under global change. Funct. Ecol. 22, 955–963.

Taylor, D.L., Bruns, T.D., 1999. Community structure of ectomycorrhizal fungi in a *Pinus muricata* forest: minimal overlap between the mature forest and resistant propagule communities. Mol. Ecol. 8, 1837–1850.

Taylor, A.F.S., Martin, F., Read, D.J., 2000. Fungal diversity in ectomycorrhizal communities of Norway spruce [*Picea abies* (L.) Karst.] and beech (*Fagus sylvatica* L.) along north-south transects in Europe. In: Schulze, E. (Ed.), Carbon and Nitrogen Cycling in European Forest Ecosystems. Springer Verlag, Heidelberg, pp. 343–365.

Taylor, J.W., Turner, E., Townsend, J.P., Dettman, J.R., Jacobson, D., 2006. Eukaryotic microbes, species recognition and the geographic limits of species: examples from the kingdom Fungi. Philos. Trans. R. Soc. B Biol. Sci. 361, 1947–1963.

Taylor, L., Leake, J., Quirk, J., Hardy, K., Banwart, S., Beerling, D., 2009. Biological weathering and the long-term carbon cycle: integrating mycorrhizal evolution and function into the current paradigm. Geobiology 7, 171–191.

Taylor, D.L., Herriott, I.C., Stone, K.E., McFarland, J.W., Booth, M.G., Leigh, M.B., 2010. Structure and resilience of fungal communities in Alaskan boreal forest soils. Can. J. For. Res. 40, 1288–1301.

Taylor, D.L., Hollingsworth, T.N., McFarland, J., Lennon, N.J., Nusbaum, C., Ruess, R.W., 2013. A first comprehensive census of fungi in soil reveals both hyperdiversity and fine-scale niche partitioning. Ecol. Monogr. 84, 3–20.

Tedersoo, L., May, T.W., Smith, M.E., 2009a. Ectomycorrhizal lifestyle in fungi: global diversity, distribution, and evolution of phylogenetic lineages. Mycorrhiza 20, 217–263.

Tedersoo, L., Suvi, T., Jairus, T., Ostonen, I., Polme, S., 2009b. Revisiting ectomycorrhizal fungi of the genus *Alnus*: differential host specificity, diversity and determinants of the fungal community. New Phytol. 182, 727–735.

Tedersoo, L., Bahram, M., Toots, M., Diédhiou, A., Henkel, T., Kjøller, R., Morris, M., Nara, K., Nouhra, E., Peay, K., et al., 2012. Towards global patterns in the diversity and community structure of ectomycorrhizal fungi. Mol. Ecol. 21, 4160–4170.

Thorn, R., Barron, G., 1984. Carnivorous mushrooms. Science 224, 76–78.

Toljander, J., Eberhardt, U., Toljander, Y., Paul, L., Taylor, A., 2006. Species composition of an ectomycorrhizal fungal community along a local nutrient gradient in a boreal forest. New Phytol. 170, 873–883.

Treseder, K.K., 2008. Nitrogen additions and microbial biomass: a meta-analysis of ecosystem studies. Ecol. Lett. 11, 1111–1120.

Treseder, K., Mack, M., Cross, A., 2004. Relationships among fires, fungi, and soil dynamics in Alaskan Boreal Forests. Ecol. Appl. 14, 1826–1838.

Trum, F., Titeux, H., Cornelis, J.-T., Delvaux, B., 2011. Effects of manganese addition on carbon release from forest floor horizons. Can. J. For. Res. 41, 643–648.

Van der Heijden, M.G., Klironomos, J.N., Ursic, M., Moutoglis, P., Streitwolf-Engel, R., Boller, T., Wiemken, A., Sanders, I.R., 1998a. Mycorrhizal fungal diversity determines plant biodiversity, ecosystem variability and productivity. Nature 396, 69–72.

Van der Heijden, M.G.A., Boiler, T., Wiemken, A., Sanders, I.R., 1998b. Different arbuscular mycorrhizal fungal species are potential determinants of plant community structure. Ecology 79, 2082–2091.

Van Der Heijden, M.G., Bardgett, R.D., Van Straalen, N.M., 2008. The unseen majority: soil microbes as drivers of plant diversity and productivity in terrestrial ecosystems. Ecol. Lett. 11, 296–310.

Veresoglou, S.D., Chen, B., Rillig, M.C., 2012. Arbuscular mycorrhiza and soil nitrogen cycling. Soil Biol. Biochem. 46, 53–62.

Visser, S., 1995. Ectomycorrhizal fungal succession in jack pine stands following wildfire. New Phytol. 129, 389–401.

Vitt, D.H., 2007. Estimating moss and lichen ground layer net primary production in tundra, peatlands, and forests. In: Fahey, T.J., Knapp, A.K. (Eds.), Principles and Standards for Measuring Primary Production. Oxford University Press, New York, pp. 82–105.

Waldrop, M., White, R., Douglas, T., 2008. Isolation and identification of cold-adapted fungi in the Fox Permafrost Tunnel, Alaska. Proc. Ninth Int. Conf. Permafrost 2, 1887–1891.

Wallander, H., Ekblad, A., Godbold, D., Johnson, D., Bahr, A., Baldrian, P., Bjork, R., Kieliszewska-Rokicka, B., Kjoller, R., Kraigher, H., et al., 2013. Evaluation of methods to estimate production, biomass and turnover of ectomycorrhizal mycelium in forests soils: a review. Soil Biol. Biochem. 57, 1034–1047.

Willis, A., Rodrigues, B., Harris, P., 2013. The ecology of arbuscular mycorrhizal fungi. Crit. Rev. Plant Sci. 32, 1–20.

Wilson, G.W., Rice, C.W., Rillig, M.C., Springer, A., Hartnett, D.C., 2009. Soil aggregation and carbon sequestration are tightly correlated with the abundance of arbuscular mycorrhizal fungi: results from long-term field experiments. Ecol. Lett. 12, 452–461.

Zalar, P., Gostincar, C., De Hoog, G., Ursic, V., Sudhadham, M., Gunde-Cimerman, N., 2008. Redefinition of *Aureobasidium pullulans* and its varieties. Stud. Mycol. 61, 21–38.

Chapter 5

Soil Fauna: Occurrence, Biodiversity, and Roles in Ecosystem Function

David C. Coleman[1] and Diana H. Wall[2]

[1]*Odum School of Ecology, University of Georgia, Athens, GA, USA*
[2]*Department of Biology and Natural Resource Ecology Laboratory, Colorado State University, Fort Collins, CO, USA*

Chapter Contents

Soil Microbiology, Ecology, and Biochemistry. http://dx.doi.org/10.1016/B978-0-12-415955-6.00005-0

I INTRODUCTION

Animals, a key group of major heterotrophs in soil systems, facilitate bacterial and fungal structure and activity and diversity in soils (Hättenschwiler et al., 2005). Many invertebrates regulate nutrient cycling by feeding directly on plant materials and organic substrates. The fragmentation or comminution of these materials enhances their decomposition. Comminution increases the surface areas of plant structural materials and exposes cytoplasm, thereby enabling greater access by microbes. Decomposition is further accelerated, as the feeding activity often results in the translocation of nitrogen (N) from the soil to the substrate in the form of fecal material and through fungal hyphae. Grazing by invertebrates disseminates microbes from one organic source to another as many microbes adhere to invertebrate exoskeletons and cuticles and survive passage through their digestive tracts (Meier and Honegger, 2002; Coleman et al., 2012). Soil fauna have been found to have a consistent positive effect on litter decomposition at global and biome scales (Garcia-Palacios et al., 2013).

Soil animals exist in food webs containing several trophic levels (Moore and de Ruiter, 2012). Some are herbivores because they feed directly on roots of living plants, but most subsist on dead plant matter (saprophytes), living microbes associated with it, or a combination of the two. Still others are carnivores, parasites, or top predators. Analyses of food webs in the soil have emphasized numbers of the various organisms, traits (Orwin et al., 2011; Wurst et al., 2012), and trophic resources. The structure of these food webs is complex, with many "missing links" poorly described or unknown (Walter et al., 1991; Scheu and Setälä, 2002; Scharroba et al., 2012), although tools such as stable isotopes are revealing C and N transfer through trophic and species levels within the soil foodweb (Crotty et al., 2012a).

II OVERVIEW OF FAUNAL BIODIVERSITY IN SOILS

The nature and extent of biodiversity in soils is impressively large (Whitman et al., 1998; Coleman and Whitman, 2005; Jeffery et al., 2010; Fierer and Lennon, 2011). Discoveries of new species and phyla in the archaea and bacteria are increasing rapidly (Pace, 2009). Discoveries among the eukarya of ever more diverse arrays of fungi, unicellular eukarya (Coleman, 2008; Buée et al., 2009; Bates et al., 2013), and animals (Wu et al., 2011) are proceeding apace as well. Animal members of the soil biota are numerous and diverse and include representatives of all terrestrial phyla. Many groups of species are not described taxonomically, and details of their natural history and biology are unknown. Only about 10% of microarthropod populations have been explored, and perhaps 10% of species have been described (André et al., 2002). Estimates of species richness based on morphological characteristics have been historically difficult for soil fauna, including ants, but molecular tools are advancing our knowledge of biodiversity in soils. Global scale studies are showing few cosmopolitan groups of soil fauna even at the family level (Wu et al., 2011) perhaps due to their high degree of functional specialization (Heemsbergen et al., 2004; Wurst and van der Putten, 2007).

When research focuses at the level of the soil ecosystem, two things are required: the cooperation of multiple disciplines (soil scientists, biogeochemists, zoologists, and microbiologists) and the lumping of animals into functional groups. These groups are often taxonomic, but species with similar biologies and morphologies are grouped together for purposes of integration at the system level (Hendrix et al., 1986; Hunt et al., 1987; Coleman et al., 1993, 2004).

The soil fauna also may be characterized by the degree of presence in the soil or microhabitat utilization by different life forms. There are transient species, exemplified by the ladybird beetle, which hibernates in the soil but otherwise lives in the plant stratum. Gnats (Diptera) are temporary residents of the soil because the adult stages live aboveground. Their eggs are laid in the soil, and their larvae feed on decomposing organic debris. In some soil situations, dipteran larvae are important scavengers. Cutworms are temporary soil residents, whose larvae feed on seedlings by night. Some nematodes that parasitize insects and beetles spend all or part of their life cycle in soil. Periodic residents spend their life histories belowground, with adults, such as the velvet mites, emerging to reproduce. The soil food webs are linked to aboveground systems, making trophic analyses much more complicated than in either subsystem alone (Wardle et al., 2004; van der Putten et al., 2013). Even permanent residents of the soil may be adapted to life at various depths in the soil.

Among the microarthropods, collembolans are examples of permanent soil residents. The morphology of collembolans reveals their adaptations for life in different soil strata. Species that dwell on the soil surface or in the litter layer may be large, pigmented, and equipped with long antennae and a well-developed jumping apparatus (furcula). Collembolans living within mineral soil tend to be smaller, with unpigmented, elongated bodies, and they possess a much reduced furcula.

A generalized classification by length illustrates a commonly used device for separating the soil fauna into size classes: microfauna, mesofauna, macrofauna, and megafauna. This classification encompasses the range from smallest to largest (i.e., from ca. 1-2 μm for the microflagellates to >2 m for giant Australian earthworms). Body width of the fauna is related to their microhabitats (Fig. 5.1). The microfauna (protozoa, rotifers, tardigrades, nematodes) inhabit water films. The mesofauna inhabit existing air-filled pore spaces and are largely restricted to existing spaces. The macrofauna have the ability to create their own spaces through their burrowing activities, and like the megafauna, they can have large influences on gross soil structure (Lavelle and Spain, 2001; van Vliet and Hendrix, 2003). Methods for studying these faunal groups are mostly size dependent. The macrofauna may be sampled as field collections, often by hand sorting, and populations of individuals are usually measured.

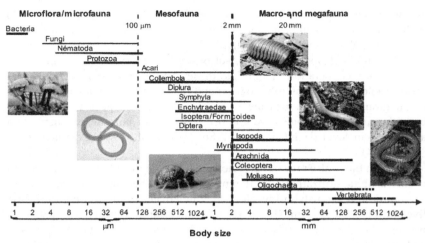

FIG. 5.1 Size (body width) classification of organisms by Swift et al. (1979), illustrated by Decaëns (2010). *With permission from John Wiley & Sons; in Lavelle (2012).*

There is considerable gradation in the classification based on body width. The smaller mesofauna exhibit characteristics of the microfauna, and so forth. The vast range of body sizes among the soil fauna emphasizes their effects on soil processes at a range of spatial scales. Three levels of participation have been suggested (Lavelle et al., 1995; Wardle, 2002): (1) "Ecosystem engineers," such as earthworms, termites, or ants, alter the physical structure of the soil itself, influencing rates of nutrient and energy flow (Jones et al., 1994); (2) "Litter transformers," the microarthropods, fragment decomposing litter and improve its availability to microbes; and (3) "Micro-food webs" include the microbial groups and their direct microfaunal predators (nematodes and protozoans). These three levels operate on different size, spatial, and time scales (Fig. 5.2; Wardle, 2002).

III MICROFAUNA

A Protozoa

The free-living protozoa of litter and soils belong to two phyla, the Sarcomastigophora and the Ciliophora. They are considered in four ecological groups: the flagellates, naked amoebae, testacea, and ciliates (Lousier and Bamforth, 1990).

A recent revision of classification of the Eukarya (Adl et al., 2005) has six major categories: (1) Amoebozoa, including all of the amoeba-like creatures; (2) Opisthokonta, the fungi through metazoa; (3) Rhizaria, including Cercozoa, Foraminifera, and Radiolaria; (4) Archaeplastida, including Glaucophyta, Rhodophyta, and Chloroplastida; (5) Chromalveolata, including Cryptophyceae and Phaeophyceae; and (6) Excavata, including many of the flagellated protozoa.

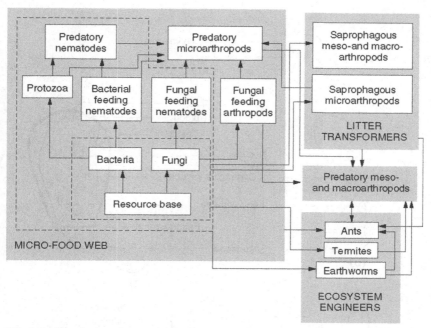

FIG. 5.2 Organization of the soil food web into three categories—ecosystem engineers, litter transformers, and micro-food webs. *(After Wardle, 2002, and Lavelle et al., 1995.)*

The number of supergroups of Eukaryotic phyla range from four to seven in other estimates (Pawlowski, 2013). By Pawlowski's recent estimate, the classical multicellular kingdoms of animals and fungi are placed in the supergroup Opisthokonta, whereas green plants together with red algae form a supergroup of Archaeplastida. All other supergroups are composed of typically unicellular eukaryotes.

A general comparison of body plans of the four ecological groups is given in Fig. 5.3, showing representatives of the four major types.

1. Flagellates (named for their one or more flagella, or whip-like propulsive organs) are among the more numerous and active of the protozoa. They play a significant role in nutrient turnover, with bacteria as their principal prey items (Kuikman and Van Veen, 1989; Zwart and Darbyshire, 1991). Numbers vary from 10^2 g^{-1} in desert soils to more than 10^5 g^{-1} in forest soils (Bamforth, 1980).

2. Naked amoebae are among the more voracious of the soil protozoa and are very numerous and active in a wide range of agricultural, grassland, and forested soils (Clarholm, 1981, 1985; Gupta and Germida, 1989). The dominant mode of feeding for the amoebae, as for the larger forms such as Ciliates, is phagotrophic (engulfing), with bacteria, fungi, algae, and other

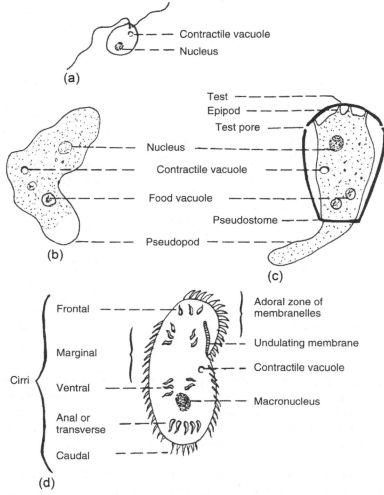

FIG. 5.3 Morphology of four types of soil Protozoa: (a) flagellate (*Bodo*); (b) naked amoeba (*Naegleria*); (c) testacean (*Hyalosphenia*); and (d) ciliate (*Oxytricha*). *(From Lousier and Bamforth, 1990.)*

fine particulate organic matter being the majority of the ingested material. They are highly plastic, in terms of their ability to explore very small cavities or pores in soil aggregates and to feed on bacteria that would otherwise be considered inaccessible to predators (Foster and Dormaar, 1991).

3. Testate amoebae: Compared with the naked amoebae, testate amoebae are often less numerous, except in moist, forested systems. They are more easily enumerated by a range of direct filtration and staining procedures. Lousier

and Parkinson (1984) noted a mean annual biomass of 0.07 g dry wt m^{-2} of aspen woodland soil, much smaller than the average annual mass for bacteria or fungi, 23 and 40 g dry wt m^{-2}, respectively. However, the testacean annual secondary production (new tissue per year) was 21 g dry wt m^{-2}, which they calculated to be essentially the entire average standing crop of the bacteria in that site.

4. Ciliates have unusual life cycles and complex reproductive patterns and tend to be restricted to very moist or seasonally moist habitats. Their numbers are lower than those of other groups, with a general range of 10-500 g^{-1} of litter/ soil. Ciliates can be very active in entering soil cavities and pores and exploiting bacterial food sources (Foissner, 1987). Ciliates, like other protozoa, have resistant or encysted forms and emerge when conditions, such as food availability, are favorable for growth and reproduction. Ciliates, flagellates, and naked and testate amoebae reproduce asexually by fission. The flagellates, naked amoebae, and testacea reproduce by syngamy, or fusion of two cells. For the ciliates, sexual reproduction occurs by conjugation, with the micronucleus undergoing meiosis in two individuals and the two cells joining at the region of the cytostome and exchanging haploid "gametic" nuclei. Each ciliate cell then undergoes fission to produce individuals, which are genetically different from the preconjugant parents (Lousier and Bamforth, 1990).

1 Methods for Extracting and Counting Protozoa

Researchers have favored the culture technique (Singh, 1946) in which small quantities of soil or soil suspensions from dilution series are incubated in small wells and inoculated with a single species of bacteria as a food source. Based on the presence or absence in each well, one can calculate the overall population density ("most probable number"). Coûteaux (1972), Foissner (1987), and Adl (2003) favor the direct count approach, in which one examines soil samples in water to see the organisms present in the subsample. The advantage of direct counting is that it is possible to observe the organisms present and not rely on the palatability of the bacterium as substrate in the series of wells in the culture technique. The disadvantage of the direct count method is that one usually employs only 5-30 mg of soil so as not to be overwhelmed with total numbers (Foissner, 1987). This discriminates against some of the rarer forms of testaceans or ciliates that may have a significant impact on an ecosystem process, if they happen to be large. The culture technique attempts to differentiate between active (trophozoite forms) and inactive (cystic) forms by treatment of replicate samples with 2% hydrochloric acid overnight. The acid kills off the trophozoite forms. After a wash in dilute NaCl, the counting continues. This assumes that all cysts will excyst after this drastic process, an assumption not always met.

2   Impacts of Protozoa on Ecosystem Function

Several investigators have noted the obvious parallel between the protozoan–microbe interaction in water films in soil and on root surfaces and in open-water aquatic systems (Stout, 1963; Clarholm, 1994; Coleman, 1994). The "microbial loop," defined by Pomeroy (1974) as the rapid recycling of nutrients by protozoan grazers, is a powerful conceptual tool. In a fashion similar to that occurring in aquatic systems, rapidly feeding protozoa may consume one or more standing crops of bacteria in soil every year (Clarholm, 1985; Coleman, 1994). This tendency is particularly marked in the rhizosphere, which provides a ready food source for microbial prey.

The population dynamics of bacteria, naked amoebae, and flagellates were followed in the humus layer of a pine forest in Sweden (Clarholm, 1994). Bacteria and flagellates began increasing in number immediately after a rainfall event and rose to a peak after two to three days. Naked amoebae rose more slowly, peaked after four to five days, and then tracked the bacterial decrease downward. The flagellates followed a pattern similar to that of the naked amoebae. Further information on protozoan feeding activities and their impacts on other organisms and ecosystem function is given in Darbyshire (1994). Bonkowski et al. (2000) suggested that protozoa, and the bacteria they feed on in the rhizosphere, produce plant-growth-promoting compounds that stimulate plant growth. The extent of protozoan influence on plant growth, particularly inducing the additional amounts of Indole acetic acid (IAA), is contentious. Certainly under specified experimental conditions, protozoans interacting with plant roots seem to cause considerable lateral root growth. Sorting out the hormonal effects from direct trophic effects, with protozoan ingestion of bacterial and fungal tissues causing enhanced uptake of N by the plants, is an ongoing subject of research in soil ecology (Bonkowski and Clarholm, 2012).

Few studies have investigated the impact of protists on other fauna in soils under field conditions. In a recent study using ciliates and flagellates that were heavily labeled with ^{13}C and ^{15}N, Crotty et al. (2012b) found that orders that were the most enriched in ^{13}C and ^{15}N of protozoan origin in both grassland and forest habitats were Nematodes, Collembola: Entomobryomorpha, and Acari: Mesostigmata, all of which had significantly greater than natural abundance for ^{13}C and/or ^{15}N. In the grassland soil, other invertebrates that were also significantly enriched were Coleoptera larvae (all combined) and Lumbricid earthworms, whereas in the woodland soil, other invertebrates that were significantly enriched were (1) Diplopoda: Polydesmidae, (2) Collembola: Poduromorpha, and (3) Acari: Oribatida (Fig. 5.4).

Protozoa and other microfauna are quite sensitive to environmental insults, and changes in the distribution and activities are diagnostic of changes in soil health (Gupta and Yeates, 1997).

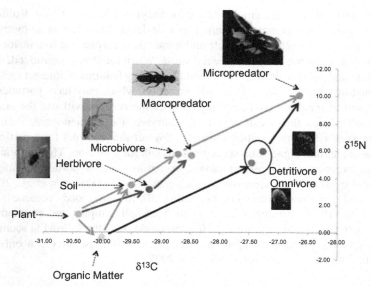

FIG. 5.4 Stable isotope analysis illustrating allocation of carbon and nitrogen through a soil food-web. *Drawing courtesy of F. V. Crotty and P. J. Murray.*

3 Distribution of Protozoa in Soil Profiles

Although protozoa are distributed principally in the upper few centimeters of a soil profile, they are also found at depth, over 200 m deep in groundwater environments (Sinclair and Ghiorse, 1989). Small (2-3 μm cell size) microflagellates decreased 10-fold in numbers during movement through 1 m in a sandy matrix under a trickling-filter facility (in dilute sewage), compared to a 10-fold reduction in bacterial transport over a 10-m distance (Harvey et al., 1995).

B Rotifera

These small (0.05- to 3-mm long) fauna are often only found when a significant proportion of water films exist in soils (Segers, 2007). While they may not be listed in major compendia of soil biota (Dindal, 1990), they are a genuine, albeit secondary, component of the soil fauna (Wallwork, 1976). Rotifers exist in leaf litter, mosses, lichens, and even more extreme environments, such as the soils of the Antarctic Dry Valleys (Treonis et al., 1999; Segers, 2007).

Rotifers are characterized by being transparent and having a three-part body: (1) an anterior ciliary structure or a "crown" with cilia, (2) a body wall that appears to be pseudosegmented (lorica) and (3) a feeding apparatus that has strong muscles and jaws. More than 90% of soil rotifers are in the order Bdelloidea, or worm-like rotifers. In these creeping forms, the suctorial rostral cilia and the adhesive disc are employed for locomotion (Donner, 1966). Rotifers, like tardigrades and nematodes, can enter a desiccated resistant state (anhyrobiosis) at any stage in their life cycle and retract their corona and foot inside the body wall in response to environmental stress. When the stress is removed, they rehydrate and become active. Additional life history features of interest include the construction of shells from a body secretion, which may have particles of debris and/or fecal material adhering to it. Some rotifers will use the empty shells of Testacea, the thecate amoebae, to survive. The parthenogenic Bdelloidea are vortex feeders, creating currents of water that conduct food particles, such as unicellular algae or bacteria, to the mouth for ingestion. The importance of these organisms is largely unknown, except as being important in biogeochemical cycling. They are extremely diverse globally (Robeson et al., 2009, 2011) and may reach numbers exceeding $10^5 \, m^{-2}$ in moist, organic soils (Wallwork, 1970). Rotifers are extracted from soil samples and enumerated using methods similar to those used for nematodes (see the following section), but morphological identification has been stymied because they can only be identified while alive and active (Segers, 2007).

IV MESOFAUNA

A Nematoda

The phylum Nematoda contains nematodes or roundworms, which are among the most numerous and diverse of the multicellular organisms found in any ecosystem. It has been estimated that four of every five animals on earth are nematodes (Bongers and Ferris, 1999). The few cosmopolitan species have been more intensively studied than the large numbers of endemic species. Nematodes are important in ecosystem functioning as belowground parasites and pathogens of plants with effects on net primary productivity (NPP) and as regulators of decomposition (Wall et al., 2012). As with the protozoa, rotifers, and tardigrades, nematodes live in water films or water-filled pore spaces in soils. Nematodes have a very early phylogenetic origin among the Eukarya (Blaxter et al., 1998), but as with other invertebrate groups, the fossil record is fragmentary. Nematodes are most closely related to the other molting animals (Ecdysozoa), such as Nematophora. The Ecdysozoa have no external cilia and similar cuticular layers (Edgecombe et al., 2011).

The overall body shape is cylindrical, tapering at the ends (Fig. 5.5). Nematode body plans are characterized by a "tube within a tube" (alimentary tract/the body wall). They have a complete digestive system or an alimentary tract, consisting of a stoma or stylet, pharynx (or esophagus), and intestine and rectum, which opens externally at the anus. The reproductive structures are complex, and sexes are generally dimorphic. Some species are parthenogenetic, producing only females. Nematodes are highly diverse, but can be identified to genus, order, or family by examining specific morphological characteristics

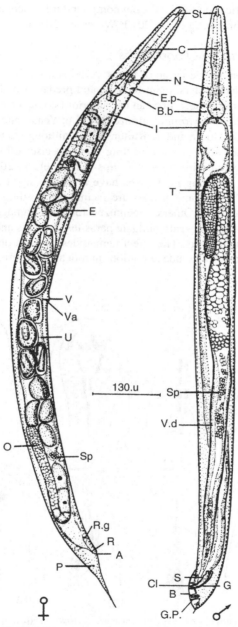

FIG. 5.5 Structures of a *Rhabditis* sp., a secernentean microbotrophic nematode of the order Rhab-ditida. (Left) Female. (Right) Male. St, stoma; C, corpus area of the pharynx; N, nerve ring; E.p, excretory pore; B.b, basal bulb of the pharynx; I, intestine; T, testis; E, eggs; V, vulva; Va, vagina; U, uterus; O, ovary; Sp, sperm; V.d, vas deferens; R.g, rectal glands; R, rectum; A, anus; S, spicules; G, gubernaculum; B, bursa; P, phasmids; G.P., genital papillae; Cl, cloaca. *(Courtesy of Proceedings of the Helminthological Society of Washington. From Poinar, 1983.)*

under high magnification (400 ×) using compound microscopes or by molecular diagnostics (Porazinska et al., 2010; Wu et al., 2011).

1 Nematodes in Soil Food Webs

Nematodes feed on a wide range of foods. A general trophic grouping is bacterial feeders, fungal feeders, plant feeders, and predators and omnivores. For the purposes of our overview, one can use anterior (stomal or mouth) structures to differentiate feeding, or trophic, groups (Fig. 5.6; Yeates and Coleman, 1982; Yeates et al., 1993; Moore and de Ruiter, 2012), although tools such as stable isotopes and metagenomics are revealing more precise information on food sources and affiliation with trophic groups (Crotty et al., 2012a; Porazinska et al., 2012). Plant-feeding nematodes have a hollow stylet that pierces cell walls of higher plants. Some species are facultative, feeding occasionally on plant roots or root hairs. Others, recognized for their damage to agricultural crops and forest plantations, are obligate parasites of plants and feed internally or externally on plant roots. The effect nematodes have on plants is generally species-specific and can include alterations in root architecture, water transport,

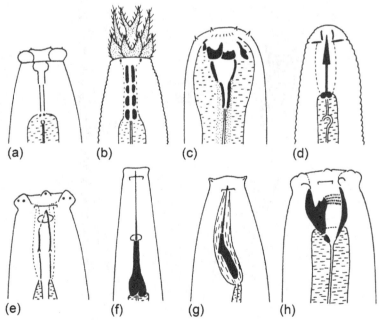

FIG. 5.6 Head structures of a range of soil nematodes. (a) *Rhabditis* (bacterial feeding); (b) *Acrobeles* (bacterial feeding); (c) *Diplogaster* (bacterial feeding, predator); (d) tylenchid (plant feeding, fungal feeding, predator); (e) *Dorylaimus* (feeding poorly known, omnivore); (f) *Xiphinema* (plant feeding); (g) *Trichodorus* (plant feeding); and (h) *Mononchus* (predator). *(From Yeates and Coleman, 1982.)*

plant metabolism, or all of these. Recently, plant parasites in the Tylenchida and some fungal feeding Aphelenchida were found to have cell wall degrading enzymes. Some of the genes for secretion of endogluconases (cellulases) and pectase lyase genes appear to play direct roles in the nematode parasitic process (Karim et al., 2009). Their enzyme products modify plant cell walls and cell metabolism (Davis et al., 2000, 2004), information that, with genome sequencing of plant parasitic nematodes such as *Meloidogyne incognita* (Abad et al., 2008), has increased development of anti-plant parasite strategies for improved crop production. These genes have greatest similarity to microbial genes for cellulases, and it appears that horizontal gene transfer has occurred from microbes to these plant parasites (Sommer and Streit, 2011).

Some of the stylet-bearing nematodes (e.g., the family Neotylenchidae) may feed on roots, root hairs, and fungal hyphae (Yeates and Coleman, 1982). Some bacterial feeders (e.g., *Alaimus*) may ingest 10-μm-wide cyanobacterial cells (*Oscillatoria*) despite the mouth of the nematode being only 1 μm wide. This indicates that the cyanobacterial cells can be markedly compressed by the nematode (Yeates, 1998). The population growth of bacterial-feeding nematodes is strongly dependent on the species of bacteria ingested (Venette and Ferris, 1998). Immature forms of certain nematodes may be bacterial feeders and then become predators or parasites on other fauna once they have matured.

The feeding habits and impacts of entomopathogenic nematodes, nematodes carrying symbiotic bacteria that are lethal to their insect host, are distributed worldwide, except for Antarctica. They have been cultured and sold commercially to control garden pests and mosquitoes (Gaugler, 2002; Hominick, 2002). The nonfeeding, infective juveniles, or third instar (dauer) larvae, of nematodes in the family Heterorhabditidae and Steinernematidae live in the soil and search for insect hosts (Gaugler, 2002). The infective juvenile enters the insect host (which it senses along a CO_2 gradient and other molecules; Dillman et al., 2012) through a body opening, punctures a membrane, and releases its symbiotic bacteria, which releases toxins that kills the host within 24-48 hours. A rapidly growing bacterial population then digests the insect cadaver and provides food for the exponentially growing adult nematode population. The symbiotic bacteria produce antibiotics and other antimicrobial substances that protect the host cadaver and adult nematodes inside from invasion by alien bacteria and fungi from the soil (Dillman et al., 2012). When the cadaver is exhausted of resources, reproduction shunts to infective juveniles, which break through the host integument and disperse into the soil. As many as 410,000 *Heterorhabditis hepialus* infective juveniles are produced in a large ghost moth caterpillar. Symbiotic bacteria occur in some plant parasitic nematodes in the Dorylaimida and Tylenchida (Haegeman et al., 2009).

2 Food Sources for Nematodes

Ruess et al. (2004) have traced the fatty acids specific to fungi to the body tissues of fungal-feeding nematodes. This technique shows considerable promise for more detailed biochemical delineation of food sources of specific feeding groups of nematodes and for groups of other soil fauna. Fungal-feeding nematodes are known to feed preferentially on different fungal species (Mankau and Mankau, 1963), including mycorrhizas and yeasts. Because of the wide range of feeding types and the fact that they seem to reflect ages of the systems in which they occur (i.e., annual vs. perennial crops, old fields, pastures, and more mature forests), nematodes have been used as indicators of overall ecological condition (Bongers, 1990; Ferris et al., 2001; Ferris and Bongers, 2006; Vervoort et al., 2012). This is a growing area of research in land management and soil ecology, one in which the intersection between nematode and other soil faunal community analysis and ecosystem function could prove to be useful for assessing land damage and restoration (Wilson and Kakouli-Duarte, 2009; Domene et al., 2011).

3   Zones of Nematode Activity in Soil

Soil nematode abundance generally decreases with increasing depth and distance from plants, as many soil nematodes are largely concentrated in the rhizosphere. Ingham et al. (1985) found up to 70% of the bacterial- and fungal-feeding nematodes in the 4-5% of the total soil that was rhizosphere, namely the amount of soil 1-2 mm from the root surface (the rhizoplane). Griffiths and Caul (1993) found that nematodes migrated to packets of decomposing grass residues, where there were considerable amounts of labile substrates and microbial food sources. They concluded that nematodes seek out these "hot spots" of concentrated organic matter and that protozoa do not. Nematodes also move and occur vertically in soils toward plant roots, but distance moved is dependent on species, soil temperature, soil type, and soil moisture. In deserts, nematodes are associated with plant roots to depths of 15 m as are mites and other biota (Freckman and Virginia, 1989), and the nematode *Halicephalus mephisto* was recently recovered from soils 3 km deep (Borgonie et al., 2011).

Nematodes have several dispersal methods, some of which involve survival mechanisms. Wallace (1959) noted that movement of nematodes locally was optimum when soil pores were half drained of free water. Using pressure plates, Demeure et al. (1979) showed that nematode movement was found to cease and anhydrobiosis to begin when the water film thickness surrounding a soil particle is between six and nine monomolecular layers of water. This is equivalent to soil pores being drained of free water. However, a nematode species from desert habitats tolerated drier soils with less pore water than a species from a moist tropical habitat, suggesting adaptation to habitat by different species. Nematodes, rotifers, and tardigrades are dispersed long distances by wind when in anhydrobiosis (Nkem et al., 2006). Elliott et al. (1980) noted that the limiting factor for nematode survival often hinges on the availability and size of soil pore necks, which enable passage between soil pores. Yeates et al. (2002) measured the movements, growth, and survival of three genera of bacterial-feeding soil nematodes in undisturbed soil cores maintained on soil pressure plates. Interestingly, the nematodes showed significant reproduction even when diameters of water-filled pores were approximately 1 μm. The anhydrobiosis gene is only one of the many genes upregulated for survival in polar soils (Adhikari et al., 2010).

4 Nematode Extraction Techniques

Nematodes may be extracted by a variety of techniques, either active or passive in nature, all of which may have different extraction efficiencies. The principal advantage of the oldest, active method, namely the Baermann funnel method, is that it is simple, requiring no sophisticated equipment or electricity. It is based on the animal's movement and gravity. Samples are placed on coarse tissue paper, on a coarse mesh screen, and then placed in the cone of a funnel and immersed in water. After crawling through the moist soil and filter paper, the nematodes fall into the neck of the funnel down to the bottom of the funnel stem, which is closed off with a screw clamp on a rubber hose. At the conclusion of the extraction (typically 48-72 hours), the nematodes in solution are drawn off into a vial and kept preserved for a later examination. Drawbacks to the technique are that only active nematodes are extracted. It also allows dormant nematodes to become active and eggs to hatch into juveniles and be extracted, yielding a slightly inflated estimate of the true, "active" population at a given time. For more accuracy in determination of populations, the passive, or flotation, techniques are generally preferred. Passive methods include filtration, or decanting and sieving, and flotation/centrifugation (Coleman et al., 1999) to remove the nematodes from the soil suspension. Elutriation methods can be employed for handling larger quantities of soil, usually greater than 500 g, or to recover large amounts and a greater diversity of nematodes. Elutriation methods rely on fast mixing of soil and water in funnels. Semiautomatic elutriators, which enhance the number of soil samples to be extracted, are available and used for assessing root mycorrhizae, rhizobia, root mass, and nematode species (Byrd et al., 1976). There are many references comparing methods, including Whitehead and Hemming (1965), Schouten and Arp (1991), and McSorley and Frederick (2004). Anhydrobiotic nematodes can be extracted in a high-molarity solution, such as sucrose, which prevents the nematodes from rehydrating (Freckman et al., 1977).

B Microarthropods

Large numbers of the microarthropods (mainly mites and collembolans) are found in most types of soils, and like nematodes, a square meter of forest floor may contain hundreds of thousands of individuals representing thousands of species. Microarthropods have a significant impact on the decomposition processes in the forest floor and are important reservoirs of biodiversity in forest ecosystems. Many microarthropods feed on fungi and nematodes, thereby linking the microfauna and microbes with the mesofauna (Chamberlain et al., 2006). Microarthropods in turn are prey for macroarthropods, such as spiders, beetles, ants, and centipedes, thus bridging a connection to the macrofauna and to identification of food webs in soils.

In the size spectrum of soil fauna, the mites and collembolans are mesofauna. Members of the microarthropod group are unique, not so much because of their body size but because of the methods used for sampling them. Small pieces of habitat (soil, leaf litter, or similar materials) are collected and the microarthropods are extracted from them in the laboratory. Most of the methods used for microarthropod extraction are either variations of the Tullgren funnel, which uses heat to desiccate the sample and force the arthropods into a collection fluid, or flotation in solvents or saturated sugar solutions followed by filtration (Edwards, 1991). Generally, flotation methods work well in low-organic, sandy soils, while Tullgren funnels perform best in soils with high organic matter content. Flotation procedures are more laborious than the Tullgren extraction. Better estimates of species number may be achieved using fewer, larger samples. However, valid comparisons of microarthropod abundance in different habitats may be obtained even if extraction efficiencies, though unknown, are similar.

Microarthropod densities vary during seasons within and between different ecosystems. Temperate forest floors with large organic matter content support high numbers ($33\text{-}88.10^3 \, m^{-2}$) of microarthropods, and coniferous forests may have in excess of $130.10^3 \, m^{-2}$. Tropical forests, where the organic layer is thin, contain fewer microarthropods (Seastedt, 1984; Coleman et al., 2004). Tillage, fire, and pesticide applications typically reduce populations, but recovery may be rapid, and microarthropod groups respond differently. These effects of disturbance on abundance and diversity can be seen for many groups of soil animals, such as nematodes and earthworms. Soil mites usually outnumber collembolans, but the latter become more abundant in some situations. In the springtime, forest leaf litter may develop large populations of "snow fleas" (*Hypogastrura nivicola* and related species). Among the mites themselves, the oribatids usually dominate, but the delicate Prostigmata may develop large populations in cultivated soils with a surface crust of algae.

1   Determining the Trophic Position of Microarthropods

Considerable progress has been made in determining the details of trophic (feeding) interactions in microarthropods, particularly for the Oribatid Mites. Stable isotope techniques have provided powerful new information on the diet of mites over time in the field. The relative positions of mite gut contents and tissues in the amount of ^{13}C and ^{15}N stable isotope signatures has enabled the assignment of Oribatid Mites into feeding guilds (Schneider et al., 2004;

Pollierer et al., 2009). Unfortunately, this requires a minimum mass of mites to enable the analyses to be performed. An alternative approach uses an analysis of mite chelicerae (the chewing mouth parts) along with measurement, where possible, of the stable isotope ratios of Oribatid Mites present in moss communities in the forest floor. Using this approach, cheliceral morphology was found to be an inexpensive and quick filter for estimation of the mites' feeding preferences. In cases of ambiguous trophic relationships, the dietary preferences were then resolved through isotope analyses (Perdomo et al., 2012). Chamberlain et al. (2006) used a combination of fatty acid gut analysis with compound specific C isotope analysis to show that two collembolan species consumed nematodes, confirming earlier observations (Ruess et al., 2004). Studying the fates of DNA from two prevalent species of soil nematodes, Heidemann et al. (2011) found that eight species of Oribatid Mites and one Mesostigmatid would act as either predators or scavengers on soil nematodes, thus unraveling the complexity of microarthropod food webs.

C Enchytraeids

In addition to earthworms (discussed under Macrofauna), another important family of terrestrial Oligochaeta is the Enchytraeidae. This group of small, unpigmented worms, also known as "potworms," is classified within the "microdrile" oligochaetes and consists of some 700 species. Species from 19 of 28 genera are found in soil; the remainder occurs primarily in marine and freshwater habitats (Dash, 1990; van Vliet, 2000; Jeffery et al., 2010; Schmelz and Collado, 2010). The Enchytraeidae are thought to have arisen in cool temperate climates where they are commonly found in moist, forest soils rich in organic matter. Various species of enchytraeids are now distributed globally from subarctic to tropical regions. Keys to the common genera were presented by Dash (1990) and Schmelz and Collado (2010). Identification of enchytraeid species is difficult, but genera may be identified by observing internal structures through the transparent body wall of specimens mounted on slides.

The Enchytraeidae are typically 10-20 mm in length and are anatomically similar to the earthworms, except for the miniaturization and rearrangement of features. They possess setae (with the exception of one genus) and a clitellum in segments XII and XIII, which contain both male and female pores. Sexual reproduction in enchytraeids is hermaphroditic and functions similar to that in earthworms. Cocoons may contain one or more eggs, and maturation of newly hatched individuals ranges from 65 to 120 days, depending on species and environmental temperature (van Vliet, 2000). Enchytraeids also display asexual strategies of parthenogenesis and fragmentation, which enhance their probability of colonization of new habitats (Dósza-Farkas, 1996).

1 Food Sources of Enchytraeids

Enchytraeids ingest both mineral and organic particles, although typically of smaller size ranges than those of earthworms. Numerous investigators have noted that finely divided plant materials, often enriched with fungal hyphae and bacteria, are a principal portion of the diet of enchytraeids. Microbial tissues are probably the fraction most readily assimilated because enchytraeids lack the gut enzymes to digest more recalcitrant soil organic matter (SOM; van Vliet, 2000; Jeffery et al., 2010). Didden (1990, 1993) suggested that enchytraeids feed predominantly on fungi, at least in arable soils, and classified a community as 80% microbivorous and 20% saprovorous. As with several other members of the soil mesofauna, the mixed microbiota that occur on decaying organic matter, either litter or roots, are probably an important part of the diet of these creatures. The remaining portions of organic matter, after the processes of ingestion, digestion, and assimilation, become part of the slow-turnover pool of soil organic matter. Zachariae (1964) suggested that so-called collembolan soil, said to be dominated by collembolan feces (particularly low-pH mor soils), were actually formed by Enchytraeidae. Mycorrhizal hyphae have been found in the fecal pellets of enchytraeids from pine litter (Ponge, 1991). Enchytraeids probably consume and further process larger fecal pellets and castings of soil macrofauna, such as collembolans and earthworms (Zachariae, 1964; Rusek, 1985). Thus, it is clear that fecal contributions to soil by soil-dwelling invertebrates provide feedback mechanisms affecting the abundance and diversity of other soil-dwelling animals.

2 Distribution and Abundance of Enchytraeids

Enchytraeids occur in moist soils from the Arctic to the tropic regions, but are rarely found in dry soils of deserts. Enchytraeid densities range from 1000 individuals m^{-2}, in intensively cultivated agricultural soil in Japan, to 140,000 individuals m^{-2} in a peat moor in the United Kingdom. In a subtropical climate, enchytraeid densities of 4000-14,000 individuals m^{-2} occur in agricultural plots in the Piedmont of Georgia, USA, whereas higher densities (20,000-30,000 individuals m^{-2}) are found in surface layers of deciduous forest soils in the southern Appalachian mountains of North Carolina (van Vliet et al., 1995). Although enchytraeid densities are typically highest in acid soils with high organic content, Didden (1995) found no statistical relationship over a broad range of data between average enchytraeid density and several environmental variables such as annual precipitation, annual temperature, or soil pH. It appears that local variability may be at least as great as variation on a wider scale, as enchytraeid densities show both spatial and seasonal variations. Vertical distributions of enchytraeids in soil are related to organic matter horizons. Up to 90% of populations may occur in the upper layers in forest and no-tillage agricultural soils, but densities may be higher in the Ah horizon of grasslands (Davidson et al., 2002). Seasonal trends in enchytraeid population densities appear to be associated with moisture and temperature regimes (van Vliet, 2000).

Enchytraeids have significant effects on SOM dynamics and on soil physical structure. Litter decomposition and nutrient mineralization are influenced primarily by interactions with soil microbial communities. Enchytraeid feeding on fungi and bacteria can increase microbial metabolic activity and turnover, accelerate the release of nutrients from microbial biomass, and change species composition of the microbial community through selective grazing. However, Wolters (1988) found that enchytraeids decreased mineralization rates by reducing microbial populations and possibly by occluding organic substrates in their feces. The influence of enchytraeids on SOM dynamics is, therefore, the net result of both enhancement and inhibition of microbial activity depending on soil texture and population densities of the animals (Wolters, 1988; van Vliet, 2000).

Enchytraeids affect soil structure by producing fecal pellets, which, depending on the animal size distribution, may enhance aggregate stability in the 600-1000 μm aggregate size fraction. In forest floors, these pellets are composed mainly of fine humus particles, but in mineral soils, organic matter and mineral particles may be mixed into fecal pellets with a loamy texture. Davidson et al. (2002) estimated that enchytraeid fecal pellets constituted nearly 30% of the volume of the Ah horizon in a Scottish grassland soil (Fig. 5.7). Encapsulation or occlusion of organic matter into these structures may reduce decomposition rates. Burrowing activities of enchytraeids have not been well studied, but there is evidence that soil porosity and pore continuity can increase in proportion to enchytraeid body size (Rusek, 1985; Didden, 1990).

3 Extraction of Enchytraeids

Enchytraeids are typically sampled in the field using cylindrical soil cores of 5-7.5 cm diameter. Large numbers of replicates may be needed for a sufficient sampling due to the clustered distribution of enchytraeid populations

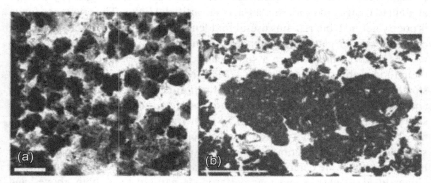

FIG. 5.7 Thin-section micrographs of fecal pellets in a grassland soil. (a) Derived from enchytraeids (scale bar, 0.5 mm). (b) Derived from earthworms (scale bar, 1.0 mm). *(From Davidson et al., 2002.)*

(van Vliet, 2000). Extractions are often done with a wet-funnel technique, similar to the Baermann funnel extraction used for nematodes. In this case, soil cores are submerged in water on the funnel and exposed for several hours to a heat and light source from above; enchytraeids move downward and are collected in the water below. Van Vliet (2000) provides a comparison of modifications of this technique.

V MACROFAUNA

A Macroarthropods

Larger insects, spiders, myriapods, and others are considered together under the appellation "macroarthropods." Typical body lengths range from about 10 mm to as much as 15 cm for centipedes (Shelley, 2002). The group includes a mixture of various arthropod classes, orders, and families. Like the microarthropods, the macroarthropods are defined more by the methods used to sample them rather than by measurements of body size. Large soil cores (10 cm diameter or greater) may be appropriate for euedaphic (dwelling within the soil) species. Arthropods can be recovered from them using flotation techniques (Edwards, 1991). Hand sorting of soils and litter is time consuming, but yields better estimates of population size. Capture–mark–recapture methods have been used to estimate population sizes of selected macroarthropod species, but the assumptions for this procedure are violated more often than not (Southwood, 1978). Pitfall traps have been widely used to sample litter- and surface-dwelling macroarthropods. This method collects arthropods that fall into cups filled with preservative. Absolute population estimates are difficult to obtain with pitfall traps, but the method yields comparative estimates when used with caution. Many of the macroarthropods are members of the group termed "cryptozoa," a group consisting of animals that dwell beneath stones or logs, under bark, or in cracks and crevices. Cryptozoans typically emerge at night to forage, and some are attracted to artificial lights. The cryptozoa fauna is poorly defined, but remains useful for identifying a group of invertebrate species with similar patterns of habitat utilization.

The macroarthropods are a significant component of soil ecosystems and their food webs (Brussaard et al., 2012). Macroarthropods differ from their smaller relatives in that they may have direct effects on soil structure. Termites and ants in particular are important movers of soil, depositing parts of lower strata on top of the litter layer (Fig. 5.8). Emerging nymphal stages of cicadas may be numerous enough to disturb soil structure. Larval stages of soil-dwelling scarabaeid beetles sometimes churn the soil in grasslands. These and other macroarthropods are part of the group that has been termed ecological engineers (Jones et al., 1994). Some macroarthropods participate in both above- and belowground parts of terrestrial ecosystems. Many macroarthropods are transient or temporary soil residents and thus form a connection between food

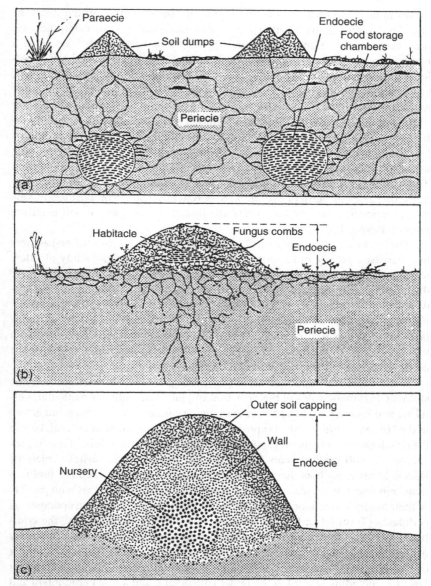

FIG. 5.8 Termite mounds. Diagrammatic representation of different types of concentrated nest systems: (a) *Hodotermes mossambicu*, (b) *Macrotermes subhyalinus*, and (c) *Nasutitermes exitiosus*. *(From Lee and Wood, 1971.)*

chains in the "green world" of foliage and the "brown world" of the soil. Caterpillars descending to the soil to pupate or migrating armyworm caterpillars are prey to ground-dwelling spiders and beetles. Macroarthropods may have a major influence on the microarthropod portion of belowground food webs. Collembola, among other microarthropods, are important food items for spiders, especially immature stadia, thus providing a macro- to microconnection. Other macroarthropods, such as cicadas, emerging from soil may serve as prey for some vertebrate animals (Lloyd and Dybas, 1966), including endangered vertebrate species (Decaëns et al., 2006), thus, providing a link to the larger megafauna. Among the macroarthropods, there are many litter-feeding species, such as the millipedes, that are important consumers of leaf, grass, and wood litter. These arthropods have major influences on the decomposition process, thereby impacting rates of nutrient cycling in soil systems. The decomposition of vertebrate carrion is largely accomplished through the actions of soil-dwelling insects (Payne, 1965).

The interaction between microbial ecologists and soil arthropod researchers has increased greatly in the last several years. For example, a study of litter-feeding millipedes has shown that they harbor a gut microbiota that is distinctly different from the ingested organic substrate of leaf litter. Using denaturing gradient gel electrophoresis (DGGE) gels, Knapp et al. (2009) found a stable, indigenous microbial community in the gut of the millipede *Cylindroiulus fulviceps* that was studied in abandoned alpine pastureland. Its gut microbiota was dominated by Gamma- and Delta-proteobacteria and was found to resist dietary changes even during a varied dietary intake.

A more general aspect of arthropod feeding biology is addressed by the study of enzymatic roles in the digestion of plant cell walls, in particular, cellulose and lignin. Considered over evolutionary time, insects were more active in the "brown world," of decomposition of plant tissues on or in the soil, before the development of herbivory, possibly coinciding with the origin of the Angiosperms ca. 160 million years before present. Some of the earliest orders of insects to arise, such as the Isoptera (Termites) have approached the problem of development of cellulases by developing symbiotic associations with protists in their hindguts (the more primitive termite families) or their own endogenous cellulases (Termitidae; Calderón-Cortés et al., 2012). Significantly, the combined synthesis and activity of cellobiohydrolases and xylanases effectively degrade lignocellulose in the termite hindgut (Gilbert, 2010).The development of endogenous cellulases, and in some cases laccases, occurs in numerous insect families, including several in the Order Coleoptera, and a few Lepidoptera and Diptera (Calderón-Cortés et al., 2012).

B   Oligochaeta (Earthworms)

Earthworms are the most familiar and, with respect to soil processes, often the most important of the soil fauna. The importance of earthworms arises from their influence on soil structure (e.g., aggregate or crumb formation, soil pore formation) and on the breakdown of organic matter applied to soil (e.g., fragmentation, burial, and mixing of plant residues; Carpenter et al., 2008). The modern era of earthworm research began with Darwin's (1881) book, *The Formation of Vegetable Mould through the Actions of Worms, with Observations of Their Habits*, which called attention to the beneficial effects of earthworms. Since then, a vast literature has established the importance of earthworms as biological agents in soil formation, organic litter decomposition, and redistribution of organic matter in the soil (Hendrix, 1995; Edwards, 1998; Lavelle & Spain, 2001).

Earthworms are classified within the phylum Annelida, class Oligochaeta. Species within the families Lumbricidae and Megascolecidae are ecologically the most important in North America, Europe, Australia, and Asia. Some of these species have been introduced worldwide by human activities and now dominate the earthworm fauna in many temperate areas. Any given locality may be inhabited by all native species, all exotic species, a combination of native and exotic species, or no earthworms at all. Relative abundance and species composition of local fauna depend greatly on soil, climate, vegetation, topography, land use history, and especially, past invasions by exotic species.

1 Earthworm Distribution and Abundance

Earthworms occur worldwide in habitats where soil, water, and temperature are favorable for at least part of the year. They are most abundant in forests and grasslands of temperate and tropical regions and least so in arid and frigid environments, such as deserts, tundra, or polar regions. Earthworm densities in a variety of habitats worldwide range from 10 to 2000 individuals m^{-2}, the highest values occurring in fertilized pastures and the lowest in acid or arid soils (coniferous or sclerophyll forests). Typical densities from temperate deciduous or tropical forests and certain arable systems range from 100 to over 400 individuals m^{-2}, representing a range of from 4 to 16 g dry mass m^{-2}. Intensive land management (especially soil tillage and application of toxic chemicals such as common soil and plant pesticides) often reduces the density of earthworms or may completely eliminate them. Conversely, degraded soils converted to conservation management often show increased earthworm densities after a suitable period of time (Curry et al., 1995; Edwards and Bohlen, 1996).

2 Biology and Ecology

Earthworms are often grouped into functional categories based on their morphology, behavior, and feeding ecology and their microhabitats within the soil (Lavelle, 1983; Lee, 1985). Methods for extraction and identification of various groups of earthworms in tropics can apply to other soils (Moreira et al., 2009). Epigeic and epi-endogeic species are often polyhumic, meaning they prefer organically enriched substrates and utilize plant litter on the soil surface and

C-rich upper layers of mineral soil. Polyhumic endogeic species inhabit mineral soil with high organic matter content (3%), such as the rhizosphere, whereas meso- and oligohumic endogeic species inhabit soil with moderate (1-3%) and low (1%) organic matter contents, respectively. Anecic species exploit both the surface litter as a source of food and the mineral soil as a refuge. The familiar *Lumbricus terrestris* is an example of an anecic species, constructing burrows and pulling leaf litter down into them. The American log worm (*Bimastos parvus*) exploits leaf litter and decaying logs with little involvement in the soil, making it an epigeic species. Epigeic species promote the breakdown and mineralization of surface litter, whereas anecic species incorporate organic matter deeper into the soil profile and facilitate aeration and water infiltration through their formation of burrows.

3   Influence on Soil Processes

Earthworms, as ecosystem engineers (Lavelle et al., 1998), have pronounced effects on soil structure as a consequence of their burrowing activities as well as their ingestion of soil and production of castings (Lavelle and Spain, 2001; van Vliet and Hendrix, 2003). Casts are produced after earthworms ingest mineral soil or particulate organic matter, mix and enrich them with organic secretions in the gut, and then deposit the material as a slurry lining their burrows or as discrete fecal pellets. Excretion of fecal pellets can occur within or on the soil, depending on earthworm species. Turnover rates of soil through earthworm casting range from 40-70 t ha^{-1} y^{-1} in temperate grasslands (Bouché, 1983) to 500-1000 t ha^{-1} y^{-1} in tropical savannas (Lavelle et al., 1992). While in the earthworm gut, casts are colonized by microbes that begin to break down SOM. As casts are deposited in the soil, microbial colonization and activity continue until readily decomposable compounds are depleted. Mechanisms of cast stabilization include organic bonding of particles by polymers secreted by earthworms and microbes, mechanical stabilization by plant fibers and fungal hyphae, and stabilization due to wetting and drying cycles and age-hardening effects (Tomlin et al., 1995). Mineralization of organic matter in earthworm casts and burrow linings produces zones of nutrient enrichment compared to bulk soil. These zones, referred to as the "drilosphere," are often sites of enhanced activity of plant roots and other soil biota (Lavelle et al., 1998). Plant growth-promoting substances have also been suggested as constituents of earthworm casts. Many earthworm castings from commercial vermicomposting operations are sold commercially as soil amendments to improve soil physical properties and enhance plant growth (Edwards, 1998).

Earthworm burrowing in soil creates macropores of various sizes, depths, and orientations, depending on species and soil type. Burrows range from about 1 to 10 mm in diameter and are among the largest of soil pores. Continuous macropores resulting from earthworm burrowing may enhance water infiltration by functioning as bypass flow pathways through soils. These pores may or may not be important in solute transport, depending on soil water content, the nature of the solute, and chemical exchange properties of the burrow linings (Edwards and Shipitalo, 1998).

4   Earthworm Effects on Ecosystems

Despite the many beneficial effects of earthworms on soil processes, some aspects of earthworm activities may be undesirable (Lavelle et al., 1998; Parmelee et al., 1998). Detrimental effects include (1) removing and burying of surface residues that would otherwise protect soil surfaces from erosion; (2) producing fresh casts that increase erosion and surface sealing; (3) increasing compaction of surface soils by decreasing SOM, particularly for some tropical species; (4) riddling irrigation ditches, making them leaky; (5) increasing losses of soil N through leaching and denitrification; and (6) increasing soil C loss through enhanced microbial respiration. Earthworms may transmit pathogens, either as passive carriers or as intermediate hosts, raising concerns that some earthworm species could be a vector for the spread of certain plant and animal diseases. The net result of positive and negative effects of earthworms, or any other soil biota, determines whether they have detrimental impacts on ecosystems (Lavelle et al., 1998). An effect, such as mixing of O and A horizons, may be considered beneficial in one setting (e.g., urban gardens) and detrimental in another (e.g., native forests). Edwards (1998) provides a review of the potential benefits of earthworms in agriculture, waste management, and land remediation.

C   Formicidae (Ants)

Formicidae, the ants, are probably the most significant family of soil insects, due to the very large influence they have on soil structure. Ants are numerous, diverse, and widely distributed from arctic to tropical ecosystems. Ant communities contain many species, even in desert areas (Whitford, 2000), and local species diversity is especially large in tropical areas. Populations of ants are equally numerous. About one-third of the animal biomass of the Amazonian rain forest is composed entirely of ants and termites, with each hectare containing in excess of 8 million ants and 1 million termites (Hölldobler and Wilson, 1990). Furthermore, ants are social insects, living in colonies with several castes.

Ants have a large impact on their ecosystems. They are major predators of small invertebrates. Their activities reduce the abundance of other predators such as spiders and carabid beetles (Wilson, 1987). Ants are ecosystem engineers, moving large volumes of soil, as much as earthworms do (Hölldobler and Wilson, 1990). Ant influences on soil structure are particularly important in deserts, where earthworm densities are low. Given the large diversity of ants, identification to species is problematic for any, but Wheeler and Wheeler (1990) offer keys to subfamilies and genera of the Nearctic ant fauna.

D   Termitidae (Termites)

Along with earthworms and ants, termites are the third major earthmoving group of invertebrates. Termites are social insects with a well-developed caste system. Through their ability to digest wood they have become economic pests of major importance in some regions of the world (Lee and Wood, 1971; Bignell and Eggleton, 2000). Termites are highly successful, constituting up to 75% of the insect biomass and 10% of all terrestrial animal biomass in the tropics (Wilson, 1992; Bignell, 2000). Although termites are mainly tropical in distribution, they occur in temperate zones as well. Termites have been called the tropical analogs of earthworms because they reach a large abundance in the tropics and process large amounts of litter. Termites in the primitive families, such as Kalotermitidae, possess a gut flora of protozoans, which enables them to digest cellulose. Their normal food is wood that has come into contact with soil. Most species of termites construct runways of soil, and some are builders of spectacular mounds. Members of the phylogenetically advanced family Termitidae do not have protozoan symbionts, but possess a formidable array of microbial symbionts (bacteria and fungi) that enable them to process and digest the humified organic matter in tropical soils (Bignell, 1984; Breznak, 1984; Pearce, 1997). A generalized sequence of events in a typical Termitinae soil-feeder gut is illustrated in Fig. 5.9 (Brauman et al., 2000).

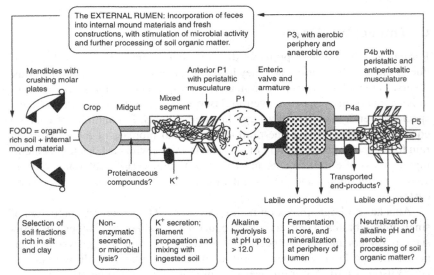

FIG. 5.9 Hypothesis of gut organization and sequential processing in soil-feeding Cubitermes-clade termites. The model emphasizes the role of filamentous prokaryotes, the extremely high pH reached in the P1, and the existence of both aerobic and anaerobic zones within the hindgut. Question marks indicate major uncertainties. Not to scale. *(From Brauman et al., 2000.)*

Three nutritional categories include wood-feeding species, plant- and humus-feeding species, and fungus growers. The last group lacks intestinal symbionts and depends on cultured fungus for nutrition. Termites have an abundance of unique microbes living in their guts. One recent study of bacterial microbiota in the gut of the wood-feeding termite *Reticulitermes speratus* found 268 phylotypes of bacteria (16S rRNA genes, amplified by PCR), including 100 clostridial, 61 spirochaetal, and 31 Bacteroides-related phylotypes (Hongoh et al., 2003). More than 90% of the phylotypes were found for the first time, but we do not know if they are active and participating in wood decay. Other phylotypes were monophyletic clusters with sequences recovered from the gut of other termite species. Cellulose digestion in termites, which was once considered to be solely due to the activities of fungi and protists and occasionally bacteria, has now been demonstrated to be endogenous to termites. Endogenous cellulose-degrading enzymes occur in the midguts of two species of higher termites in the genus Nasutitermes and in the Macrotermitinae (which cultivate basidiomycete fungi in elaborately constructed gardens) as well (Bignell, 2000).

In contrast to the C-degradation situation, only prokaryotes are capable of producing nitrogenase to fix N_2. This process occurs in the organic-matter-rich, microaerophilic milieu of termite guts. Some termite genera have bacteria that fix relatively small amounts of N, but others, including *Mastotermes* and *Nasutitermes*, fix from 0.7 to 21 g N g^{-1} fresh wt day^{-1}. This equals 20-61 µg N per colony per day, which would double the N content if N_2 fixation was the sole source of N and the rate per termite remained constant (N content of termites assumed to be 11% on a dry weight basis) (Breznak, 2000). Some insight into the roles of soil-feeding termites in the terrestrial N cycle has been provided by Ngugi and Brune (2012), who measured significant denitrification in the hindguts of two genera of soil feeding termites ranging from 0.4 to 3.9 nmol h^{-1} (g fresh wt.)$^{-1}$ N_2O, providing direct evidence that soil-feeding termites are a hitherto unrecognized source of this greenhouse gas in tropical soils. For an extensive exposition of the role of termites in the dynamics of SOM and nutrient cycling in ecosystems worldwide, refer to Bignell and Eggleton (2000).

VI ROLES OF SOIL FAUNA IN ECOSYSTEMS

Evidence on how the role of soil fauna in food webs will respond to environmental change and influence aboveground processes has advanced considerably (see Cheeke et al., 2012; de Vries et al., 2012; Wall et al., 2012; Bragazza et al., 2013). Loss of species due to varying management practices, erosion, pollution, urbanization, and resulting effects on ecosystem function and services are becoming widely recognized and related to larger issues of biodiversity loss, desertification, and elevated greenhouse gas concentrations (Koch et al., 2013). This has resulted in international attention to the importance of soils and, in particular, to soil biodiversity. An analysis (see the *European Atlas of*

Soil Biodiversity, Jeffery et al., 2010) includes maps showing the potential vulnerability of soil biodiversity and ecosystem services to various environmental changes, including land use change. We are now able to answer basic questions about soil fauna including such topics as whether there are cosmopolitan vs. endemic fauna species; what is the local and global biogeographic distribution and range of faunal species; what are the factors influencing the distribution of key species; and whether the loss of species affects ecosystem function (Cock et al., 2012; Wall et al., 2012). Food web ecology, with its emphasis on community assembly and disassembly, has the potential to act as an integrating concept across conservation biology and community and ecosystem ecology as well as for provision of ecosystem services (de Vries et al., 2012; Moore and de Ruiter, 2012; Thompson et al., 2012; Wall et al., 2012).

Global experiments and syntheses have continued to address the quantification of the role of soil fauna in ecosystem processes and, in particular, have led to increased evidence for their contribution to C cycling. Global multisite experiments show that soil fauna are key regulators of decomposition rates at biome and global scales (Wall et al., 2008; Powers et al., 2009; Makkonen et al., 2012). Garcia-Palacios et al. (2013) conducted a meta-analysis on 440 litterbag case studies across 129 sites to assess how climate, litter quality, and soil invertebrates affect decomposition. This analysis showed fauna were responsible for ~27% average enhancement of litter decomposition across global and biome scales.

Agricultural practices affect many of the key functional and structural attributes of ecosystems in several ways: the transformation of mature ecosystems into ones that are in a managed developmental state are induced by tillage operations and other activities such as applying fertilizers and pesticides. These manipulations have the potential to shift the elemental balance of a system and decrease species diversity and alter the soil food web (Cheeke et al., 2012; Moore & de Ruiter, 2012). Conventional tillage practices alter the distribution of organic material and affect the rate of formation of micro- and macroaggregates in the soil profile. This has a profound effect on the turnover rates of organic matter that is associated with the aggregates (Elliott and Coleman, 1988; Six et al., 2004; O'Brien and Jastrow, 2013) as well as affecting ecosystem services (Cheeke et al., 2012). De Vries et al. (2012) showed that grassland, fungal-based, food webs were more resilient than agricultural fields with bacterial food webs and provided evidence for management options that enhanced ecosystem services. Cock et al. (2012) provide evidence for manipulating soil invertebrates to benefit agriculture and to enhance ecosystem services, such as biological control and C sequestration. Their focus includes successful case studies in sustainable agriculture. Koch et al. (2013) bring attention to the global policy impact of land degradation and loss of soils, biodiversity and ecosystem services and its implications for food security. Soil fauna continue to be an exciting research field, linking aboveground systems to belowground diversity and function, as well as to environmental issues at local to global scales (Wall, 2004; Cock et al., 2012; Wall et al., 2012).

VII SUMMARY

Soil fauna may be considered as very efficient means to assist microbes in colonizing and extending their reach into the horizons of soils worldwide. Their roles as colonizers, comminutors, and engineers within soils have been emphasized, but new technologies and global environmental issues are yielding new questions about how to manipulate soil fauna for the long-term sustainability of soils. The demand for taxonomic specialists for all groups of soil biota is increasing as we currently recognize that molecular information alone is insufficient for many studies. Stable isotope technologies are revealing transfer of C and N through soil food webs with more precision as to the role of each trophic group. In addition, this technique is clarifying that soil faunal species hitherto thought to be within one trophic group are not, thus leading to research about the structure and resilience of soil food webs. Information on the biogeography of soil fauna, their latitudinal gradient patterns, their relationship to aboveground hot spots and to land management strategies, as well as their taxonomic status and natural history, will be critical for understanding how bacteria, archaea, fungi, protozoa, and soil invertebrates will interact and respond to multiple global changes (Wall et al., 2001, 2008; Wu et al., 2011; Fierer et al., 2012). For example, soil invertebrates can be invasive species, which, depending on the species, can affect soil C sequestration, soil fertility, and plant and animal health and result in economic and ecosystem change. We feel protection of known biodiversity in ecosystems clearly must include the rich pool of soil species. This is because data for some of these species individually and collectively indicate tight connections to biodiversity aboveground, major roles in ecosystem processes, and provision of ecosystem benefits for human well-being (Wall, 2004; Wardle et al., 2004; Wall et al., 2005). In addition, climate envelopes can be developed as more is known about soil species identity, geographic ranges, and activities, aiding in options for land management and projections for the functioning of tomorrow's ecosystems under climate change. For further reading on the roles of fauna in soil processes, see Coleman et al. (2004, 2012), Wall (2004), and Wall et al. (2012).

ACKNOWLEDGMENTS

We would like to thank Drs. Matt Knox and Zach Sylvain for their careful editing and helpful comments.

REFERENCES

Abad, P., Gouzy, J., Aury, J.M., et al., 2008. Genome sequence of the metazoan plant-parasitic nematode Meloidogyne incognita. Nat. Biotechnol. 26, 909–915.

Adhikari, B.N., Wall, D.H., Adams, B.J., 2010. Effect of slow desiccation and freezing on gene transcription and stress survival of an Antarctic nematode. J. Exp. Biol. 213, 1803–1812.

Adl, S.M., 2003. The Ecology of Soil Decomposition. CAB International, Wallingford.

Adl, S.M., Simpson, A.G.B., Farmer, M.A., et al., 2005. The new higher level classification of eukaryotes with emphasis on the taxonomy of protists. J. Eukaryot. Microbiol. 52, 399–451.

André, H.M., Ducarme, X., Lebrun, P., 2002. Soil biodiversity: myth, reality or conning? Oikos 96, 3–24.

Bamforth, S.S., 1980. Terrestrial protozoa. J. Protozool. 27, 33–36.

Bates, S.T., Clemente, J.C., Flores, G.E., Walters, W.A., Parfrey, L.W., Knight, R., Fierer, N., 2013. Global biogeography of highly diverse protistan communities in soil. ISME J. 7, 652–659.

Bignell, D.E., 1984. The arthropod gut as an environment for microorganisms. In: Anderson, J.M., Rayner, A.D.M., Walton, D.W.H. (Eds.), Invertebrate-Microbial Interactions. Cambridge University Press, Cambridge, pp. 205–227.

Bignell, D.E., 2000. Introduction to symbiosis. In: Abe, T., Bignell, D.E., Higashi, M. (Eds.), Termites: Evolution, Sociality, Symbioses, Ecology. Kluwer Academic, Dordrecht, pp. 189–208.

Bignell, D.E., Eggleton, P., 2000. Termites in ecosystems. In: Abe, T., Bignell, D.E., Higashi, M. (Eds.), Termites: Evolution, Sociality, Symbioses, Ecology. Kluwer Academic, Dordrecht, pp. 363–387.

Blaxter, M.L., De Ley, P., Garey, J.R., Liu, L.X., Scheldeman, P., Vierstraete, A., Vanfleteren, J.R., Mackey, L.Y., Dorris, M., Frisse, L.M., Vida, J.T., Thomas, W.K., 1998. A molecular evolutionary framework for the phylum Nematoda. Nature 392, 71–75.

Bongers, T., 1990. The maturity index—an ecological measure of environmental disturbance based on nematode species composition. Oecologia 83, 14–19.

Bongers, T., Ferris, H., 1999. Nematode community structure as a bioindicator in environmental monitoring. Trends Ecol. Evol. 14, 224–228.

Bonkowski, M., Clarholm, M., 2012. Stimulation of plant growth through interactions of bacteria and protozoa: testing the auxiliary microbial loop hypothesis. Acta Protozool. 51, 237–247.

Bonkowski, M., Griffiths, B., Scrimgeour, C., 2000. Substrate heterogeneity and microfauna in soil organic 'hotspots' as determinants of nitrogen capture and growth of ryegrass. Appl. Soil Ecol. 14, 37–53.

Borgonie, G., Garcia-Moyano, A., Litthauer, D., Bert, W., Bester, A., van Heerden, E., Moller, C., Erasmus, M., Onstott, T.C., 2011. Nematoda from the terrestrial deep subsurface of South Africa. Nature 474, 79–82.

Bouché, M.B., 1983. The establishment of earthworm communities. In: Satchell, J.E. (Ed.), Earthworm Ecology: From Darwin to Vermiculture. Chapman and Hall, London, pp. 431–448.

Bragazza, L., Parisod, J., Buttler, A., Bardgett, R.D., 2013. Biogeochemical plant-soil microbe feedback in response to climate warming in peatlands. Nat. Climate Change 3, 273–277.

Brauman, A., Bignell, D.E., Tayasu, I., 2000. Soil-feeding termites: biology, microbial associations and digestive mechanisms. In: Abe, T., Bignell, D.E., Higashi, M. (Eds.), Termites: Evolution, Sociality, Symbioses, Ecology. Kluwer Academic, Dordrecht, pp. 233–260.

Breznak, J.A., 1984. Biochemical aspects of symbiosis between termites and their intestinal microbiota. In: Anderson, J.M., Rayner, A.D.M., Walton, D.W.H. (Eds.), Invertebrate-Microbial Interactions. Cambridge University Press, Cambridge, pp. 173–203.

Breznak, J., 2000. Ecology of prokaryotic microbes in the guts of wood- and litter-feeding termites. In: Abe, T., Bignell, D.E., Higashi, M. (Eds.), Termites: Evolution, Sociality, Symbioses, Ecology. Kluwer Academic, Dordrecht, pp. 209–231.

Brussaard, L., Aanen, D.K., Briones, M.J.I., Decaëns, T., De Deyn, G.B., Fayle, T.M., James, S.W., Nobre, T., 2012. Biogeography and phylogenetic community structure of soil invertebrate ecosystem engineers: global to local patterns, implications for ecosystem functioning and services and global environmental change impacts. In: Wall, D.H., Bardgett, R.D., Behan-Pelletier, V., Herrick, J.E., Jones, H., Ritz, K., Six, J., Strong, D.R., van der Putten, W.H. (Eds.), Soil Ecology and Ecosystem Services. Oxford University Press, Oxford, UK, pp. 201–232.

Buée, M., Reich, M., Murat, C., Morin, E., Nilsson, R.H., Uroz, S., Martin, F., 2009. 454 Pyrose-quencing analyses of forest soils reveal an unexpectedly high fungal diversity. New Phytol. 184, 449–456.

Byrd, D.W.J., Barker, K.R., Ferris, H., Nusbaum, C.J., Griffin, W.E., Small, R.H., Stone, C.A., 1976. Two semi-automatic elutriators for extracting nematodes and certain fungi from soil. J. Nematol. 8, 206–212.

Calderón-Cortés, N., Quesada, M., Watanabe, H., Cano-Camacho, H., Oyama, K., 2012. Endogenous plant cell wall digestion: a key mechanism in insect evolution. Annu. Rev. Ecol. Evol. Syst. 43, 45–71.

Carpenter, D., Hodson, M.E., Eggleton, P., Kirk, C., 2008. The role of earthworm communities in soil mineral weathering: a field experiment. Mineral. Mag. 72, 33–36.

Chamberlain, P.M., Bull, I.D., Black, H.I.J., Ineson, P., Evershed, R.P., 2006. Collembolan trophic preferences determined using fatty acid distributions and compound-specific stable carbon isotope values. Soil Biol. Biochem. 38, 1275–1281.

Cheeke, T., Coleman, D.C., Wall, D.H., 2012. Microbial Ecology in Sustainable Agroecosystems. CRC Press, Boca Raton.

Clarholm, M., 1981. Protozoan grazing of bacteria in soil—impact and importance. Microb. Ecol. 7, 343–350.

Clarholm, M., 1985. Possible roles for roots, bacteria, protozoa and fungi in supplying nitrogen to plants. In: Fitter, A.H., Atkinson, D., Read, D.J., Usher, M.B. (Eds.), Ecological Interactions in Soil: Plants, Microbes and Animals. Blackwell, Oxford, pp. 355–365.

Clarholm, M., 1994. The microbial loop in soil. In: Ritz, K., Dighton, J., Giller, K.E. (Eds.), Beyond the Biomass. Wiley/Sayce, Chichester, pp. 221–230.

Cock, M.J., Biesmeijer, J.C., Cannon, R.J., Gerard, P.J., Gillespie, D., Jiménez, J.J., Lavelle, P.M., Raina, S.K., 2012. The positive contribution of invertebrates to sustainable agriculture and food security. CAB Reviews Perspect. Agriculture, Vet. Sci. Nutr. Nat. Res. 7, 1–27.

Coleman, D.C., 1994. The microbial loop concept as used in terrestrial soil ecology studies. Microb. Ecol. 28, 245–250.

Coleman, D.C., 2008. From peds to paradoxes: linkages between soil biota and their influences on ecological processes. Soil Biol. Biochem. 40, 271–289.

Coleman, D.C., Whitman, W.B., 2005. Linking species richness, biodiversity and ecosystem function in soil systems. Pedobiologia 49, 479–497.

Coleman, D.C., Hendrix, P.F., Beare, M.H., Cheng, W., Crossley Jr., D.A., 1993. Microbial and faunal dynamics as they affect soil organic matter dynamics in subtropical agroecosystems. In: Paoletti, M.G., Foissner, W., Coleman, D.C. (Eds.), Soil Biota and Nutrient Cycling Farming Systems. CRC Press, Boca Raton, pp. 1–14.

Coleman, D.C., Blair, J.M., Elliott, E.T., Wall, D.H., 1999. Soil invertebrates. In: Robertson, G.P., Coleman, D.C., Bledsoe, C.S., Sollins, P. (Eds.), Standard Soil Methods for Long-Term Ecological Research. Oxford University Press, New York, pp. 349–377.

Coleman, D.C., Crossley Jr., D.A., Hendrix, P.F., 2004. Fundamentals of Soil Ecology. Elsevier, San Diego.

Coleman, D.C., Vadakattu, G., Moore, J.C., 2012. Soil ecology and agroecosystem studies: a dynamic world. In: Cheeke, T., Coleman, D.C., Wall, D.H. (Eds.), Microbial Ecology in Sustainable Agroecosystems. CRC Press, Boca Raton, pp. 1–21.

Coûteaux, M.M., 1972. Distribution des thécamoebiens de la litiére et de l'humus de deux sols forestier d'humus brut. Pedobiologia 12, 237–243.

Crotty, F.V., Adl, S.M., Blackshaw, R.P., Murray, P.J., 2012a. Using stable isotopes to differentiate trophic feeding channels within soil food webs. J. Eukaryot. Microbiol. 59, 520–526.

Crotty, F.V., Adl, S.M., Blackshaw, R.P., Murray, P.J., 2012b. Protozoan pulses unveil their pivotal position within the soil food web. Microb. Ecol. 63, 905–918.

Curry, J.P., Byrne, D., Boyle, K.E., 1995. The earthworm population of a winter cereal field and its effects on soil and nitrogen turnover. Biol. Fertil. Soils 19, 166–172.

Darbyshire, J.F., 1994. Soil Protozoa. CAB International, Wallingford.

Darwin, C., 1881. The Formation of Vegetable Mould, Through the Action of Worms. John Murray, London.

Dash, M.C., 1990. Oligochaeta: Enchytraeidae. In: Dindal, D.L. (Ed.), Soil Biology Guide. Wiley, New York, pp. 311–340.

Davidson, D.A., Bruneau, P.M.C., Grieve, I.C., Young, I.M., 2002. Impacts of fauna on an upland grassland soil as determined by micromorphological analysis. Appl. Soil Ecol. 20, 133–143.

Davis, E.L., Hussey, R.S., Baum, T.J., Bakker, J., Schots, A., Rosso, M., Abad, P., 2000. Nematode parasitism genes. Annu. Rev. Phytopathol. 38, 365–396.

Davis, E.L., Hussey, R.S., Baum, T.J., 2004. Getting to the roots of parasitism. Trends Parasitol. 20, 134–141.

Decaëns, T., Jimenez, J.J., Gioia, C., Measey, G.J., Lavelle, P., 2006. The values of soil animals for conservation biology. Eur. J. Soil Biol. 42, S23–S38.

Decaëns, T., 2010. Macroecological patterns in soil communities. Glob. Ecol. Biogeogr. 19, 287–302.

Demeure, Y., Freckman, D.W., Gundy, S.D.V., 1979. Anhydrobiotic coiling of nematodes in soil. Nematology 11, 189–195.

de Vries, F.T., Liiri, M.E., Bjørnlund, L., Bowker, M.A., Christensen, S., Setälä, H.M., Bardgett, R.-D., 2012. Land use alters the resistance and resilience of soil food webs to drought. Nat. Climate Change 2, 276–280.

Didden, W.A.M., 1990. Involvement of Enchytraeidae (Oligochaeta) in soil structure evolution in agricultural fields. Biol. Fertil. Soils 9, 152–158.

Didden, W.A.M., 1993. Ecology of Enchytraeidae. Pedobiologia 37, 2–29.

Didden, W.A.M., 1995. The effect of nitrogen deposition on enchytraeid-mediated decomposition and mobilization—a laboratory experiment. Acta Zool. Fenn. 196, 60–64.

Dillman, A.R., Guillermin, M.L., Lee, J.H., Kim, B., Sternberg, P.W., Hallem, E.A., 2012. Olfaction shapes host-parasite interactions in parasitic nematodes. Proc. Natl. Acad. Sci. U. S. A. 109, E2324–E2333.

Dindal, D.L., 1990. Soil Biology Guide. Wiley, New York.

Domene, X., Chelinho, S., Campana, P., Natal-da-Luz, T., Alcañiz, J.M., Andrés, P., Römbke, J., Sousa, J.P., 2011. Influence of soil properties on the performance of Folsomia candida: implications for its use in soil ecotoxicology testing. Environ. Toxicol. Chem. 30, 1497–1505.

Donner, J., 1966. Rotifers. Warne, London.

Dósza-Farkas, K., 1996. Reproduction strategies in some enchytraeid species. In: Dósza-Farkas, K. (Ed.), Newsletter on Enchytraeidae. Eötvös Lorand University, Budapest, pp. 25–33.

Edgecombe, G.D., Giribet, G., Dunn, C.W., Hejnol, A., Kristensen, R.M., Neves, R.C., Rouse, G.W., Worsaae, K., Sorensen, M.V., 2011. Higher-level metazoan relationships: recent progress and remaining questions. Organ. Divers. Evol. 11, 151–172.

Edwards, C.A., 1991. The assessment of populations of soil-inhabiting invertebrates. Agric. Ecosyst. Environ. 34, 145–176.

Edwards, C.A., 1998. Earthworm Ecology. St. Lucie Press, Boca Raton.

Edwards, C.A., Bohlen, P.J., 1996. Earthworm Biology and Ecology. Chapman and Hall, London.

Edwards, W.M., Shipitalo, M.J., 1998. Consequences of earthworms in agricultural soils: aggregation and porosity. In: Edwards, C.A. (Ed.), Earthworm Ecology. CRC Press, Boca Raton, pp. 147–161.

Elliott, E.T., Coleman, D.C., 1988. Let the soil work for us. Ecol. Bull. 39, 23–32.

Elliott, E.T., Anderson, R.V., Coleman, D.C., Cole, C.V., 1980. Habitable pore-space and microbial trophic interactions. Oikos 35, 327–335.

Ferris, H., Bongers, T., 2006. Nematode indicators of organic enrichment. J. Nemat. 38, 3–12.

Ferris, H., Bongers, T., de Goede, R.G.M., 2001. A framework for soil food web diagnostics: extension of the nematode faunal analysis concept. Appl. Soil Ecol. 18, 13–29.

Fierer, N., Lennon, J.T., 2011. The generation and maintenance of diversity in microbial communities. Am. J. Bot. 98, 439–448.

Fierer, N., Leff, J.W., Adams, B.J., Nielsen, U.N., Bates, S.T., Lauber, C.L., Owens, S., Gilbert, J.A., Wall, D.H., Caporaso, J.G., 2012. Cross-biome metagenomic analyses of soil microbial communities and their functional attributes. Proc. Natl. Acad. Sci. U. S. A. 109, 21390–21395.

Foissner, W., 1987. Soil protozoa: fundamental problems, ecological significance, adaptations in ciliates and testaceans, bioindicators, and guide to the literature. Prog. Protozool. 2, 69–212.

Foster, R.C., Dormaar, J.F., 1991. Bacteria-grazing amebas in situ in the rhizosphere. Biol. Fertil. Soils 11, 83–87.

Freckman, D.W., Virginia, R.A., 1989. Plant-feeding nematodes in deep-rooting desert ecosystems. Ecology 70, 1665–1678.

Freckman, D.W., Kaplan, D.T., van Gundy, S.D., 1977. A comparison of techniques for extraction and study of anhydrobiotic nematodes from dry soils. J. Nematol. 9, 176–181.

García-Palacios, P., Maestre, F.T., Kattge, J., Wall, D.H., 2013. Climate and litter quality differently modulate the effects of soil fauna on litter decomposition across biomes. Ecol. Lett. 16, 1045–1053.

Gaugler, R. (Ed.), 2002. Entomopathogenic Nematology. CAB International, Wallingford.

Gilbert, H.J., 2010. The biochemistry and structural biology of plant cell wall decomposition. Plant Physiol. 153, 444–455.

Griffiths, B.S., Caul, S., 1993. Migration of bacterial-feeding nematodes, but not protozoa, to decomposing grass residues. Biol. Fertil. Soils 15, 201–207.

Gupta, V.V.S.R., Germida, J.J., 1989. Influence of bacterial-amoebal interactions on sulfur transformations in soil. Soil Biol. Biochem. 21, 921–930.

Gupta, V.V.S.R., Yeates, G.W., 1997. Soil microfauna as bioindicators of soil health. In: Pankhurst, C., Doube, B.M., Gupta, V.V.S.R. (Eds.), Biological Indicators of Soil Health. CAB International, Wallingford, pp. 201–233.

Haegeman, A., Vanholme, B., Jacob, J., Vandekerckhove, T.T.M., Claeys, M., Borgonie, G., Gheysen, G., 2009. An endosymbiotic bacterium in a plant-parasitic nematode: member of a new *Wolbachia* supergroup. Int. J. Parasitol. 39, 1045–1054.

Harvey, R.W., Kinner, N.E., Bunn, A., MacDonald, D., Metge, D., 1995. Transport behavior of groundwater protozoa and protozoan-shaped micro-spheres in sandy aquifer sediments. Appl. Environ. Microbiol. 61, 209–271.

Hättenschwiler, S., Tiunov, A.V., Scheu, S., 2005. Biodiversity and litter decomposition in terrestrial ecosystems. Annu. Rev. Ecol. Evol. Syst. 36, 191–218.

Heemsbergen, D., Berg, M.P., van Hall, J., Faber, J.H., Verhoef, H.A., 2004. Biodiversity effects on soil processes explained by inter-specific functional trait dissimilarity. Science 306, 1019–1020.

Heidemann, K., Scheu, S., Ruess, L., Maraun, M., 2011. Molecular detection of nematode predation and scavenging in oribatid mites: laboratory and field experiments. Soil Biol. Biochem. 43, 2229–2236.

Hendrix, P.F., 1995. Earthworm Ecology and Biogeography in North America. CRC Press, Boca Raton.

Hendrix, P.F., Parmelee, R.W., Crossley Jr., D.A., Coleman, D.C., Odum, E.P., Groffman, P., 1986. Detritus food webs in conventional and no-tillage agroecosystems. Bioscience 36, 374–380.

Hölldobler, B., Wilson, E.O., 1990. The Ants. Belknap Press, Cambridge, MA.

Hominick, W.M., 2002. Biogeography. In: Gaugler, R. (Ed.), Entomopathogenic Nematology. CAB International, Wallingford, pp. 115–143.

Hongoh, Y., Ohkuma, M., Kudo, T., 2003. Molecular analysis of bacterial microbiota in the gut of the termite *Reticulitermes speratus* (Isoptera; Rhinotermitidae). FEMS Microbiol. Ecol. 44, 231–242.

Hunt, H.W., Coleman, D.C., Ingham, E.R., Ingham, R.E., Elliott, E.T., Moore, J.C., Rose, S.L., Reid, C.P.P., Morley, C.R., 1987. The detrital food web in a shortgrass prairie. Biol. Fertil. Soils 3, 57–68.

Ingham, R.E., Trofymow, J.A., Ingham, E.R., Coleman, D.C., 1985. Interactions of bacteria, fungi, and their nematode grazers—effects on nutrient cycling and plant-growth. Ecolog. Monogr. 55, 119–140.

Jeffery, S., Gardi, C., Jones, A., Montanarella, L., Marmo, L., Miko, L., Ritz, K., Peres, G., Römbke, J., van der Putten, W.H., 2010. European Atlas of Soil Biodiversity. European Commission, Publications Office of the European Union, Luxembourg.

Jones, C.G., Lawton, J.H., Shachak, M., 1994. Organisms as ecosystem engineers. Oikos 69, 373–386.

Karim, N., Jones, J.T., Okada, H., Kikuchi, T., 2009. Analysis of expressed sequence tags and identification of genes encoding cell-wall-degrading enzymes from the fungivorous nematode *Aphelenchus avenae*. BMC Genomics 10, 525.

Knapp, B.A., Seeber, J., Podmirseg, S.M., Rief, A., Meyer, E., Insam, H., 2009. Molecular fingerprinting analysis of the gut microbiota of *Cylindroiulus fulviceps* (Diplopoda). Pedobiologia 52, 325–336.

Koch, A., McBratney, A., Adams, M., Field, D., Hill, R., Lal, R., Abbott, L., Angers, D., Baldock, J., Barbier, E., Binkley, D., Bird, M., Bouma, J., Chenu, C., Crawford, J., Butler Flora, C., Goulding, K., Gunwald, S., Hempel, J., Jastrow, J., Lehmann, J., Lorenz, K., Minasny, B., Morgan, C., O'Donnell, A., Parton, W., Rice, C., Wall, D., Whitehead, D., Young, I., Zimmermann, M., 2013. Soil Security: Solving the Global Soil Crisis. Global Policy 4, 434–441.

Kuikman, P.J., Van Veen, J.A., 1989. The impact of protozoa on the availability of bacterial nitrogen to plants. Biol. Fertil. Soils 8, 13–18.

Lavelle, P., 1983. The structure of earthworm communities. In: Satchell, J.E. (Ed.), Earthworm Ecology: From Darwin to Vermiculture. Chapman and Hall, London, pp. 449–466.

Lavelle, P., 2012. Soil as a habitat. In: Wall, D.H., Bardgett, R.D., Behan-Pelletier, V., Herrick, J.E., Jones, H., Ritz, K., Six, J., Strong, D.R., van der Putten, W.H. (Eds.), Soil Ecology and Ecosystem Services. Oxford University Press, Oxford, pp. 28–44.

Lavelle, P., Spain, A.V., 2001. Soil Ecology. Kluwer Academic, Dordrecht.

Lavelle, P., Blanchart, E., Martin, A., Spain, A.V., Martin, S., 1992. Impact of soil fauna on the properties of soils in the humid tropics. In: Myths and Science of Soils of the Tropics. Soil Science Society of America, Madison, pp. 157–185.

Lavelle, P., Lattaud, C., Trigo, D., Barois, I., 1995. Mutualism and biodiversity in soils. Plant Soil 170, 23–33.

Lavelle, P., Pashanasi, B., Charpentier, F., Gilot, C., Rossi, J., Derouard, L., Andre, J., Ponge, J., Bernier, N., 1998. Influence of earthworms on soil organic matter dynamics, nutrient dynamics and microbiological ecology. In: Edwards, C.A. (Ed.), Earthworm Ecology. CRC Press, Boca Raton, pp. 103–122.

Lee, K.E., 1985. Earthworms: Their Ecology and Relationships with Soils and Land Use. Academic Press, Sydney.

Lee, K.E., Wood, T.G., 1971. Termites and Soils. Academic Press, London.

Lloyd, M., Dybas, H.S., 1966. The periodical cicada problem I. Population ecology. Evolution 20, 133–149.

Lousier, J.D., Parkinson, J., 1984. Annual population dynamics and production ecology of Testacea (Protozoa, Rhizopoda) in an aspen woodland soil. Soil Biol. Biochem. 16, 103–114.

Lousier, J.D., Bamforth, S.S., 1990. Soil Protozoa. In: Dindal, D.L. (Ed.), Soil Biology Guide. Wiley, New York, pp. 97–136.

Makkonen, M., Berg, M.P., Handa, I.T., Hättenschwiler, S., van Ruijven, J., van Bodegom, P.M., Aerts, R., 2012. Highly consistent effects of plant litter identity and functional traits on decomposition across a latitudinal gradient. Ecol. Lett. 15, 1033–1041.

Mankau, R., Mankau, S.K., 1963. The role of mycophagous nematodes in the soil. I. The relationships of Aphelenchus avenae to phytopathogenic soil fungi. In: Doeksen, J., Drift, J.V. (Eds.), Soil Organisms. North-Holland, Amsterdam, pp. 271–280.

McSorley, R., Frederick, J.J., 2004. Effects of extraction method on perceived composition of the soil nematode community. Appl. Soil Ecol. 27, 55–63.

Meier, F.A., Honegger, R., 2002. Faecal pellets of lichenivorous mites contain viable cells of the lichen-forming ascomycete Xanthoria parietina and its green algal photobiont, Tebouxia arboricola. Biol. J. Linn. Soc. 76, 259–268.

Moore, J.C., de Ruiter, P.C., 2012. Soil food webs in agricultural ecosystems. In: Cheeke, T., Coleman, D.C., Wall, D.H. (Eds.), Microbial Ecology of Sustainable Agroecosystems. CRC Press, Boca Raton, pp. 63–88.

Moreira, F.M.S., Huising, E.J., Bignell, D.E., 2009. A Handbook of Tropical Soil Biology: Sampling and Characterization of Belowground Biodiversity. Earthscan, London, UK.

Ngugi, D.K., Brune, A., 2012. Nitrate reduction, nitrous oxide formation, and anaerobic ammonia oxidation to nitrite in the gut of soil-feeding termites (Cubitermes and Ophiotermes spp.). Environ. Microbiol. 14, 860–871.

Nkem, J.N., Wall, D.H., Virginia, R.A., Barrett, J.E., Broos, E.J., Porazinska, D.L., Adams, B.J., 2006. Wind dispersal of soil invertebrates in the McMurdo Dry valleys, Antarctica. Polar Biol. 29, 346–352.

O'Brien, S.L., Jastrow, J.D., 2013. Physical and chemical protection in hierarchical soil aggregates regulates soil carbon and nitrogen recovery in restored perennial grasslands. Soil Biol. Biochem. 61, 1–13.

Orwin, K.H., Kirschbaum, M.U.F., St John, M.G., Dickie, I.A., 2011. Organic nutrient uptake by mycorrhizal fungi enhances ecosystem carbon storage: a model-based assessment. Ecol. Lett. 14, 493–502.

Pace, N.R., 2009. Mapping the tree of life: progress and prospects. Microbiol. Mol. Biol. Rev. 73, 565–576.

Parmelee, R.W., Bohlen, P.J., Blair, J.M., 1998. Earthworms and nutrient cycling processes: integrating across the ecological hierarchy. In: Edwards, C.A. (Ed.), Earthworm Ecology. St. Lucie Press, Boca Raton, pp. 123–143.

Pawlowski, J., 2013. The new micro-kingdoms of eukaryotes. BMC Biol. 11, 40.

Payne, J.A., 1965. A summer carrion study of the baby pig, *Sus scrofa* Linnaeus. Ecology 46, 592–602.

Pearce, M.J., 1997. Termites: Biology and Pest Management. CAB International, Wallingford.

Perdomo, G., Evans, A., Maraun, M., Sunnucks, P., Thompson, R., 2012. Mouthpart morphology and trophic position of microarthropods from soils and mosses are strongly correlated. Soil Biol. Biochem. 53, 56–63.

Poinar Jr., G.O., 1983. The Natural History of Nematodes. Prentice Hall International, Englewood Cliffs.

Pollierer, M.M., Langel, R., Scheu, S., Maraun, M., 2009. Compartmentalization of the soil animal food web as indicated by dual analysis of stable isotope ratios (N-15/N-14 and C-13/C-12). Soil Biol. Biochem. 41, 1221–1226.

Pomeroy, L.R., 1974. The ocean's food web: a changing paradigm. Bioscience 24, 499–504.

Ponge, J.F., 1991. Food resources and diets of soil animals in a small area of Scots pine litter. Geoderma 49, 33–62.

Porazinska, D.L., Sung, W., Giblin-Davis, R.M., Thomas, W.K., 2010. Reproducibility of read numbers in high-throughput sequencing analysis of nematode community composition and structure. Mol. Ecol. Resour. 10, 666–676.

Porazinska, D.L., Creer, S., Caporaso, J.G., Knight, R., Thomas, W.K., 2012. Sequencing our way towards understanding global eukaryotic biodiversity. Trends Ecol. Evol. 27, 234–244.

Powers, J.S., Montgomery, R.A., Adair, E.C., Brearley, F.Q., DeWalt, S.J., Castanho, C.T., Chave, J., Deinert, E., Ganzhorn, J.U., Gilbert, M.E., González-Iturbe, J.A., Bunyavejchewin, S., Grau, H.R., Harms, K.E., Hiremath, A., Iriarte-Vivar, S., Manzane, E., De Oliveira, A.A., Poorter, L., Ramanamanjato, J.-B., Salk, C., Varela, A., Weiblen, G.D., Lerdau, M.T., 2009. Decomposition in tropical forests: a pan-tropical study of the effects of litter type, litter placement and mesofaunal exclusion across a precipitation gradient. J. Ecol. 97, 801–811.

Robeson, M.S., Costello, E.K., Freeman, K.R., Whiting, J., Adams, B., Martin, A.P., Schmidt, S.K., 2009. Environmental DNA sequencing primers for eutardigrades and bdelloid rotifers. BMC Ecol. 9, 25.

Robeson, M.S., King, A.J., Freeman, K.R., Birky, C.W., Martin, A.P., Schmidt, S.K., 2011. Soil rotifer communities are extremely diverse globally but spatially autocorrelated locally. Proc. Natl. Acad. Sci. U. S. A. 108, 4406–4410.

Ruess, L., Häggblom, M.M., Langel, R., Scheu, S., 2004. Nitrogen isotope ratios and fatty acid composition as indicators of animal diets in belowground systems. Oecologia 139, 336–346.

Rusek, J., 1985. Soil microstructures—contributions on specific soil organisms. Quaestiones Entomologicae 21, 497–514.

Scharroba, A., Dibbern, D., Hünninghaus, M., Kramer, S., Moll, J., Butenschoen, O., Bonkowski, M., Buscot, F., Kandeler, E., Koller, R., Krüge, D., Lueders, T., Scheu, S., Ruess, L., 2012. Effects of resource availability and quality on the structure of the micro-food web of an arable soil across depth. Soil Biol. Biochem. 50, 1–11.

Scheu, S., Setälä, H., 2002. Multitrophic interactions in decomposer food-webs. In: Tscharntke, B., Hawkins, B.A. (Eds.), Multitrophic Level Interactions. Cambridge University Press, Cambridge, pp. 223–264.

Schmelz, R.M., Collado, R., 2010. A guide to European terrestrial and freshwater species of Enchytraeidae (Oligochaeta). Soil Organ. 82, 1–176.

Schneider, K., Migge, S., Norton, R.A., Scheu, S., Langel, R., Reineking, A., Maraun, M., 2004. Trophic niche differentiation in soil microarthropods (Oribatida, Acari): evidence from stable isotope ratios (15N/14N). Soil Biol. Biochem. 36, 1769–1774.

Schouten, A.J., Arp, K.K.M., 1991. A comparative study on the efficiency of extraction methods for nematodes from different forest litters. Pedobiologia 35, 393–400.

Seastedt, T.R., 1984. The role of microarthropods in decomposition and mineralization processes. Annu. Rev. Entomol. 29, 25–46.

Segers, H., 2007. Annotated checklist of the rotifers (Phylum Rotifera), with notes on nomenclature, taxonomy and distribution. Zootaxa 1564, 1–104.

Shelley, R.M., 2002. A Synopsis of the North American centipedes of the order Scolopendromorpha (Chilopoda). In: Memoir 5. Virginia Museum of Natural History, Martinsville.

Sinclair, J.L., Ghiorse, W.C., 1989. Distribution of aerobic bacteria, protozoa, algae and fungi in deep subsurface sediments. Geophys. J. Roy. Astron. Soc. 7, 15–31.

Singh, B.N., 1946. A method of estimating the numbers of soil protozoa, especially amoebae, based on their differential feeding on bacteria. Ann. Appl. Biol. 33, 112–119.

Sommer, R.J., Streit, A., 2011. Comparative genetics and genomics of nematodes: genome structure, development, and lifestyle. Annu. Rev. Genet. 45, 1–20.

Southwood, T.R.E., 1978. Ecological Methods with Particular Reference to the Study of Insect Populations. Chapman, London.

Stout, J.D., 1963. The terrestrial plankton. Tuatara 11, 57–65.

Swift, M.J., Heal, O.W., Anderson, J.M., 1979. Decomposition in Terrestrial Ecosystems. University of California Press, Berkeley.

Thompson, R.M., Brose, U., Dunne, J.A., Hall Jr., R.O., Hladyz, S., Kitching, R.L., Martinez, N.D., Rantala, H., Romanuk, T.N., Stouffer, D.B., Tylianakis, J.M., 2012. Food webs reconciling the structure and function of biodiversity. Trends Ecol. Evolut. 27, 689–697.

Tomlin, A.D., Shipitalo, M.J., Edwards, W.M., Protz, R., 1995. Earthworms and their influence on soil structure and infiltration. In: Hendrix, P.F. (Ed.), Earthworm Ecology and Biogeography in North America. CRC Press, Boca Raton, pp. 159–183.

Treonis, A.M., Wall, D.H., Virginia, R.A., 1999. Invertebrate biodiversity in Antarctic dry valley soils and sediments. Ecosystems 2, 482–492.

van der Putten, W.H., Bardgett, R.D., Bever, J.D., Bezemer, T.M., Casper, B.B., Fukami, T., Kardol, P., Klironomos, J.N., Kulmatiski, A., Schweitzer, J.A., Suding, K.N., Van de Voorde, T.F.J., Wardle, D.A., 2013. Plant-soil feedbacks: the past, the present and future challenges. J. Ecol. 101, 265–276.

van Vliet, P.C.J., 2000. Enchytraeids. In: Sumner, M. (Ed.), Handbook of Soil Science. CRC Press, Boca Raton, pp. C70–C77.

van Vliet, P.C.J., Hendrix, P.F., 2003. Role of fauna in soil physical processes. In: Abbott, L.K., Murphy, D.V. (Eds.), Soil Biological Fertility—A Key to Sustainable Land Use in Agriculture. Kluwer Academic, Dordrecht, pp. 61–80.

van Vliet, P.C.J., Beare, M.H., Coleman, D.C., 1995. Population dynamics and functional roles of Enchytraeidae (Oligochaeta) in hardwood forest and agricultural systems. Plant Soil 170, 199–207.

Venette, R.C., Ferris, H., 1998. Influence of bacterial type and density on population growth of bacterial-feeding nematodes. Soil Biol. Biochem. 30, 949–960.

Vervoort, M.T., Vonk, J.A., Mooijman, P.J., Van den Elsen, S.J., Van Megen, H.H., Veenhuizen, P., Landeweert, R., Bakker, J., Mulder, C., Helder, J., 2012. SSU ribosomal DNA-based monitoring of nematode assemblages reveals distinct seasonal fluctuations within evolutionary heterogeneous feeding guilds. PLoS ONE 7, e47555.

Wall, D.H., 2004. Sustaining Biodiversity and Ecosystem Services in Soil and Sediments. Island Press, Washington, DC.

Wall, D.H., Snelgrove, V.R., Covich, A.P., 2001. Conservation priorities for soil and sediment invertebrates. In: Soule, M.E., Orians, G.H. (Eds.), Conservation Biology: Research Priorities for the Next Decade. Island Press, Washington, DC, pp. 99–123.

Wall, D.H., Fitter, A., Paul, E., 2005. Developing new perspectives from advances in soil biodiversity research. In: Bardgett, R.D., Usher, M.B., Hopkins, D.W. (Eds.), Biological Diversity and Function in Soils. Cambridge University Press, Cambridge, pp. 3–30.

Wall, D.H., Bradford, M.A., St John, M.G., Trofymow, J.A., Behan-Pelletier, V., Bignell, D.D.E., Dangerfield, J.M., Parton, W.J., Rusek, J., Voigt, W., Wolters, V., Gardel, H.Z., Ayuke, F.O., Bashford, R., Beljakova, O.I., Bohlen, P.J., Brauman, A., Flemming, S., Henschel, J.R., Johnson, D.L., Jones, T.H., Kovarova, M., Kranabetter, J.M., Kutny, L., Lin, K.C., Maryati, M., Masse, D., Pokarzhevskii, A., Rahman, H., Sabara, M.G., Salamon, J.A., Swift, M.-J., Varela, A., Vasconcelos, H.L., White, D., Zou, X.M., 2008. Global decomposition experiment shows soil animal impacts on decomposition are climate-dependent. Glob. Chang. Biol. 14, 2661–2677.

Wall, D.H., Bardgett, R.D., Behan-Pelletier, V., Herrick, J.E., Jones, H., Ritz, K., Six, J., Strong, D.-R., van der Putten, W.H., 2012. Soil Ecology and Ecosystem Services. Oxford University Press, Oxford, 406.

Wallace, H.R., 1959. The movement of eelworms in water films. Ann. Appl. Biol. 47, 366–370.

Wallwork, J.A., 1970. Ecology of Soil Animals. McGraw-Hill, London.

Wallwork, J.A., 1976. The Distribution and Diversity of Soil Fauna. Academic Press, London.

Walter, D.E., Kaplan, D.T., Permar, T.A., 1991. Missing links: a review of methods used to estimate trophic links in soil food webs. Agric. Ecosyst. Environ. 34, 399–405.

Wardle, D.A., 2002. Communities and Ecosystems: Linking the Aboveground and Belowground Components. Princeton University Press, Princeton.

Wardle, D.A., Bardgett, R.D., Klironomos, J.N., Setala, H., van der Putten, W.H., Wall, D.H., 2004. Ecological linkages between aboveground and belowground biota. Science 304, 1629–1633.

Wheeler, G.C., Wheeler, J., 1990. Insecta: Hymenoptera Formicidae. In: Dindal, D.L. (Ed.), Soil Biology Guide. Wiley, New York, pp. 1277–1294.

Whitehead, A.G., Hemming, J.R., 1965. A comparison of some quantitative methods of extracting small vermiform nematodes from soil. Ann. Appl. Biol. 55, 25–38.

Whitford, W.G., 2000. Keystone arthropods as webmasters in desert ecosystems. In: Coleman, D.C., Hendrix, P.F. (Eds.), Invertebrates as Webmasters in Ecosystems. CAB International, Wallingford, pp. 25–41.

Whitman, W.B., Coleman, D.C., Wiebe, W.J., 1998. Prokaryotes: the unseen majority. Proc. Natl. Acad. Sci. U. S. A. 95, 6578–6583.

Wilson, E.O., 1987. Causes of ecological success—the case of the ants—the 6th Tansley lecture. J. Anim. Ecol. 56, 1–9.

Wilson, E.O., 1992. The Diversity of Life. Norton, New York.

Wilson, M.J., Kakouli-Duarte, T., 2009. Nematodes as Environmental Indicators. CABI Press, United Kingdom p. 315.

Wolters, V., 1988. Effects of *Mesenchytraeus glandulosus* (Oligochaeta, Enchytraeidae) on decomposition processes. Pedobiologia 32, 387–398.

Wu, T., Ayres, E., Bardgett, R., Wall, D., Garey, J., 2011. Molecular study of worldwide distribution and diversity of soil animals. Proc. Natl. Acad. Sci. U. S. A. 108, 17720–17725.

Wurst, S., van der Putten, W.H., 2007. Root herbivore identity matters in plant-mediated interactions between root and shoot herbivores. Basic Appl. Ecol. 8, 491–499.

Wurst, S., De Deyn, G.B., Orwin, K., 2012. Soil biodiversity and functions. In: Wall, D.H., Bardgett, R.D., Behan-Pelletier, V., Herrick, J.E., Jones, H., Ritz, K., Six, J., Strong, D.R., van der Putten, W.H. (Eds.), Soil Ecology and Ecosystem Services. Oxford University Press, Oxford, pp. 28–44.

Yeates, G.W., 1998. Feeding in free-living soil nematodes: a functional approach. In: Perry, R.N., Wright, D.J. (Eds.), The Physiology and Biochemistry of Free-Living and Plant-Parasitic Nematodes. CAB International, Wallingford, pp. 245–269.

Yeates, G.W., Coleman, D.C., 1982. Role of nematodes in decomposition. In: Freckman, D.W. (Ed.), Nematodes in Soil Ecosystems. University of Texas Press, Austin, pp. 55–80.

Yeates, G.W., Bongers, T., Degoede, R.G.M., Freckman, D.W., Georgieva, S.S., 1993. Feeding-habits in soil nematode families and genera—an outline for soil ecologists. J. Nematol. 25, 315–331.

Yeates, G.W., Dando, J.L., Shepherd, T.G., 2002. Pressure plate studies to determine how moisture affects access of bacterial-feeding nematodes to food in soil. Eur. J. Soil Sci. 53, 355–365.

Zachariae, G., 1964. Welche Bedeutung haben Enchyträus in Waldboden? In: Jongerius, A. (Ed.), Soil Micromorphology. Elsevier, Amsterdam, pp. 57–68.

Zwart, K.B., Darbyshire, J.F., 1991. Growth and nitrogenous excretion of a common soil flagellate, *Spumella* sp. J. Soil Sci. 43, 145–157.

FURTHER READING

Six, J., Bossuyt, H., Degryze, S., Denef, K., 2004. A history of research on the link between (micro) aggregates, soil biota, and soil organic matter dynamics. Soil Tillage Res. 79, 7–31.

Chapter 6

Molecular Approaches to Studying the Soil Biota

Janice E. Thies
Department of Crop and Soil Sciences, Cornell University, Ithaca, NY, USA

Chapter Contents

I INTRODUCTION

Rapid technological advances in high-throughput, nucleic acid sequencing, mass spectroscopy, microscopy, and microfluidics have enabled the fields of genomics, transcriptomics, and proteomics to begin to be applied more readily to soil ecological studies. They are swiftly changing our access to and understanding of the diversity, activity, and functions of soil biotic communities. Molecular approaches are being used with increasing frequency to open the so-called black box of microbial life in soil.

Soil Microbiology, Ecology, and Biochemistry. http://dx.doi.org/10.1016/B978-0-12-415955-6.00006-2

151

Molecular microbial ecology relies on extracting and characterizing nucleic acids and other cellular components, such as phospholipid fatty acids and proteins (Chapter 7), from soil organisms. It is now common to extract deoxyribonucleic acids (DNA) and ribonucleic acids (RNA) from cells contained within soil samples and analyze them directly in hybridization experiments, use them in polymerase chain reaction (PCR) amplification experiments, and sequence them in high throughput, next-generation sequencing reactions. Molecular approaches have allowed us to detect and begin to characterize the vast diversity of soil microbes previously unimagined. Direct microscopic counts of soil bacteria are typically one to two orders of magnitude higher than counts obtained by culturing. Molecular methods have allowed us access to the largely undescribed 90-99% of the soil biological community.

The aim of many molecular community analyses is to describe population diversity as described by taxon richness and evenness. However, virtually every step—from soil sampling to sequence analysis—may introduce biases (Fig. 6.1; Lombard et al., 2011), which must be taken into account when interpreting the results of subsequent analyses. Due to these potential biases, it is difficult, if not impossible, to assess the true abundance of different taxa using these approaches. These methods alone, although very powerful, cannot be used to assign function unambiguously to different taxa in complex soil communities. Hence, molecular methods should be used in concert with other approaches (termed a polyphasic or multiphasic approach) to achieve a more holistic understanding of the structure and function of soil biotic communities.

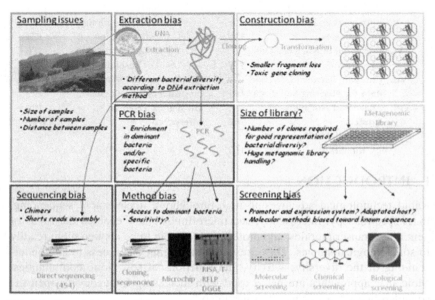

FIG. 6.1 Potential sources of bias associated with different steps in the molecular analysis of soil microbial communities. *Lombard et al. (2011).*

II TYPES AND STRUCTURES OF NUCLEIC ACIDS

The two types of nucleic acids present in all cells, DNA and RNA, are the target molecules for most molecular analyses. Watson and Crick in 1953 described a double helix of nucleotide bases that could "unzip" to make copies of itself. DNA was known to contain equimolar ratios of adenine (A) and thymine (T) and of cytosine (C) and guanine (G). They recognized that the adenine-thymine pair, held together by two hydrogen bonds, and the cytosine-guanine pair, held together by three hydrogen bonds, could fit together to form the rungs of a ladder of nucleotides. Molecular analyses of nucleic acids take advantage of these base-pairing rules (Fig. 6.2).

The DNA backbone is composed of deoxyribose (sugar), phosphates, and the associated purine (A and G) and pyrimidine (T and C) bases. The base-pairing specificity between the nucleotides leads directly to the faithful copying of both strands of the DNA double helix during replication and can be exploited to make copies of selected genes (or regions of the DNA molecule) *in vitro* by use of PCR. Base-pairing specificity in DNA allows for the transcription of genes coding for ribosomal RNA (rRNA) used to produce ribosomes that

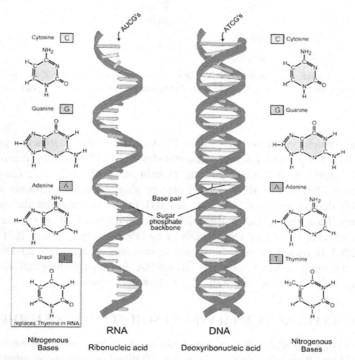

FIG. 6.2 The basic structure of DNA and RNA. DNA is double-stranded, whereas RNA is single-stranded. The nucleotide thymine in DNA is replaced with uracil in RNA. *Courtesy of the National Human Genome Research Institute.*

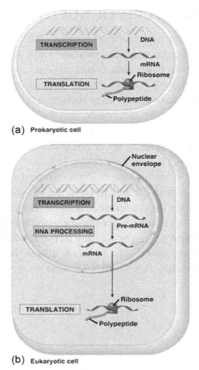

(a) Prokaryotic cell

(b) Eukaryotic cell

FIG. 6.3   The relationship between DNA gene sequence, transcription into mRNA, and translation into the coded protein sequence in (a) prokaryotes and (b) eukaryotes. *Campbell and Reece (2005).*

facilitate protein synthesis, messenger RNA (mRNA) that carries the genetic instructions for protein assembly, and transfer RNA (tRNA) used to transport and link amino acids together during protein assembly (Fig. 6.3; Campbell and Reece, 2005). The RNA transcripts are single stranded, contain ribose instead of deoxyribose, and substitute uracil (U) for thymine in the nucleic acid sequence. To analyze RNA extracted from soils, it must first be reverse transcribed to complementary DNA (cDNA) by reverse transcriptase (RT) PCR. The cDNA is then analyzed. Most molecular analyses target either DNA or rRNA, although recently, analysis of mRNA from environmental samples has become possible (see Section IX. D Metatranscriptomics).

III NUCLEIC ACID ANALYSES IN SOIL ECOLOGY STUDIES

The techniques available for analyzing nucleic acids fall into three basic categories: (1) analysis of nucleic acids *in situ*, (2) direct analysis or sequencing of nucleic acid extracts, or (3) analysis of PCR amplified segments of the DNA

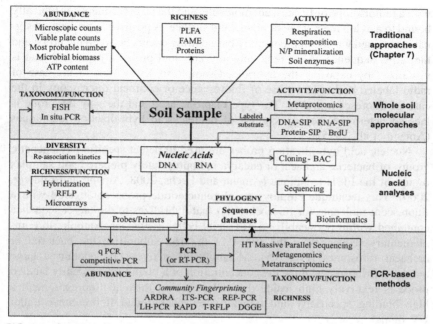

FIG. 6.4 Overview of traditional and molecular approaches used in soil ecology studies.

molecule. In PCR-based approaches, the primers chosen dictate the target for amplification, such as rRNA genes or genes that code for proteins with functions of ecological interest, such as those involved in nitrogen fixation (*nif*), ammonia (*amo*A) or methane (*pmo*A) oxidation, or denitrification (*nar*G, *nap*A, *nir*S, *nir*K, *nor*B, *nos*Z). The techniques commonly used for microbial analysis (Fig. 6.4) require that target molecules be separated from the soil matrix prior to analysis. A few techniques, such as fluorescent *in situ* hybridization (FISH), do not. Our abilities to interpret molecular analysis results and design more robust methods depend on the continuing development of nucleic acid and protein sequence databases and associated bioinformatics analytical tools. Targeted primers for use in PCR assays and probes for use in hybridization studies can be designed through comparative sequence analyses that have the required level of specificity for a given study.

IV DIRECT MOLECULAR ANALYSIS OF SOIL BIOTA

A Nucleic Acid Hybridization

Nucleic acid hybridization involves the bonding of a short, complementary nucleic acid strand (probe) to a target sequence. The probe is generally labeled with a radioisotope or fluorescent molecule, and the target sequence is typically bound to a nylon membrane or other solid surface. A positive hybridization signal is obtained when complementary base pairing occurs between the probe and the target sequence. After removing any unbound probe, a positive signal is visualized by exposing the hybridized sample to X-ray film in the case of radio-labeled probes or by use of fluorescence or confocal microscopy in the case of fluorescent probes. The type of probe used and the way the probe is labeled determine the applications of nucleic acid hybridization techniques (Amann et al., 1995).

Nucleic acid hybridization probes are used to detect specific phylogenetic groups of bacteria, archaea, or eucarya in appropriately prepared soil samples by use of the FISH technique (Amann and Fuchs, 2008; Amann and Ludwig, 2000). This technique employs an oligonucleotide probe conjugated with a fluorescent molecule (or fluorochrome) that hybridizes to the target sequence contained within permeabilized microbial cells *in situ*, for example, to complementary sequences in the rRNA of the 16S subunit of the bacterial or archaeal ribosomes (Fig. 6.5). Metabolically active cells contain a large number of ribosomes; thus, the concentration of a bound fluorescently labeled probe is relatively high inside the cells, causing them to fluoresce, with a high binding specificity and typically low background fluorescence under UV light.

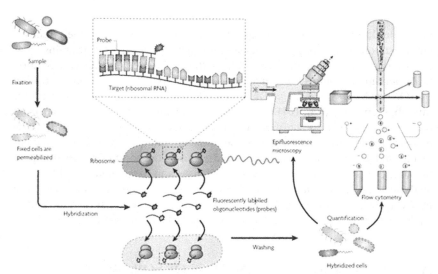

FIG. 6.5 Basic steps in fluorescence *in situ* hybridization analysis. The sample is first fixed to stabilize the cells and permeabilize the cell membranes. The labeled oligonucleotide probe is then added and allowed to hybridize to its intracellular targets before the excess probe is washed away. The sample is then ready for single-cell identification and quantification by either epifluorescence microscopy or flow cytometry. *Amann and Fuchs (2008).*

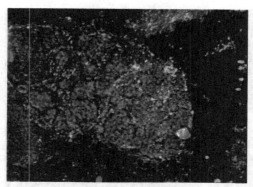

FIG. 6.6   Fluorescence *in situ* hybridization of bacteria (EUB 338 probe labeled with Oregon green) and archaea (ARCH 915 probe labeled with cy3) colonies isolated from a Cold Seep in the Black Sea. *Courtesy of Prof. Joachim Reitner, Univ. of Goettingen.*

For simultaneous counting of subpopulations in a given sample, probes can be designed that bind to specific sequences in rRNA that are found only in a particular group of organisms (archaea, bacteria, or eukarya) and used in conjunction with each other (Fig. 6.6). FISH can also be combined with microautoradiography to determine specific substrate-uptake profiles for individual cells within complex microbial communities (e.g., STARFISH, substrate-tracking autoradiographic fluorescent *in situ* hybridization; Lee et al., 1999; Ouverney and Fuhrman, 1999). The FISH approach is especially useful when used in conjunction with confocal laser scanning microscopy (CLSM), which allows for three-dimensional visualization of the relative positions of diverse populations within complex communities, such as biofilms (Fig. 6.7a) and the surfaces of soil aggregates (Binnerup et al., 2001). FISH can also be used to identify and quantify single cells by use of epifluorescence microscopy or flow cytometry (Fig. 6.5; Amann and Fuchs, 2008). The key advantage of FISH is the ability to visualize and identify organisms in their natural environment on a microscale. Such techniques have enormous potential for studying microbial interactions with plants and the ecology of target microbial populations in soil. However, the binding of fluorescent dyes to organic matter, resulting in nonspecific fluorescence, is a common problem in high SOM soils, such as peats, or other particles with high surface charge, such as black C. Available image analysis software may be "trained" to detect only those aspects of an image that meet specified criteria.

The probeBase website (http://www.microbial-ecology.net/probebase/) provides comprehensive lists and information about probes that are commonly used to address specific questions about the presence and location of selected organisms in soil communities. This includes lists of probes used specifically for FISH and DNA microarrays (see Section VIII. B; Loy et al., 2003).

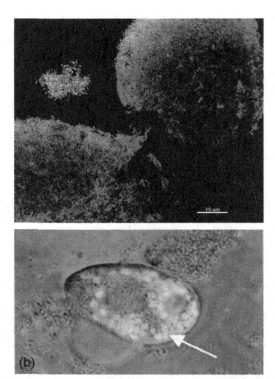

FIG. 6.7   (a) Confocal scanning laser microscopy image of *gfp* transformed *Psuedomonas aeruginosa* PA01 (green) embedded in its biofilm matrix (red). *(Courtesy of the Center for Biofilm Engineering, Montana State University; photo credit, Melis Penic.)* (b) The protozoan *Tetrahymena* sp. after ingesting *gfp*-tagged *Moraxelia* sp. G21 cells. *Errampali et al. (1999).*

B   Confocal Microscopy

Confocal laser scanning microscopy (CLSM), combined with *in situ* hybridization techniques, has been applied with considerable success to visualize the structure of soil microbial communities. The basic principle of CLSM is to create an image that is composed only of emitted fluorescence signals from a single plane of focus. This is done using a pinhole aperture that eliminates signals that may be coming from portions of the field that are out of focus. A series of these optical sections is scanned at specific depths, and then each section is "stacked" using imaging software. This gives rise to either a two-dimensional image that includes all planes of focus in the specimen or a computer-generated three-dimensional image. This gives unprecedented resolution in viewing environmental samples, enabling organisms to be better differentiated from particulate matter, as well as giving insight into the three-dimensional spatial relationships of microbial communities in their environment (Fig. 6.7a).

V BIOSENSORS AND MARKER GENE TECHNOLOGIES

Introduced marker genes, such as *lux*AB (luminescence), *lac*Z (β galactosidase), and *xyl*E (catechol 2, 3-dioxygenase) have been used successfully in soil microbial ecology studies. One such gene that has attracted a lot of attention in rhizosphere studies is *gfp*, which encodes the green fluorescent protein (GFP). This bioluminescent genetic marker can be used to identify, track, and count specific organisms into which the gene has been cloned and that have been reintroduced into the environment (Chalfie et al., 1994). The *gfp* gene was discovered in, and is derived from, the bioluminescent jelly fish, *Aequorea victoria* (Prasher et al., 1992). Once cloned into the organism of interest, GFP methods require no exogenous substrates, complex media, or expensive equipment to monitor and, hence, are favored for environmental applications (Errampalli et al., 1999). The GFP-marked cells can be identified using a standard fluorescence microscope fitted with excitation and emission filters of the appropriate wavelengths. There is no GFP activity in plants or the bacteria and fungi that interact with them, making *gfp* an excellent target gene that can be introduced into selected bacterial or fungal strains and used to study plant-microbe interactions (Errampalli et al., 1999). Basically, *gfp* is transformed into either the chromosome (for fungi) or a plasmid in a bacterial strain, where it is subsequently replicated. Various gene constructs have been made that differ in the type of promoters or terminators used and some contain repressor genes such as *lac*I for control of *gfp* expression. Once key populations in a sample are known and isolates obtained, they can be subsequently transformed with *gfp* or other genes producing detectable products to track them and assess their functions and interactions in soil and the rhizosphere. Red-shifted and yellow-shifted variants have been described. Development of *gfp* transformants that fluoresce at different excitation and emission wavelengths makes it possible to identify multiple bacterial populations simultaneously. The *gfp* gene has been introduced into *Sinorhizobium meliloti* (Alphaproteobacteria) and *Pseudomonas aeruginosa* (PA01, Gammaproteobacteria), among other common soil bacteria (Fig. 6.7b). Marked strains can be visualized in rhizobium infection threads, root nodules (Stuurman et al., 2000), colonized roots, biofilms, and even inside digestive vacuoles of protozoa (Fig. 6.7b). If the *gfp* genes are cloned along with specific promoters, such as the *mel*A (α galactosidase) promoter, then they can be used as biosensors to determine if the inducers, in this case galactosides, are present and at what relative concentration in the surrounding environment (Bringhurst et al., 2002). Marker gene approaches are restricted to use in organisms that can be cultured and transformed.

VI EXTRACTION OF NUCLEIC ACIDS (DNA/RNA)

Both whole community nucleic acid analyses and those based on PCR amplification of target genes require that the nucleic acids contained in a soil sample be separated from the solid phase. A variety of methods to extract nucleic acids from soils of varying texture have been developed (Bruns and Buckley, 2002). There are two main approaches: (1) cell fractionation and (2) direct lysis. In cell fractionation, intact microbial cells are extracted from the soil matrix. After extraction, the cells are chemically lysed, and the DNA is separated from the cell wall debris and other cell contents by a series of precipitation, binding, and elution steps. In the direct lysis methods, microbial cells are lysed directly in the soil or sediment and then the nucleic acids are separated from the soil matrix by similar means as described earlier. The main concerns when choosing a suitable protocol are extraction efficiency, obtaining a sample that is representative of the resident community, and obtaining an extract free of contaminants that interfere with subsequent analyses, such as PCR or hybridization with nucleic acid probes (Lombard et al., 2011).

Extraction efficiency is a key concern for obtaining a nucleic acid extract that is representative of the soil community. Gram-positive (Gr+) organisms, resting spores, fungal hyphae, and cysts of protozoa and nematodes are more difficult to lyse than cells of Gr⁻ organisms or vegetative stages of various soil fauna. Hence, unless lysis procedures are robust, a nucleic acid extract from soil may be biased toward organisms that are more readily ruptured or dislodged from the soil matrix. Feinstein et al. (2009) examined the extraction efficiency of a popular, commercial soil DNA extraction kit by repeating the bead-beating lysis procedure six consecutive times, removing the supernatant, and replacing the lysis solution each time. They tested a clay, sand, and organic soil and found for all three soil types that they continued to extract substantial amounts of DNA with each round of bead-beating, most notably in the first three rounds of extraction.

Extraction efficiency of both cell fractionation and direct lysis procedures can be assessed by direct microscopy, wherein extracted soil is examined using fluorescent stains for intact microbial cells. Alternatively, soil samples may be spiked with a known quantity of labeled DNA and then recovery of added DNA assessed. Feinstein et al. (2009) used quantitative PCR (qPCR) to show that repeated rounds of DNA extraction from the same soil sample continued to yield increasing numbers of amplifiable ribosomal RNA genes, illustrating the initially low lysis and recovery efficiency.

Obtaining a sample that is representative of the resident community is challenging. Microbes in soil are frequently in a resting state or near starvation, making them more difficult to lyse than cells growing rapidly in culture media. Direct cell extraction protocols must ensure that cells are released from soil without bias, and direct lysis procedures must ensure that nucleic acids are not adsorbed to clays or SOM, and thus not recovered. Failure to recover a representative sample of nucleic acids from soil may affect later data interpretation (Fig. 6.1, Lombard et al., 2011). Feinstein et al. (2009) analyzed community composition by T-RFLP and pyrosequencing following each of six rounds of lysis and extraction. They found that bias associated with incomplete extraction could be minimized by pooling the first three extractions prior to downstream analyses.

Coextraction of contaminants, such as humic substances, is also a common problem. Such contaminants may interfere with PCR amplification or hybridization experiments to the point where the reactions may fail entirely. Several methods have been suggested to eliminate or reduce contaminants. One approach is to use a commercial, post-PCR DNA cleanup kit. Although these kits are normally designed to remove salts from post-PCR amplification reactions, they may also work to reduce contaminants in DNA extracts, thereby reducing PCR inhibition. Alternatively, the nucleic acid extract can be rinsed with dilute EDTA or be gel purified prior to analysis. A simple dilution of the DNA extract may alleviate problems with PCR amplification, but may also dilute out soil community members of interest. All manipulations of nucleic acid extracts can lead to loss of material and, hence, sparsely represented members of the community may be lost from subsequent analyses. All postextraction cleanup procedures add expense and processing time, thus reducing the number of samples that can be analyzed. Most postextraction analyses require that the concentration of extracted nucleic acids and their quality be known. Sample nucleic acid concentration can be estimated by treating the DNA sample with Hoescht 33258 or SYBR Green I dye and exposing it to UV light. Absorbance of the sample is measured by fluorimetry and compared to a calf-thymus DNA standard curve. The absorbance of the extract at 260 nm (DNA) and at 280 nm (contaminating proteins) can be measured in a UV spectrophotometer. The ratio between these two readings is used as an indicator of the purity (quality) of the DNA extract.

VII CHOOSING BETWEEN DNA AND RNA FOR STUDYING SOIL BIOTA

A key decision a researcher must make prior to molecular analysis of soil microbial communities is whether to extract microbial DNA, RNA, or both types of nucleic acids. DNA analysis has been used most frequently because DNA is more stable and easier and less costly to extract from soil. Postextraction analyses are straightforward and yield considerable information about the presence of various organisms in a given sample. The key problem with DNA analysis is that it does not necessarily reflect either the abundance or potential activity of different organisms in a sample. When cells die, the nucleic acids released into the soil solution are hydrolyzed rapidly by nucleases. However, DNA contained in dead, unruptured cells within soil aggregates or otherwise protected from decomposition will be extracted along with that from moribund and active cells. On the other hand, RNA is highly labile, making it very difficult to extract from soil before it is degraded.

Commercial nucleic acid extraction kits are now readily available. For DNA extraction, examples include the PowerSoil® DNA Isolation kit (MoBio Laboratories, Inc., Carlsbad, CA) and the FastDNA® SPIN kit for Soil (MP Biomedicals, Santa Ana, CA). Dineen et al. (2010) tested the ability of six commercial DNA extraction kits to recover bacterial spore DNA from soil. They found that the PowerSoil® kit was most efficient at removing PCR inhibitors, whereas the FastDNA® kit provided the highest DNA yield from the loam soil they tested. For RNA extraction, the RNA PowerSoil® Total RNA Isolation kit (MoBio Laboratories), and FastRNA Pro™ Soil-Direct kit (MP Biomedicals) are commonly used. A relatively straightforward method for ribosome extraction from soil is given in Felske et al. (1999).

VIII ANALYSIS OF NUCLEIC ACID EXTRACTS

A DNA:DNA Reassociation Kinetics

When DNA is denatured by either heating or use of a denaturant (e.g., urea), the double-helix structure is lost as the two strands, held together by hydrogen bonds between complementary base pairs (A:T and G:C), come apart. When the denaturant is removed or the temperature is lowered, complementary strands will reanneal. When genome complexity is low, the time it takes for all single strands to find their complement is brief. As complexity increases, the time it takes for complementary strands to reanneal increases. This is referred to as a C_0t curve, where C_0 is the initial molar concentration of nucleotides in single-stranded DNA, and t is time. This measure reflects both the total amount of information in the system (richness or number of unique genomes) and the distribution of that information (evenness or the relative abundance of each unique genome; Liu and Stahl, 2002), making it among the more robust methods for estimating extant diversity in a given sample. Yet, DNA:DNA reassociation kinetics provide no information on the identity or function of any member of the microbial community. Torsvik et al. (1998), using this approach, estimated that the community genome size in an undisturbed, organic soil was equivalent to 6000-10,000 *E. coli* genomes. A soil polluted with heavy metals contained 350-1500 genome equivalents. Culturing produced less than 40 genome equivalents from both soils (Torsvik and Øvreås, 2002). Such experiments, in addition to those employing direct counts using epifluorescence microscopy, are what lead us to understand that culturing methods have only shown us the tip of the iceberg with regard to the actual diversity of soil microbial communities. The study by Torsvik et al. (1998) was reanalyzed by Gans et al. (2005) using improved analytical methods. Their analysis yielded an estimate of the extant diversity contained in the undisturbed soil sample of 8.3 million distinct genomes in 30 g of soil. This was two orders of magnitude higher than soil with heavy metal contamination, which was estimated to contain only 7900 genome equivalents, 99.9% fewer than in the unpolluted soil.

B Microarrays

Microarrays, based on nucleic acid hybridization, represent an exciting approach to microbial community analysis. In microarrays, the oligonucleotide

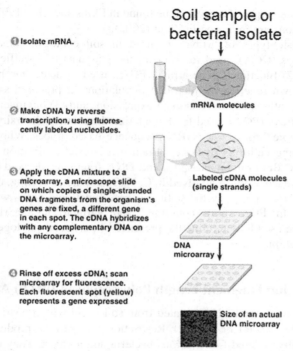

FIG. 6.8 Flowchart depicting the steps in developing and using a microarray, from extracting mRNA from a sample to reading the microarray with a laser scanner. *Campbell and Reese (2005).*

probes, rather than the extracted DNA or RNA targets, are immobilized on a solid surface in a miniaturized matrix. Thousands of probes can be tested simultaneously for hybridization with sample DNA or RNA. In contrast to other hybridization techniques, the sample nucleic acids to be probed, rather than the probes themselves, are fluorescently labeled. After the labeled sample nucleic acids are hybridized to the probes contained on the microarray, positive signals are detected by use of CSLM or other laser microarray scanning devices (Fig. 6.8). A fully developed DNA microarray could include a set of probes encompassing virtually all known natural microbial groupings and thereby serve to monitor the population structure at multiple levels of resolution (Ekins and Chu, 1999; Wu et al., 2001). Such arrays could allow for an enormous increase in sample throughput. A major drawback of microarrays, for use in soil ecology studies, is their need for a high copy number of target DNA/RNA to obtain a signal that is detectable with current technologies. Techniques to improve the sensitivity have been reported by Denef et al. (2003). Nonspecific binding of target nucleic acids to the probes is also a serious issue that needs to be overcome (Zhou and Thompson, 2002). The details of microarray construction and types of arrays can be found in Ekins and Chu (1999). Ecological applications are reviewed in Zhou (2003).

Three basic types of arrays are used in soil ecology: (1) community genome arrays (CGA), used to compare the genomes of specific groups of organisms; (2) functional gene arrays (FGA), used to detect the presence of genes of known function in microbial populations in prepared soil samples and more recently used to detect gene expression; and (3) phylogenetic oligonucleotide arrays (POA), used to characterize the relative diversity of organisms in a sample through use of rRNA sequence-based probes. Whole genomic DNA, requiring culturing of target organisms, is used to develop the probes used in CGA. Both oligonucleotides and DNA fragments derived from functional genes, such as those involved in C, N, S, and metal cycling, can be used to prepare FGAs and query the status of these functional genes within a soil community. In POA, both conserved and variable region rRNA gene sequences are used to determine the presence of particular phylogenic groups in a given sample.

C   Restriction Fragment Length Polymorphism (RFLP) Analysis

In RFLP analysis, total DNA isolated from soil is cut with a restriction endonuclease (often *Eco*R1 or *Hin*dIII). Restriction enzymes are produced by and were originally isolated from various bacteria and archaea. They cut doublestranded DNA at palindromic sequences (those that read in the same order both backward and forward) and can be selected to cut either frequently or infrequently along the isolated DNA strands. The variation (polymorphisms) in the length of resulting DNA fragments is visualized by running the DNA fragments on an electrophoretic gel and staining the gel with ethidium bromide or SYBR Green I, which intercalate in the DNA molecule and fluoresce under UV light. These variations in fragment lengths are then used as a "fingerprint" to differentiate between soil communities. Discrimination between communities is often difficult because of the large number of fragments generated and the difficulty in resolving closely spaced bands in a gel. The RFLP analysis is rarely used on its own for diversity studies. It is most commonly used in conjunction with the Southern method of transferring (blotting) the DNA fragments from the electrophoretic gel onto a nitrocellulose or other membrane. The blot is then probed with appropriately labeled oligonucleotide gene probes to test for the presence of specific sequences (Brown, 2001) or used in conjunction with PCR of amplified ribosomal genes in a technique called amplified ribosomal DNA restriction analysis (ARDRA). The ultimate aim of RFLP-based methods is to compare differences between DNA fingerprints obtained from different communities. Observed differences can then be characterized more fully using other methods, such as those that involve cloning and sequencing of DNA fragments of interest.

D Cloning and Sequencing

DNA sequence information is obtained from environmental samples in three main ways: (1) cloning DNA extracted from soil directly, (2) cloning of PCR-amplified DNA, followed in both cases by sequencing of the cloned DNA, and more recently (3) by high-throughput massive, parallel sequencing (pyrosequencing). Direct cloning involves isolating DNA from the soil, ligating the DNA into a vector (most frequently a self-replicating plasmid), and transforming (moving) the vector into a competent host bacterium, such as commercially available *E. coli*-competent cells, where it can be maintained and multiplied as a recombinant DNA clone library. Once a clone library is obtained, DNA inserts contained in the clones can be reisolated from the host cells, purified, and sequenced. The clone library can also be screened for biological activity expressed directly in *E. coli* or probed for sequences of interest using various genomics applications (Green and Sambrook, 2012). This approach circumvents the need to culture microorganisms from environmental samples and provides a relatively unbiased sampling of the genetic diversity in the sample.

Large fragments (100-300 kb) of genomic DNA can be isolated directly from soil into bacterial, artificial chromosome (BAC) vectors (Rondon et al., 2000). The BAC vectors are low-copy number plasmids that can readily maintain large DNA inserts. Sequences homologous to the low-G+C Gr^+ Acidobacterium, Cytophagales, and Proteobacteria were found in two BAC libraries analyzed by Rondon et al. (2000). They also identified clones that expressed amylase, nuclease, lipase, hemolytic, and antibacterial activities. Their study initiated the field of metagenomics, which is the analysis of all the genomes recovered from a population of organisms in a given environmental sample (Handelsman, 2004). Treusch et al. (2005) probed metagenomic libraries derived from a range of environments and discovered that uncultivated members of the Crenarchaeota contained gene sequences homologous to the ammonia monooxygenase (*amo*A) gene in nitrifying bacteria. Leininger et al. (2006) examined 12 soils from different climatic zones from both agricultural and unmanaged systems and found that the number of Crenarchaeota *amo*A sequences ammonia-oxidizing archaea (AOA) was consistently higher than that of ammonia-oxidizing bacteria (AOB), with the ratio of AOA to AOB ranging from 1.5 to 232. This suggests that the Crenarchaeota play a more significant role in global N cycling than previously thought (Nicol and Schleper, 2006).

An alternative method for creating large clone libraries from soil sequences that allows subsequent profiling of microbial communities is called serial analysis of ribosomal sequence tags (SARST). A region of the 16S rRNA gene, such as the V1-region, is amplified by PCR. The various V1-region amplicons are joined together through a series of enzymatic and ligation (linking) steps. The resulting concatemers are then purified, cloned, screened, and sequenced. The sequences of the individual V1 amplicons are deduced by ignoring the linking sequences and analyzing each sequence tag individually (Neufeld et al., 2004). Neufeld and Mohn (2006) used SARST to analyze arctic tundra and boreal forest soils and found that overall diversity was higher in the arctic tundra soil. They suggested that the high C flux and low pH characteristics of the boreal forest soils might contribute to lower bacterial diversity or that the high diversity in the arctic soils may be influenced by allochthonous organisms coming in via air currents and being preserved by low temperature. The comparative diversity between the two systems did not change when singleton sequence tags were eliminated from the analysis, suggesting that the Arctic may serve as an unrecognized reservoir of microbial diversity and biochemical potential.

Significant progress has been made since the dideoxy chain termination method for DNA sequencing was first described by Sanger et al. (1977), particularly for working with environmental samples. The advent of fluorescent dyes, improvements in gel matrix technology, cycle sequencing using a thermal cycler, capillary systems, use of lasers, and automated gel analysis and microfluidics allow up to 1100 bases of sequence to be generated or deduced in a single reaction. It is now cost-effective to send sample DNA to a commercial sequencing facility for analysis.

Part of the usefulness of DNA sequencing lies in determining gene sequences of unique or important members of a soil community for use in developing specific primers and gene probes to address ecological questions. Gene sequences, once obtained, are submitted to and maintained within various databases, such as GenBank (http://www.ncbi.nlm.nih.gov/) and its collaborating databases, the European Molecular Biology Laboratory (EMBL, http://www.ebi.ac.uk/embl/), the DNA databank of Japan (DDBJ), and the Ribosomal Database Project II (http://rdp.cme.msu.edu/html/; Cole et al., 2014). The growth of the GenBank database over the last 30 years can be appreciated from Fig. 6.9. The whole genome shotgun (WGS) sequence database was developed in 2002. Continued development of databases through DNA sequencing is essential and is a prerequisite to good primer and probe design.

Several PCR-based community analysis methods described later, such as denaturing gradient gel electrophoresis (DGGE) and two-dimensional polyacrylamide gel electrophoresis (2D-PAGE), allow DNA fragments to be retrieved in a selective manner.

E Stable Isotope Probing

Stable isotope probing (SIP) allows microbial identity to be linked to functional activity through the use of substrates labeled with stable isotopes. It has been used to its best advantage by labeling substrates that are used almost exclusively by the population of interest (Dumont and Murrell, 2005; Radajewski et al., 2000; Wellington et al., 2003). A selected substrate is highly enriched with a stable isotope, primarily ^{13}C, or more recently, ^{15}N (Buckley et al., 2007),

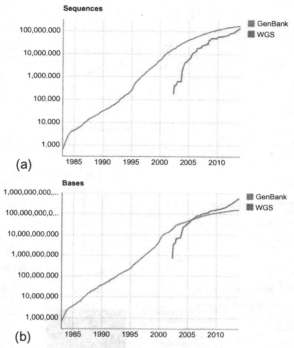

FIG. 6.9 Growth in (a) number of sequence and (b) number of base accessions in the GenBank database since its inception in 1982 and in the genome shotgun (WGS) database beginning in 2002. *http://www.ncbi.nlm.nih.gov/genbank/statistics.*

and then incorporated into the soil. After incubating briefly, cellular components of interest are recovered from the soil sample and analyzed for the incorporated, stable isotope label. Microbes that are actively using the added substrate can then be identified. DNA (DNA-SIP) and rRNA (RNA-SIP) are the biomarkers used most frequently (Radajewski et al., 2003), although phospholipid fatty acids (PLFAs; Treonis et al., 2004) and proteins (Jehmlich et al., 2010) have also been used successfully (Chapter 7). The isotopically labeled nucleic acids are separated from unlabeled nucleic acids by centrifugation in a cesium chloride (CsCl) density-gradient. Once separated, labeled nucleic acids can be amplified using PCR and universal primers to bacteria, archaea, or eucarya. Analysis of the PCR products through cloning and sequencing allows the microbes that have assimilated the labeled substrate to be identified. The stable isotope labeled nucleic acids can also be assessed by using either phylogenetic microarrays or metagenomic analyses (Fig. 6.10). This approach has been applied successfully to study methanotrophs and methylotrophs (McDonald et al., 2005). Information about which microbes were assimilating root exudates was obtained in a study of active rhizosphere communities using

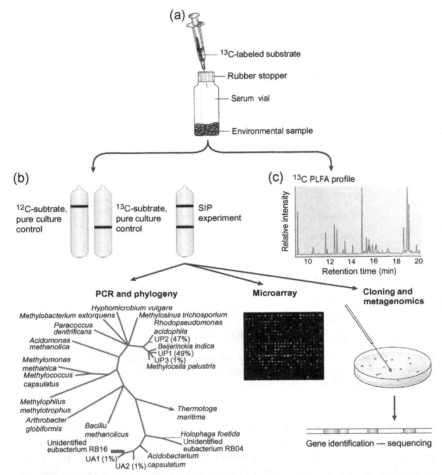

FIG. 6.10 Stable isotope probing procedure with associated assays for assessing incorporation of isotopically labeled substrate into cellular constituents and determining the sequence(s) of bacteria that have incorporated the label. *Dumont and Murrell (2005).*

^{13}C-CO_2 labeling of host plants (Griffiths et al., 2004). Rangel-Castro et al. (2005) used ^{13}C-CO_2 pulse-labeling, followed by RNA-SIP, to study the effect of liming on the structure of the rhizosphere community that was metabolizing root exudates in a grassland. Their results indicated that limed soils contained a microbial community that was more complex and more active in using ^{13}C-labeled compounds in root exudates than were those in unlimed soils.

The SIP-based approaches hold great potential for linking microbial identity with function (Dumont and Murrell, 2005), but at present, a high degree of labeling is necessary to separate labeled from unlabeled marker molecules.

The need for high substrate concentrations may bias community responses. Use of long incubation times to ensure that sufficient label is incorporated increases the risk of cross-feeding ^{13}C from the primary consumers to the rest of the community. It can also be difficult to identify the enriched nucleic acids within the density gradient. The point at which a given nucleic acid molecule is retrieved from the CsCl gradient is a function of both the incorporation of the heavy isotope and the G+C content of the nucleic acids.

IX PARTIAL COMMUNITY ANALYSES—PCR-BASED ASSAYS

The PCR involves (1) the separation of a double-stranded DNA template into two strands (denaturation), (2) the hybridization (annealing) of oligonucleotide primers (short strands of nucleotides of a known sequence) to the template DNA, and (3) the elongation of the primer-template hybrid by a DNA polymerase enzyme. Each of these steps is accomplished by regulating the heat of the reaction. The temperature is raised to 92-96 °C to denature the template DNA, and then lowered to 42-65 °C to allow the primers to anneal to the template. The temperature is then raised to 72 °C, or the ideal temperature for the activity of the DNA polymerase used in the reaction. This cycle is repeated from 25 to 30 times, each cycle doubling the number of products (amplicons) in the reaction (Fig. 6.11). The discovery of thermal-stable DNA polymerases from organisms, such as *Thermus aquaticus* (Deinococcus-Thermus, *Taq* polymerase), has made PCR possible as a standard protocol (Mullis and Faloona, 1987; Saiki et al., 1988) and led to the award of a Nobel Prize to Kary Mullis in 1993.

The annealing temperature is a critical choice. Lower temperatures are less stringent and may allow base "mismatch" to occur when the primer binds to the template. Higher annealing temperatures are more stringent, and therefore, primers bind with higher fidelity to their target sequences. The potential target genes for PCR are many and varied, limited only by available sequence information. The primers used for soil ecological studies may target specific DNA sequences, such as those coding for the small subunit (SSU) rRNA genes, genes of known function, sequences that are repeated within microbial genomes (rep-PCR) or use arbitrary primers (randomly amplified polymorphic DNA, RAPD) to generate a PCR fingerprint. Primers can be selected to target different levels of taxonomic resolution. Ribosomal RNA genes are highly conserved and therefore discriminate between sequences at the genus level or above. Repeated sequence and arbitrary primers are used to discriminate at a finer scale, separating isolates at the strain level. The probeBase website (http://www.microbial-ecology.net/probebase/list.asp?list=primer) contains a comprehensive list of oligonucleotide primers used in PCR-fingerprinting and in a range of other applications.

The most common targets for characterizing microbial communities are the rRNA genes due to their importance in establishing phylogenetic and taxonomic relationships (Woese et al., 1990). These include SSU rRNA genes

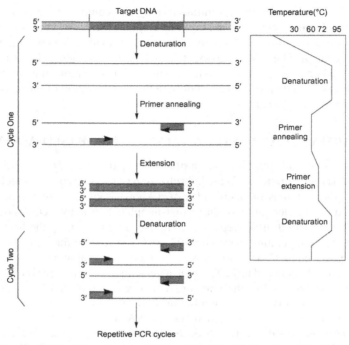

FIG. 6.11   Mechanics of the polymerase chain reaction. The right-hand panel depicts the temperature variation occurring in a thermal cycler during the reaction. The left-hand panel illustrates how the DNA template and primers interact to copy the target DNA during the changes in temperature as cycling proceeds.

(16S in bacteria and archaea or 18S in eucarya), the large subunit (LSU) rRNA genes (23S in bacteria and archaea or 28S in eucarya), or the internal transcribed spacer (ITS) region (sequences that lie between the SSU and LSU genes). Other defined targets are genes that code for ecologically significant functions, such as those that code for proteins involved in N_2 fixation (e.g., *nif*H which encodes the iron protein of nitrogenase reductase; *amo*A, which codes for ammonium monooxygenase, a key enzyme in nitrification reactions; and *nir*S, which codes for nitrite reductase, a key enzyme in denitrification reactions).

Sources of bias must be considered where PCR is used (Fig. 6.1, Wintzingerode et al., 1997; Acinas et al., 2005; Lombard et al., 2011). The main sources of bias in amplifying soil community DNA are (1) the use of very small sample sizes (typically only 500 mg of soil), which may represent only a small fraction of the soil community; (2) preferential amplification of some DNA templates due to the greater ease of DNA polymerase binding; and in the (3) amplification of the rRNA genes, the fact that many bacteria contain multiple copies of these operons. For example, *Bacillus* and *Clostridium* species (Firmicutes) contain 15 rRNA gene copies; hence, sequences from such species will be overrepresented among the amplification products. In addition, chimeras—amplicons composed of double-stranded DNA where each strand is derived from a different organism rather than a single organism—may be generated. This problem is sometimes a consequence of using too many cycles in the PCR.

An advance in PCR analysis that allows specific gene targets to be quantified is quantitative PCR (qPCR), also called real-time PCR. The qPCR employs fluorogenic probes or dyes bound to the primers, which are used to quantify the number of copies of a target DNA sequence in a sample. This approach has been used successfully to quantify specific functional genes in soil samples, including ammonia monooxygenase (*amo*A), nitrite reductase (*nir*S or *nir*K), and particulate methane monooxygenase (*pmo*A) genes to quantify ammonia oxidizing (Hermansson and Lindgren, 2001), denitrifying (Henry et al., 2004), and methanotrophic (Kolb et al., 2003) bacteria, respectively. The qPCR coupled with primers to specific internal transcribed spacer (ITS) or rRNA gene sequences has also been used to quantify ectomycorrhizal (Landeweert et al., 2003) and arbuscular mycorrhizal fungi (Filion et al., 2003), as well as cyst nematodes (Madani et al., 2005) in soil.

A Electrophoresis of Nucleic Acids

Amplified PCR products are visualized most often by running samples in an electrophoretic gel, staining the DNA within the gel with ethidium bromide, SYBR Green I, or another fluorescent dye with high-affinity binding to DNA and viewing the stained, separated PCR products under UV light. Nucleic acids are negatively charged and will run to the positive pole in an electric field. The gel matrix provides resistance to the movement of nucleic acids by virtue of the pore sizes within the gel, such that DNA fragments of smaller size will move through the matrix faster than those of a larger size. A standard, molecular weight marker is run along with the samples to enable the sizes of PCR products to be assigned. Analysis of the amplified products is based on the presence and pattern of DNA bands of various sizes contained in the gel matrix.

Agarose, the most popular medium for electrophoretic separation of medium and large-sized nucleic acids, has a large working range, but poor resolution. Depending on the agarose concentration used, nucleic acids between 0.1 kilobase (kb) and 70 kb in size can be separated. Polyacrylamide is the preferred matrix for separating proteins, single-stranded DNA fragments up to 2000 bases in length, or double-stranded DNA fragments of less than 1 kb. Polyacrylamide gels separate macromolecules on the basis of configuration in addition to the characteristics of size, charge, and G+C content and is used in denaturing (or temperature) gradient gel electrophoresis (DGGE or TGGE). This shape-dependent mobility forms the basis of a suite of techniques that exploit inter- and intra-strand nucleotide interactions, can be used to rapidly screen amplified DNA for very fine-scale sequence differences, and is thus used in studies of genetic diversity. These techniques include single-strand conformation polymorphisms (SSCP; Dewit and Klatser, 1994), DGGE (Muyzer and Smalla, 1998), and heteroduplex mobility assays (HMA; Espejo and Romero, 1998).

B   PCR Fingerprinting

PCR fingerprinting can be accomplished by several different methods, all of which are aimed at distinguishing differences in the genetic makeup of biotic populations from different samples. The advantages of these techniques is that they are rapid in comparison with sequencing methods, thus enabling high sample throughput, and they can be used to target sequences that are phylogenetically or functionally significant. Depending on the primers chosen, PCR fingerprints can be used to distinguish between isolates at the strain level or to characterize target microbes at the community level. The more common PCR fingerprinting techniques used for characterizing soil microbial community composition are DGGE or TGGE (Muyzer and Smalla, 1998) and terminal restriction fragment length polymorphism (T-RFLP) analysis (Clement et al., 1998; Liu et al., 1997). Both techniques can be used to separate PCR products that are initially of a similar length by employing additional methods to separate the amplicons into a greater number of bands (or fragment lengths) that are then used for community comparisons. For DGGE, only about 20-40 bands can be resolved clearly in the denaturing gel; thus, the target sequences amplified are normally $\sim$100-200 bases in length. Hypervariable (V) regions of the 16S rRNA gene are typically targeted because they can also provide phylogenetic information; however, many other gene targets have been used successfully (Nakatsu, 2007). In T-RFLP, nearly the full length of the 16S rRNA genes ($\sim$1500 bases) is amplified. The amplicons are then cut with a restriction enzyme, and the terminal fragments are sized by detection of the fluorescent label attached to the primer used in the PCR reaction. Applications of these approaches are reviewed in Nakatsu (2007) for DGGE and in Thies (2007) for T-RFLP.

Additional PCR fingerprinting techniques that target the ribosomal gene sequences (ribotyping) include amplified ribosomal DNA restriction analysis (ARDRA, Smit et al., 1997), automated ribosomal intergenic spacer analysis (ARISA, Ranjard et al., 2001), and two-dimensional polyacrylamide gel electrophoresis (2D-PAGE) of the ITS (Jones and Thies, 2007).

The successful application of PCR fingerprinting to population studies relies heavily on the correct interpretation of the banding or spot patterns observed on electrophoretic gels. Gel images are typically digitized, and band detection software is used to mark the band locations in the gel. The resulting band pattern is then exported to a statistical software package for analysis. Some analyses require that the fingerprint patterns obtained must first be converted to presence/absence matrices, although average band density data are also used. The matrices generated are then compared using cluster analysis, multidimensional scaling, principal component analysis, redundancy analysis, canonical correspondence analysis, or the additive main effects multiplicative interaction model (Culman et al., 2008). Each analysis will allow community comparisons, yet each has associated strengths and weaknesses. There are several software packages available that will enable one to compare and score PCR fingerprints and produce similarity values for a given set of samples. Software packages, such as BioNumerics and GelCompar (Applied Maths, Kortrijk, Belgium), Canoco (Microcomputing, Ithaca, NY), and PHYLIP (freeware via GenBank and the RDPII), among others, are commonly used. Culman et al. (2009) offer the online tool, T-REX, for processing and analyzing T-RFLP data.

C Metagenomics

The burgeoning field of meta-*omics* involves the study of all the biological molecules extracted from an environmental sample (Fig. 6.12). Metagenomics involves the large-scale analysis of microbial genomes extracted from soil. Metatranscriptomics is the study of the genes being expressed by the active members of the soil biotic community and is based on the analysis of RNA transcripts (see Section D). Study of proteins extracted from the environmental sample is the province of metaproteomics (Chapter 7); whereas metabolomics encompasses the study of metabolites, including sugars, lipids, amino acids, and nucleotides. Recently, the soil volatilome (volatile organic compounds with biological activity) has begun to be studied (Insam, 2014). These new approaches are allowing us to identify and comprehensively assess the activity and function of members of the soil biological community (Nannipieri et al., 2014).

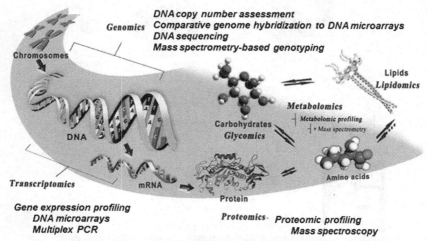

FIG. 6.12 Illustration of the array of -omics approaches developed, many of which are being used in soil ecology studies. *Adapted from Wu et al. (2011).*

Metagenomics has been enabled by the advent of: (1) high throughput, massive parallel sequencing (Margulies et al., 2005), also referred to as pyrosequencing; (2) next-generation sequencing (NGS); and (3) sequencing by synthesis (Nyren et al., 1993; Ronaghi et al., 1996, 1998). In sequencing by synthesis, the pyrophosphate (PPi) released on base incorporation during amplification is converted to ATP by ATP sulfurylase, the amount of ATP is subsequently determined by the luciferase assay. Automated 454 pyrosequencing, commercialized by 454 Life Sciences (Roche Diagnostics Corp., Branford, CT) involves isolating and fragmenting genomic DNA, ligating the fragments to adapters, and then separating them into single strands. The single-stranded fragments are bound to beads, with one fragment per bead, with the beads then captured in droplets of a "PCR-reaction-in-oil emulsion." DNA amplification takes place inside each droplet, such that the bead contained in the droplet has 10 million copies of a unique DNA template attached to it. The emulsion is dispersed, the DNA strands are denatured, and the beads carrying the single-stranded DNA copies are placed into individual wells on a fiberoptic slide. Smaller beads that have the enzymes needed for pyrophosphate sequencing immobilized on them are then added to the wells. The fiberoptic slide is placed into a chamber through which the sequencing reagents flow. The base of the slide comes into optical contact with a second fiberoptic bundle that is fused to a charge-coupled device (CCD) sensor. Reagents are delivered cyclically to the chamber and flow into the wells of the fiberoptic slide. Simultaneous extension reactions occur on the template-carrying beads, such that each time a nucleotide is incorporated, inorganic pyrophosphate is released and photons are generated (Fig. 6.13). After an

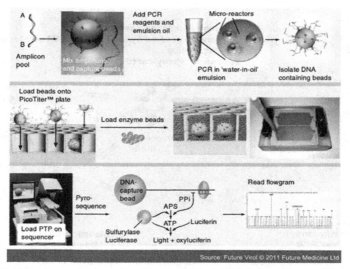

FIG. 6.13   Flow diagram for pyrosequencing using the Roche 454 platform. *Gega and Kozal (2011)*.

individual nucleotide is pulsed into the chamber, a wash containing apyrase is used to prevent nucleotides from remaining in the wells before the next nucleotide is introduced. Raw signals are captured, and the background is subtracted, normalized, and corrected. Postrun analyses are used for base calling and sequence alignments. Pyrosequencing circumvents cloning of DNA fragments into bacterial vectors and handling of individual clones and holds great promise for characterizing population diversity.

Data quality is a major issue in any *omics* analysis, particularly with soil microbial community analyses (Hazen et al., 2013; Lombard et al., 2011). The potential biases associated with each step in any soil metagenomics project (Fig. 6.1) are not trivial. They include sampling bias during sample collection and processing (particularly acute for soil sampling and nucleic acid extraction), PCR bias, and biases associated with pyrosequencing, genome assembly, and library construction. Větrovský and Baldrian (2013) introduced the SEED software pipeline (http://www.biomed.cas.cz/mbu/lbwrf/seed/main.html) to help control method-dependent errors in the analysis of pyrosequencing data. Their recommended data analysis workflow includes removing sequences of inferior quality, sequence denoising, deleting chimeric sequences, trimming sequences so they contain the same part of the template, and clustering sequences into groups based on their similarity. This process is followed by sequence alignment, sequence identification, community composition analysis, and random resampling prior to calculating diversity parameters.

Data quantity must also be managed. The Roche 454 platform has the capacity for over one million base reads per run, with 85% of the bases from reads longer than 500 base pairs (http://www.454.com/). Competing platforms, including the Illumina/Solexa Genome Analyzer System (Illumina, Inc., San Diego, CA) and the Applied Biosystems SOLiD^{TM} System (Life Technologies-Thermo Fisher Scientific, Grand Island, NY), have significantly higher output, but much shorter read lengths, which make genome assembly more difficult and potentially more prone to bias. Luo et al. (2012) compared the ability of the Roche 454 and Illumina platforms to characterize the same DNA sample extracted from a freshwater planktonic community. They found that the gene assemblies derived overlapped in ~90% of their total sequences and that they estimated gene and genome abundances similarly. The Roche 454 produced a 502 Mbp (450 bp long reads) sequence compared to 2460 Mbp (100 bp pair-ended reads) obtained from the Illumina sequencing, but once assembled, these resulted in 46 Mbp and 57 Mbp of total unique sequences, respectively. The Illumina platform produced longer and more accurate contigs (overlap between any two DNA sequences that result from the sequencing effort with a common string of bases), despite the substantially shorter read lengths, at 25% of the cost of the 454 platform.

Bioinformatics approaches, their tools, and algorithms must be improved continually to ensure emergence of the highest quality and quantity of information from next-generation sequencing experiments. Approaches have been published in an issue of the journal *Bioinformatics* entitled "Bioinformatics for Next Generation Sequencing" and are updated regularly (http://www.oxfordjournals.org/our_journals/bioinformatics/nextgenerationsequencing.html; Bateman and Quackenbush, 2009).

An ambitious soil metagenome project, The Earth Microbiome Project (EMP; http://www.earthmicrobiome.org/), aims to characterize soil metagenomes sampled from Earth's major biomes and to enable comparisons between them. The EMP aims to generate shotgun metagenomic sequence data of 16S rRNA amplicons from 10,000 environmental samples (Gilbert and Meyer, 2012). Participating researchers follow standard DNA extraction and amplification protocols, use the same primers and brand of DNA polymerase, and follow standard sequencing protocols using the same sequencing platform to reduce the potential bias associated with all the steps in sample analyses. Other objectives of the EMP are to produce a global atlas of protein functions, a catalogue of the taxonomic distribution of reassembled genomes, and to then use the data generated to develop predictive ecosystems models (Gilbert and Meyer, 2012).

D   Metatranscriptomics

The metatranscriptome is composed of all the RNA transcripts (rRNA, mRNA, miRNA, and tRNA) produced by the biota in a sampled soil. Cell processes, particularly RNA transcription, slow down as substrate becomes limiting. Poulsen et al. (1993) have shown that the RNA content of cells under substrate-limited conditions can decrease by up to 60% within a few hours. Thus, analysis of RNA is reflective of the portion of the soil microbial community that is active at the time of sampling (Blagodatskaya and Kuzyakov, 2013) and thus provides a robust means to examine microbial responses to soil management.

The RNA extracted from soil is dominated by rRNA and tRNA and includes eukaryotic RNA along with that of bacteria and archaea. Metatranscriptomics surveys based on reverse transcription and sequencing of rRNA (RNA-Seq, Wang et al., 2009) extracted from soil have been very successful in identifying active populations of both eukaryotes and prokaryotes. Turner et al. (2013) used comparative metatranscriptomics to examine bulk soil and rhizosphere communities of wheat, pea, oat, and an oat mutant (*sad*1) unable to produce antifungal avenacins. They found profound differences—at the kingdom level—in the communities that developed in the rhizosphere of the different plant species growing in the same soil, confirming a significant amount of previous work showing plant species as a key driver in shaping soil microbial communities. The largest proportion of the sequences they recovered was from bacteria, which ranged from 91% for bulk soil to 73.7% of total sequences for the pea rhizosphere. Proteobacteria, Actinobacteria, Firmicutes,

Acidobacteria, Planctomycetes, and Bacteroidetes were all well represented in recovered sequences. Bacterial communities in the plant rhizospheres differed from those in the bulk soil and differed between the plant species at the genus level; however, at the phylum level, prokaryotic communities in the wheat rhizosphere community did not differ from those in the bulk soil. Archaea consistently represented approximately 5% of the sequences recovered overall, whereas eukaryotes made up 2.8% of the bulk soil community and 3.3% of the wheat rhizosphere community, but were highly represented in the pea (20.7%) and oat (16.6%) rhizosphere communities. The pea rhizosphere was highly enriched for fungi, and nematodes and bacterivorous protozoa were enriched in all rhizospheres.

Once thought nearly impossible, extraction, reverse transcription, amplification, and sequencing of prokaryotic mRNA transcripts is now ongoing, but remains technically challenging, as mRNA typically represents less than 5% of the RNA extracted (Carvalhais et al., 2012), is often extremely short-lived, and is frequently transcribed and translated simultaneously in prokaryotes (Fig. 6.2). To assess the functional activities of the soil microbial community, the mRNA in the extract must be enriched and the rRNA removed. Subtractive hybridization, treatment with endonucleases that preferentially degrade rRNA or duplex specific nuclease treatment are used to remove rRNA. The latter was found to be the most efficient at enriching mRNA (Yi et al., 2011). Depending on the aims of the study, eukaryotic mRNA must also be removed. Eukaryotic mRNA transcripts contain 3′-poly-A tails that can be captured by use of surfaces coated with poly(dt) probes, thus enriching mRNA from bacteria and archaea in the extract. The mRNA must be reverse-transcribed (RT) into cDNA, which is then sequenced using one of the sequencing platforms described earlier. Several life sciences companies produce kits that facilitate each step in this process. Every step has associated methodological challenges, which are summarized and discussed in Carvalhais et al. (2012). Details of sample processing and cDNA preparation are given in Carvalhais and Schenk (2013).

Myrold et al. (2014) provide a summary of recent soil metagenomic (Table S6.1), metatranscriptomic (Table S6.2), and metaproteomic (Table S6.3) studies and discuss their potential for improving our understanding of how C, N, and other nutrients cycle in soil. (See Supplemental Tables S6.1, S6.2, and S6.3 on the online Supplemental Material at http://booksite.elsevier.com/9780124159556). The study by Urich et al. (2008) stands out as the first shotgun metatranscriptomics study of a soil environment. They extracted the total RNA pool and analyzed both the rRNA and mRNA recovered without amplification, allowing them to examine both the structure and function of the soil community. In their analyses, the eukarya represented 10.3% (SSU) and 13.3% (LSU) of the ribo-tags recovered, whereas 87.2% (SSU) and 83.8% (LSU) were from the bacteria and 1.5% (SSU) and 1.4% (LSU) were from the archaea. These proportions, at the domain level, agree closely with those reported by Turner et al. (2013) discussed earlier. Urich et al. (2008) found that the Crenarcheota were represented by ~0.01% of the rRNA sequence tags, with the Crenarchaeal candidate division Group I.1b dominating among the archaeal ribo-tags recovered. They went on to infer activity *in situ* by confirming the presence of mRNA tags specific for enzymes involved in ammonia oxidation and CO_2 fixation. This finding corroborates those of Treusch et al. (2005) and Nicol and Schleper (2006); see Section VIII. D who, through early cloning and sequencing work, identified the significant role of the Crenarcheota in ammonia oxidation in a diversity of soils. Among the bacteria, Proteobacteria and Actinobacteria were most abundant, followed by the Firmicutes, Acidobacteria, and Planctomycetes; again, in close agreement with the findings of Turner et al. (2013). Dominant among the eukarya ribo-tags identified were the fungi (primarily Ascomycota and Glomeromycota), plants (Viridiplantae), and Metazoa (protists); with ~50%, ~20%, and ~10% of the ribo-tags assigned to these kingdoms, respectively. The 1370 ribo-tags classified as protists were particularly interesting and important as they represented the largest molecular dataset of a protist community that had been generated to date and provided the first window into their community structure in soil. Slime molds (Mycetozoa, 25%) dominated the metazoan community, followed by the Cercozoa (amoeba and flagellates, 17%), Plasmodiophorida (largely plant parasites, 16%), and Alveolata (e.g., ciliates and dinoflagellates, 15%). All are known to have members with a voracious appetite for soil bacteria and are thus integral to organic matter turnover and N, P, and S mineralization in soil.

X   LEVEL OF RESOLUTION

Genes change as they acquire fixed mutations over time. The number of differences between two homologous sequences reflects both the evolutionary rate of the sequences and the time separating them, in other words—how long it has been since they had a common ancestor. Consequently, different sequences need to be selected to resolve variation at different taxonomic levels. In general, noncoding DNA evolves faster than transcribed DNA because it is under no selection pressure to remain unchanged; therefore, intergenic spacer regions evolve more rapidly than other sequences. Slowest to change are the structural rRNA genes, although these too have variable regions that are exploited regularly to characterize soil biological communities. For example, the V3 region of 16S rRNA genes is a common target in DGGE analysis of bacterial diversity, whereas the internal transcribed spacer (ITS) is used frequently for studies of fungal diversity as the 18S rRNA genes are not sufficiently variable. The information that can be obtained from the use of molecular approaches depends on the analysis technique chosen. The level of resolution required, coupled with study aims, will largely guide the choice of technique used for a given study (Fig. 6.4).

XI FACTORS THAT MAY AFFECT MOLECULAR ANALYSES

Any study of soil microbial ecology requires that the spatial and temporal inter-actions of potential reactants, such as soil type and moisture content, be consid-ered. Hence, the time, location, season of sampling, sample volume, mixing, compositing, and replication are all important for deriving meaningful data (Chapter 7). Because all samples are by their nature small volumes on which measurements are taken to represent the whole soil, scaling errors can occur frequently. For molecular analyses, it is critical that representative samples are taken, the samples are protected from change between the field and the time they are analyzed in the laboratory, and the samples are not contaminated by either inadvertent mixing with each other or by coming in contact with other samples during handling, transport, and analysis (Fig. 6.1). To avoid alter-ing the composition of the microflora prior to analysis, soil samples should be stabilized as soon as practicable. For molecular measures, soil should be frozen immediately at -20 °C and processed within several weeks of sampling. If analyses will be significantly delayed, soil should be stored at -80 °C. For RNA analyses, soil is snap-frozen in liquid N in the field to prevent changes in the metatranscriptome prior to analysis. Soil drying should be avoided entirely when intended for microbial population analyses using molecular methods.

Soil constituents with cation and anion exchange capacity, such as humic acids, clays, soil organic matter (SOM), and pyrogenic C are the key chemical factors that may interfere with molecular analysis of soil communities. Negatively charged clay particles in a moist soil will be surrounded by hydrated cations, which create a localized zone of positive charge. This will attract microorganisms, which possess a net negative charge at the pH of most soil habitats. Binding of bacteria to solid surfaces through such ionic interactions makes it difficult to separate cells and DNA/RNA released from cells from the soil matrix (Feinstein et al., 2009). Hence, soil type has a significant influence on DNA/RNA extraction efficiency. Nucleic acid extraction is partic-ularly difficult from soils with high clay, organic matter, and pyrogenic C contents.

Humic substances inhibit *Taq* DNA polymerase in the PCR, interfere with restriction enzyme digestion, and reduce transformation efficiency during cloning and DNA hybridization. Effective removal of humic acids is often required prior to quantifying or amplifying DNA. Humic substances are difficult to remove as they remain soluble under similar conditions to that of DNA; hence, direct extraction of DNA may require an additional purification step to obtain DNA of sufficient purity for downstream assays. Commercial DNA extraction kits are available that include improved DNA cleanup steps and yield high-quality DNA extracts with few impurities that affect down-stream DNA analyses, such as the PowerSoil® DNA Isolation Kit (MoBio Laboratories, Inc.).

XII FUTURE PROMISE

We have come a long way in developing our understanding of soil microbiology, ecology, and biochemistry, but have many milestones yet to meet. Molecular tools offer unparalleled opportunities to characterize soil biota in culture and directly from field soils. These tools are allowing us to ask questions at much larger geographic scales than has been previously possible (e.g., The Earth Microbiome Project, http://www.earthmicrobiome.org/). We are now able to examine such issues as how microbial populations vary across soil types and climatic zones, in association with plant roots and between various plant species, and in response to soil management or soil pollution. The "meta-*omics*" approaches are rapidly developing and evolving technologies that are expanding our access to both biotic diversity and gene expression in the soil environment. However, Myrold and Nannipieri (2014) caution that ". . . omics research should go beyond being descriptive and be more hypothesis driven, with a greater focus on experimental design rather than on the technological prowess" and remind us that a polyphasic approach remains necessary for a comprehensive understanding of soil biogeochemistry. Techniques such as RFLP analysis of isotopically labeled, amplified *nif*H, *amo*A, *nir*S, and *pmo*A sequences, and DNA-, RNA-, and protein-SIP allow us to target, with high specificity, organisms or groups of organisms responsible for specific functions in soil, particularly those involved in key transformations in the C, N, and S nutrient cycles.

The amount of work that remains is daunting, yet exciting as so much remains to be discovered. These rapidly expanding technical developments open new horizons of research that allow a far more detailed view of microbial diversity and function in soils that should help to guide us in improving the management of our precious soil resources.

REFERENCES

Acinas, S.G., Sarma-Ruptavtarm, R., Klepac-Ceraj, V., Polz, M.F., 2005. PCR-induced sequence artifacts and bias: insights from comparison of two 16S rRNA clone libraries constructed from the same sample. Appl. Environ. Microbiol. 71, 8966–8969.

Amann, R.I., Fuchs, B.M., 2008. Single-cell identification in microbial communities by improved fluorescence *in situ* hybridization techniques. Nat. Rev. Microbiol. 6, 329–348.

Amann, R.I., Ludwig, W., 2000. Ribosomal RNA-targeted nucleic acid probes for studies in microbial ecology. FEMS Microbiol. Rev. 24, 555–565.

Amann, R.I., Ludwig, W., Schleifer, K.H., 1995. Phylogenetic identification and *in situ* detection of individual microbial cells without cultivation. Microbiol. Rev. 59, 143–169.

Bateman, A., Quackenbush, J., 2009. Bioinformatics for next generation sequencing. Bioinformatics 25, 429.

Binnerup, S.J., Bloem, J., Hansen, B.M., Wolters, W., Veninga, M., Hansen, M., 2001. Ribosomal RNA content in microcolony forming soil bacteria measured by quantitative 16S rRNA hybridization and image analysis. FEMS Microbiol. Ecol. 37, 231–237.

Blagodatskaya, E., Kuzuakov, Y., 2013. Active microorganisms in soil: critical review of estimation criteria and approaches. Soil Biol. Biochem. 67, 192–211.

Bringhurst, R.M., Cardon, Z.G., Gage, D.J., 2002. Galactosides in the rhizosphere: utilization by *Sinorhizobium meliloti* and development of a biosensor. Proc. Natl. Acad. Sci. U. S. A. 98, 4540–4545.

Brown, T.S., 2001. Southern blotting and related DNA detection techniques. Encyclopedia of Life Sciences. Nature Publishing Group, London, United Kingdom.

Bruns, M.A., Buckley, D.H., 2002. Isolation and purification of microbial community nucleic acids from environmental samples. In: Hurst, C.J. (Ed.), Manual of Environmental Microbiology, second ed. ASM Press, Washington, DC.

Buckley, D.H., Huangyutitham, V., Hsu, S.-F., Nelson, T.A., 2007. Stable isotope probing with [15] N achieved by disentangling the effects of genome G+C content and isotope enrichment on DNA density. Appl. Environ. Microbiol. 73, 3189–3195.

Campbell, N.A., Reece, J.B., 2005. Biology, seventh ed. Pearson, Benjamin Cummings, San Francisco.

Carvalhais, L.C., Schenk, P.M., 2013. Sample processing and cDNA preparation for microbial meta-transcriptomics in complex soil communities. Methods Enzymol. 351, 251–267.

Carvalhais, L.C., Dennis, P.G., Tyson, G.W., Schenk, P.M., 2012. Application of metatranscriptomics to soil environments. J. Microbiol. Methods 91, 246–251.

Chalfie, M., Tu, Y., Euskirchen, G., Ward, W.W., Prasher, D.C., 1994. Green fluorescent protein as a marker for gene expression. Science 263, 802–895.

Clement, B.G., Kehl, L.E., DeBord, L., Kitts, C.L., 1998. Terminal restriction fragment patterns (TRFPs), a rapid, PCR-based method for the comparison of complex bacterial communities. J. Microbiol. Methods 31, 135–142.

Cole, J.R., Wang, Q., Fish, J.A., Chai, B., McGarrell, D.M., Sun, Y., Brown, C.T., Porras-Alfaro, A., Kuske, C.R., Tiedje, J.M., 2014. Ribosomal Database Project: data and tools for high throughput rRNA analysis. Nucleic Acids Res. 41. http://dx.doi.org/10.1093/nar/gkt1244 Database issue, first published online 27 Nov. 2013.

Culman, S.W., Gauch, H.G., Blackwood, C.B., Thies, J.E., 2008. Analysis of T-RFLP data using AOV and ordination methods: a comparative study. J. Microbiol. Methods 75, 55–63.

Culman, S.W., Bukowski, R., Gaugh, H.G., Cadillo-Quiroz, H., Buckley, D.H., 2009. *T-REX*: Software for the processing and analysis of T-RFLP data. BMC Informatics 10, 171–180.

Denef, V.J., Park, J., Rodriques, J.L.M., Tsoi, T.V., Hashsham, S.A., Tiedje, J.M., 2003. Validation of a more sensitive method for using spotted oligonucleotide DNA microarrays for functional genomics studies on bacterial communities. Environ. Microbiol. 5, 933–943.

Dewit, M.Y.L., Klatser, P.R., 1994. *Mycobacterium leprae* isolates from different sources have identical sequences of the spacer regions between the 16S and 23S ribosomal RNA genes. Microbiology 140, 1983–1987.

Dineen, S.M., Aranda, R., Anders, D.L., Robertson, J.M., 2010. An evaluation of commercial DNA extraction kits for the isolation of bacterial spore DNA from soil. J. Appl. Microbiol. 109, 1886–1896.

Dumont, M.G., Murrell, J.C., 2005. Stable isotope probing—linking microbial identity to function. Nat. Rev. 3, 499–504.

Ekins, R., Chu, F.W., 1999. Microarrays: their origins and applications. Trends Biotechnol. 17, 217–218.

Errampalli, D., Leung, K., Cassidy, M.B., Kostrzynska, M., Blears, M., Lee, H., Trevors, J.T., 1999. Applications of the green fluorescent protein as a molecular marker in environmental microorganisms. J. Microbiol. Methods 35, 187–199.

Espejo, R.T., Romero, J., 1998. PAGE analysis of the heteroduplexes formed between PCR-amplified 16S rRNA genes: estimation of sequence similarity and rDNA complexity. Microbiology 144, 1611–1617.

Feinstein, L.M., Sul, W.J., Blackwood, C.B., 2009. Assessment of bias associated with incomplete extraction of microbial DNA from soil. Appl. Environ. Microbiol. 75, 5428–5433.

Felske, A., Engelen, B., Nübel, U., Backhaus, H., 1999. Direct ribosome isolation from soil to extract bacterial rRNA for community analysis. Appl. Environ. Microbiol. 62, 4162–4167.

Filion, M., St-Arnaud, M., Jabaji-Hare, S.H., 2003. Direct quantification of fungal DNA from soil substrate using real-time PCR. J. Microbiol. Methods 53, 67–76.

Gans, J., Wolinsky, M., Dunbar, J., 2005. Computational improvements reveal great bacterial diversity and high metal toxicity in soil. Science 309, 1387–1390.

Gega, A., Kozal, M.J., 2011. New technology to detect low-level drug-resistant HIV variants. Future Virol. 6, 1–26.

Gilbert, J.A., Meyer, F., 2012. Modeling the Earth's microbiome: a real world deliverable for microbial ecology. Am. Soc. Microbiol. Microbe Mag. 7, 64–69.

Green, M.R., Sambrook, J., 2012. Molecular Cloning: A Laboratory Manual, fourth ed. Cold Spring Harbor Laboratory, Cold Spring Harbor, NY.

Griffiths, R.I., Manefield, M., Ostle, N., McNamara, N., O'Donnell, A.G., Bailey, M.J., Whiteley, A.S., 2004. $^{13}CO_2$ pulse labelling of plants in tandem with stable isotope probing: methodological considerations for examining microbial function in the rhizosphere. J. Microbiol. Methods 58, 119–129.

Handelsman, J., 2004. Soils—the metagenomics approach. In: Bull, A.T. (Ed.), Microbial Diversity and Bioprospecting. ASM Press, Washington, DC, pp. 109–119.

Hazen, T.C., Rocha, A.M., Techtmann, S.M., 2013. Advances in monitoring environmental microbes. Curr. Opin. Biotechnol. 24, 526–533.

Henry, S., Baudoin, E., Lopez-Gutierrez, J.C., Martin-Laurent, F., Baumann, A., Philippot, L., 2004. Quantification of denitrifying bacteria in soils by *nir*K gene targeted real-time PCR. J. Microbiol. Methods 59, 327–335.

Hermansson, A., Lindgren, P.E., 2001. Quantification of ammonia-oxidizing bacteria in arable soil by real-time PCR. Appl. Environ. Microbiol. 67, 972–976.

Insam, H., 2014. Soil volatile organic compounds as tracers for microbial activities in soils. In: Nannipieri, P., Pietramellara, G., Renella, G. (Eds.), Omics in Soil Science. Caister Academic Press, Portland, OR.

Jehmlich, N., Schmidt, F., Taubert, M., Seifert, J., Bastida, F., von Bergen, M., Richnow, H.-H., Vogt, C., 2010. Protein-based stable isotope probing. Nat. Protoc. 5, 1957–1966.

Jones, C.M., Thies, J.E., 2007. Soil microbial community analysis using two-dimensional polyacrylamide gel electrophoresis of the bacterial ribosomal internal transcribed spacer regions. J. Microbiol. Methods 69, 256–267.

Kolb, S., Knief, C., Stubner, S., Conrad, R., 2003. Quantitative detection of methonotrophs in soil by novel *pmo*A-targeted real-time PCR assays. Appl. Environ. Microbiol. 69, 2423–2429.

Landeweert, R., Veenman, C., Kuyper, T.W., Fritze, H., Wernars, K., Smit, E., 2003. Quantification of ectomycorrhizal mycelium in soil by real-time PCR compared to conventional quantification techniques. FEMS Microbiol. Ecol. 45, 283–292.

Lee, N., Nielsen, P.H., Andreasen, K.H., Juretschko, S., Nielsen, J.L., Schleifer, K.-H., Wagner, M., 1999. Combination of fluorescent *in situ* hybridization and microautoradiography - a new tool for structure-function analyses in microbial ecology. Appl. Environ. Microbiol. 65, 1289–1297.

Leininger, S., Urich, T., Schloter, M., Schwark, L., Qi, J., Nicol, G.W., Prosser, J.I., Schuster, S.C., Schleper, C., 2006. Archaea predominate among ammonia-oxidizing prokaryotes in soils. Nature 442, 806–809.

Liu, W.-T., Stahl, D.A., 2002. Molecular approaches for the measurement of density, diversity and phylogeny. In: Hurst, C.J. (Ed.), Manual Environmental Microbiology, second ed. ASM Press, Washington, DC.

Liu, W.T., Marsh, T.L., Cheng, H., Forney, L., 1997. Characterization of microbial diversity by determining Terminal Restriction Fragment Length Polymorphisms of genes encoding 16S rRNA. Appl. Environ. Microbiol. 63, 4516–4522.

Lombard, N., Prestat, E., van Elsas, J.D., Simonet, P., 2011. Soil-specific limitations for access and analysis of soil microbial communities by metagenomics. FEMS Microbiol. Ecol. 78, 31–49.

Loy, A., Horn, M., Wagner, M., 2003. ProbeBase: an online resource for rRNA-targeted oligonucleotide probes. Nucleic Acids Res. 31, 514–516.

Luo, C., Tsementzi, D., Kyrpides, N., Read, T., Konstantinidis, K.T., 2012. Direct comparisons of Illumina vs. Roche 454 sequencing technologies on the same microbial community DNA sample. PLoS ONE 7 (2), e30087. http://dx.doi.org/10.1371/journal.pone.0030087.

Madani, M., Subbotin, S.A., Moens, M., 2005. Quantitative detection of the potato cyst nematode, *Globodera pallida*, and the beet cyst nematode, *Heterodera schachtii*, using real-time PCR with SYBR green I dye. Mol. Cell. Probes 19, 81–86.

Margulies, M., Egholm, M., Altman, W.E., Attiya, S., Bader, J.S., Bemben, L.A., Berka, J., Braverman, M.S., Chen, Y.-J., Chen, Z., Dewell, S.B., Du, L., Fierro, J.M., Gomes, X.V., Godwin, B.C., He, W., Helgesen, S., Ho, C.H., Irzyk, G.P., Jando, S.C., Alenquer, M.L.I., Jarvie, T.P., Jirage, K.B., Kim, J.B., Knight, J.R., Lanza, J.R., Leamon, J.H., Lefkowitz, S.M., Lei, M., Li, J., Lohman, K.L., Lu, H., Makhijani, V.B., McDade, K.E., McKenna, M.P., Myers, E.W., Nickerson, E., Nobile, J.R., Plant, R., Puc, B.P., Ronan, M.T., Roth, G.T., Sarkis, G.J., Simons, J.F., Simpson, J.W., Srinivasan, M., Tartaro, K.R., Tomasz, A., Vogt, K.A., Volkmer, G.A., Wang, S.H., Wang, Y., Weiner, M.P., Yu, P.G., Begley, R.F., Rothberg, J.M., 2005. Genome sequencing in microfabricated high-density picolitre reactors. Nature 437, 376–380.

McDonald, I.R., Radajewski, S., Murrell, J.C., 2005. Stable isotope probing of nucleic acids in methanotrophs and methylotrophs: a review. Org. Geochem. 36, 779–787.

Mullis, K.B., Faloona, F.A., 1987. Specific synthesis of DNA *in vitro* via a polymerase catalysed chain reaction. Methods Enzymol. 155, 335–350.

Muyzer, G., Smalla, K., 1998. Application of denaturing gradient gel electrophoresis (DGGE) and temperature gradient gel electrophoresis (TGGE) in microbial ecology. Anton. Leeuw. Int. J. Gen. Mol. Microbiol. 73, 127–141.

Myrold, D.D., Nannipieri, P., 2014. Classical techniques versus omics approaches. In: Nannipieri, P., Pietramellara, G., Renella, G. (Eds.), Omics in Soil Science. Caister Academic Press, Portland, OR.

Myrold, D.D., Zeglin, L.H., Jansson, J.K., 2014. The potential of metagenomic approaches for understanding soil microbial processes. Soil Sci. Soc. Am. J. 78, 3–10.

Nakatsu, C.H., 2007. Soil microbial community analysis using denaturing gradient gel electrophoresis. Soil Sci. Soc. Am. J. 71, 562–571.

Nannipieri, P., Pietramellara, G., Renella, G., 2014. Omics in Soil Science. Caister Academic Press, Portland, OR.

Neufeld, J.D., Mohn, W.W., 2006. Unexpectedly high bacterial diversity in arctic tundra relative to boreal forest soils revealed with serial analysis of ribosomal sequence tags. Appl. Environ. Microbiol. 71, 5710–5718.

Neufeld, J.D., Yu, Z., Lam, W., Mohn, W.W., 2004. Serial analysis of ribosomal sequence tags (SARST): a high-throughput method for profiling complex microbial communities. Environ. Microbiol. 6, 131–144.

Nicol, G.W., Schleper, C., 2006. Ammonia-oxidising Chrenarchaeota: important players in the nitrogen cycle? Trends Microbiol. 14, 207–212.

Nyren, P., Pettersson, B., Uhlen, M., 1993. Solid phase DNA minisequencing by an enzymatic luminometric inorganic pyrophosphate detection assay. Anal. Biochem. 208, 171–175.

Ouverney, C.C., Fuhrman, J.A., 1999. Combined microautoradiography-16S rRNA probe technique for determination of radioisotope uptake by specific microbial cell types in situ. Appl. Environ. Microbiol. 65, 1746–1752.

Poulsen, L.K., Ballard, G., Stahl, D.A., 1993. Use of rRNA fluorescence in situ hybridization for measuring the activity of single cells in young and established biofilms. Appl. Environ. Microbiol. 59, 1354–1360.

Prasher, D.C., Eckenrode, V.K., Ward, W.W., Prendergast, F.G., Cormier, M.J., 1992. Primary structure of the *Aequorea victoria* green fluorescent protein. Gene 111, 229–233.

Radajewski, S., Ineson, P., Parekh, N.R., Murrell, J.C., 2000. Stable-isotope probing as a tool in microbial ecology. Nature 403, 646–649.

Radajewski, S., McDonald, I.R., Murrell, J.C., 2003. Stable-isotope probing of nucleic acids: a window to the function of uncultured microorganisms. Curr. Opin. Biotechnol. 14, 296–302.

Rangel-Castro, J.I., Killham, K., Ostle, N., Nicol, G.W., Anderson, I.C., Scrimgeour, C.M., Ineson, P., Meharg, A., Prosser, J.I., 2005. Stable isotope probing analysis of the influence of liming on root exudate utilization by soil microorganisms. Environ. Microbiol. 7, 828–838.

Ranjard, L., Poly, F., Lata, J.-C., Mougel, C., Thioulouse, J., Nazaret, S., 2001. Characterization of bacterial and fungal soil communities by automated ribosomal intergenic spacer analysis fingerprints: biological and methodological variability. Appl. Environ. Microbiol. 67, 4479–4487.

Ronaghi, M., Karamohamed, S., Pettersson, B., Uhlén, M., Nyrén, P., 1996. Real-time DNA sequencing using detection of pyrophosphate release. Anal. Biochem. 242, 84–89.

Ronaghi, M., Uhlen, M., Nyren, P., 1998. A sequencing method based on real-time pyrophosphate. Science 281, 363–365.

Rondon, M.R., August, P.R., Betterman, A.D., Brady, S.F., Grossman, T.H., Liles, M.R., Loiacono, K.A., Lynch, B.A., MacNiel, I.A., Minor, C., Tiong, C.L., Gilman, M., Osburne, M.-S., Clardy, J., Handelsman, J., Goodman, R.M., 2000. Cloning the soil metagenome: a strategy for accessing the genetic and functional diversity of uncultured microorganisms. Appl. Environ. Microbiol. 66, 2541–2547.

Saiki, R.K., Gelfand, D.H., Stoffel, S., Scharf, S.J., Higuchi, R., Horn, G.T., Mullis, K.B., Erlich, H.-A., 1988. Primer-directed enxymatic amplification of DNA with a thermostable DNA-polymerase. Science 239, 487–491.

Sanger, F., Nicklen, S., Coulson, A.R., 1977. DNA sequencing with chain-terminating inhibitors. Proc. Natl. Acad. Sci. U. S. A. 74, 5463–5467.

Smit, E., Leeflang, O., Wernars, K., 1997. Detection of shifts in microbial community structure and diversity in soil caused by copper contamination using amplified ribosomal DNA restriction analysis. FEMS Microbiol. Ecol. 23, 249–261.

Stuurman, N., Bras, C.P., Schlaman, H.R.M., Wijfjes, A.H.M., Bloemberg, G., Spaink, H.P., 2000. Use of green fluorescent protein color variants expressed on stable broad-host-range vectors to visualize rhizobia interacting with plants. Mol. Plant Microbe Interact. 13, 1163–1169.

Thies, J.E., 2007. Soil microbial community analysis using terminal restriction fragment length polymorphisms. Soil Sci. Soc. Am. J. 71, 579–591.

Torsvik, V., Øvreås, L., 2002. Microbial diversity and function in soil: from genes to ecosystems. Curr. Opin. Microbiol. 5, 240–245.

Torsvik, V., Daae, F.L., Sandaa, R.-A., Øvreås, L., 1998. Novel techniques for analysing microbial diversity in natural and perturbed environments. J. Biotechnol. 64, 53–62.

Treonis, A.M., Ostle, N.J., Stott, A.W., Primrose, R., Grayston, S.J., Ineson, P., 2004. Identification of groups of metabolically-active rhizosphere microorganisms by stable isotope probing of PLFAs. Soil Biol. Biochem. 36, 533–537.

Treusch, A.H., Leininger, S., Kletzin, A., Schuster, S.C., Klenk, H.P., Schleper, C., 2005. Novel genes for nitrite reductase and Amo-related proteins indicate a role of uncultivated mesophilic Crenarchaeota in nitrogen cycling. Environ. Microbiol. 7, 1985–1995.

Turner, T.R., Ramakrishnan, K., Walshaw, J., Heavens, D., Alston, M., Swarbreck, D., Osbourn, A., Grant, A., Poole, P.S., 2013. Comparative metatranscriptomics reveals kingdom level changes in the rhizosphere microbiome of plants. ISME J. 7, 2248–2258.

Urich, T., Lanzén, A., Qi, J., Huson, D.H., Schleper, C., Schuster, S.C., 2008. Simultaneous assessment of soil microbial community structure and function through analysis of the meta-transcriptome. PLoS ONE 3, e2527.

Větrovský, T., Baldrian, P., 2013. Analysis of soil fungal communities by amplicon pyrosequencing: current approaches to data analysis and the introduction of the pipeline SEED. Biol. Fertil. Soils 49, 1027–1037.

Wang, Z., Gerstein, M., Snyder, M., 2009. RNA-Seq: a revolutionary tool for transcriptomics. Nat. Rev. Genet. 10, 57–63.

Wellington, E.M.H., Berry, A., Krsek, M., 2003. Resolving functional diversity in relation to microbial community structure in soil: exploiting genomics and stable isotope probing. Curr. Opin. Microbiol. 6, 295–301.

Wintzingerode, F.V., Gobel, U.V., Stackebrandt, E., 1997. Determination of microbial diversity in environmental samples: pitfalls of PCR-based rRNA analysis. FEMS Microbiol. Rev. 21, 213–229.

Woese, C.R., Kandler, O., Wheelis, M.L., 1990. Towards a natural system of organisms: proposal for the domains Archaea, Bacteria and Eucarya. Proc. Natl. Acad. Sci. U. S. A. 87, 4576–4579.

Wu, L.Y., Thompson, D.K., Li, G.S., Hurt, R.A., Tiedje, J.M., Zhou, J.Z., 2001. Development and evaluation of functional gene arrays for detection of selected genes in the environment. Appl. Environ. Microbiol. 67, 5780–5790.

Wu, R.Q., Zhao, X.F., Wang, Z.Y., Zhou, M., Chen, Q.M., 2011. Novel molecular events in oral carcinogenesis via integrative approaches. J. Dent. Res. 90, 561–572.

Yi, H., Cho, Y.J., Won, S., Lee, J.E., Yu, H.J., Kim, S., Schroth, G.P., Luo, S., Chun, J., 2011. Duplex-specific nuclease efficiently removes rRNA for prokaryotic RNA-seq. Nucleic Acids Res. 39.

Zhou, J.Z., 2003. Microarrays for bacterial detection and microbial community analysis. Curr. Opin. Biotechnol. 6, 288–294.

Zhou, J., Thompson, D.K., 2002. Challenges in applying microarrays to environmental studies. Curr. Opin. Biotechnol. 13, 204–207.

Chapter 7

Physiological and Biochemical Methods for Studying Soil Biota and Their Functions

Ellen Kandeler

Institute of Soil Science and Land Evaluation, Soil Biology, University of Hohenheim, Stuttgart, Baden-Württemberg, Germany

Chapter Contents

Soil Microbiology, Ecology, and Biochemistry. http://dx.doi.org/10.1016/B978-0-12-415955-6.00007-4

I   INTRODUCTION

Biological and biochemically mediated processes in soils are fundamental to terrestrial ecosystem function because members of all trophic levels in ecosystems depend on the soil both as a source of nutrients and energy and for transformation of complex organic compounds. These biotic decomposition processes are studied at multiple levels of resolution (Sinsabaugh et al., 2002): (1) at the molecular level, plant fiber structure and enzymatic characteristics of degradation are investigated; (2) at the organismal level, the focus is on functional gene analyses, regulation of enzyme expression, and growth kinetics; and (3) at the community level, research concentrates on metabolism, microbial succession, and competition between microbial and faunal communities. Results from multiple levels of resolution must be integrated to fully understand microbial functions in soils (Sinsabaugh et al., 2002). The functions of soil biota are investigated by a range of methods focusing either on broad physiological properties (e.g., soil respiration, N-mineralization) or specific reactions of soil microorganisms (e.g., ammonia monoxygenase production by bacterial and archaeal nitrifiers). The activities of approximately 100 enzymes have been identified in soils (Tabatabai and Dick, 2002). A challenge for the future is to locate these enzymes in the three-dimensional network of soils and relate their activity to soil processes.

This chapter will focus on the important biochemical and physiological methods applied in soil microbiology and soil biochemistry. Biochemical techniques are used to determine the distribution and diversity of soil microorganisms, whereas, physiological methods are applied to understand the physiology of single cells, soil microbial communities, and biogeochemical cycling in terrestrial ecosystems. Information about methods focusing specifically on faunal abundance and activity can be found in Chapters 4 and 6. Material about the development and application of biochemical and physiological methods to study soil microorganisms is summarized in Table 7.1. Before starting any analysis in soil microbiology, it is important to select an adequate experimental design and sampling strategy.

II   SCALE OF INVESTIGATIONS AND COLLECTION OF SAMPLES

Pico- and nano-scale investigations are used to reveal the structure and chemical composition of organic substances and microorganisms, as well as the interactions between biota and humic substances. These can identify organisms, characterize their relationships, determine their numbers, and measure the rates of physiological processes. Tremendous progress has recently been made at the nano-scale in visualizing nutrient uptake by individual soil microbes, element storage in microorganisms, and the clarification of root-microbe interactions using secondary ion mass spectrometry (SIMS, see the Section IX;

TABLE 7.1 Books and Book Chapters of Methods in Soil Microbiology

Methods in Soil Microbiology	References
Soil microbiology and soil biochemistry	Levin et al. (1992), Alef and Nannipieri (1995), Weaver et al. (1994), Schinner et al. (1996)
Fungi	Frankland et al. (1991), Newell and Fallon (1991), Horwitz et al. (2013)
Actinomycetes	McCarthy and Williams (1991)
Digital analysis of soil microorganisms	Wilkinson and Schut (1998)
Soil enzymes	Burns and Dick (2002), Dick (2011), Sinsabaugh and Shah (2012), Burns et al. (2013)
Tracer techniques (^{13}C, ^{14}C, ^{11}C, ^{15}N)	Coleman and Fry (1991), Knowles and Blackburn (1993), Schimel (1993), Boutton and Yamasaki (1996), Amelung et al. (2008)
Gross nitrogen fluxes (^{15}N pool dilution)	Murphy et al. (2003)
Soil biological processes and soil organisms	Robertson et al. (1999)
Stable isotope probing techniques	Neufeld et al. (2007)
PLFAs	Ruess and Chamberlain (2010)
Metaproteomics	Becher et al. (2013)

Clode et al., 2009; Wagner, 2009). Such results boost our understanding of chemical and biological processes and structures at larger scales.

Microscale investigations concentrate either on soil aggregates or on different microhabitats characterized by high turnover of organic material (e.g., rhizosphere, drilosphere, and soil litter interface). For example, microbial abundance and C dynamics at the soil litter interface can be studied in microcosms where moisture content and solute transport through soil are regulated by irrigation through fine needles and through suction plates on the bottom of the soil cores. These microcosms have been used to study the metabolism and cometabolism of bacterial and fungal degradation of different organic substances (Poll et al., 2010). Because high-activity areas are heterogeneously distributed within the soil matrix, hot spots of activity may make up < 10% of the total soil volume, but may represent > 90% of the total biological activity (Beare et al., 1995). As a consequence, upscaling of data from the small to the plot or regional scale remains difficult because spatial distribution patterns are still incompletely known.

Investigations at the plot scale are the dominant sampling strategy for soil chemical and biological studies. A representative number of soil samples are taken from the study site (arable land, grassland, and forest) and either combined into bulk samples or treated as separate samples. Usually, random samples are combined from representative areas that are described by uniform soil type, soil texture, and habitat characteristics. Samples of agricultural soils are often taken from specific soil depths (e.g., 0-20 cm or 0-30 cm layers), and samples of forest soils are taken from specific soil horizons (e.g., litter horizon, A horizon). If soil biological data are distributed normally, the number of samples that are necessary for a given level of accuracy can be found by using the relationship

$$N = t_2 C^2 / E^2, \tag{7.1}$$

where N = the number of samples to be collected; t = Student's statistic that is appropriate for the level of confidence and number of samples collected; C = the coefficient of variation (standard deviation divided by mean); and E = the acceptable error as a proportion of the mean.

Descriptions of sampling time, frequency, and intensity, as well as preparation, archiving, and quality control, are given by Robertson et al. (1999). Soil microbiological data obtained from soil samples become more informative if supplemented by information about corresponding soil physical, chemical, and biotic factors (Table 7.2).

Spatial patterns are strongly dependent on the organisms studied, the characteristics of the study area, and the spacing of the samples. When topography and soil chemical and physical properties are relatively uniform, spatial patterns of soil biota are primarily structured by plants (plant size, growth form, and spacing). Therefore, simple *a priori* sampling designs are often inappropriate. A nested spatial sampling design is useful to explore spatial aggregation across a range of scales. For patch size estimation and mapping at a particular scale, the spatial sampling design can be optimized using simulation. To increase the statistical power in belowground field experiments and monitoring programs, exploratory spatial sampling and geostatistical analysis can be used to design a hotspot stratified sampling scheme (Klironomos et al., 1999; Robertson, 1994).

Knowledge of the spatial dependency of soil biota attributes helps to interpret their ecological significance at the ecosystem level. Similarity of ranges between soil physicochemical and biological attributes points to ecological identity. Figure 7.1 gives an example of the spatial distribution of microbial biomass in the top 10 cm of a grassland soil at the scale of 6×6 m^2. Kriged maps of soil microorganisms can be used to relate the spatial distribution of soil

TABLE 7.2 Physical, Chemical, and Biological Properties that Help Interpret Data on the Function and Abundance of Soil Biota

Physical and Chemical Soil Properties		Biological Soil Properties
Topography	Particle size and type	Plant cover and productivity
Parent material	CO_2 and O_2 status	Vegetation history
Soil type, soil pH	Bulk density	Abundance of soil animals
Moisture status	Temperature: range and variation	Microbial biomass
Water infiltration	Rainfall: amount and distribution	Organic matter inputs and roots present

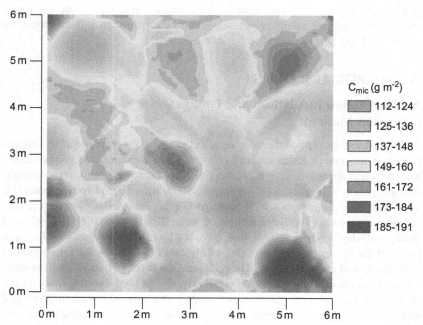

FIG. 7.1 Spatial distribution of microbial carbon (g C_{mic} m^{-2}) in a Vertic Stagnosol under grassland. The kriged map illustrates the heterogeneity of C_{mic} at the scale of 6×6 m^{-2}.

microorganisms to physical and chemical soil properties of the same site. Nevertheless, biochemical processes in the soil are dynamic, leading to variation in both space and time. Landscape-scale analyses by geostatistical methods are a useful tool for identifying and explaining spatial relationships between soil biochemical processes and site properties. Further model improvements, however, should focus on identifying and mapping time-space patterns using approaches such as fuzzy classification and geostatistical interpolation.

III STORAGE AND PRETREATMENT OF SAMPLES

Biological analyses are performed as soon as possible after soil sampling to minimize the effects of storage on soil microbial communities. Moist soil can be stored for up to three weeks at 4 °C when samples cannot be processed immediately. If longer storage periods are necessary, the samples taken to measure most soil physiological properties (soil microbial biomass, enzyme activities, etc.) can be stored at -20 °C. The soil is then allowed to thaw at 4 °C for about two days before analysis. The soil disturbance incidental to sampling may in itself trigger changes in the soil population during the storage interval. Observations on a stored sample may not be representative of the undisturbed field soil. If samples are stored, care should be taken to ensure that samples do not dry out and that anaerobic conditions do not develop. Analyses are usually preceded by sieving the samples through a 2-mm mesh to remove stones, roots, and debris. Wet soil samples can be sieved either through a 5-mm mesh or gently predried before using the 2-mm mesh.

IV MICROBIAL BIOMASS

A Fumigation Incubation and Fumigation Extraction Methods

Microbial biomass is measured to give an indication of the response of soil microbiota to management, environmental change, site disturbance, and soil pollution. Two different approaches are based on CO_2 evolution. The chloroform fumigation incubation (CFI) method (Jenkinson and Powlson, 1976) exposes moist soil to ethanol-free chloroform for 24 hours to kill the indigenous microorganisms. After removal of the fumigant, a flush of mineralized CO_2 and NH_4^+ is released during a 10-day incubation. This flush is caused by soil microorganisms that have survived the fumigation and are using cell lysates produced by the fumigation as available C and energy sources. The released CO_2 is trapped in an alkaline solution and quantified by titration. Alternatively, the CO_2 of the headspace of the samples is measured by gas chromatography. An assay of nonfumigated soil serves as a control. The amount of microbial biomass C is calculated as:

$$\text{Biomass C} = (\text{Fc} - \text{Ufc}) : \text{Kc}, \tag{7.2}$$

where biomass C = the amount of C trapped in the microbial biomass, Fc = CO_2 produced by the fumigated soil, Ufc = CO_2 produced by the nonfumigated soil sample, and Kc = fraction of the biomass C mineralized to CO_2. The Kc value is a constant, representative of the cell utilization efficiency of the fumigation procedure. This efficiency is considered to be about 40-45% for many soils (e.g., a constant of 0.41-0.45). Deviations from this range are found for subsurface and tropical soils. The Kc factor of soil samples can be estimated by measuring the $^{14}CO_2$ release of soil microorganisms isolated from different soils that use radiolabeled bacterial cells as substrates.

The chloroform fumigation extraction (CFE) method involves the extraction and quantification of microbial constituents (C, N, S, and P) immediately following $CHCl_3$ fumigation of the soil (Brookes et al., 1985). The efficiency of soil microbial biomass extraction has to be taken into account. To convert extracted organic C to biomass C, the calibration factor of k_{EC} 0.45 is recommended for surface soils (Joergensen, 1996; Joergensen et al., 2011); the k_{EC} factor of soils from other environments (subsurface soils, peat soils, etc.) should be experimentally derived. The CFE method can be applied to a wide range of soils. Soils containing large amounts of living roots require a preextraction procedure of roots because these cells are also affected by the fumigation procedure. Direct extraction of C from nonfumigated and chloroform fumigated soils with 0.5 M K_2SO_4 solution for 30 minutes could be a rapid alternative method for quantifying microbial biomass C in soils (Gregorich et al., 1990). Results of the direct extraction method are comparable with results derived from the CFE method (Setia et al., 2012).

B Substrate-Induced Respiration

The substrate-induced respiration method (SIR) estimates the amount of C held in living, heterotrophic microorganisms by measuring the initial respiration after the addition of an available substrate (Anderson and Domsch, 1978). In general, soil samples are amended with glucose, and respiration is followed for several hours (no increase in the population of microorganisms should occur). The initial respiratory response is proportional to the amount of microbial C present in the soil sample. Results can be converted to mg biomass C by applying the following conversion factor, derived from the calibration of substrate-induced respiration to the chloroform fumigation incubation technique:

$$y = 40.04x + 0.37, \qquad (7.3)$$

where y is biomass C (mg 100 g^{-1} dry weight soil) and x is the respiration rate (ml CO_2 100 g^{-1} soil h^{-1}). Respired CO_2 can be measured by titration of the CO_2 trapped in the headspace, or by gas chromatographic analysis of the headspace. The addition of glucose at levels of optimum concentration to generate the maximal initial release of CO_2 must be independently determined for each soil type and should be applied accordingly to standardize the substrate induction method between different soil types. The SIR method using titrimetric measurement of CO_2 is frequently applied because it is simple, fast, and inexpensive. A disadvantage of static systems, such as SIR that use alkaline absorption of evolved CO_2, is that the O_2 partial pressure may change, causing overestimations in neutral or alkaline soils. Nevertheless, most versions of the three methods for estimating microbial biomass (CFI, CFE, and SIR) gave very similar results from a range of 20 arable and forest sites in an interlaboratory comparison (Beck et al., 1997).

By using selective antibiotics, the SIR approach can also be applied to measure the relative contributions of fungi and bacteria to the soil microbial community. Glucose-induced respiration is determined in the presence of streptomycin to inhibit prokaryotes and in the presence of cyclohexamide to inhibit eukaryotes. An automated infrared gas analyzer system is used to continuously measure CO_2 production, and a computer program is used to calculate the bacterial/fungal respiration based on the following criteria: (1) proof of no unselective inhibition, and (2) proof of no shifts in the biosynthesis rates of bacteria and fungi (Bååth and Anderson, 2003).

V COMPOUND-SPECIFIC ANALYSES OF MICROBIAL BIOMASS AND MICROBIAL COMMUNITY STRUCTURE

Signal molecules differ in their specificity. The amount of ATP extracted from soil gives a signature of all living microorganisms; the ergosterol content is attributable to the fungal biomass. Signal molecules such as phospholipid fatty acids or respiratory quinones are used as indicators of the microbial community's structural diversity. A prerequisite for the use of a compound as a signal molecule is that it be unstable outside the cell because the compound extracted from the soil should represent living organisms only.

A ATP as a Measure of Active Microbial Biomass

All biosynthetic and catabolic reactions within cells require the participation of adenosine triphosphate (ATP), which is sensitive to phosphatases and does not persist in soil in a free state. The substrate luciferin, an aromatic N- and S-containing molecule, reacts with ATP and luciferase in the presence of Mg to yield an enzyme-luciferin-adenosine monophosphate intermediate. This, in the presence of O_2, breaks down to produce free adenosine monophosphate, inorganic P, and light. The light emitted is measured by a photometer or scintillation counter and plotted against ATP content to form a standard curve. The measurement of ATP content relative to AMP and ADP plus ATP (AEC = $(ATP + 0.5 \times ADP)/(AMP + ADP + ATP)$ gives a measure of the energy charge of the soil biota.

B   Microbial Membrane Components and Fatty Acids

Lipids play an important role as sources of energy (i.e., neutral lipids), as structural components of membranes (i.e., phospholipids) and/or as storage products of cells. To date, more than 1000 different lipids have been identified. Fatty acids are the major fraction of lipids; they contain either a fully saturated or unsaturated C chain containing, in most cases, one or two double bonds. Their classification is based on the number of C atoms counted from the carboxyl group (delta end) to the nearest double bond (Fig. 7.2). Alternatively, a single fatty acid is expressed as the number of C atoms counted from the terminal methyl group (ω end). The latter classification is frequently used in soil microbiology because the position of the double bond in the fatty acid depends on the pathway of biosynthesis, revealing characteristic "ω families" (ω3, ω6, and ω9; Ruess and Chamberlain, 2010).

Lipids can be extracted from the soil biota with a one-phase chloroform-methane extraction. Separation of the extracted lipids on silicic acid columns yields neutral lipids, glycolipids, and polar lipids. The neutral lipids can be further separated by HPLC, derivatization, and gas chromatography to yield sterols and triglycerides. Glycolipids yield poly-β-hydroxybutyrate through the processes of hydrolysis and derivatization, followed by gas chromatography. The mixture of polymers can be analyzed to determine the nutritional status of bacteria, such as those associated with plant roots. Polar lipids are separated by hydrolysis, derivatization, and gas chromatography to yield phospholipid phosphate, phospholipid glycerol, phospholipid fatty acids, and ether lipids.

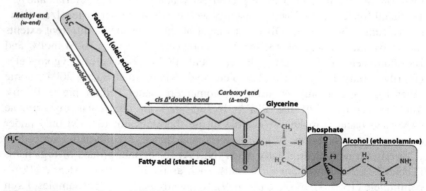

FIG. 7.2   Structure of phospholipid fatty acids. Each phospholipid consists of a polar hydrophilic head (glycerine, phosphate, alcohol) and two hydrophobic fatty acid tails. The fatty acid structure affects the bilayer structure of the cell membrane. Unsaturated fatty acids (green) are bent, whereas saturated fatty acids (blue) are not bent, making it possible for them to be packed into the cell membrane more tightly than unsaturated fatty acids.

A simple approach to characterize microbial communities is the extraction and subsequent methylation of lipids from soils to release their respective fatty acid methyl esters (FAMEs). Because the extracted lipids derive not only from cellular storage compounds and membranes of living organisms, but also from plant tissues in various stages of decomposition, it is difficult to draw conclusions about changes in soil microbial community structure from these patterns (Drenovsky et al., 2004). However, FAME analysis is more rapid than phospholipid fatty acids (PLFA) analysis and has been applied to describe microbial communities in agricultural soils.

Phospholipids are found in the membranes of all living cells, but not in their storage products. Due to their taxonomic specificity and their rapid degradation after cell death, they are frequently used as signature molecules for microbial abundance and microbial community structure under different environmental conditions (Frostegård et al., 2011; Ruess and Chamberlain, 2010). PLFAs are extracted in single-phase solvent extractions and can be analyzed by: (1) colorimetric analysis of the phosphate after hydrolysis, (2) colorimetric analysis or gas chromatography (GC) after esterification, (3) capillary GC, and (4) GC-mass spectrometry or triple-quadruple mass spectrometry. The mass profile on mass spectrometry yields information on phospholipid classes present and on their relative intensities.

Characteristic fatty acids in phospholipid and neutral lipid fractions of bacteria, fungi, plants, and animals in soils are shown in Table 7.3. The presence of indicator PLFAs unique to certain taxa is inferred from pure culture studies. There is still a debate on whether or not some biomarkers common to a certain group of microorganisms might also be found in smaller amounts in other organisms. The following biomarkers, however, are widely accepted: (1) branched chain fatty acids (iso, anteiso) as indicators for Gr$^+$ bacteria; (2) cyclopropyl fatty acids for Gr$^-$ bacteria; and (3) 18:2ω6,9 for fungi. 18:2ω6,9 is usually well correlated to ergosterol, an alternative fungal biomarker. Different fungal phyla contribute at different extents to the overall amount of 18:2ω6,9: Ascomycetes comprise 36-61 mol% and Basidiomycetes 45-57 mol% of linoleic acid (18:2ω6,9), whereas Zygomycetes contribute only 12-22 mol% of linoleic acid (Klamer and Bååth, 2004). Some other PLFAs can only be used as biomarkers under certain preconditions: The co-occurrence of the PLFA 16:1ω5 in both arbuscular mycorrhizae (AM) and bacteria makes it possible to use this indicator for AM only under environmental conditions of low bacterial abundance. An alternative procedure is to rely on neutral lipid fatty acids (NLFA) for certain groups of organisms (Table 7.3). Straight-chain fatty acids, such as palmitic (16:0), stearic (18:0), palmitoleic (16:1ω7), or oleic (18:1ω9), frequently occur in lipid samples. Even though of limited taxonomic value, they can be used as indicators of microbial biomass.

TABLE 7.3 Characteristic Ester-Linked Fatty Acids in the Lipids of Common Soil Biota

Fatty Acid Type	Frequently Found	Lipid Fraction	Predominant Origin
Saturated			
> C20, straight	22:0, 24:0	PLFA, NLFA	Plants
Iso/anteiso methyl-branched	i, a in C14-C18	PLFA	Gram-positive bacteria
10-Methyl-branched	10ME in C15-C18	PLFA	Sulfate reducing bacteria
Cyclopropyl ring	cy17:0, cy19:0	PLFA	Gram-negative bacteria
Hydroxy substituted	OH in C10-C18	PLFA	Gram-negative bacteria, actinomycetes
Monounsaturated			
Double bond C5	16:1ω5	PLFA	AM fungi, bacteria
		NLFA	AM fungi
Double bond C7	16:1 ω7	PLFA	Bacteria widespread
	18:1 ω7	PLFA	Bacteria, AM fungi
		NLFA	AM fungi
Double bond C8	18:1 ω8	PLFA	Methane-oxidising bacteria
Double bond C9	18:1 ω9	PLFA	Fungi
		PLFA	Gram-positive bacteria
		NLFA	Nematodes
	20:1 ω9	PLFA	AM fungi (*Gigaspora*)
Polyunsaturated			
ω6 family	18:2 ω6,9	PLFA	Fungi (saprophytic, EM)
		NLFA	Animals
	18:3 ω6,9,12	PLFA	Zygomycetes
	20:4 ω6,9,12,15	PLFA, NLFA	Animals widespread
ω3 family	18:3 ω3,9,12	PLFA	Higher fungi
	20:5 ω3,6,9,12,15	PLFA	Algea
		PLFA, NLFA	Collembola

Adapted from Ruess and Chamberlain (2010).
NLFA, neutral lipid fatty acid; PLFA, phospholipid fatty acid.

C Phospholipid Etherlipids

Analyses of archaeal membrane lipids (e.g., phospholipid etherlipids [PLEL]) are increasingly being included in ecological studies as a comparatively unbiased complement to gene-based microbiological approaches (Gattinger et al., 2003; Pitcher et al., 2010). The PLEL-derived isoprenoid side chains are measured by GC/MS and provide a broad picture of the archaeal community in a soil extract because lipids identified in isolates belonging to the subkingdom Eury- and Crenarchaeota are covered. Monomethyl-branched alkanes are dominant and account for 43% of the total identified ether-linked hydrocarbons, followed by straight-chain (unbranched) and isoprenoid hydrocarbons, which account for 34.6% and 15.5%, respectively. The ether lipid crenarchaeol has been postulated as a specific biomarker for all Archaea carrying out ammonia oxidation (AOA) (Pitcher et al., 2010). Because not many cultivated representatives outside the AOA lineages of the Group I Crenarchaeota are available, it is not yet clear how widespread crenarchaeol synthesis is among all Archaea.

D Respiratory Quinones

The quinone profile, which is represented as a molar fraction of each quinone type in a soil, is a simple and useful tool for analyzing population dynamics in soils. The total amount of quinones can be used as an indicator of microbial biomass (Fujie et al., 1998). These are essential components of the electron transport systems of most organisms and are present in membranes of mitochondria or chloroplasts. Isoprenoid quinones are chemically composed of benzoquinone (or naphthoquinone) and an isoprenoid side chain. There are two major groups of quinones in soils: (1) ubiquinones (1-methyl-2-isoprenyl-3,4-dimethoxypara-benzoquinone), and (2) menaquinones (1-isoprenyl-2-methyl-naphthoquinone). Most microorganisms contain only one species of quinone as their major quinone, which remains unchanged by physiological conditions. The half-lives of those quinones released by dead soil microorganisms are very short (in the range of several days). The diversity of quinone species can be interpreted directly as an indicator of microbial diversity.

E Ergosterol as a Measure of Fungal Biomass

Sterols in fungi typically exist as a mixture of several sterols, with one comprising more than 50% of the total sterol composition (Weete et al., 2010). Ergosterol is the predominant sterol of many fungi (ascomycetes and basidiomycetes), yet it does not occur in plants (Fig. 7.3). In addition to ergosterol, there appear to be at least five taxon-specific end-products of sterol biosynthesis that occur as dominant sterols in certain lineages, including some subclades of ascomycetes and basidiomycetes. For example, Glomeromycota, arbuscular mycorrhizal fungi, contain 24-ethyl cholesterol, and members of the family Mortierellaceae contain desmosterol (Grandmougin-Ferjani et al., 1999).

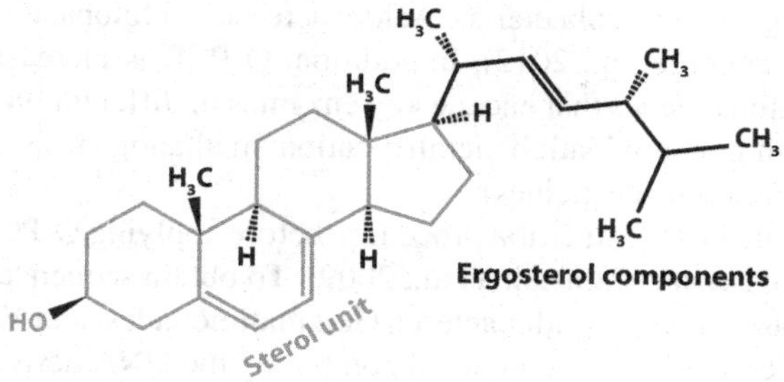

FIG. 7.3 Chemical structure of ergosterol. The sterol unit, which is common to all sterols, is marked in green.

Ergosterol is extracted by methanol and detected using high-pressure liquid chromatography with a UV detector (HPLC-UV). Because chromatographic coelution might be a problem, reversed-phase liquid chromatography with positive-ion atmospheric pressure chemical ionization tandem mass spectrometry (LC/APCI/MS/MS) can be used for full quantification and confirmation of ergosterol (Verma et al., 2002). The ergosterol content varies from 0.75 to 12.9 $\mu g\ g^{-1}$ soil in arable, grassland, and forest soils. These values correspond to 5 to 31 mg ergosterol g^{-1} fungal dry weight depending on species and growth conditions. The ratio of ergosterol to microbial biomass C is used as an index of fungal biomass to total soil microbial biomass. Shifts in microbial community structure due to soil contamination or change in vegetation can be detected using the ergosterol to microbial biomass C ratio. The content of coprostanol, which is a sterol present after sewage sludge disposal and contamination by municipal wastes, is a useful marker of human fecal matter contamination of soils.

F   Gene Abundance as a Measure of Biomass of Specific Groups of Soil Microorganisms

Quantitative PCR (Q-PCR or real-time PCR) is increasingly applied to quantify the abundance and expression of taxonomic and functional gene markers within soils. The Q-PCR-based analyses combine end-point detection PCR with fluorescent detection to measure the accumulation of amplicons in "real time" during each cycle of the PCR amplification (Smith and Osborn, 2009). The quantification of gene or transcript numbers is performed by detection of amplicons during the early exponential phase of the PCR when they are proportional to the starting template concentration. This procedure can be coupled with a reverse transcription reaction to determine gene expression (RT-Q-PCR). The Q-PCR approaches targeting highly conserved regions of the 16S rRNA gene are currently used to quantify bacterial and/or archaeal copy numbers. The use of taxa-specific sequences within hypervariable regions of genes makes it possible to target specific groups of soil organisms at higher taxonomic resolution (e.g., ß Proteobacteria, Acidobacteria, Actinobacteria; Philippot et al., 2011; Lennon et al., 2012). In addition, Q-PCR is increasingly used to quantify functional genes that encode key enzymes in different biogeochemical cycles (e.g., ammonia oxidation, denitrification, methanogenesis, methane oxidation, degradation of pesticides).

The adequate DNA extraction procedure before applying Q-PCR methods is still a matter of debate (Feinstein et al., 2009). To obtain sequences from rarely cultivated groups such as Acidobacteria, Gemmatimonades, and Verrucomicrobia, Feinstein et al. (2009) recommend combining the DNA derived from three sequential extraction procedures. A detailed description of different DNA extraction procedures and PCR reactions, as well as application of these molecular techniques, can be found in Chapter 6.

G   Component-Specific Analyses of Microbial Products

After cell death, many microbial products are not immediately decomposed. Therefore, several lipids and N-containing compounds (e.g., lipopolysaccharides, glycoproteins, amino sugars) can be used to track back the bacterial or bacterial plus fungal residues of different fungi and bacteria (Amelung et al., 2008). The outer cell membrane of Gr⁻ bacteria contains unique lipopolysaccharide (LPS) polymers. The peptidoglycan of bacterial cell walls contains N-acetylmuramic acid and diaminopimelic acid. Chitin, a polymer of N-acetylglucosamine, is found in many fungi, but is also present in the exoskeleton of invertebrates. Three amino sugars that are mainly derived from dead microbial biomass can be used to trace the fate of C and N within residues of bacteria and fungi. These amino sugars in soils comprise about 1-5% of total SOC. Muramic acid and galactosamine mainly originate from bacterial peptidoglycan, whereas glucosamine is mainly derived from fungal cell walls.

The enantiomers of amino acids have been used as markers for both bacterial residues and cell aging. D-amino acids are produced by racemization from their respective L-enantiomers during cell aging. Because aged and dead cells do not have any D-amino acid oxidase to decompose D-amino acids, these compounds may accumulate in soils. D-lysine is an especially promising age marker. It is devoid of any nutritional value, and its formation in soil is independent from microbial activity and soil organic matter (SOM) turnover.

Arbuscular mycorrhizal fungi (AMF) produce a recalcitrant AM-specific glycoprotein, glomalin. It is extracted from soil by applying several cycles of autoclaving, quantified using a Bradford assay, and evaluated for immunoreactivity using a monoclonal antibody.

Metaproteomic studies of whole microbial communities from soils and sediments comprise three basic steps: (1) sample preparation including protein extraction, purification, and concentration; (2) protein or peptide separation and MS analysis; and (3) peptide identification based on the obtained

MS/MS data, followed by an assignment of the peptides to proteins in an appropriate database (Becher et al., 2013). Amino acid sequences and relative abundance information for detected peptides allow characterizing expressed protein functions within communities with high resolution (Müller and Pan, 2013). However, a critical evaluation of the extraction protocol is crucial for the quality of the metaproteomics data of soils (Keiblinger et al., 2012).

VI ISOTOPIC COMPOSITION OF MICROBIAL BIOMASS AND SIGNAL MOLECULES

Radioactive and stable isotope tracers are used to follow the flow of C and nutrients (N, P, and S) into total microbial biomass, specific groups of microorganisms, or components of the soil microbial community. Studies of ^{13}C have increasingly gained interest due to improved sensitivity of ^{13}C measurements and its nonradioactive nature (in contrast to ^{14}C studies). Carbon use of total microbial biomass or specific groups of microorganisms are followed by labeling techniques that differ in their actual ^{13}C enrichment: natural abundance, near natural abundance (50-500‰) or highly enriched (> 500‰; Amelung et al., 2008). Natural abundance ^{13}C labeling tracer approaches are based on the physiological difference during the photosynthetic fixation of CO_2 between C_3 and C_4 plants. Switching C_3 to C_4 plants, ^{13}C pulse labeling or continuous labeling introduces a distinct C signal to the soil system as well as to the soil microbial community (Kramer et al., 2012). The combined application of ^{14}C plant labeling and ^{13}C natural abundance allows linking aboveground C fixation by photosynthesis to belowground C dynamics of rhizosphere microorganisms.

A Isotopic Composition of Microbial Biomass

Determining the isotopic composition of microbial biomass C is an important tool for the study of soil microbial ecology and the decomposition and transformation of soil organic C. The use of the fumigation incubation (FI) method is restricted to isotopes ^{14}C, ^{13}C, ^{15}N, whereas the fumigation extraction (FE) method can be used with a larger range of isotopes (i.e., ^{14}C, ^{13}C, ^{15}N, ^{32}P, and ^{35}S). Because C_3 plants (e.g., wheat) and C_4 plants (e.g., maize) differ in their natural abundance of ^{13}C, these materials provide a natural label that can be used for decomposition studies in both microcosm and field experiments. Determinations of ^{13}C/^{12}C are performed with an offline sample preparation technique combined with isotope analysis by a dual-inlet isotope ratio mass spectrometer (IRMS) or an online analysis using an element analyzer connected to an isotope ratio mass spectrometer (EA-IRMS). An alternative method is based on the UV-catalyzed liquid oxidation of fumigated and nonfumigated soil extracts combined with trapping of the released CO_2 in liquid N; $\delta^{13}CO_2$-C is subsequently determined with a gas chromatograph connected to an IRMS (Potthoff et al., 2003). The ^{13}C analysis can also be done using an automated continuous-flow IRMS. In addition, isotopic analysis of $\delta^{15}N$ and $\delta^{13}C$ of DNA extracted from soil provides a way to determine the natural abundance of this important fraction of soil microbial biomass (Schwartz et al., 2007). This is a powerful tool for elucidating C and N cycling processes through soil microorganisms.

B Stable-Isotope Probing of Fatty Acids, Ergosterol, and Nucleic Acids

Stable isotope probing techniques (SIP) introduce a stable isotope-labeled substrate into a microbial community, follow the fate of the substrate by extracting signal molecules such as phospholipid fatty acids and nucleic acids, and determine which specific molecules have incorporated the isotope (Kreuzer-Martin, 2007). These techniques allow for the study of substrate assimilation in minimally disturbed communities and provide a tool for linking functional activity to specific members of microbial communities.

The PLFA-SIP techniques are characterized by higher sensitivity than RNA-SIP or DNA-SIP techniques. Due to the high sensitivity of the PLFA-SIP method, incubations can be carried out at near *in situ* concentrations of labeled substrates. After incubation of the soil with labeled substrates, all lipids are extracted. The isotopic compositions of fatty acids are analyzed by gas chromatography-combustion-isotopic ratio mass spectrometry (GC-IRMS). The enrichment of ^{13}C in single biomarkers (Table 7.3) is used to identify the groups of microorganisms feeding on the added substrate. In contrast to other SIP-based methods (e.g., RNA-SIP), it is not necessary to purify unlabeled from labeled biomarkers. If two C sources with different isotopic signatures are available (e.g., ^{13}C litter and SOC), PLFA-SIP allows the quantification of the relative assimilation of each of these food resources. Alternatively, the natural abundance of a C_4 plant (e.g., maize), introduced to a soil previously planted with solely C_3 crops, can be used to trace C routing from the food into the soil microbial community. Both PLFA- and NLFA-SIP techniques can also be used to study the trophic transfer of C from primary decomposers to higher levels of the soil food web (e.g., nematodes, collembolans, and earthworms; Ruess and Chamberlain, 2010). For example, Murase et al. (2011) studied the strain-specific incorporation of methanotrophic biomass into eukaryotic grazers in a rice field soil with a PLFA-SIP technique.

Carbon flow from a given substrate into fungi can be followed by compound-specific analysis of ergosterol, combined with a stable isotopic probing technique. Soils are incubated with a substrate (e.g., ^{13}C labeled litter or litter of a C_4 plant) for a period of days to months. After the incubation, ergosterol is extracted, purified by a preparative HPLC, followed by the determination of ^{13}C ergosterol using a GC-C-IRMS or tandem mass spectrometry (GC-MS-MS; Kramer et al., 2012).

The DNA- and RNA-SIP techniques involved labeling of soils with a substrate (e.g., ^{13}C glucose, ^{13}C cellulose, maize-derived litter) for a period of hours up to several weeks, extraction of DNA or RNA, and buoyant density gradient centrifugation (Neufeld et al., 2007). The nucleic acids of microorganisms having assimilated the ^{13}C have a higher buoyant density than the nucleic acids of organisms that have not. The ^{13}C-containing nucleic acid is separated from unlabeled nucleic acid during the centrifugation, and labeled nucleic acid is concentrated in heavier gradient fractions. Nucleic acid fractions at various positions in the gradient are collected and amplified by the polymerase chain reaction (PCR) or reverse transcription (RT) PCR using primers complementary to 16S rDNA or rRNA. Alternatively, specific functional genes are used as amplification targets. Amplified products can be separated by denaturing gradient gel electrophoresis (DGGE), cloned directly, or subjected to terminal restriction fragment length polymorphism analysis for phylogenetic characterization.

C Growth Rates from Signature Molecules and Leucine/ Thymidine Incorporation

Growth and turnover rates can be determined by incorporating tracers into precursors of cytoplasmic constituents, membranes, or cell wall components. The increase in the tracer over short time periods yields estimates of growth. For example, a technique to estimate relative fungal growth rates is based on the addition of ^{14}C-acetate to a soil slurry and measurement of the subsequent uptake and incorporation by fungi of the labeled acetate into the fungus-specific ergosterol (Newell and Fallon, 1991). The specificity of this incorporation was shown by using fungal and bacterial inhibitors. Incorporation rates were linear up to 18 hours after the acetate addition, but absolute growth rates cannot be calculated due to the uncertainty of conversion factors and problems associated with the incorporation of the added acetate. Similar techniques have also been developed for C isotopes in microbial lipids. In combination with their role as biomarkers, fatty acids can also be used to monitor the C flux in a bacterial community by measuring the ratios of their C isotopes. Before fatty acids can be used as chemotaxonomic markers and indicators of substrate usage in microbial communities, calibration studies on the degree and strain specificity of isotopic ^{13}C fractionation with regard to the growth substrate are necessary.

Bacterial and fungal growth rates can also be separately determined by using leucine/thymidine incorporation to estimate bacterial growth and acetate incorporation into ergosterol to estimate fungal growth (Rousk and Bååth, 2011). A parameter to quantify how C is partitioned between growth and respiration is carbon-use efficiency (CUE), defined as the ratio of the amount of C employed in new biomass (excluding C excreted in the form of metabolites and enzymes) relative to the amount of C that has been consumed (Manzoni et al., 2012). High values of carbon-use efficiency indicate efficient growth and relatively low release of C to the atmosphere.

VII   PHYSIOLOGICAL ANALYSES

A wide array of physiologically based soil processes are known (e.g., decomposition, ammonification, nitrification, denitrification, N_2 fixation, P mineralization, and S transformations). These are described in later chapters, which detail these processes. Here we primarily cover culture studies, respiration techniques, N-mineralization, and enzyme measurements.

A   Culture Studies

Culturing a soil organism involves transferring its propagules to a nutrient medium conducive to growth. These techniques allow isolation of specific soil organisms from a wide range of soils. Culture techniques are selective and are designed to detect organisms with particular growth forms or biochemical capabilities. After liberating cells from surfaces of soil particles and aggregates in sterile water, Windogradsky's solution, or physiological saline (0.85% NaCl), soil suspensions are diluted to an appropriate concentration. The degree of dilution required is related to the initial number of organisms in the soil. Serial dilutions, usually 1:10, stepwise, are made on a known weight of wet soil. Aggregates are broken in the Waring blender, often in the presence of a dispersing agent such as $Na_4P_2O_7$. Replicate 0.1 mL portions of appropriate suspensions are transferred to solid or liquid media in which individual propagules develop to the point of visible growth. After appropriate incubation, single colonies on solid media are counted, and each is equated with a single propagule in the soil suspension (CFU: colony forming unit).

The plate count of bacteria in soils, waters, and sediments usually represents 1-5% of the number determined by direct microscopy, leading to many discussions concerning nonculturable bacteria in nature (Kjelleberg, 1993) and the applicability of this method for quantitative soil microbiology. The plate count methods are not suitable for enumerating fungal populations or densities because spores and fragments of mycelia develop as single colonies that are counted. Due to these strong limitations, many researchers doubt whether plate counts can be used for quantification of soil microbiota.

Population densities of various groups of bacteria can be estimated by the most probable number (MPN) technique. This method uses a liquid medium to support growth of soil microorganisms. The dilution count is based on determination of the highest soil dilution that will still provide growth in a suitable medium. An actual count of single cells or colonies is not necessary. Employment of replicate inocula (ten usually, five minimum) from each of three successive serial dilutions at the estimated extinction boundary for growth enables resultant visible growth to be converted to numbers within statistical limits of reliability (Gerhardt et al., 1994). A desktop computer program has been written to determine MPN and related confidence limits as well as to correct for biases (Klee, 1993). This methodology is useful because it allows a bacterial population to be estimated based on process-related attributes, such as nitrification, denitrification by denitrifiers, or N fixation by free-living aerobic and microaerophilic N-fixing bacteria. The MPN technique requires an appropriate choice of growth medium and accurate serial dilution to obtain quantitative data. In general, the results are usually less precise than those obtained with direct plating methods.

B   Isolation and Characterization of Specific Organisms

Specific organisms are usually isolated from soil by using liquid or solid substrates. These can involve very general substrates from which individual colonies are picked for further characterization. Also useful are media containing inhibitors, such as chloramphenicol, tetracycline, and streptomycin, which inhibit protein synthesis by binding to the 50S ribosomal subunit or nalidixic acid, which inhibits DNA synthesis of Gr$^-$ bacteria. Specific growth environments include high or low pH, anaerobiosis, or high salinity. Plants are used to identify and enrich specific soil populations. Growth of compatible legumes makes it possible to isolate and identify rhizobia. Root or leaf pathogens and mycorrhizae are often similarly enriched and identified. Isolation can be performed without growth media by microscopic examination combined with micromanipulators or optical laser tweezers, which can separate single cells, spores, or hyphal fragments.

Genetic markers that are incorporated into isolated organisms before soil reinoculation include antibiotic resistance, the lux operon for light emission, and enzyme-activity genes. Resistance to most presently used antibiotics is usually carried on a plasmid. The use of an appropriate vector makes it possible to transfer this plasmid to soil isolates. These isolates can then be reintroduced into soil and recovered by growth on the appropriate medium containing that antibiotic. Only organisms carrying the antibiotic resistance gene(s) will be capable of growth.

Automated, multisubstrate approaches have been used as a metabolic fingerprinting system. For example, the "Biolog System" makes it possible to test growth reactions of many isolates on a broad range of media and to analyze community composition. This, combined with the use of a broad range of known organisms and mathematical clustering techniques, allows characterization of unknowns by sorting them into affinity types occurring in that habitat. A community-level physiological approach is based on the C utilization profiles of soil dilutions inoculated onto the redox-based plates used in multisubstrate systems such as Biolog. This bypasses the need to work with isolated culturable organisms, but still requires growth (reduction) in the appropriate substrate. Because of their generally oxidative and mycelial nature, fungi are not amenable to approaches by substrate utilizing-clustering analyses. Multiple substrate finger-printing also has limitations (i.e., measuring the enzyme activity of a growing microbial population or the selectivity of the microbial response).

C   Microbial Growth and Respiration

Microbial communities oxidize naturally occurring organic material such as carbohydrates according to the generalized equation:

$$CH_2O + O_2 \rightarrow CO_2 + H_2O + \text{intermediates} + \text{cellular material} + \text{energy}. \quad (7.5)$$

Under anaerobic conditions, the most common heterotrophic reaction is that of fermentation, which in its simplest form is

$$C_6H_{12}O_6 \rightarrow 2\,CH_3CH_2OH + 2\,CO_2 + \text{intermediates} + \text{cellular material} + \text{energy}. \quad (7.6)$$

Measurement of microbial activity is complex under anaerobic conditions. Fermentation products and CH_4 produced within anaerobic microsites can diffuse to aerobic areas, where oxidation to CO_2 and H_2O can occur. Microbial respiration is determined by measuring either the release of CO_2 or the uptake of O_2. Because the atmospheric CO_2 concentration is only 0.036% versus 20% for O_2, measurements of CO_2 production are more sensitive than those for O_2. One method of CO_2 measurement involves aeration trains. Here, NaOH is used to trap evolved CO_2 in an air stream from which CO_2 is removed before the air is exposed to the soil sample. The reaction occurs as follows:

$$2\,NaOH + CO_2 \rightarrow Na_2CO_3 + H_2O. \quad (7.7)$$

Before titration, $BaCl_2$ or $SrCl_2$ is added to precipitate the CO_3^{2-} as $BaCO_3$, and excess NaOH is back titrated with acid. The use of carbonic anhydrase and a double endpoint titration provides greater accuracy when CO_2 concentrations are low. In the laboratory, NaOH containers placed in sealed jars are convenient and effective for CO_2 absorption. The jars must be opened at intervals so that the O_2 concentration does not drop below 10%. Gas chromatography, with thermal conductivity detectors, can be used to measure the CO_2 concentration after CO_2 is separated from other constituents on column materials such as Poropak Q. Computer-operated valves in conjunction with GC allow time-sequence studies to be automatically conducted. Infrared gas analyzers (IRGA) are sensitive to CO_2 and can be used for both static and flow systems in the laboratory and in the field after H_2O, which absorbs in the same general wavelength, has been removed. The results are expressed either per unit of soil dry weight ($\mu g\ CO_2\text{-C}\ g^{-1}\ \text{soil}\ h^{-1}$) or per unit of microbial biomass ($mg\ CO_2\text{-C}\ g^{-1}\ C_{mic}\ h^{-1}$). The ratio of respiration to microbial biomass is termed the metabolic quotient (qCO_2) and is in the range of 0.5-3 mg $CO_2\text{-C}\ g^{-1}\ C_{mic}\ h^{-1}$. The metabolic quotient is particularly useful in differentiating the responses of soil biota to soil management techniques. For example, stress, heavy metal pollution, and nutrient deficiency increase qCO_2 because microbial biomass decreases and respiration increases.

The measurement of CO_2 can be augmented by incorporating ^{14}C or ^{13}C into chosen substrates. The tracer may be molecules such as glucose, cellulose, amino acids, and herbicides, or complex materials, such as microbial cells or plant residues. Substituting $SrCl_2$ for $BaCl_2$ allows the precipitate to be measured in mass spectrometers. The measurement of $^{14}CO_2$ or $^{13}CO_2$ makes it possible to calculate the decomposition rate of soil organic matter as well as to establish a balance of the C used in growth relative to substrate decomposition and microbial by-products. Mass spectrometers capable of directly analyzing a gaseous sample for $^{13}CO_2$ are preferable to the precipitation procedure.

A variety of methods exist to measure soil CO_2 efflux under field conditions (Sullivan et al., 2010). Static or dynamic chambers placed on the soil surface allow for measurement of soil CO_2 efflux by quantifying the increase of headspace gas concentration within the chamber headspace over a certain time. This procedure does not alter the soil structure, and therefore, field respiration rates of the indigenous microbial population are reliably measured. Gas samples are taken from the static field chamber using a gastight syringe and are then injected into a gas chromatograph. Measurements of other gases such as N_2O and CH_4 from the same samples are possible. A dynamic chamber is coupled with an IRGA in the field. This procedure allows for instantaneous soil CO_2 efflux measurements. An alternative option is to model CO_2 efflux from soil CO_2 gradient profiles. Solid-state IRGA probes (e.g., GMM 222; Vaisala, Inc., Helsinki, Finland) are buried vertically at different soil depths (e.g., 2, 10, and 20 cm into the mineral soil). The probes are connected to a data logger and multiplexer. The diffusivity of CO_2 in soil can be calculated by models (Sullivan et al., 2010).

The eddy covariance technique ascertains the net ecosystem exchange (NEE) rate of CO_2 (the NEE represents the balance of gross primary productivity and respiration of an ecosystem). The covariance between fluctuations in vertical wind velocity and CO_2 mixing ratio across the interface between the atmosphere and a plant canopy is measured at a flux tower (Baldocchi, 2003). Although flux tower data represent point measurements with a footprint of typically 1 km^2, they are especially useful for validating models and to spatialize biospheric fluxes at the regional scale (Papale and Valentini, 2003).

Soil organic matter decomposition in the field can be examined by following the decay process (i.e., weight loss) of added litter. Site-specific litter or standard litter (e.g., wheat straw) is placed in nylon mesh bags and exposed on or just below the soil surface. Organic matter decomposition can also be followed by the minicontainer system (Fig. 7.4). The system consists of polyvinylchloride bars (PVC bars) as carriers and minicontainers enclosing the straw material, which can be exposed horizontally or vertically in soil, as well as placed on the soil surface. The minicontainers are filled with 150 to 300 mg of organic substrates and closed by nylon mesh of variable sizes (20, 250, or 500 μm, or 2 mm) to exclude or include

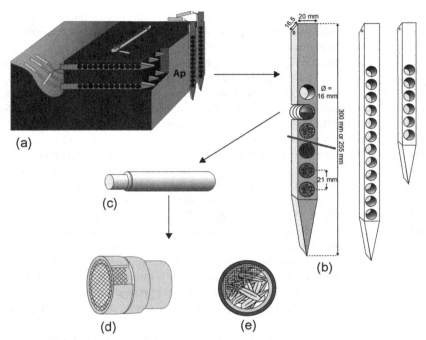

FIG. 7.4   The decomposition of organic substrates can be estimated by exposing minicontainers filled with site-specific or standard material (e.g., maize straw) to particular conditions for a certain period of time (from weeks to several months; from Eisenbeis et al., 1999). (a) Minicontainer can be exposed to the top layers of agricultural soils (Ap = ploughed A horizon) and to forest soils for several weeks up to years. Vertical insertion of the bars will give information about gradients of decomposition within a soil profile; horizontal exposure of the bars help to explain spatial variation of decomposition within one horizon. (b) Polyvinylchloride bars are 6 or 12 minicontainers that can be removed from the bar by a rod after exposure. (c) A minicontainer in side view (d) where the left end is closed with a gauze disc held in place with a ring. The mesh size of the gauze will inhibit and allow colonization of the organic substrates by mesofauna (20 μm, 250 μm, 500 μm, or 2 mm). Minicontainers can be filled with organic material of different qualities (straw, litter, cellulose, etc.)

the faunal contribution to organic matter decomposition. After an exposure time of several weeks to months, decomposition is calculated based on the weight loss of the oven-dried material taking into consideration the ash content of the substrate. A time series analysis enables the dynamics of decomposition processes to be investigated.

D   Nitrogen Mineralization

Nitrogen (N) mineralization is estimated in field or laboratory experiments as the release of inorganic N from soil. Alternatively, specific steps of the N mineralization process can be estimated (e.g., arginine deaminase, urease, ammonia monooxygenase). Nitrogen availability is measured using aerobic and anaerobic incubation tests as well as soil inorganic N measurements. The recommended methods differ in incubation time and temperature, moisture content, and extraction of ammonium and nitrate. Frequently, soils are incubated under aerobic conditions and analyzed for ammonium, nitrite, and nitrate. Because mineral N is partly immobilized into the microbial biomass during incubation, these incubation methods yield the net production of ammonium and nitrate. Isotope pool dilution techniques enable gross rates of nitrification (or mineralization) to be determined by monitoring the decline in the ^{15}N abundance in a nitrate or ammonium pool, labeled at $t = 0$ and receiving unlabeled N via nitrification or mineralization (Murphy et al., 2003). Labeled N can be applied as $^{15}NH_4^+$ solution or injected as $^{15}NH_3$ gas into soil. The ^{15}N pool dilution and enrichment can also be used to separate the heterotrophic and autotrophic pathways of nitrification. An isotopic dilution experiment using $^{14}NH_4$ $^{15}NO_3$ yields rates of nitrification by the combined autotrophic and heterotrophic paths. A parallel isotope dilution experiment with $^{15}NH_4$ $^{15}NO_3$ provides the gross mineralization rate and the size and ^{15}N abundance of the nitrate pool at different time intervals. Spatial variability of the tracer addition and extraction must be taken into account when interpreting such data.

Taylor et al. (2010) published an approach to estimate the archaeal and bacterial contributions to the nitrification potential of different soils. In a first step, ammonia monooxygenase of all soil microorganisms is irreversibly inactivated by acetylene; in a second step, acetylene is removed. Archaeal and bacterial contribution to the ammonia-oxidizing potential can be measured from the recovery of nitrification potential (RNP) in the presence and absence of kanamycin, an inhibitor of bacterial protein synthesis (Taylor et al., 2012). The inhibitor prevents resynthesis of ammonia monooxygenase by bacteria. Any RNP that recovers in the presence of a bacterial protein synthesis inhibitor is likely to be contributed by archaea.

VIII ACTIVITIES OF ENZYMES

Enzymes are specialized proteins that combine with a specific substrate and act to catalyze a biochemical reaction. In soils, enzyme activities are essential for energy transformation and nutrient cycling. The enzymes commonly extracted from soil, and their range of activities, are given in Table 7.4. Some enzymes (e.g., urease) are constitutive and routinely produced by cells; others, such as cellulase, are adaptive or induced and formed only in the presence of a susceptible substrate, some other initiator, or in the absence of an inhibitor. Dehydrogenases are often measured because they are not constitutive and also because they are found only in living systems. Enzymes associated with proliferating cells occur in the cytoplasm, the periplasmic membrane, and the cell membrane. Figure 7.5 illustrates that soil enzymes are not only associated with proliferating cells, but also with humic colloids and clay minerals as extracellular enzymes.

TABLE 7.4 Some Enzymes Extracted from Soils, the Reactions They Catalyze, and Their Range of Activities

Enzyme	Reaction	Range of Activity
Cellulase	Endohydrolysis of 1,4-β-glucosidic linkages in cellulose, lichenin, and ceral β-glucose	0.02-3.33 μM glucose g^{-1} h^{-1}
Xylanase	Hydolysis of 1,4-β-glucosidic linkages in hemicellulose	0.06-130 μM glucose g^{-1} h^{-1}
β-Fructo-furanosidase (Invertase)	Hydrolysis of terminal nonreducing β-D-fructo-furanoside residues in β-fructofuranosides	0.61-130 μM glucose g^{-1} h^{-1}
β-Glucosidase	Hydrolysis of terminal, nonreducing β-D-glucose residues with release of β-D-glucose	0.09-405 μM p-nitrophenol g^{-1} h^{-1}
Chitinase	Random hydrolysis of N-acetyl-β-D-glucosaminide 1,4-β linkages in chitin and chitodextrins	31-213 nM MUF g^{-1} h^{-1}
Proteinase	Hydrolysis of proteins to peptides and amino acids	0.5-2.7 μM tyrosine g^{-1} h^{-1}
Leucine-aminopeptidase	Metallopeptidase cleaving N-terminal residues from proteins and peptides	3-380 nM MUF g^{-1} h^{-1}
Urease	Hydrolysis of urea to CO_2 and NH_4^+	0.14-14.3 μM N-NH_3 g^{-1} h^{-1}
Alkaline phosphatase	Orthophosphoric monoester + H_2O → an alcohol + orthophosphate	6.76-27.3 μM p-nitrophenol g^{-1} h^{-1}
Acid phosphatase	Orthophosphoric monoester + H_2O → an alcohol + orthophosphate	0.05-86.3 μM p-nitrophenol g^{-1} h^{-1}
Arylsulfatase	A phenol sulfate + H_2O → a phenol + sulfate	0.01-42.5 μM p-nitrophenol g^{-1} h^{-1}
Catalase	$2H_2O_2 \rightarrow O_2 + 2H_2O$	2.55-3.08 μM O_2 g^{-1} h^{-1}

Adapted from Tabatabai and Fung (1992), Nannipieri et al. (2002), and Dick (2011). MUF, methylumbelliferyl.

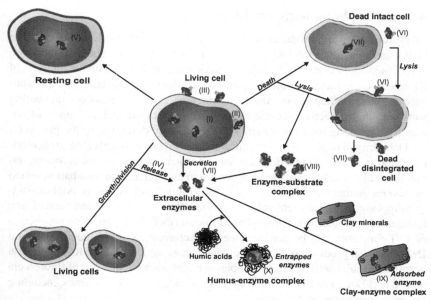

FIG. 7.5   Locations of enzymes. (From Burns, 1982 redrawn by R. S. Boeddinghaus.) I, Intracellular enzymes; II, Periplasmatic enzymes; III, Enzymes attached to outer surface of cell membranes; IV, Enzymes released during cell growth and division; V, Enzymes within nonproliferating cells (spores, cysts, seeds, endospores); VI, Enzymes attached to dead cells and cell debris; VII, Enzymes leaking from intact cells or released from lysed cells; VIII, Enzymes temporarily associated in enzyme-substrate complexes; IX, Enzymes absorbed to surfaces of clay minerals; X, Enzymes complexed with humic colloids.

Because many of the enzymes that are frequently measured can be intracellular, extracellular, bound, and/or stabilized within their microhabitat, most assays determine enzymatic potential, but not necessarily the activity of proliferating microorganisms.

Methods for analysis of a broad range of enzymes are described by Tabatabai (1994), Alef and Nannipieri (1995), Schinner et al. (1996), and Dick (2011). A general introduction to enzymes in the environment, their activity, ecology, and applications, is given by Burns and Dick (2002). This section will primarily focus on colorimetric and fluorimetric techniques to measure enzyme activities in soils and on some approaches for visualizing the locations of soil enzymes. Methods of soil enzyme activities using spectroscopic and fluorometric approaches are available in bench scale and microplate format (Deng et al., 2013; Dick, 2011). Use of the microplate format offers the advantage of simultaneous analysis of multiple enzymes using a small amount of soil (Vepsäläinen et al., 2001). A microplate reader can measure the absorbance or fluorescence up to 96 wells simultaneously.

A Spectrophotometric Methods

Many substrates and products of enzymatic reactions absorb light either in the visible or in the ultraviolet region of the spectrum, or they can be measured by a color reaction. Due to their higher sensitivity, methods based on the analysis of the released product are more frequently used than methods based on the analysis of substrate depletion. Analyzing enzyme activities involves incubating soils with the respective substrate at a fixed temperature, pH, and time, subsequently extracting the product and colorimetrically determining the product.

Dehydrogenases, which are intracellular enzymes catalyzing oxidation-reduction reactions, can be detected using a water-soluble, almost colorless, tetrazolium salt, which forms a reddish formazan product after an incubation period of several hours. A commonly used tetrazolium salt is 2-(p-iodophenyl)-3-(p-nitrophenyl)-5-phenyltetrazolium chloride (INT), which is transformed into an intensely colored, water-insoluble formazan (INT-formazan). The production of INT-formazan can be detected spectrophotometrically by quantifying total INT-formazan production. Another enzyme assay system used to reflect general or total microbial activity in soil samples is based on the hydrolysis of fluorescein diacetate (FDA). The FDA is hydrolyzed by a wide variety of enzymes, including esterases, proteases, and lipases.

Enzymes involved in C-cycling (xylanase, cellulase, invertase, and trehalase) are measured based on the release of sugars after soils are incubated with a buffered solution (pH 5.5) containing their corresponding substrates (xylan, carboxymethylcellulose, sucrose, or trehalose). The incubation period depends on the substrate used: high-molecular-weight substrates are incubated for 24 hours, whereas low-molecular-weight substrates are incubated for 1 to 3 hours. Reducing sugars released during the incubation period cause the reduction of potassium hexacyanoferrate (III) in an alkaline solution. Reduced potassium hexacyanoferrate (II) reacts with ferric ammonium sulfate in an acid solution to form a complex of ferric hexacyanoferrate (II; Prussian blue), which is determined colorimetrically.

Enzymes involved in P-cycling (phosphomonoesterase, phosphodiesterase, phosphotriesterase) are preferably determined after the addition of a substrate analog. Phosphomonoesterase hydrolyzes the phosphate-ester-bond of the p-nitrophenylphosphate, and the p-nitrophenol released is measured as a yellow compound under alkaline conditions. For determining arylsulfatase, which catalyzes the hydrolysis of organic sulfate ester, p-nitrophenyl sulfate is used as a substrate analog. p-nitrophenyl substrates are widely used for different enzyme assays because the released product p-nitrophenol can be quantitatively extracted. Urease is an important extracellular enzyme that hydrolyzes urea into CO_2 and NH_3. The urease assay involves measuring the released ammonia either by a colorimetric procedure or by steam distillation followed by a titration assay of ammonia.

Although the described potential enzyme activities are based on the addition of an excess amount of suitable substrate and subsequent determination of product release, Bünemann (2008) suggests reversing the approach by adding excess of an enzyme. In this case, substrate availability becomes rate limiting, and the maximum release of product indicates the availability of a given substrate. The method is mainly applied to characterize the potential bioavailability of organic P compounds after adding different phosphatases to environmental samples (soil, manure, and sediment extracts; Bünemann, 2008).

B Fluorescence Methods

Fluorogenic substrates are used to assay extracellular enzymes in aquatic and terrestrial environments. These substrates contain an artificial fluorescent molecule and one or more natural molecules (e.g., glucose, amino acids), and they are linked by a specific type of binding (e.g., peptide binding, ester binding). The substrates used are conjugates of the highly fluorescent compounds 4-methylumbelliferone (MUB) and 7-amino-4-methyl coumarin (AMC). Marx et al. (2001) used this method to measure the activities of enzymes involved in C-cycling (ß-D-glucosidase, ß-D-galactosidase, ß-cellobiase, ß-xylosidase), N-cycling (leucine aminopeptidase, alanine amino peptidase, lysine-alanine aminopeptidase), P-cycling (acid phosphatase), and S-cycling (arylsulfatase). Fluorogenic model substrates are not toxic and are supplied to soil suspensions in high or increasing quantities to measure the maximum velocity of hydrolysis (v_{max}). Fluorescence is observed after enzymatic splitting of the complex molecules (Fig. 7.6). All calibrations are

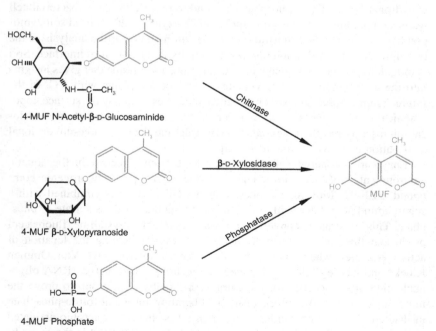

4-MUF N-Acetyl-β-D-Glucosaminide

4-MUF β-D-Xylopyranoside

4-MUF Phosphate

Chitinase

β-D-Xylosidase

Phosphatase

MUF

FIG. 7.6 Enzyme assays using different nonfluorescent substrates (left), and the highly fluorescent product (right) after a short-term incubation during which the fluorescent product is cleaved from the original substrates.

done with soil suspensions because of interference in the detection from soil particles and quenching for fluorescence (Dick, 2011).

The increasing interest of soil microbiologists in the use of fluorogenic substrates to measure soil enzyme activities is mainly due to their high sensitivity. A comparative study between a fluorometric and a standard colorimetric enzyme assay based on p-nitrophenyl substrates generated similar values for the maximum rate of phosphatase and ß-glucosidase (v_{max}), but the affinity for their respective substrates (as indicated by K_m values, Michaelis-Menten constant) was up to two orders of magnitude greater for the 4-methylumbelliferyl substrates compared to the p-nitrophenyl substrates (Marx et al., 2001). This high sensitivity of fluorometric enzyme assays provides an opportunity to detect enzyme activities in small samples (e.g., microaggregates and rhizosphere samples) and/or low activity samples (subsoil, peat, and soil solutions).

IX   IMAGING MICROBIAL ACTIVITIES

Thin-section techniques combined with histochemical and imaging techniques can visualize the location of enzyme proteins and their activities. Papers by Foster et al. (1983), using either transmission electron microscopy (TEM) or scanning electron microscopy (SEM), showed peroxidase, succinic dehydrogenase, and acid phosphatase bound to root rhizosphere, bacterial cell walls, and organic matter in soil. Spohn and Kuzyakov (2013) used soil zymography with fluorescent substrates as an *in situ* method for the analysis of the two-dimensional distribution of enzyme activity at the soil–root interface. Soil zymography can be combined with ^{14}C imaging, a technique that gives insights into the distribution of photosynthates after labeling plants with ^{14}C. In the future, atomic force microscopy, which measures small-scale surface topographical features, will help us to understand enzyme-clay interactions. Confocal microscopy has the potential to reconstruct the detailed three-dimensional distribution of enzyme proteins in soils.

Enzymatic properties of single cells have been screened in the aquatic environment, biofilms, and activated sludge by a type of fluorogenic compound, ELF-97 (Molecular Probes, Eugene, OR), which is combined with a sugar, amino acid, fatty acid, or inorganic compound, such as sulfate or phosphate. This substrate is converted to a water-insoluble, crystalline, fluorescent product at the site of enzymatic hydrolysis, thus indicating the location of active enzymes when viewed by fluorescence microscopy. Van Ommen Kloeke and Geesey (1999) combined this technique with a 16S rRNA oligonucleotide probe specific for the cytophaga-flavobacteria group to prove the importance of these microorganisms in liberating inorganic ortho-phosphate in discrete, bacteria-containing areas of the floc matrix in aerobic activated sludge (Fig. 7.7). Phosphatase activity was primarily localized in the immediate vicinity of the bacterial cells rather than dispersed throughout the floc or

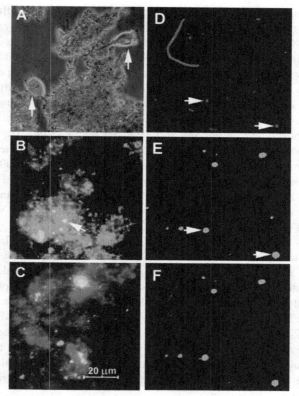

FIG. 7.7   (a) Phase contrast photomicrograph image of activated sludge floc material and associated microorganisms. Arrows indicate protozoa grazing on the floc particles. (b) Epifluorescence photomicrograph image of the same field of view as in (a) of whole floc material, revealing the areas of intense phosphatase activity (yellow spots where ELF crystals have precipitated, arrow). (c) Epifluorescence photomicrograph image of thin section of a floc particle stained red-orange with acridine orange, revealing discrete regions within the floc displaying phosphatase activity (green spots) as a result of incubation in the presence of ELP-P. Bar = 20 μm. (d) Epifluorescence photomicrograph image of a homogenized activated sludge filtrate showing floc bacteria from the cytophaga-flavobacteria group by using a filter combination that reveals cells which react with CF319a oligonucleotide probe (red). (e) Epifluorescence photomicrograph image of the same field of view as (d) using a filter combination that resolves crystals of ELF (green spots) indicating areas of phosphatase activity. (f) Merged images of (d) and (e), revealing the subpopulation of cells that react positively with the CF319a and also display phosphatase activity. *According to Van Ommen Kloeke and Geesey (1999).*

associated with protozoa (see Fig. 7.7). These new substrates have the potential to be used for visualizing the location of enzyme activities in microenvironments of soil.

Imaging and enumeration of specific soil microorganisms can be performed by fluorescence *in situ* hybridization (FISH) techniques. However, the detection of FISH-stained cells is often affected by strong autofluorescence of the background. If the FISH approach is coupled with catalyzed reporter deposition (CARD), CARD-FISH-stained cells are suitable for automated counting using digital image analysis (Neufeld et al., 2007). In addition, very sensitive mass spectrometric techniques (secondary ion mass spectrometry: SIMS) have made possible the visualization of single-cell ecophysiology of microbes in environmental samples (Wagner, 2009). The SIMS uses an energetic primary ion beam to produce and expel secondary particles (primarily atoms or molecules) from the soil surface under high vacuum. The elemental, isotopic, and molecular compositions of these particles are determined by mass spectrometry. Using this technique, Pumphrey et al. (2009) imaged phenol-degrading microorganisms in soils that had been exposed to ^{12}C- or ^{13}C-phenol and showed that about 30% of the soil microorganisms lived exclusively on phenol. This method has also been applied to visualize N assimilation of single cells in soils.

Whereas the SIMS techniques can provide results at the single cell level, the technique by Stiehl-Braun et al. (2011) makes the study of the spatial distribution of active microorganisms in soil at larger scales (cm scale) possible. When undisturbed soil cores were exposed to ^{14}CH$_4$, methanotrophs incorporated ^{14}C into their cells. The spatial distribution of ^{14}C incorporation was characterized by a combined micromorphological and autoradiographical technique (Stiehl-Braun et al., 2011). A ^{14}C phosphor imaging approach can also be used to detect the temporal release of root exudates from plants after ^{14}C pulse labeling (Pausch and Kuzyakov, 2011). This method has the potential to elucidate C transfer from plant to rhizosphere microorganisms in the future.

X   FUNCTIONAL DIVERSITY

Different aspects of functional diversity have advanced our understanding of the significance of biodiversity to biochemical cycling. This includes several levels of resolution: (1) the importance of biodiversity to specific biogenic transformations; (2) the complexity and specificity of biotic interactions in soils that regulate biogeochemical cycling; and (3) how biodiversity may operate at different hierarchically arranged spatial and temporal scales to influence ecosystem structure and function (Beare et al., 1995). Most methods for measuring functional diversity consider only the biodiversity of those groups that regulate biochemical cycling. Several approaches enable functional diversity to be measured in situations where taxonomic information is poor. These include the use of binary biochemical and physiological descriptors to characterize isolates, evaluation of enzymatic capabilities for utilization of particular substrates, extraction of DNA and RNA from the soil, and the application of gene probes that code for functional enzymes. Recent advances in genomic analysis and stable isotope probing are the first steps toward a better understanding of linkages between structure and function in microbial communities.

Commercially available BIOLOG bacterial identification system plates or community-level physiological profiles (CLPP) have been used to assess functional diversity of microorganisms based on utilization patterns of a wide range (up to 128) of individual C sources. Members of the microbial community that can be cultured, and which exhibit the fast growth rates typical for r-strategists, primarily contribute to CLPP analysis. Preston-Mafham et al. (2002) assessed the pros and cons of its use and pointed out inherent biases and limitations as well as possible ways of overcoming certain difficulties. Today, the original CLPP approach using BIOLOG microtiter plates, due to severe limitations of this method, has been replaced by alternative approaches based on CO_2 or O_2 detection (Degens and Harris, 1997). The CO_2 in the headspace of multiple flasks can be measured by infrared detection or with sealed microtiter plates containing pH-sensitive dye. The microtiter system, based on O_2-sensitive fluorophore, is used to detect basal soil respiration and substrate-induced responses at amendment levels 10- to 100-fold less than that used with any other CLPP approaches (Garland et al., 2010).

An alternative approach was proposed by Kandeler et al. (1996) using the prognostic potential of 16 soil microbial properties (microbial biomass, respiration, N-mineralization, and 13 soil enzymes involved in cycling of C, N, P, and S). Multivariate statistical analysis is used to calculate the functional diversity from measured soil microbial properties. The approach is based on the following assumptions. The composition of the microbial species assemblage (taxonomic diversity) determines the community's potential for enzyme synthesis. The actual rate of enzyme production and the fate of produced enzymes are modified by environmental effects as well as by ecological interactions. The spectrum and amount of active enzymes are responsible for the functional capability of a microbial community irrespective of whether they are active inside or outside the cell. According to the diversity concept in ecology, presence or absence of a certain function, as well as quantification of the potential of the community to realize this function are considered. This approach may permit evaluation of the status of a changed ecosystem (e.g., by soil pollution, soil management, global change) while providing insight into the functional diversity of the soil microbial community of the undisturbed habitat.

Physiological methods are applied to better understand the physiology of single cells, soil biological communities, food webs, and biogeochemical cycling in terrestrial ecosystems. Small-scale studies explain biological reactions in aggregates, the rhizosphere, or at the soil–litter interface. Combining physiological and molecular methods helps us to understand gene expression, protein synthesis, and enzyme activities at both micro- and nano-scale. Linking these methods can also help explain whether the abundance and/or function of organisms are affected by soil management, global change, and soil pollution.

At the field scale, researchers use biochemical and physiological methods to investigate the functional response of soil organisms to the manipulation or preservation of soils. These applications include microbe-plant interactions and control of plant pathogens, as well as better characterization of organic matter decomposition and its impact on local and global C and N cycling. Soil biologists investigate the effect of soil management (tillage, fertilizer, pesticides, crop rotation) or disturbance on the function of soil organisms. In many cases, soil microbial biomass and/or soil microbial processes can be early predictors of the effects of soil management on soil quality and can also indicate the expected rapidity of these changes. Monitoring of soil microbial properties is also necessary in environmental studies that test the use of soil microorganisms in bioremediation and composting. Future challenges in functional soil microbiology are to use our present knowledge to upscale these data to regional and global scales.

REFERENCES

Alef, K., Nannipieri, P., 1995. Methods in Applied Soil Microbiology and Biochemistry. Academic Press, San Diego.

Amelung, W., Brodowski, S., Sandhage-Hofmann, A., Bol, R., 2008. Combining biomarker with stable isotope analyses for assessing the transformation and turnover of soil organic matter. Adv. Agron. 100, 155–250.

Anderson, J.P.E., Domsch, K.H., 1978. A physiological method for the quantitative measurement of microbial biomass in soils. Soil Biol. Biochem. 10, 215–221.

Bååth, E., Anderson, T.H., 2003. Comparison of soil fungal/bacterial ratios in a pH gradient using physiological and PLFA-based techniques. Soil Biol. Biochem. 35, 955–963.

Baldocchi, D.D., 2003. Assessing the eddy covariance technique for evaluating carbon dioxide exchange rates of ecosystems: past, present and future. Glob. Chang. Biol. 9, 479–492.

Beare, M.H., Coleman, D.C., Crossley, D.A., Hendrix, P.F., Odum, E.P., 1995. A hierarchical approach to evaluating the significance of soil biodiversity to biochemical cycling. Plant Soil 170, 5–22.

Becher, D., Bernhardt, J., Fuchs, S., Riedel, K., 2013. Metaproteomics to unravel major microbial players in leaf litter and soil environments: challenges and perspectives. Proteomics 13, 2895–2909.

Beck, T., Joergensen, R.G., Kandeler, E., Makeschin, F., Nuss, E., Oberholzer, H.R., Scheu, S., 1997. An inter-laboratory comparison of ten different ways of measuring soil microbial biomass C. Soil Biol. Biochem. 29, 1023–1032.

Boutton, T.W., Yamasaki, S.-I., 1996. Mass Spectrometry of Soils. Marcel Dekker, New York.

Brookes, P.C., Landman, A., Puden, G., Jenkinson, D.S., 1985. Chloroform fumigation and the release of soil nitrogen: a rapid direct extraction method to measure microbial biomass nitrogen in soil. Soil Biol. Biochem. 17, 837–842.

Bünemann, E.K., 2008. Enzyme additions as a tool to assess the potential bioavailability of organically bound nutrients. Soil Biol. Biochem. 40, 2116–2129.

Burns, R.G., 1982. Enzyme activity in soil: location and possible role in microbial ecology. Soil Biol. Biochem. 14, 423–427.

Burns, R.G., Dick, R.P., 2002. Enzymes in the Environment—Activity, Ecology, and Applications. Marcel Dekker, New York.

Burns, R.G., DeForest, J.L., Marxsen, J., Sinsabaugh, R.L., Stromberger, M.E., Wallenstein, M.D., Weintraub, M.N., Zoppini, A., 2013. Soil enzymes in a changing environment: current knowledge and future directions. Soil Biol. Biochem. 58, 216–234.

Clode, P.L., Kilburn, M.R., Jones, D.L., Stockdale, E.A., Cliff III, J.B., Herrmann, A.M., Murphy, D.V., 2009. *In situ* mapping of nutrient uptake in the rhizosphere using nanoscale secondary ion mass spectrometry. Plant Physiol. 151, 1751–1757.

Coleman, D.C., Fry, B. (Eds.), 1991. Carbon Isotope Techniques. Academic Press, San Diego.

Degens, B.P., Harris, J.A., 1997. Development of a physiological approach to measuring the catabolic diversity of soil microbial communities. Soil Biol. Biochem. 29, 1320–1509.

Deng, S., Popova, I.E., Dick, L., Dick, R., 2013. Bench scale and microplate format assay of soil enzyme activities using spectroscopic and fluorometric approaches. Appl. Soil Ecol. 64, 84–90.

Dick, R., 2011. Methods of Soil Enzymology. In: Soil Science of America Book Series 9. Soil Science Society of America, Inc, Madison, Wisconsin.

Drenovsky, R.E., Elliott, G.N., Graham, K.J., Scow, K.M., 2004. Comparison of phospholipid fatty acid (PLFA) and total soil fatty acid methyl esters (TSFAME) for characterizing soil microbial communities. Soil Biol. Biochem. 36, 1793–1800.

Eisenbeis, G., Lenz, R., Heiber, T., 1999. Organic residue decomposition: the Minicontainer-system—a multifunctional tool in decomposition studies. Environ. Sci. Pollut. Res. 6, 220–224.

Feinstein, L.M., Sul, W.J., Blackwood, C.B., 2009. Assessment of bias associated with incomplete extraction of microbial DNA from soil. Appl. Environ. Microbiol. 75, 5428–5433.

Foster, R.C., Rovira, A.D., Cock, T.W., 1983. Ultrastructure of the Root-Soil Interface. American Phytopathological Society, St. Paul, MN.

Frankland, J.C., Dighton, J., Boddy, L., 1991. Fungi in soil and forest litter. In: Grigorova, R., Norris, J.R. (Eds.), Methods in Microbiology, vol. 22. Academic Press, London, pp. 344–404.

Frostegård, Å., Tunlid, A., Bååth, E., 2011. Use and misuse of PLFA measurements in soils. Soil Biol. Biochem. 43, 1621–1625.

Fujie, K., Hu, H.K., Tanaka, H., Urano, K., Saitou, K., Katayama, A., 1998. Analysis of respiratory quinones in soil for characterization of microbiota. Soil Sci. Plant Nutr. 44, 393–404.

Garland, J.L., Mackowiak, C.L., Zaboloy, M.C., 2010. Organic waste amendment effects on soil microbial activity in a corn-rye rotation: application of a new approach to community-level physiological profiling. Appl. Soil Ecol. 44, 262–269.

Gattinger, A., Gunther, A., Schloter, M., Munch, J.C., 2003. Characterisation of Archaea by polar lipid analysis. Acta Biotechnol. 23, 21–28.

Gerhardt, P.G., Murray, R.G.E., Wood, W.A., Krieg, N.R., 1994. Methods for General and Molecular Bacteriology. American Society of Microbiology, Washington, D.C.

Grandmougin-Ferjani, A., Dalpe, Y., Hartmann, M.A., Laruellea, F., Sancholle, M., 1999. Sterol distribution in arbuscular mycorrhizal fungi. Phytochemistry 50, 1027–1031.

Gregorich, E., Wen, G., Voroney, R., Kachanoski, R., 1990. Calibration of a rapid direct chloroform extraction method for measuring soil microbial biomass C. Soil Biol. Biochem. 22, 1009–1011.

Horwitz, B.A., Mukherjee, P.K., Kubicek, C.P., 2013. Genomics of Soil- and Plant-associated Fungi. In: Series: Soil Biology, vol. 36. Springer, Heidelberg, Berlin.

Jenkinson, D.S., Powlson, D.S., 1976. The effects of biocidal treatments on metabolism in soil. V. A method for measuring soil biomass. Soil Biol. Biochem. 8, 209–213.

Joergensen, R.G., 1996. The fumigation-extraction method to estimate soil microbial biomass: calibration of the k_{EC} value. Soil Biol. Biochem. 28, 25–31.

Joergensen, R.G., Wu, J., Brookes, P.C., 2011. Measuring soil microbial biomass using an automated procedure. Soil Biol. Biochem. 43, 873–876.

Kandeler, E., Kampichler, C., Horak, O., 1996. Influence of heavy metals on the functional diversity of soil microbial communities. Biol. Fertil. Soils 23, 299–306.

Keiblinger, K.M., Wilhartitz, I.C., Schneider, T., Roschitzki, B., Schmid, E., Eberl, L., Riedel, K., Zechmeister-Boltenstern, S., 2012. Soil metaproteomics—comparative evaluation of protein extraction protocols. Soil Biol. Biochem. 54, 14–24.

Kjelleberg, S., 1993. Starvation in Bacteria. Plenum, New York.

Klamer, A.J., Bååth, E., 2004. Estimation of conversion factors for fungal biomass determination in compost using ergosterol and PLFA 18:2ω6,9. Soil Biol. Biochem. 36, 57–65.

Klee, A.J., 1993. A computer program for the determination of most probable number and its confidence limits. J. Microb. Methods 18, 91–98.

Klironomos, J.N., Rilling, M.C., Allen, M.F., 1999. Designing belowground field experiments with the help of semi-variance and power analyses. Appl. Soil Ecol. 12, 227–238.

Knowles, R., Blackburn, T.H. (Eds.), 1993. Nitrogen Isotope Techniques. Academic Press, San Diego.

Kramer, S., Marhan, S., Ruess, L., Armbruster, W., Butenschoen, O., Haslwimmer, H., Kuzyakov, Y., Pausch, J., Scheunemann, N., Schoene, J., Schmalwasser, A., Totsche, K.U., Walker, F., Scheu, S., Kandeler, E., 2012. Carbon flow into microbial and fungal biomass as a basis for the belowground food web of agroecosystems. Pedobiologia 55, 111–119.

Kreuzer-Martin, H.W., 2007. Stable isotope probing: linking functional activity to specific members of microbial communities. Soil Sci. Soc. Am. J. 71, 611–619.

Lennon, J.T., Aanderud, Z.T., Lehmkuhl, B.K., Schoolmaster Jr., D.R., 2012. Mapping the niche space of soil microorganisms using taxonomy and traits. Ecology 93, 1867–1879.

Levin, M.A., Seidler, R.J., Rogul, M., 1992. Microbial Ecology: Principles, Methods and Applications. McGraw-Hill, New York.

Manzoni, S., Taylor, P., Richter, A., Porporato, A., Agren, G.I., 2012. Environmental and stoichiometric controls on microbial carbon-use efficiency in soils. New Phytol. 196, 79–91.

Marx, M.C., Wood, M., Jarvis, S.C., 2001. A microplate fluorimetric assay for the study of enzyme diversity in soils. Soil Biol. Biochem. 33, 1633–1640.

McCarthy, A.J., Williams, S.T., 1991. Methods for studying the ecology of actinomycetes. In: Grigorova, R., Norris, J.R. (Eds.), Methods in Microbiology, vol. 22. Academic Press, London, pp. 533–563.

Müller, R.S., Pan, C., 2013. Sample handling and mass spectrometry for microbial metaproteomic analyses. Methods Enzymol. 531, 289–303.

Murase, J., Kees Hordijk, K., Tayasu, I., Bodelier, P.L.E., 2011. Strain-specific incorporation of methanotrophic biomass into eukaryotic grazers in a rice field soil revealed by PLFA-SIP. FEMS Microbiol. Ecol. 75, 284–290.

Murphy, D.V., Recous, S., Stockdale, E.A., Fillery, I.R.P., Jensen, L.S., Hatch, D.J., Goulding, K.W.T., 2003. Gross nitrogen fluxes in soil: theory, measurement and application of N-15 pool dilution techniques. Adv. Agron. 79, 69–118.

Nannipieri, P., Kandeler, E., Ruggiero, P., 2002. Enzyme activities and microbiological and biochemical processes in soil. In: Burns, R.G., Dick, R.P. (Eds.), Enzymes in the Environment—Activity, Ecology and Applications. Marcel Dekker, New York, pp. 1–34.

Neufeld, J.D., Wagner, M., Murrell, J.C., 2007. Who eats what, where and when? Isotope-labeling experiments are coming of age. ISME J. 1, 103–110.

Newell, S.Y., Fallon, R.D., 1991. Toward a method for measuring instantaneous fungal growth rates in field samples. Ecology 72, 1547–1559.

Papale, D., Valentini, R., 2003. A new assessment of European forests carbon exchange by eddy fluxes and artificial neural network spatialization. Glob. Chang. Biol. 9, 525–535.

Pausch, J., Kuzyakov, Y., 2011. Photoassimilate allocation and dynamics of hotspots in roots visualized by ^{14}C phosphor imaging. J. Plant Nutr. Soil Sci. 174, 12–19.

Philippot, L., Tscherko, D., Bru, D., Kandeler, E., 2011. Distribution of high bacterial taxa across the chronosequence of two alpine glacier forelands. Microb. Ecol. 61, 303–312.

Pitcher, A., Rychlik, N., Hopmans, E.C., Spieck, E., Rijpstra, W.I.C., Ossebaar, J., Schouten, S., Wagner, M., Sinninghe Damste, J.S.S., 2010. Crenarchaeol dominates the membrane lipids of *Candidatus Nitrososphaera gargensis*, a thermophilic Group I.1b Archaeon. ISME J. 4, 542–552.

Poll, C., Pagel, H., Devers, M., Martin-Laurent, F., Ingwersen, J., Streck, T., Kandeler, E., 2010. Regulation of bacterial and fungal MCPA degradation at the soil-litter interface. Soil Biol. Biochem. 42, 1879–1887.

Potthoff, M., Loftfield, N., Buegger, F., Wick, B., Jahn, B., Joergensen, R.G., Flessa, H., 2003. The determination of δ^{13}C in soil microbial biomass using fumigation-extraction. Soil Biol. Biochem. 35, 947–954.

Preston-Mafham, J., Boddy, L., Randerson, P.F., 2002. Analysis of microbial community functional diversity using sole-carbon-source utilisation profiles—a critique. FEMS Microbiol. Ecol. 42, 1–14.

Pumphrey, G.M., Hanson, B.T., Chandra, S., Madsen, E.L., 2009. Dynamic secondary ion mass spectrometry imaging of microbial populations utilizing ^{13}C-labeled substrates in pure culture and in soil. Environ. Microbiol. 11, 220–229.

Robertson, J.P., 1994. The impact of soil and crop management practices on soil spatial heterogeneity. In: Pankhurst, C.E., Doube, B.M., Gupta, V., Grace, P. (Eds.), Soil Biota. CSIRO, East Melbourne, pp. 156–161.

Robertson, G.P., Coleman, D.C., Bledsoe, C.S., Sollins, P., 1999. Standard Soil Methods for Long-Term Ecological Research. Oxford University Press, New York.

Rousk, J., Bååth, E., 2011. Growth of saprotrophic fungi and bacteria in soil. FEMS Microbiol. Ecol. 78, 17–30.

Ruess, L., Chamberlain, P.M., 2010. The fat that matters: soil food web analysis using fatty acids and their carbon stable isotope signate. Soil Biol. Biochem. 42, 1898–1910.

Schimel, D.S., 1993. Theory and Application of Tracers. Academic Press, San Diego.

Schinner, F., Öhlinger, R., Kandeler, E., Margesin, R. (Eds.), 1996. Methods in Soil Biology. Springer, Berlin.

Schwartz, E., Blazewicz, S., Doucett, R., Hungate, B.A., Hart, S.C., Dijkstra, P., 2007. Natural abundance δ^{15}N and δ^{13}C of DNA extracted from soil. Soil Biol. Biochem. 39, 3101–3107.

Setia, R., Verma, S.L., Marschner, P., 2012. Measuring microbial biomass carbon by direct extraction-comparison with chloroform fumigation-extraction. Eur. J. Soil Biol. 53, 103–106.

Sinsabaugh, R.L., Shah, J.J.F., 2012. Ecoenzymatic stoichiometry and ecological theory. Annu. Rev. Ecol. Evol. Syst. 43, 313–343.

Sinsabaugh, R.L., Carreiro, M.M., Alvarez, S., 2002. Enzymes and microbial dynamics of litter decomposition. In: Burns, R.G., Dick, R.P. (Eds.), Enzymes in the Environment—Activity, Ecology, and Application. Decker, New York, pp. 249–266.

Smith, C.J., Osborn, A.M., 2009. Advantages and limitations of quantitative PCR (Q-PCR)-based approaches in microbial ecology. FEMS Microbiol. Ecol. 67, 6–20.

Spohn, M., Kuzyakov, Y., 2013. Distribution of microbial- and root-derived phosphatase activities in the rhizosphere depending on P availability and C allocation—coupling soil zymography with ^{14}C imaging. Soil Biol. Biochem. 67, 106–113.

Stiehl-Braun, P.A., Hartmann, A., Kandeler, E., Buchmann, N., Pascal, N., 2011. Interactive effects of drought and N fertilisation on the spatial distribution of methane assimilation in grassland soils. Glob. Chang. Biol. 17, 2629–2639.

Sullivan, B.W., Dore, S., Kolb, T.E., Hart, S.C., Montes-Helu, M.C., 2010. Evaluation of methods for estimating soil carbon dioxide efflux across a gradient of forest disturbance. Glob. Chang. Biol. 16, 2449–2460.

Tabatabai, M.A., 1994. Soil enzymes. In: Weaver, R.W., Angel, J.S., Bottomley, P.S. (Eds.), Methods of Soil Analysis. Part 2. Microbiological and Biochemical Properties. Book Series No. 5. Soil Science Society of America, Madison, Wisconsin, pp. 775–833.

Tabatabai, M.A., Dick, W.A., 2002. Enzymes in soil—research and development in measuring activities. In: Burns, R.G., Dick, R.P. (Eds.), Enzymes in the Environment—Activity, Ecology and Applications. Marcel Dekker, New York, pp. 567–596.

Tabatabai, M., Fung, M., 1992. Extraction of enzymes from soil. In: Stotzky, G., Bollag, J.-M. (Eds.), Soil Biochemistry, vol. 7. Dekker, New York, pp. 197–227.

Taylor, A.E., Zeglin, L.H., Dooley, S., Myrold, D.D., Bottomley, P.J., 2010. Evidence of different contributions of Archaea and Bacteria to the ammonia-oxidizing potential of divers Oregon soils. Appl. Environ. Microbiol. 76, 7691–7698.

Taylor, A.E., Zeglin, L.H., Wanzek, T.A., Myrold, D.D., Bottomley, P., 2012. Dynamics of ammonia-oxidizing archaea and bacteria populations and contributions to soil nitrification potentials. ISME J. 6, 2024–2032.

Van Ommen Kloeke, F., Geesey, G.G., 1999. Localization and identification of populations of phosphatase-active bacterial cells associated with activated sludge flocs. Microbiol. Ecol. 38, 201–214.

Vepsäläinen, M., Kukkonen, S., Vestberg, M., Sirviö, H., Niemi, R.M., 2001. Application of soil enzyme activity test kit in a field experiment. Soil Biol. Biochem. 33, 1665–1672.

Verma, B., Robarts, R.D., Headley, J.V., Peru, K.M., Christofi, N., 2002. Extraction efficiencies and determination of ergosterol in a variety of environmental matrices. Commun. Soil Sci. Plant Anal. 33, 3261–3275.

Wagner, M., 2009. Single-cell ecophysiology of microbes as revealed by Raman microspectroscopy or secondary ion mass spectrometry imaging. Annu. Rev. Microbiol. 63, 411–429.

Weaver, R., Angele, S., Bottomley, P., 1994. Methods of Soil Analyses. Part 2. Biochemical and Biological Properties of Soil. American Society of Agronomy, Madison, WI.

Weete, J.D., Abril, M., Blackwell, M., 2010. Phylogenetic distribution of fungal sterols. PLoS One 5, e10899.

Wilkinson, M.H.F., Schut, F. (Eds.), 1998. Digital Image Analysis of Microbes—Imaging, Morphology, Fluorometry and Motility Techniques and Applications. John Wiley & Sons, Chichester.

Chapter 8

The Spatial Distribution of Soil Biota

Serita D. Frey

Department of Natural Resources & the Environment, University of New Hampshire, Durham, NH, USA

Chapter Contents

I INTRODUCTION

Biota that occur in soil represent a large proportion of Earth's biodiversity, have worldwide distribution, and are known to survive and grow in some seemingly unlikely and often inhospitable places. This includes the canopies of tropical trees, deep subsurface environments, recently deposited volcanic materials, under deep snow in alpine systems, Antarctic Dry Valley soils, and in cryoconite holes, pockets of meltwater containing windblown soil on the surface of glaciers (Fig. 8.1). Although there are still large gaps in our understanding of how soil organisms are distributed, there has been a dramatic increase in information obtained in this area over the past decade. This chapter summarizes what is known about the distribution of soil biota, from geographic differences at the landscape, regional, and continental scales, down to variability in microbial populations at the aggregate and pore level.

II THE BIOGEOGRAPHY OF SOIL BIOTA

Biogeography, the science of documenting spatial patterns of biological diversity from local to continental scales, examines variation in genetic, phenotypic, and physiological characteristics at different scales (e.g., distantly located

Soil Microbiology, Ecology, and Biochemistry. http://dx.doi.org/10.1016/B978-0-12-415955-6.00008-6

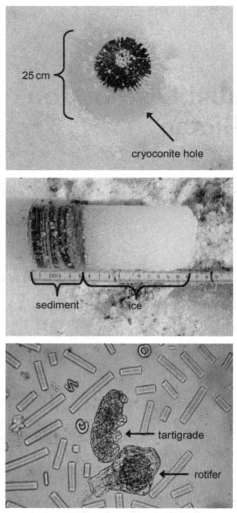

FIG. 8.1 Cryoconite hole formed on the surface of an Antarctic glacier. (Top) The ice of the cryoconite hole reflects light differently than the surrounding ice, making it easy to spot on the glacial surface. (Middle) Ice core from the cryoconite hole showing the accumulation of windblown sediment at the bottom of the hole. (Bottom) Organisms associated with sediments in the cryoconite hole. The green rods are cyanobacteria. *Photos courtesy of Dorota Porazinska, University of Florida, and Thomas Nylen, Portland State University; used with permission.*

sampling sites or along environmental gradients). It emphasizes understanding the factors that generate and maintain organism distributions (Ramette and Tiedje, 2006). Questions addressed include (1) why global biodiversity is so great; (2) why organisms live where they do; (3) how many taxa can coexist;

and (4) how organisms will respond to environmental change. Biogeography is a relatively mature science, with the study of plant and animal diversity patterns going back more than two centuries to Carl Linnaeus's early plant and animal surveys. The global distributions of most of the world's flora and fauna are generally known. From this work, it is well established that most macroscopic plant and animal species are not globally distributed, but instead have restricted geographic distributions because of climate sensitivity and natural barriers to migration. This isolation has, over geological time, led to the evolution of new species and the development of geographically distinct plant and animal communities.

Until recently, there has been less emphasis on understanding and mapping the biogeography of microscopic organisms, due primarily to methodological constraints. The recent advances in molecular sequencing technologies have revealed that microbial diversity far exceeds that of macroscopic organisms; however, the geographic patterns and factors controlling this diversity are only beginning to be examined. The field of microbial biogeography is thus in its infancy, but the number of papers published in just the past few years has increased dramatically. It is now possible, by combining molecular sequencing and environmental data with advanced analytical approaches (e.g., neural networks), to generate predictive models and maps of microbial diversity with which to build a better understanding of microbial biogeography and ultimately, the relationships between microbial diversity, community composition, and ecosystem function (Fierer and Ladau, 2012; Larsen et al., 2012). These approaches will bring new and exciting insights to our understanding of microbial biogeography, and it is quite likely that our current ideas about microbial diversity patterns will be challenged.

A common perception was that microorganisms are cosmopolitan in their distribution, being capable of growth in many different places worldwide. This idea goes back more than a century to Baas Becking, a Dutch scientist, who suggested that *"everything is everywhere*, but, *the environment selects"* (de Wit and Bouvier, 2006). Microorganisms, due to their small size and large numbers, are continuously being moved around, often across continental-scale distances. Dispersal mechanisms include (1) water transport via rivers, groundwater, and ocean currents; (2) airborne transport in association with dust particles and aerosols, especially during extreme weather events, such as hurricanes and dust storms; (3) transport on or in the intestinal tracts of migratory birds, insects, and aquatic organisms; and (4) human transport through air travel and shipping. Model simulations of aerial transport suggest that microbial taxa less than 20 μm in size can disperse across continents on an annual timescale (Wilkinson et al., 2012). Microbes are abundant in the upper atmosphere, where thousands of distinct bacterial taxa representing a wide range of phyla have been identified (Smith et al., 2013). A high abundance of Actinobacteria and Firmucutes contain spore-forming and G^+ taxa capable of surviving the extreme conditions associated with long-range transport. Even isolated environments, like those found in the Antarctic, show a wide range of microbial species that appear to have been introduced from other places, often by human visitors (Azmi and Seppelt, 1998). It is not known how many transported microbes survive their journey and the degree to which they become established in their new environment. There are numerous potential barriers to success following microbial dispersal to a new habitat, including limited resource availability, physicochemical requirements (e.g., temperature, salinity, pH), and ecological constraints (e.g., competition with native taxa).

Finlay (2002) has argued that anything smaller than about 1 mm will likely be ubiquitous in the environment. This argument was initially supported by research on protozoa that, based on morphological analysis, are represented by a relatively limited number of species with a global distribution. There are only about 4000 morphologically described species of free-living soil ciliates. This is in comparison to 5 million insect species, many of which are known to have geographically restricted ranges. The thought that protozoa have global distributions was supported by a study of protozoa living in the sediments of a geographically isolated crater lake in Australia (Finlay and Clarke, 1999). Of 85 protozoan species collected from the Australian system, all had been previously described and known from northern Europe. They apparently reached the crater by dispersal from other freshwater, soil, and marine environments. The conclusion drawn from this work was that high rates of dispersal limit the geographic isolation necessary for speciation to occur, leading to a cosmopolitan distribution (Finlay, 2002).

This interpretation is now challenged by the application of high throughput sequencing methods for the assessment of soil microbial diversity. A recent deep, sequencing analysis of soil protistan communities, at sites representing a broad geographic range and variety of biome types, revealed a high level of diversity and distribution patterns that suggest that most taxa in this group are not globally distributed (Bates et al., 2013). More than 1000 taxa representing 13 phyla were identified. A high number of rare taxa were observed, along with several taxa rarely found in soil environments. Only one taxon was found in more than 75% of the samples, and the majority of taxa (84%) were found at five or fewer sites.

Similar results have been observed for soil rotifers, a ubiquitously distributed group of microbial eukaryotes that are important contributors to biogeochemical cycling (Robeson et al., 2011). Rotifer diversity was found to be high, with global sampling revealing close to 800 taxa. Distant communities were different from one another even in similar environments, and there was almost no overlap in taxon occurrence between soil communities sampled at distances of ~150 m apart. Many novel phylotypes were also observed at each sampling location, suggesting a high degree of endemism (where taxa are unique to a particular site).

Soil prokaryotes, particularly bacteria, have been the most extensively studied from a microbial biogeography perspective and provide the most compelling evidence refuting a cosmopolitan distribution for soil microorganisms. Certainly soil bacteria can be considered to have a cosmopolitan distribution at the phylum level given that several bacterial phyla (Acidobacteria, Actinobacteria, Proteobacteria, and Bacteroides) have a worldwide distribution in terrestrial ecosystems and occur in roughly equivalent proportions across biomes (Fierer et al., 2009). However, this is akin to saying that flowering plants (Magnoliophyta phylum) have a global distribution, a rather obvious statement. If we look at the 16S rRNA gene level (using the commonly used bacterial species definition of 97% sequence similarity), there are examples where the same taxa have been identified in similar habitats located in different geographic regions (Ramette and Tiedje, 2006).

Existing data suggest that endemism is more the rule than the exception. Fluorescent *Pseudomonas* strains isolated from soil samples collected at ten sites on four continents showed no overlap between sites (Cho and Tiedje, 2000). More recent analyses suggest that most bacterial phylotypes are rare and the most dominant taxa represent, at most, 2-5% of sequences in a given sample (Fierer and Lennon, 2011). A small proportion of bacterial taxa overlap among sites, with the majority of taxa (often >75%) being unique at the site level (Chu et al., 2010; Fulthorpe et al., 2008; Lauber et al., 2009; Nemergut et al., 2011).

Deep sequencing analyses indicate that meaningful biogeographic patterns do exist for microorganisms, suggesting that microbes are not exempt from the fundamental evolutionary processes (i.e., geographic isolation and natural selection) that shape plant and animal communities (Rout and Callaway, 2012). Perhaps everything is not everywhere. There is a caveat to this statement, however. The use of cultivation-independent, high-throughput sequencing methodologies yields thousands of sequences and suggests a much higher level of diversity than previously anticipated. Deep sequencing produces large numbers of taxa represented by only a few sequences, data often interpreted as evidence of the "rare microbial biosphere." Sequencing technologies, like any other analytical approach, are subject to bias, including sequencing errors, poor alignment quality, use of inappropriate denoising and clustering approaches, and inconsistent results among targeted regions. These biases can greatly influence the resulting diversity estimates, and there are concerns that many rare sequences may actually be artifacts. Several research groups have started to investigate the potential significance of this problem. Bachy et al. (2013) conducted a comparative analysis using both morphological and molecular approaches for a species-rich order of planktonic protists (tintinnid ciliates) that are easy to distinguish morphologically. They found that molecular approaches were efficient at detecting the species identified morphologically. At the same time, the sequencing results often indicated that there were more species in a sample than the total number of cells enumerated by microscopy. Before generalizations regarding microbial biogeography can be made, a careful analysis of sequencing approaches is needed to minimize errors associated with sequencing bias.

III VERTICAL DISTRIBUTION WITHIN THE SOIL PROFILE

The abundance and biomass of most soil organisms is highest in the top 10 cm of soil and declines with depth in parallel with organic matter content and prey availability. Approximately 65% of the total microbial biomass is found in the top 25 cm of the soil profile. Below that depth, microbial densities typically decline by one to three orders of magnitude (Fig. 8.2). Abundances of G- bacteria, fungi, and protozoa are typically highest at the soil surface, whereas G+ bacteria, Actinobacteria, and archaea tend to increase in proportional abundance with increasing depth (Fierer et al., 2003). In forest soils, the litter layer is dominated by fungi, with fungal biomass being up to three times higher than that of bacterial biomass (Baldrian et al., 2012). The relative importance of bacteria increases in the organic and mineral soil horizons. Hyphal density of and root colonization by mycorrhzial fungi decrease substantially below 20 cm. Microbial grazers (e.g., protozoa, collembola) also decrease with depth, often more rapidly than either their bacterial or fungal prey. Collembolan numbers peak at 1-5 cm below the soil surface and drop to almost none below 10 cm.

Although generally low, the numbers and activities of soil organisms at depth vary spatially depending on gradients in texture, pH, temperature, water availability, and organic matter content. Interfaces between layers often generate localized regions of greater water availability where microorganisms may exhibit increased numbers or activity due to improved access to water and nutrients. There are active cells in deeper soil horizons, and on a depth-weighted basis, microbial biomass below 25 cm represents up to 30-40% of the microbial biomass in the soil profile.

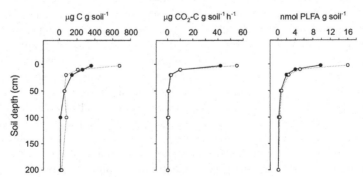

FIG. 8.2 Microbial biomass with depth for a valley (open circles) and terrace (closed circles) soil profile as determined by three methods. (Left) Chloroform fumigation extraction. (Middle) Substrate-induced respiration. (Right) Phospholipid fatty acid (PLFA) analysis. *Adapted from Fierer et al. (2003).*

Microbial diversity and community composition are also vertically strati-fied. Like microbial biomass, microbial diversity is typically highest in the top 10 cm of soil and declines with depth. Bacterial diversity dropped by 20-40% from surface soil to deeper horizons in a montane forest in Colorado, U.S.A. (Eilers et al., 2012). There are distinct differences in the microbial com-munity colonizing the litter layer versus the organic horizon of forest soils. The litter layer in a spruce forest in Central Europe was enriched in Acidobacteria and Firmicutes, whereas Actinobacteria were more abundant in the organic horizon (Baldrian et al., 2012). Surface mineral soils harbor a different micro-bial community than that found at depth, with the transition typically occurring between 10 and 25 cm (Eilers et al., 2012). The depth distribution of the bac-terial community in a montane forest was primarily driven by a decline in the relative abundance of Bacteroidetes and a peak in the relative abundance of Ver-rucomicrobia between 10 and 50 cm (Fig. 8.3). The Actinobacteria were less strongly structured by depth, showing similar relative abundances down to 150 cm. The depth distribution pattern of other bacterial groups (e.g., Acidobacteria, Alphaproteobacteria) was less consistent and differed depending on sampling location.

Fungal community composition differs substantially between the litter and organic horizons, with up to 40% of the most abundant taxa found in either the litter layer or the organic horizon (Baldrian et al., 2012). The litter layer in forest soils typically exhibits a high abundance of saprotrophic fungi, whereas

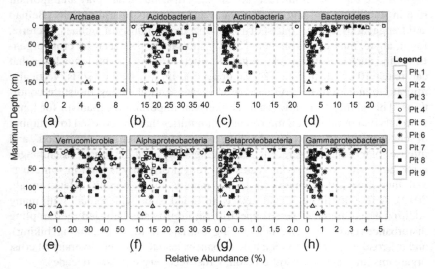

FIG. 8.3 Relative abundance of archaea and bacterial taxa with depth at nine locations in a mon-tane forest in Colorado, U.S.A. *From Eilers et al. (2012); used with permission.*

mycorrhizal species dominate the organic horizon and surface mineral soil. Of 22 ectomycorrhizal fungal taxa identified in a coniferous forest soil, 2 taxa were restricted to the organic horizon, whereas 11 were found only in the mineral soil (Rosling et al., 2003). Observed differences in fungal taxa between soil horizons are likely related to their ecological functions in the system. Baldrian et al. (2012) analyzed forest soil fungi that harbor the *cbhI* gene which encodes for cellobiohydrolase, an extracellular enzyme that mediates the degradation of cellulose. The diversity of cellulolytic fungi was higher in the litter layer compared to the organic horizon, and the fungal taxa mediating cellulose decomposition were distinct between these two horizons. Some of the most abundant *cbhI* sequences were transcribed by fungi with low relative abundance, indicating that some species, although not highly abundant, may nonetheless make an important contribution to cellulose decomposition.

Soil invertebrate communities also vary with depth, particularly in the organic and upper mineral soil horizons. There is a succession from litter-dwelling to soil-dwelling species of collembola at the interface between organic and mineral soil layers. Gut content analysis has shown that collembola near the top of the organic horizon feed preferentially on pollen grains, whereas those at the bottom of this layer feed mainly on fungal material and highly decomposed organic matter (Ponge, 2000).

Soil biota are widely distributed as biological soil crusts (also called cryptogamic, microbiotic, microphytic crusts), which form a protective soil covering in the interspaces between the patchy plant cover in arid and semiarid regions (see Fig. 8.4). Soil crusts are globally distributed and play an important role in stabilizing soil and cycling water and nutrients in arid regions (Belnap and Lange, 2001). They consist of a specialized community of archaea, bacteria, fungi, algae, mosses, and lichens. Sticky substances produced by crust organisms bind surface soil particles together, often forming a continuous layer that can reach 10 cm in thickness. Cyanobacteria, in arid surface soils, vertically migrate to and from the soil surface in response to changing moisture conditions. This ability to follow water is likely an important mechanism for long-term survival of desert soil microbial communities that are exposed to a number of challenging conditions, including water stress, nutrient limitation, and intense solar radiation. A recent, sequencing analysis of the fungal assemblages associated with soil crusts indicates that fungal diversity is high in soil crusts and that diversity increases as crusts mature successionally (Bates et al., 2012). A few dominant fungal taxa were found to be widespread, with many additional taxa found to be site specific. Soil crusts are damaged by trampling disturbances caused by livestock grazing, tourist activities (e.g., hiking, biking), and off-road vehicle traffic. Such disturbances lead to reduced diversity of crust organisms and increased soil erosion. Crust recovery can take decades.

The deep subsurface environment, that region below the top few meters of soil, as well as deep aquifers, caves, bedrock, and unconsolidated sediments, was once considered hostile to and devoid of living organisms. It is now known

FIG. 8.4 Biological soil crusts. (Top) Landscape with healthy soil crusts. (Middle) Close-up of mature crusts on the Colorado Plateau, U.S.A. (Bottom) Cyanobacteria adhered to sand grains. *Courtesy of Jayne Belnap, USGS Canyonlands Field Station, Moab, UT, U.S.A.; used with permission.*

that large numbers of microorganisms reside in the "deep biosphere," often to depths of tens to hundreds of meters. Subsurface microbiology emerged as a discipline largely in response to groundwater quality issues and because modern drilling and coring techniques have made possible the retrieval and characterization of undisturbed, uncontaminated samples collected from deep environments. Scientists from many disciplines (e.g., geology, hydrology, geochemistry, and environmental engineering) are now interested in understanding the role of microorganisms in soil genesis, contaminant degradation, maintenance of groundwater quality, and the evolution of geological formations (e.g., caves). Many novel microorganisms with unique biochemical and genetic traits have been isolated from subsurface environments, and these organisms may have important industrial and pharmaceutical uses.

Estimates from a recent census suggest that there are 3×10^{29} microbial cells residing in the deep biosphere globally (Hoehlert and Jorgensen, 2013). Jansson (2013) has estimated a total of 10^{30} bacteria alone on our planet relative to the 10^{34} stars in the universe. Deep subsurface microbial populations are dominated by bacteria and archaea, and nearly all of the major taxonomic and physiological prokaryotic groups have been found. Bacterial densities range from less than 10^1 to 10^8 cells g^{-1} material, depending on the method of enumeration and the depth of sample collection. Groundwater sampled from aquifers or unconsolidated sediments have bacterial concentrations ranging from 10^3 to 10^6 cells mL^{-1}. Microbial biomass in the deep subsurface is typically several to many orders of magnitude lower than that observed for surface soils, where populations often reach 10^9 bacteria and archaea per gram soil. The turnover of the subsurface biomass is typically on the timescale of centuries to millennia.

Subsurface microbial communities have a high proportion of inactive, nonviable, and nonculturable cells. Most subsurface microbes have no cultivated relatives, making it difficult to determine the physiological characteristics of even the dominant taxa (Hoehlert and Jorgensen, 2013). It is known that metabolic rates are very slow compared to terrestrial surface environments, with microbial activity at these depths being limited by water availability, temperature, and the availability of energy sources. It is estimated that the energy flux available to the deep biosphere is $<1\%$ of the C fixed photosynthetically at Earth's surface (Hoehlert and Jorgensen, 2013). Energy sources include low-concentration organic substrates or reduced inorganic substrates such as H_2, CH_4, or S_2. As in surface soils, the size and interconnectivity of pores is an important factor regulating microbial growth. In the subsurface environment, pores exist in unconsolidated materials or as fractures or fissures in consolidated rock. Microbial activity occurs in pores 0.2 to 15 μm in size, whereas little to no activity has been observed in pores with openings of less than 0.2 μm.

Although bacteria and archaea dominate most subsurface microbial communities, protozoa and fungi have been observed at depth under certain conditions. Protozoan populations, typically at or below the level of detection at unpolluted sites, have been observed in the subsurface of contaminated sites. Protozoan presence may thus represent a useful indicator of environmental contamination. Protozoa have been shown to stimulate subsurface nitrification and bacterial degradation of dissolved organic C. Both ectomycorrhizal and arbuscular mycorrhizal hyphae have also been recovered in fractured, granitic bedrock at depths greater than 2 m (Egerton-Warburton et al., 2003). The ability of mycorrhiza to

FIG. 8.5 Gypsum associated with biofilms containing sulfur-oxidizing bacteria in Lower Kane Cave, Wyoming, U.S.A. *Photo courtesy of Annette Engel, University of Tennessee–Knoxville; used with permission.*

grow and function in subsurface environments may enable their host plants living in dry climates to survive drought conditions by enhancing water and nutrient uptake from bedrock sources.

The study of subsurface microorganisms is transforming our ideas regarding the extent of the biosphere and the role that microorganisms have and continue to play in the evolution of the subsurface environment. As one example, scientists have found evidence suggesting that caves and aquifers in karst environments are formed, in part, by microbial activity rather than exclusively by abiotic, geochemical reactions as was previously thought (Fig. 8.5). Sulfur-oxidizing bacteria colonize carbonate surfaces in the cave or aquifer, use H_2S as an energy source, and produce sulfuric acid as a by-product (Engel and Randall, 2011). This acid facilitates the conversion of limestone to gypsum, which is more easily dissolved by water.

IV MICROSCALE HETEROGENEITY IN MICROBIAL POPULATIONS

Soil is a complex, three-dimensional space comprised of mineral particles, plant roots, organic materials, pore spaces, and organisms (Fig. 8.6). The shape and arrangement of soil minerals and organic particles is such that a network of pores of various shapes and sizes exists, resulting in a highly uneven distribution of water, oxygen, and solutes. Between 45% and 60% of the total soil volume is comprised of pores that are either air or water filled, depending on moisture conditions. These pores may be open and connected to adjoining pores or closed and isolated from the surrounding soil. Pores of different shapes, sizes, and degree of continuity provide a mosaic of microbial habitats with very different physical, chemical, and biological characteristics, resulting in an uneven distribution of soil organisms. Because soil organisms themselves vary in size, structural heterogeneity at this scale determines where a particular organism can reside, the degree to which its movement is restricted, and its interactions with other organisms.

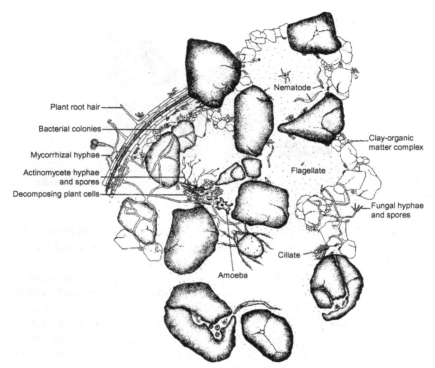

FIG. 8.6 Diagrammatic representation of a plant root and associated biota in an approximately 1 cm² area. *Adapted from S. Rose and T. Elliott, personal communication.*

The heterogeneous nature of the soil pore network plays a fundamental role in determining microbial abundance, diversity, activity, and community composition by affecting the relative proportion of air- versus water-filled pores, which in turn regulates water and nutrient availability, gas diffusion, and biotic interactions, such as competition and predation. Microbial activity, measured as respiratory output (i.e., CO_2 evolution), is dependent on the location of organic matter substrates within the pore network and on optimum concentrations of water and oxygen. Microbial activity is maximized when about 60% of the total soil pore space is water filled. As soil moisture declines below this level, pores become poorly interconnected, water circulation becomes restricted, and dissolved nutrients, which are carried by the soil solution, become less available for microbial utilization. At the other extreme, when most or all of the pores are filled with water, oxygen becomes limiting because diffusion rates are significantly greater in air than through water. Gas diffusion into micropores is particularly slow because small pores often retain water even under dry conditions. Restricted oxygen diffusion into micropores combined with biological oxygen consumption during the decomposition of organic matter can lead to the rapid development and persistence of anaerobic conditions. Thus, survival of soil biota residing in small pores depends on their ability to carry out anaerobic respiration (e.g., denitrification), replacing oxygen with an alternative electron acceptor (e.g., NO_3^-).

This chapter has focused, so far, on the spatial distribution of biota at the scale of soil profiles, sites, or continents, but we might also ask whether there is a microbiogeography at the soil aggregate or pore scale. Recent evidence suggests that microbial diversity and community composition do vary significantly across a gradient of aggregate fractions and pore size classes. Bacteria can occupy both large and small soil pores; however, 80% of bacteria are thought to reside in small pores. The maximum diameter of pores most frequently colonized by bacteria is estimated to range from 2.5 μm in fine soils to 9 μm in coarse textured soils. Few bacteria have been observed in pores <0.8 μm in diameter, which means that 20-50% of the total soil pore volume, depending on soil texture and pore size distribution, cannot be accessed and utilized by the microbial community.

Electron microscopy has revealed that bacteria often occur as isolated cells or small colonies (<10 cells) associated with decaying organic matter; however, larger colonies of several hundred cells have been observed on the surface of aggregates (Fig. 8.7). Bacterial cells are often embedded in mucilage, a sticky substance of bacterial origin to which clay particles attach. Clay encapsulation and residence in small pores may provide bacteria with protection against dessication, predation, bacteriophage attack, digestion during travel through an earthworm gut, and the deleterious effects of introduced gases such as ethylene bromide, a soil fumigant.

Fungi are generally found on aggregate surfaces and in large pores; however, they can use substrates in smaller pores if their hyphae can penetrate them (Ruamps et al., 2011). Like bacteria, fungal hyphae are often sheathed in extracellular mucilage, which not only serves as protection against predation and desiccation, but is also a gluing agent in the soil aggregation process. Extensive hyphal networks grow through the soil and over aggregate surfaces, binding soil particles together, thereby playing an important role in the formation and stabilization of aggregates.

The distribution of soil fauna is highly dependent on the size and arrangement of soil aggregates and pore spaces. Nematode abundance and community composition were found to be more highly influenced by aggregate size than by soil management practice (Briar et al., 2011). Nematode density was consistently lower in small macroaggregates (250-1000 μm) compared to large macroaggregates (>1000 μm) and interaggregate spaces. There was also a higher juvenile:adult ratio of nematodes observed in small macroaggregates, suggesting that juveniles, because of their smaller size, can more readily access nutrient resources (e.g., bacterial prey) in smaller pores. Smaller aggregates may also provide refuge for juvenile nematodes, protecting them from predation by larger soil fauna.

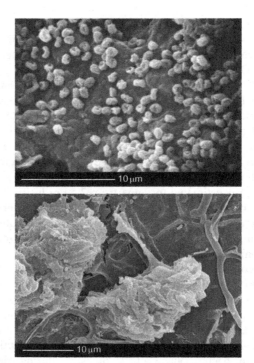

FIG. 8.7   (Top) Bacteria on the surface of an aggregate isolated from a grassland soil. (Bottom) An amoeba with its extended pseudopodia engulfing bacteria. *Photos courtesy of V. V. S. R. Gupta, CSIRO Land and Water, Glen Osmond, South Australia, Australia; used with permission.*

Soil heterogeneity influences nutrient cycling dynamics by restricting organism movement and thereby modifying the interactions between organisms. For example, small pores influence trophic relationships and nutrient mineralization by providing refuges and protection for smaller organisms, particularly bacteria, against attack from larger predators (e.g., protozoa, nematodes) that are typically unable to enter smaller pores. The location of bacteria within the pore network is a key factor in their growth and activity. Bacterial populations are consistently high in small pores, but highly variable in large pores where they are vulnerable to being consumed. This may explain why introduced bacteria (e.g., *Rhizobium* and biocontrol organisms) often exhibit poor survival relative to indigenous bacteria. When they are introduced in such a way as to be transported by water movement into small, protected pores, their ability to persist is enhanced.

V   DRIVERS OF SPATIAL HETEROGENEITY

Organism abundance, diversity, and activity are not randomly distributed in soil, but vary in a patchy fashion both horizontally across a landscape and vertically through the soil profile. Different groups of organisms exhibit different spatial patterns because they each react to soil conditions in different ways. This spatial heterogeneity in microbial properties is an inherent feature of soils and has been shown to correlate with gradients in site and soil characteristics, including abiotic factors (bulk density, texture, moisture, oxygen concentration, pH, soil organic matter content, inorganic N availability, soil aggregation), biotic factors (food web interactions, vegetation dynamics), management factors (tillage, cropping system), and global change factors (soil warming, altered precipitation regimes, land-use change, N deposition, invasive species). Some of these factors are important at the microscopic scale, whereas others act over larger distances. Microbial biomass and collembolan abundance in an agroecosystem reflected large-scale gradients in soil C content and cultivation practices (Fromm et al., 1993). In other cases, soil characteristics have been found to explain a relatively minor amount (<30%) of the spatial variation in organism abundance (Robertson and Freckman, 1995). Spatial heterogeneity can be high even in soils that appear relatively homogenous at the plot or field scale (Franklin and Mills, 2003, 2009). Nematode populations were strongly patterned in an agroecosystem despite many years of soil tillage and monoculture cropping, suggesting that spatial heterogeneity may be even higher in less-disturbed systems (Robertson and Freckman, 1995).

Information on microbial spatial patterns can be used to design more statistically powerful experiments, improve our understanding of how soil communities develop, and determine what factors are important for regulating and maintaining soil biodiversity and microbial community function. However, determining the appropriate scale of study is complicated because the factors influencing microbial community structure and function are themselves scale dependent, and there are complex interactions between scales that need to be taken into account (Berg, 2012). Historically, spatial variability in soil characteristics has been regarded as random noise in the system; however, geostatistical approaches are now commonly used to quantify spatial soil heterogeneity and to provide information on the underlying causes of observed spatial patterns. These analyses indicate that the scale at which soil organisms aggregate can vary from a few millimeters to hundreds or thousands of meters, and it is highly dependent on the body size, mobility, and dispersal ability of the organism(s) being studied.

Spatial patterns in microbial community parameters may be controlled in some instances by geographic distance, but increasingly, research suggests that microbial communities are more often structured by environmental gradients (Nemergut et al., 2011). Recent deep sequencing analyses have demonstrated that there can be as much variation in microbial community structure within a single soil pit as there is among soils collected from different biomes separated by thousands of kilometers (Eilers et al., 2012). This suggests that local environmental heterogeneity may be more important than geographic distance in determining microbial community differences. At large spatial scales, climatic factors (temperature and precipitation), which are strongly associated with changes in elevation or latitude, might be expected to play a key role in determining microbial community structure because these are primary drivers structuring plant communities at regional to global scales. Thus, we might predict that microbial diversity follows the same latitudinal and elevational diversity patterns typically observed for many plant and animal taxa; namely that species richness is highest at the equator or at lower elevations of mountains, decreasing toward the poles or higher elevations. Recently, a number of studies have been conducted to determine if, in fact, soil biota follow the same biogeographical patterns as plants and animals and whether those variables that drive plant and animal diversity are also good predictors of microbial diversity. The take-home message from these studies is that the factors structuring soil microbial communities appear to be distinctly different from those driving biogeographic patterns of plants and animals. In particular, soil organisms do not appear to follow the same elevational or latitudinal patterns observed for many plant and animal taxa (Fierer et al., 2011).

Bacterial diversity often is largely explained by soil pH (Chu et al., 2010; Lauber et al., 2008; Rousk et al., 2010), with this pattern largely driven by changes in the Acidobacteria, Actinobacteria, and Bacteroides groups (Lauber et al., 2009). Bacterial taxon richness tends to be highest in neutral soils and lowest in acid soils, but particular bacterial groups are more strongly affected by pH than others. Some acidobacterial groups strongly decline with increasing pH, whereas others increase or are not significantly affected (Rousk et al., 2010). Perhaps it is not surprising that soil bacteria are sensitive to pH given their tendency toward narrow pH tolerances, closely constrained pH optima, and activities that are reduced by deviations in this optimum range (Bååth, 1996; Fernandez-Calvino and Bååth, 2010).

Although soil pH is emerging as a strong predictor of bacterial community structure at local to continental scales, it does not appear to be the key driver for other soil microbial groups; pH was not a significant factor driving fungal community composition (Rousk et al., 2010), which was better predicted by soil nutrient status (e.g., C:N ratio; Lauber et al., 2008). Soil moisture, as influenced by gradients in mean annual temperature and precipitation, has also been identified as an important factor influencing fungal biomass (Fig. 8.8; Frey et al., 1999), growth (Meier et al., 2010), and species richness (McGuire et al., 2012). Soil protists are another group for which pH appears to be of limited importance in determining biogeographic patterns in community composition, the latter being more strongly related to soil moisture availability (Bates et al., 2013). Archaeal abundance is correlated with soil C:N ratios, being higher in soils with lower ratios (Bates et al., 2011).

Soil heterogeneity in resource availability as influenced by land-use patterns, and plant community dynamics has been hypothesized to be responsible for the high levels of soil biodiversity (Ettema and Wardle, 2002). In particular, spatial isolation may be an important determinant of microbial community structure by facilitating species coexistence. In a simulated soil environment, one bacterial species dominated the community under saturated conditions where the pore network was highly connected (Treves et al., 2003). However,

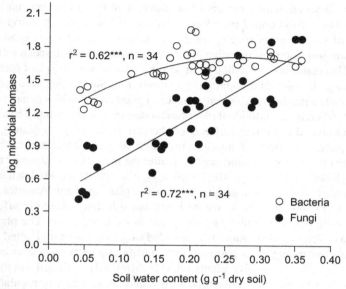

FIG. 8.8 Relationship between soil water content of the 0-5 cm depth and bacterial and fungal biomass C for soils collected from six long-term tillage comparison experiments in North America. *From Frey et al. (1999); used with permission.*

under low moisture conditions (i.e., discontinuous water films), spatial isolation of microbial populations allowed a less competitive species to become established in the community. These results concur with the observation that saturated subsurface soils have lower microbial diversity compared to unsaturated surface soils (Zhou et al., 2004).

Complex interactions and feedbacks occurring at the plant-soil interface and plant-mediated heterogeneity in soil physical and chemical characteristics also play a critical role in regulating the composition and diversity of soil microbial communities. Microbial abundance, diversity, community composition, and activity often reflect the plant species present in a given soil. For example, bacteria isolated from the rhizospheres of different grass species exhibited differential growth and C source utilization patterns (Westover et al., 1997). Plant-mediated differences in the microbial community are potentially attributable to specific variations between plant species in the quality and quantity of organic matter inputs to the soil. Plant resource quantity and quality can also be altered by disturbances such as herbivore grazing and global change (e.g., elevated atmospheric CO_2, N deposition). Plant herbivory has been shown to increase C allocation to roots, root exudation, fine root turnover, soil dissolved organic C, microbial biomass and activity, and faunal activity (Bardgett and Wardle, 2003). These changes in turn alter soil N availability, plant N acquisition, photosynthetic rates, and ultimately plant productivity.

There is accumulating evidence that plants "culture" a soil community that then feeds back to control their long-term survival and growth. Survival and growth of several grass species were significantly reduced when grown with their own soil community rather than that of another plant species (Bever, 1994). This result was attributed to an accumulation of specific plant pathogens or a change in microbial community composition. Such a negative feedback may provide a mechanism for maintenance of plant communities in natural ecosystems, whereby an individual plant cannot dominate a community because of the accumulated detrimental effects of the soil community on plant growth. Exotic plants that become invasive may have escaped local control by soil organisms at invaded sites and may even alter the microbial community to their benefit (Rout and Callaway, 2012). Callaway et al. (2004) reported that spotted knapweed (*Centaurea maculosa*), an invasive plant in North America, cultivates a soil community in its native European soil that negatively affects its own growth, possibly controlling its spread in its home range. The plant cultivates a different soil community at invaded sites in the western United States, positively enhancing its own growth and contributing to its success as an invasive species. Thus, invasive plants interact differently with soil microorganisms in their home range compared to invaded sites, escaping regulation by natural enemies and benefitting from mutualistic relationships with soil organisms in the introduced range that serve to reduce competition by suppressing the growth of native plant species and increasing soil nutrient supplies (Rout and Callaway, 2012).

Just as the plant community may be a determinant of microbial community structure, the diversity and composition of the microbial community may play a role in plant community dynamics. The diversity of arbuscular mycorrhizal fungi was observed to be a major factor contributing to plant diversity, above- and belowground plant biomass, and soil nutrient availability in a macrocosm experiment simulating North American old-field ecosystems (van der Heijden et al., 1998). An increase in arbuscular mycorrhizal fungal diversity was accompanied by a significant increase in the length of mycorrhizal hyphae in the soil leading to greater soil resource acquisition. As the number of mycorrhizal fungal species increased, plant diversity, biomass, and plant tissue phosphorus content increased, while the phosphorus concentration in the soil decreased.

A primary focus of this section has been on factors driving microbial community structure; however, it is also useful to consider regional and global patterns in soil microbial biomass, which represent a considerable fraction of living biomass on Earth, with estimates between 1% and 3% of total terrestrial soil C. A meta-analysis of belowground plant, microbial, and faunal biomass indicates that microbial biomass ranges from < 5 to ~ 800 g C m^{-2} across biomes, with the highest concentrations occurring in temperate coniferous and tropical forests (Fierer et al., 2009). Climate, vegetation, soil characteristics, and land-use patterns all interact to influence microbial abundance and biomass at a given location. Microbial biomass tends to follow global patterns of plant biomass and productivity, but soil organic C is the best predictor of microbial biomass levels at a given location. Microbial biomass is generally positively related to soil organic matter contents in most ecosystems, with peat and organic soils being an exception. Levels of microbial biomass are also typically correlated with soil clay content. Clay minerals promote microbial growth by maintaining the pH in an optimal range, buffering the nutrient supply, adsorbing metabolites that are inhibitory to microbial growth, and providing protection from desiccation and grazing through increased aggregation. Total amounts of microbial biomass are also impacted by land use, with lower levels typically observed in arable compared to undisturbed forest and grassland soils due to cultivation-induced losses of organic matter. Microbial biomass is also correlated with latitude, with microbial biomass tending to be lower but more highly variable at high latitudes. This increased variability in microbial biomass with increasing latitude is attributed to a higher interseasonal variation in temperature (Wardle, 1998).

VI SUMMARY

The spatial patterns of soil organisms and the factors that drive this heterogeneity have only begun to be explored in detail. Recent developments in molecular sequencing technologies, geostatistics, and other analytical approaches have resulted in an exponential increase in the number of studies published on the topic of microbial biogeography. Our understanding about how soil microbial communities are spatially organized across scales ranging from the soil pore to the globe has fundamentally changed in the past few years. It is becoming increasingly evident that soil microorganisms are not cosmopolitan in their distribution, but exhibit meaningful biogeographic patterns. These distribution patterns often appear to be shaped by forces other than those typically observed for plant and animal taxa. For more information on microbial biogeography, see the reviews by Ramette and Tiedje (2006), Fierer (2008), Green et al. (2008), Fierer and Lennon (2011), and Rout and Callaway (2012).

REFERENCES

Azmi, O.R., Seppelt, R.D., 1998. The broad-scale distribution of microfungi in the Windmill Islands region, continental Antarctica. Polar Biol. 19, 92–100.

Bååth, E., 1996. Adaptation of soil bacterial communities to prevailing pH in different soils. FEMS Microbial. Ecol. 19, 227–237.

Bachy, C., Dolan, J.R., Lopez-Garcia, P., Deschamps, P., Moreira, D., 2013. Accuracy of protest diversity assessments: morphology compared with cloning and direct pyrosequencing of 18S rRNA genes and ITS regions using the conspicuous tintinnid ciliates as a case study. ISME J. 7, 244–255.

Baldrian, P., Kolařík, M., Štursová, M., Kopecký, J., Valášková, V., Větrovský, T., Zifčáková, L., Snajdr, J., Rídl, J., Vlček, Č., Voříšková, J., 2012. Active and total microbial communities in forest soil are largely different and highly stratified during decomposition. ISME J. 6, 248–258.

Bardgett, R.D., Wardle, D.A., 2003. Herbivore-mediated linkages between aboveground and below-ground communities. Ecology 84, 2258–2268.

Bates, S.T., Berg-Lyons, D., Caporaso, J.G., Walters, W.A., Knight, R., Fierer, N., 2011. Examining the global distribution of dominant archaeal populations in soil. ISME J. 5, 908–917.

Bates, S.T., Nash, T.H., Garcia-Pichel, F., 2012. Patterns of diversity for fungal assemblages of biological soil crusts from the southwestern United States. Mycologia 104, 353–361.

Bates, S.T., Clemente, J.C., Flores, G.E., Walters, W.A., Parfrey, L.W., Knight, R., Fierer, N., 2013. Global biogeography of highly diverse protistan communities in soil. ISME J. 7, 652–659.

Belnap, J., Lange, O.L., 2001. Biological Soil Crusts: Structure, Function and Management. Ecological Studies. Springer-Verlag, Berlin.

Berg, M., 2012. Patterns of biodiversity at fine and small spatial scales. In: Wall, D.H. (Ed.), Soil Ecology and Ecosystem Services. Oxford University Press, Oxford, UK, pp. 136–149.

Bever, J.D., 1994. Feedback between plants and their soil communities in an old field community. Ecology 73, 1965–1977.

Briar, S.S., Fonte, S.J., Park, I., Six, J., Scow, K., Ferris, H., 2011. The distribution of nematodes and soil microbial communities across soil aggregate fractions and farm management systems. Soil Biol. Biochem. 43, 905–914.

Callaway, R.M., Thelen, G.C., Rodriguez, A., Holben, W.E., 2004. Soil biota and exotic plant invasion. Nature 427, 731–733.

Cho, J., Tiedje, J.M., 2000. Biogeography and degree of endemicity of fluorescent *Pseudomonas* strains in soil. Appl. Environ. Microbiol. 66, 5448–5456.

Chu, H., Fierer, N., Lauber, C.L., Caporaso, J.G., Knight, R., Grogan, P., 2010. Soil bacterial diversity in the Arctic is not fundamentally different from that found in other biomes. Environ. Microbiol. 12, 2998–3006.

de Wit, R., Bouvier, T., 2006. '*Everything is everywhere, but, the environment selects*'; what did Baas Becking and Beijerinck really say? Environ. Microbiol. 8, 755–758.

Egerton-Warburton, L.M., Graham, R.C., Hubbert, K.R., 2003. Spatial variability in mycorrhizal hyphae and nutrient and water availability in a soil-weathered bedrock profile. Plant Soil 249, 331–342.

Eilers, K.G., Debenport, S., Anderson, S., Fierer, N., 2012. Digging deeper to find unique microbial communities: the strong effect of depth on the structure of bacterial and archaeal communities in soil. Soil Biol. Biochem. 50, 58–65.

Engel, A.S., Randall, K.W., 2011. Experimental evidence for microbially mediated carbonate dissolution from the saline water zone of the Edwards Aquifer, Central Texas. Geomicrobiol. J. 28, 313–327.

Ettema, C.H., Wardle, D.A., 2002. Spatial soil ecology. Trends Ecol. Evol. 17, 177–183.

Fernandez-Calvino, D., Bååth, E., 2010. Growth response of the bacterial community to pH in soils differing in pH. FEMS Microbiol. Ecol. 73, 149–156.

Fierer, N., 2008. Microbial biogeography: patterns in microbial diversity across space and time. In: Zengler, K. (Ed.), Accessing Uncultivated Microorganisms: from the Environment to Organisms and Genomes and Back. ASM Press, Washington DC, pp. 95–115.

Fierer, N., Ladau, J., 2012. Predicting microbial distributions in space and time. Nat. Methods 9, 549–551.

Fierer, N., Lennon, J.T., 2011. The generation and maintenance of diversity in microbial communities. Am. J. Bot. 98, 439–448.

Fierer, N., Schimel, J.P., Holden, P.A., 2003. Variations in microbial community composition through two soil depth profiles. Soil Biol. Biochem. 35, 167–176.

Fierer, N., Strickland, M.S., Liptzin, D., Bradford, M.A., Cleveland, C.C., 2009. Global patterns in belowground communities. Ecol. Lett. 12, 1238–1249.

Fierer, N., McCain, C.M., Meir, P., Zimmermann, M., Rapp, J.M., Silman, M.R., Knight, R., 2011. Microbes do not follow the elevational diversity patterns of plants and animals. Ecology 92, 797–804.

Finlay, B.J., 2002. Global dispersal of free-living microbial eukaryote species. Science 296, 1061–1063.

Finlay, B.J., Clarke, K., 1999. Ubiquitous dispersal of microbial species. Nature 400, 828.

Franklin, R.B., Mills, A.L., 2003. Multi-scale variation in spatial heterogeneity for microbial community structure in an eastern Virginia agricultural field. FEMS Microbiol. Ecol. 44, 335–346.

Franklin, R., Mills, A., 2009. Importance of spatially structured environmental heterogeneity in controlling microbial community composition at small spatial scales in an agricultural field. Soil Biol. Biochem. 41, 1833–1840.

Frey, S.D., Elliott, E.T., Paustian, K., 1999. Bacterial and fungal abundance and biomass in conventional and no-tillage agroecosystems along two climatic gradients. Soil Biol. Biochem. 31, 573–585.

Fromm, H., Winter, K., Filser, J., Hantschel, R., Beese, F., 1993. The influence of soil type and cultivation system on the spatial distributions of the soil fauna and microorganisms and their interactions. Geoderma 60, 109–118.

Fulthorpe, R.R., Roesch, L.F.W., Riva, A., Triplett, E., 2008. Distantly sampled soils carry few species in common. ISME J. 2, 901–910.

Green, J.L., Bohannan, B.J.M., Whitaker, R.J., 2008. Microbial biogeography: from taxonomy to traits. Science 320, 1039–1043.

Hoehlert, T.M., Jorgensen, B.B., 2013. Microbial life under extreme energy limitation. Nat. Rev. Microbiol. 11, 83–94.

Jansson, J.K., 2013. The life beneath our feet. Nature 494, 40–41.

Larsen, P.E., Field, D., Gilbert, J.A., 2012. Predicting bacterial community assemblages using an artificial neural network approach. Nat. Methods 9, 621–625.

Lauber, C.L., Strickland, M.S., Bradford, M.A., Fierer, N., 2008. The influence of soil properties on the structure of bacterial and fungal communities across land-use types. Soil Biol. Biochem. 40, 2407–2415.

Lauber, C.L., Hamady, M., Knight, R., Fierer, N., 2009. Pyrosequencing-based assessment of soil pH as a predictor of soil bacterial community structure at the continental scale. Appl. Environ. Microbiol. 75, 5111–5120.

McGuire, L., Fierer, N., Bateman, C., Treseder, K.K., Turner, B.L., 2012. Fungal community composition in neotropical rain forests: the influence of tree diversity and precipitation. Microb. Ecol. 63, 804–812.

Meier, C.L., Rapp, J., Bowers, R.M., Silman, M., Fierer, N., 2010. Fungal colonization of a common wood substrate across a tropical elevation gradient: temperature sensitivity, community composition, and potential for above-ground decomposition. Soil Biol. Biochem. 42, 1083–1090.

Nemergut, D.R., Costello, E.K., Hamady, M., Lozupone, C., Jiang, L., Schmidt, S.K., Fierer, N., 2011. Global patterns in the biogeography of bacterial taxa. Environ. Microbiol. 13, 135–144.

Ponge, J., 2000. Vertical distribution of collembola (Hexapdoa) and their food resources in organic horizons of beech forests. Biol. Fertil. Soils 32, 508–522.

Ramette, A., Tiedje, J.M., 2006. Biogeography: an emerging cornerstone for understanding prokaryotic diversity, ecology, and evolution. Microb. Ecol. 53, 197–207.

Robertson, G.P., Freckman, D.W., 1995. The spatial distribution of nematode trophic groups across a cultivated ecosystem. Ecology 76, 1425–1432.

Robeson, M.S., King, A.J., Freeman, K.R., Birky, C.W., Martin, A.P., Schmidt, S.K., 2011. Soil rotifer communities are extremely diverse globally but spatially autocorrelated locally. Proc. Natl. Acad. Sci. 108, 4406–4410.

Rosling, A., Landeweert, R., Lindahl, B.D., Larsson, K.-H., Kuyper, T.W., Taylor, A.F.S., Finlay, R.D., 2003. Vertical distribution of ectomycorrhizal fungal taxa in a podzol soil profile. New Phytol. 159, 775–783.

Rousk, J., Bååth, E., Brookes, P.C., Lauber, C.L., Lozupone, C., Caporaso, J.G., Knight, R., Fierer, N., 2010. Soil bacterial and fungal communities across a pH gradient in an arable soil. ISME J. 4, 1340–1351.

Rout, M.E., Callaway, R.M., 2012. Interactions between exotic invasive plants and soil microbes in the rhizosphere suggest that 'everything is not everywhere'. Ann. Bot. 110, 213–222.

Ruamps, L.S., Nunan, N., Chenu, C., 2011. Microbial biogeography at the soil pore scale. Soil Biol. Biochem. 43, 280–286.

Smith, D.J., Timonen, H.J., Jaffe, D.A., Griffin, D.W., Birmele, M.N., Perry, K.D., Ward, P.D., Roberts, M.S., 2013. Intercontinental dispersal of bacteria and archaea in transpacific winds. Appl. Environ. Microbiol. http://dx.doi.org/10.1128/AEM.03029-12.

Treves, D.S., Xia, B., Zhou, J., Tiedje, J.M., 2003. A two-species test of the hypothesis that spatial isolation influences diversity in soil. Microb. Ecol. 45, 20–28.

Van der Heijden, M.G.A., Klironomos, J.N., Ursic, M., Moutoglis, P., Streitwolf-Engel, R., Boller, T., Wiemken, A., Sanders, I.R., 1998. Mycorrhizal fungal diversity determines plant biodiversity, ecosystem variability and productivity. Nature 396, 69–72.

Wardle, D.A., 1998. Controls of temporal variability of the soil microbial biomass: a global-scale synthesis. Soil Biol. Biochem. 13, 1627–1637.

Westover, K.M., Kennedy, A.C., Kelley, S.E., 1997. Patterns of rhizosphere microbial community structure associated with co-occurring plant species. J. Ecol. 85, 863–873.

Wilkinson, D.M., Koumoutsaris, S., Mitchell, E.A.D., Bey, I., 2012. Modelling the effect of size on the aerial dispersal of microorganisms. J. Biogeogr. 39, 89–97.

Zhou, J., Xia, B., Huang, H., Palumbo, A.V., Tiedje, J.M., 2004. Microbial diversity and heterogeneity in sandy subsurface soils. Appl. Environ. Microbiol. 70, 1723–1734.

FURTHER READING

Fierer, N., Jackson, R.B., 2006. The diversity and biogeography of soil bacterial communities. Proc. Natl. Acad. Sci. 103, 626–631.

Chapter 9

The Metabolic Physiology of Soil Microorganisms

Alain F. Plante, Maddie M. Stone and William B. McGill
Department of Earth and Environmental Science, University of Pennsylvania, Philadelphia, PA, USA

Chapter Contents

I INTRODUCTION

The soil habitat (Chapter 2) is the intersection between the atmosphere, hydrosphere, lithosphere, and biosphere and represents a highly complex environment with variable properties across space and time. This environmental heterogeneity generates a large diversity of selective pressures and available resources. Coupled with the evolutionary and metabolic flexibility of microorganisms, the result is the large phylogenetic and metabolic diversity found in the soil microbial population. Soil microorganisms can be classified on the basis of their genetic code and a set of phylogenetic markers (Chapters 3-6), dividing the population into archaea, bacteria, and eukarya. Recent advances in high-throughput molecular

Soil Microbiology, Ecology, and Biochemistry. http://dx.doi.org/10.1016/B978-0-12-415955-6.00009-8

tools have revealed soil to be the most taxonomically diverse habitat on Earth, with best estimates suggesting that there may be 10,000 species of microorganisms per cm^3 of soil (Fierer and Lennon, 2011). A different way to organize the soil microbial population is based on their metabolic capacity and physiology. In this aspect, soils are also the most metabolically diverse habitat on Earth. There is some overlap between these two classification systems. Eukaryotes have two basic metabolic strategies: autotrophy used by plants, and aerobic heterotrophy used by animals. Prokaryotes exhibit many variations and also have unique metabolic strategies not found in Eukaryotes. Despite these broad generalizations, groups of soil microorganisms defined by metabolic strategies or ecosystem function may include organisms that are only distantly related by phylogenetics, thus making the relationship between structure (i.e., phylogenetic classification) and function (i.e., metabolic classification) challenging to determine.

It is important to remember what soil microorganisms need to survive and reproduce. They require the raw materials used to build and repair their cells, a source of energy from which ATP can be synthesized, and a means of transferring electrons from one compound to another. It is also important to remember that the acquisition of resources and transfers of energy performed by the soil microbial population is subject to limitations imposed by chemical thermodynamics. The biogeochemical transformations mediated by soil organisms result from their search for energy, which is obtained by passing electrons (e^-) from donors to acceptors in multiple interconnected oxidation-reduction couples, which lead to cycles through which electrons flow. Electron donors and acceptors are often different elements, and consequently these flows of e^- unite cycles of elements, alter the mobility and functions of elements, and regulate soil biological transformations. Flow of electrons among these cycles unites activities of extremely diverse groups of soil organisms. This chapter examines the metabolic diversity of the soil microbial population and provides examples of several metabolic strategies relevant to soil function and biogeochemistry.

Information on the basis of redox reactions and bioenergetics is given in Section 1 of the online Supplemental Material (http://booksite.elsevier.com/9780124159556). A series of excellent online videos provides a refresher on the fundamental biology and chemistry discussed in this chapter. Access information provided in Sections 2 and 3 provides a glossary of some terms utilized in this chapter. Additional information about the basics of bioenergetics (including references to texts, links to online background information videos, and a glossary) are also provided in the online Supplemental Material.

II FOUNDATIONS OF MICROBIAL METABOLISM

A Stoichiometry

Recent work suggests that the elemental composition of soil microbial populations is relatively constrained and similar to the Redfield ratio of ocean systems, although substantial variability exists. Cleveland and Liptzin (2007) found the C:N:P ratio for soil microbes to be 60:7:1, which is significantly different from that of bulk soil (186:13:1). This disparity in elemental ratios between microbes and their substrates, and the microbial population's ability to maintain elemental ratios despite changes in resource availability, have profound implications for understanding controls on microbial metabolism (Sinsabaugh et al., 2008; Manzoni et al., 2010).

B Redox Reactions

As stated previously, all organisms require energy. To understand the vast diversity of microbial energy-acquisition strategies, it is important to start with several basic concepts of thermodynamics, the branch of chemistry that predicts which reactions are energetically favorable and which are not. Metabolism is, in essence, a suite of oxidation-reduction (redox) reactions. Oxidation refers to a loss of electrons by a substance and leads to a new oxidized substance with a lower level of potential energy. Conversely, reduction refers to the gain of electrons by a substance and leads to the production of a reduced substance with greater potential energy. A key concept is that oxidation and reduction reactions are always coupled; the oxidation of one compound necessarily leads to the reduction of another. Redox reactions lead to changes in the valence of elements, their oxidation state. The oxidation state of elements in a given reaction determines what is being oxidized and what is being reduced. Several general rules can be applied for determining the oxidation state of an element:

(1) Elements in the free state have an oxidation number of zero.
(2) Elements present as monoatomic ions have an oxidation state equal to the ionic charge (e.g., ferrous iron, or Fe^{2+} has an oxidation state of +2).
(3) In combination with other elements, hydrogen typically has an oxidation state of +1, and oxygen typically has an oxidation state of -2. An exception to this rule is that hydrides for hydrogen and peroxides for oxygen both have oxidation states of -1.
(4) The sum of oxidation states in the atoms comprising a compound must equal the overall net charge of the compound.

As an example, the first step in nitrification (a set of oxidation reactions performed by soil microorganisms that transform ammonium to nitrate), involves the oxidation of ammonium to nitrite. The oxidation states of the constituent elements in this reaction are written above the elements in the following reaction:

$$\overset{-3\ +1}{NH_4^+} + 1.5 \overset{0}{O_2} \rightarrow \overset{+3\ -2}{NO_2^-} + \overset{+1}{H^+} + \overset{+1\ -2}{H_2O} \tag{9.1}$$

Notice that the sum of the oxidation states of the elements in each compound equal the overall net charge. For example, a net charge of +1 can be determined for ammonium as follows:

$$(-3[N]) + (+1[H] \times 4) = +1 \tag{9.2}$$

What is the likelihood of a reaction to occur in nature? Given that all redox reactions are composed of an oxidation and reduction reaction, overall reactions can be written as two separate half reactions. The relative tendency of each half reaction to occur allows us to determine the likelihood of the overall reaction to proceed in a particular direction. For example, the reduction of CO_2 to methane that is accomplished by a class of anaerobic archaea known as methanogens is composed of two half reactions:

an oxidation:

$$4H_2 \rightarrow 8H^+ + 8e^- \tag{9.3}$$

and a reduction:

$$CO_2 + 8H^+ + 8e^- \rightarrow CH_4 + 2H_2O \tag{9.4}$$

resulting in the overall reaction:

$$CO_2 + 4H_2 \rightarrow CH_4 + 2H_2O \tag{9.5}$$

The tendency of reaction (9.5) to proceed as written will be determined by the tendencies of reactions (9.3 and 9.4) to proceed in the direction indicated. This can be determined by measuring the electrical potential (E_o) of each half reaction relative to a reference substance (H_2) under conditions of standard temperature, acidity, and pressure (25 °C, pH $= 0$, 1 atm). This potential is then corrected to pH $= 7$ to reflect the fact that most metabolic reactions occur at near-neutral conditions in the cell. To facilitate comparison among reactions, all half reactions are written as reductions, and the sign of the corresponding electrical potential is adjusted accordingly. These standard values are known as reduction potentials (E_0'). Low (more negative) reduction potentials indicate a low tendency for the reaction to occur, or a high tendency for the opposite oxidation reaction to occur. Summing the reduction potentials for two half reactions allows us to determine which direction a redox reaction tends to proceed in. For example, the formation of water from hydrogen and oxygen can be written as two half reactions with the following reduction potentials:

$$2H^+ + 2e^- \rightarrow H_2 \; E_0' = -0.42 \, V \tag{9.6}$$

$$\tfrac{1}{2}O_2 + 2H^+ + 2e^- \rightarrow H_2O \; E_0' = +0.82 \, V \tag{9.7}$$

Notice how both half reactions are written as reductions, even though for water to form reaction (9.6) would be written in reverse such that diatomic hydrogen would be oxidized. Reaction (9.6) has a more negative reduction potential than reaction (9.7), indicating that, under standard conditions, reaction (9.7) is more likely to proceed as written. For the reaction to proceed the other way, energy would have to be supplied from an external source.

C   Energetics

All compounds contain intrinsic energy, known as free energy or Gibbs free energy ($G^{0\prime}$). An important property of a metabolic reaction is the corresponding change in free energy ($\Delta G^{0\prime}$) that the reaction produces. The $\Delta G^{0\prime}$ value indicates the amount of energy released or required by a given reaction. A negative $\Delta G^{0\prime}$ value indicates that a reaction is exergonic, or energy-releasing, and will thus proceed spontaneously, whereas a positive $\Delta G^{0\prime}$ value indicates an endergonic reaction that requires external energy inputs to proceed. Most organisms obtain cellular energy through catabolism, or the exergonic breakdown of energy-rich organic molecules. By contrast, most organisms produce new organic molecules through a suite of endergonic processes collectively known as anabolism. Equilibrium conditions are established at $\Delta G^{0\prime} = 0$.

The following equation can be used to calculate the $\Delta G^{0\prime}$ for a given redox reaction:

$$\Delta G^{0\prime} = -nF\Delta E_0' \tag{9.8}$$

where n is the number of electrons transferred, F is Faraday's constant ($96.5\,\text{kJ V}^{-1}\,\text{mol}^{-1}$ or $23.1\,\text{kcal V}^{-1}\,\text{mol}^{-1}$), and $\Delta E_0'$ is the difference between the reduction potentials for two half reactions comprising a redox reaction when both are oriented to reflect the direction of the combined reaction. As an example, the first step in denitrification involves the reduction of nitrate to nitrite:

$$NO_3^- + 2H^+ + 2e^- \rightarrow NO_2^- + H_2O \tag{9.9}$$

$\Delta E_0'$ for this reaction is the difference between the two reduction potentials for the half reactions:

$$\Delta E_0' = +0.43 - (-0.37) = 0.8\,\text{V} \tag{9.10}$$

$$\Delta G^{0\prime} = -(1)(96.5)(0.8) = -77.2\,\text{kJmol}^{-1} \text{ of } NO_2 \text{ produced} \tag{9.11}$$

Equation (9.9) shows that redox pairs with higher reduction potentials will produce more negative $\Delta G^{0\prime}$ values. The more negative the $\Delta G^{0\prime}$ for a reaction, the more energy is released that can then be used to generate ATP and other cellular energy-carrying molecules. In other words, the higher the $\Delta E_0'$ value for a redox pair, the more energetically favorable that redox reaction is.

Figure 9.1 depicts the relationship between oxidized and reduced substrates as a vertically arranged hierarchy of oxidation-reduction half reactions. Compounds on the left half of the diagram are in an oxidized state and therefore accept electrons, whereas compounds on the right half of the diagram are in

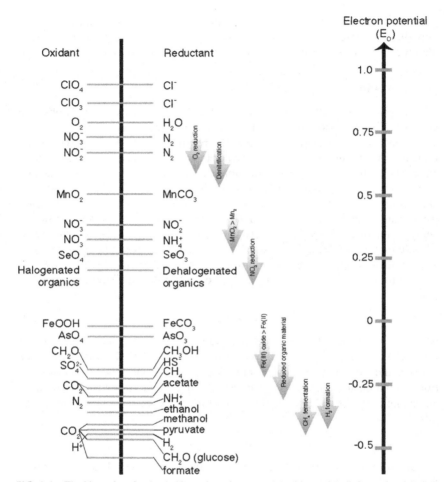

FIG. 9.1 The hierarchy of redox half reactions that control the biogeochemical reactions carried out by soil microorganisms. *(Redrawn from Madsen, E. L., 2008. "Environmental Microbiology: From genomes to biogeochemistry." Blackwell Publishing, Malden, MA.)*

a reduced state and donate electrons. Redox conditions are highly oxidizing at the top of the hierarchy and become progressively more reducing moving down the hierarchy. Reduction potentials for the various oxidant-reductant couples are listed along the right. Thermodynamically favorable half reactions are determined by linking pairs of compounds such that the lower reaction in the pair proceeds leftward, therefore producing electrons, and the upper reaction proceeds rightward, accepting electrons. The order of redox pairs listed in Fig. 9.1 reflects a hierarchical order of energy yield, which in turn dictates the metabolic strategies of soil microorganisms. Key catabolic reactions in microbial metabolism that drive the soil C cycle pair the oxidation of organic C in the lower right to a series of terminal electron acceptors in the upper left.

That a large range of redox couples are employed by soil microorganisms underscores the complexity and variability of the soil environment.

D Role of Enzymes in Metabolism

Many reactions in nature do not proceed spontaneously at rates that are compatible with cellular life. Some reactions occur too quickly, releasing energy that cannot be harvested for cellular processes. Other reactions, even reactions with very negative $\Delta G^{0\prime}$ values, may occur too slowly in nature to be of any use to organisms. Reactions often occur slowly because the existing chemical bonds require an initial energy input to be broken. This energy input is called the activation energy, or the energy input required for a reaction to proceed spontaneously. Thus for a reaction to proceed spontaneously, often an external energy source is required or the activation energy must be reduced.

A catalyst is a substance that promotes a reaction by reducing the activation energy required for that reaction to proceed, without itself being altered in the process. In biological systems, enzymes are specialized proteins that act as catalysts for a range of catabolic and anabolic reactions, thus allowing these reactions to occur at biologically useful rates. Enzymes function by temporarily binding reactants at an "active site," where they are physically oriented in a manner that facilitates the formation of the desired product. Enzymes often require other molecules known as coenzymes to carry out their functions. A coenzyme can be any nonprotein substance that binds to an enzyme. Metal ions, electron carrier molecules, and certain vitamins often serve as coenzymes. Coenzymes can either bind covalently (prosthetic group) or noncovalently (e.g., NADH). Coenzymes often supply substances such as electrons or hydrogen ions that are required for the reaction to proceed. The result of enzymatic catalysis is that the metabolic reactions of life can proceed at biologically controlled rates with minimum energy inputs (i.e., reduced activation energies). Figure 9.2 illustrates this by depicting the activation energy required for an enzyme catalyzed versus a noncatalyzed reaction.

Although most enzymes in multicellular organisms function intracellularly, it is often advantageous for soil microorganisms to release enzymes into the environment. Such enzymes are known as extracellular enzymes (or exoenzymes). In soils, extracellular enzymes play a key role in decomposing complex organic polymers into oligomeric and monomeric substances that are small enough to be transported inside the cell (Sinsabaugh and Follstad Shah, 2012). Extracellular enzymes have been widely studied by soil microbiologists and are considered key agents in the decomposition of organic matter and the biogeochemical cycling of C, N, and P in soils (Dick, 2011). The measurement of extracellular enzyme activity in soils has been used to as an index of microbial activity, soil fertility, and nutrient cycling rates (Nannipieri et al., 2012).

Extracellular enzymes can be broadly divided into two classes: hydrolases and oxidases. Hydrolases are substrate-specific; their structure enables them to catalyze reactions that cleave specific bonds (such as C-O and C-N) in organic

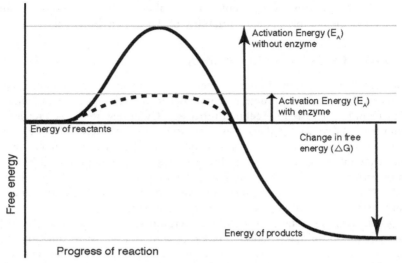

FIG. 9.2   Conceptual illustration of the role of enzymes in mediating the chemical thermodynamics of reactions.

matter. Oxidases use either oxygen (oxygenases) or hydrogen peroxide (peroxidases) to oxidize a broad suite of molecules that share similar bonds (such as C-C or C-O). Table 9.1 lists the most commonly measured extracellular enzymes in soils and their functions. All these enzymes have been demonstrated to play an important role in the cycling of at least one of the three most biotically important nutrients: C, N, and P.

III   METABOLIC CLASSIFICATION OF SOIL ORGANISMS

Classification by how microbes acquire raw materials that make up their cells (particularly C), an energy source, and a source of electrons for redox reactions provides a means to divide virtually all soil microorganisms into five groups (Fig. 9.3). Energy sources are divided into two main categories: chemicals and light. Organisms that derive energy by converting light energy into chemical energy are called phototrophs, whereas organisms that derive energy from the breakdown of energy-rich chemicals are called chemotrophs. The chemical sources of energy are further broken down into organic sources (e.g., compounds containing multiple C-C bonds) and inorganic sources (e.g., ammonia, hydrogen gas, methane, or elemental sulfur). The C sources used by soil microbes as building blocks for biosynthesis are divided into two categories: organic (or fixed) compounds and inorganic (or gaseous) compounds. Organisms that use an organic form of C are referred to as heterotrophs, whereas organisms that fix inorganic C (i.e., CO_2) are referred to as autotrophs. Finally,

TABLE 9.1 Commonly Measured Extracellular Enzymes, Their Functions in Soils, and Classification

Enzyme	Enzyme Function	Enzyme Class
Acid phosphatase	Mineralizes organic P into phosphate by hydrolyzing phosphoric (mono) ester bonds under acidic conditions.	Hydrolase
α-Glucosidase	Principally a starch-degrading enzyme that catalyzes the hydrolysis of terminal, nonreducing 1,4-linked α-D-glucose residues, releasing α-D-glucose.	Hydrolase
β-Glucosidase	Catalyzes the hydrolysis of terminal 1,4 linked β-D-glucose residues from β-D-glucosides, including short-chain cellulose oligomers.	Hydrolase
β-Xylosidase	Degrades xylooligomers (short xylan chains) into xylose.	Hydrolase
Cellobiohydrolase	Catalyzes the hydrolysis of 1,4-β-D-glucosidic linkages in cellulose and cellotetraose, releasing cellobiose.	Hydrolase
N-Acetyl glucosaminidase	Catalyzes the hydrolysis of terminal 1,4 linked N-acetyl-beta-D-glucosaminide residues in chitooligosaccharides (chitin-derived oligomers).	Hydrolase
Leucine aminopeptidase	Catalyzes the hydrolysis of leucine and other amino acid residues from the N-terminus of peptides. Amino acid amides and methyl esters are also readily hydrolyzed by this enzyme.	Hydrolase
Urease	Catalyzes the hydrolysis of urea into ammonia and carbon dioxide.	Hydrolase
Phenol oxidase	Also known as polyphenol oxidase or laccase. Oxidizes benzenediols to semiquinones with O_2.	Oxidase
Peroxidase	Catalyzes oxidation reactions via the reduction of H_2O_2. It is considered to be used by soil microorganisms as a lignolytic enzyme because it can degrade molecules without a precisely repeated structure.	Oxidase

(Modified from German et al., 2011.)

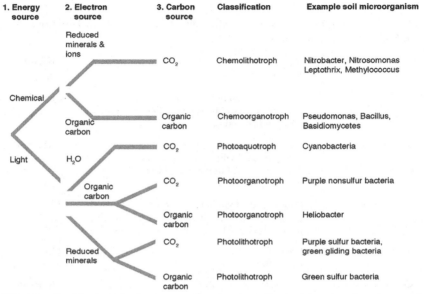

FIG. 9.3 Classification of soil microorganisms by metabolic physiology based on sources of energy, carbon, and terminal electron acceptor (i.e., reducing equivalents).

the electron source can be either organic (organotrophs), inorganic (lithotrophs), or water (aquatrophs).

According to this classification, plants could be called photosynthetic autotrophs or photoaquatrophs. Soil fungi or other microorganisms that consume plant-derived organic substrates could be called chemosynthetic organoheterotrophs or heterotrophic chemoorganotrophs. Microorganisms that oxidize inorganic compounds for energy and use CO_2 as their C source, such as nitrifying bacteria, could be called chemosynthetic lithoautotrophs. However, these full metabolic designations to describe the metabolic strategy of an organism are often abbreviated to "autotrophs," "heterotrophs" and "lithotrophs" because certain combinations of metabolic strategies typically occur together. It should be noted that the boundaries between some of these classifications can be unclear because some microorganisms may change their metabolic strategies in response to changes in environmental conditions. For instance, some soil microorganisms can switch between chemoorganotrophy and chemolithotrophy depending on substrate availability.

A key distinction between autotrophs and heterotrophs is that autotrophs make their own biosynthetic molecules directly via C fixation, whereas heterotrophs must degrade preexisting organic compounds via catabolism to synthesize new biomolecules. The use of reduced C as a carbon, energy or electron source, or some combination of the three, drives the soil C cycle.

IV CELLULAR ENERGY TRANSFORMATIONS

The ultimate result of catabolism is the production of energy storage compounds. Adenosine triphosphate (ATP) is the primary energy storage molecule used to activate the reactions needed for growth and reproduction by all living organisms. The energy released from the reaction $ATP \rightarrow ADP + Pi$ is used to fuel many intracellular reactions, both catabolic and anabolic. Formation of ATP is accomplished by two pathways: substrate-level phosphorylation and electron transport phosphorylation, which is also known as oxidative phosphorylation.

A Substrate-Level Versus Oxidative Phosphorylation

Substrate-level phosphorylation can be a significant source of ATP in anaerobic organisms but is generally insignificant in aerobic and facultatively anaerobic organisms when compared with the ATP production from oxidative phosphorylation. Substrate-level phosphorylation involves the direct production of ATP during the enzymatic oxidation of another substance. Removal of e^- and H^+ from an organic molecule allows the incorporation of inorganic phosphate and the production of a phosphorylated intermediate. Hydrolysis of the phosphoryl group of these intermediates releases enough energy to form ATP. As an example of substrate-level phosphorylation, Fig. 9.4 shows the transfer of a phosphoryl group from acetyl phosphate to $ADP \rightarrow ATP$. Yields of ATP are low, and consequently biomass production is low. Organisms that rely on this pathway tend to grow slowly. The production of fermentation intermediates is high, and these frequently become available for downstream metabolism by other microbes.

Oxidative phosphorylation is a more generalized means of ATP production because electrons from any number of oxidation reactions can be processed through the same electron transport chain, which occurs during both respiration and photosynthesis. The initial energy-yielding oxidation (of reduced organic C in chemoorganotrophs) releases electrons, which are transferred to the electron carriers NAD^+ (nicotineamide-adenine-dinucleotide) and FAD (flavin-adenine-dinucleotide), reducing these molecules to NADH and FADH. NADH and FADH shuttle electrons to an electron transport chain composed of a series of membrane-localized electron acceptors (for bacteria, either the inner

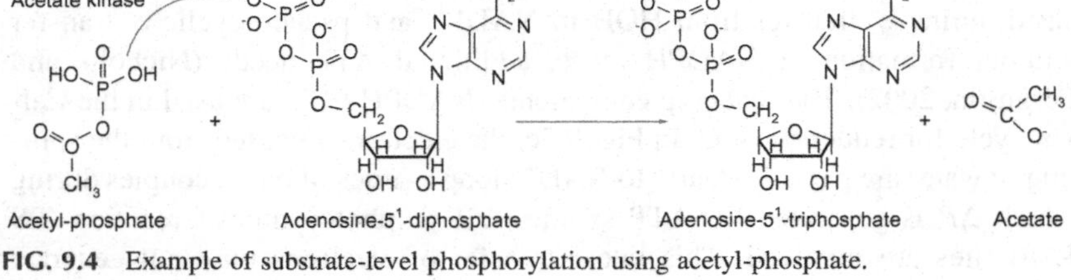

FIG. 9.4 Example of substrate-level phosphorylation using acetyl-phosphate.

mitochondrial membrane or the plasma membrane). These electron acceptors ultimately pass their electrons to a terminal acceptor. In aerobic respiration, this terminal acceptor is O_2, but any electron acceptor from Fig. 9.1 can be used provided its reduction potential is greater than that of the original, oxidized substance. Reduction of some of the membrane-localized electron acceptors draws in an H^+ ion from the cytoplasm (n-phase of the membrane). As it flips to a reduced state, the H^+ is extruded to the periplasmic space (p-phase), thereby increasing the H^+ concentration in the periplasm. As this process continues, a gradient of H^+ develops across the membrane with a high concentration in the p-phase compared to the n-phase. An electrical potential, or proton-motive force (ΔP), is consequently developed to drive H^+ from the p- to the n-phase through the ATP synthase complex. A portion of the energy released from this process is used to produce ATP from ADP.

Compared with substrate level phosphorylation, oxidative phosphorylation is a much more efficient means of producing ATP. However, it does require the presence of a terminal electron acceptor. Substrate level phosphorylation can occur under the most reducing environmental conditions when an external oxidant may not be present.

B   How ATP Production Varies for Different Metabolic Classes of Soil Microorganisms

Four variations on a theme for generating ATP and reducing equivalents in autotrophic and aerobic chemotrophic organisms are illustrated in Fig. 9.5. In Fig. 9.5a, electron transport through the respiratory chain to B_{ox} generates the ΔP (proton-motive force) needed to synthesize ATP (e.g., chemoorganotrophs such as *Pseudomonas* spp. (Proteobacteria)). Reducing equivalents (NADH) are produced by oxidation of substrates within the cytoplasm. In Fig. 9.5b, reverse electron transport is used to generate NADH, which is used to reduce CO_2 in the Calvin cycle. Electron transport through the respiratory chain to B_{ox} generates the ΔP needed both to synthesize ATP and to drive reverse electron transport (e.g., chemolithotrophs such as *Nitrobacter* (Proteobacteria)).

Photosynthesis is the ultimate source of energy to allow soil organisms to work. Photosynthesis requires no oxygen (although it may partially cycle O_2), and may or may not generate O_2. If photosynthesis uses HOH as an electron donor as it does for photoaquatrophs, then it generates O_2 and is called oxygenic photosynthesis (Table 9.2, Eq. 9.1). In oxygenic photosynthesis, ATP is synthesized during e^- transfer from HOH to $NADP^+$, and pseudo-cyclic e^- transfer without formation of NADPH meets additional ATP needs (Nicholls and Ferguson, 2002). The reducing equivalents ($NADPH + H^+$) are used in the Calvin cycle for reduction of C. In Fig. 9.5c, the electrons released from the splitting of water are passed linearly to NAD^+ along a series of redox couples during which ΔP is generated for ATP synthesis. Two photosynthesis reaction (PS Rxn) sites are involved. This linear transfer of electrons may not generate

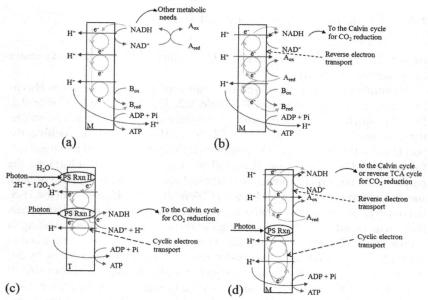

FIG. 9.5 Generation of ATP and reducing equivalents (NADH) in: (a) chemoorganotrophs where substrates are oxidized within the cytoplasm (e.g., *Pseudomonas* spp.); (b) chemolithotrophs in which substrates are oxidized at the membrane (e.g., *Nitrobacter* spp.); (c) oxygenic phototrophic cyanobacteria; or (d) anoxygenic phototrophic bacteria such as Rhodospirillaceae (purple nonsulfur bacteria), Chromatiaceae (purple sulfur bacteria), Chlorobiaceae (green sulfur bacteria), and Chloroflexaceae (green gliding bacteria), when oxidizing substrates such as H_2. Small circles within the plasma membrane (M) or thylakoid membrane (T; an intracytoplasmic membrane) represent redox complexes. PS Rxn is the photosynthesis reaction site. Red arrows indicate transformations of energy carriers occurring at the membrane. Blue arrows indicate the motion of protons, and green arrows indicate the motion of electrons.

TABLE 9.2 Anoxygenic and Oxygenic Photosynthesis

Oxygenic Photosynthesis, for example, *Panicum* (grass) *Nostoc* (cyanobacteria)

$h\upsilon + 2HOH + 3 (ADP+Pi) \rightarrow 4e^- + 4H^+ + O_2 + 3ATP$

$3ATP + CO_2 + 4e^- + 4H^+ \rightarrow CH_2O + HOH + 3 (ADP+Pi)$

Overall: $h\upsilon + CO_2 + HOH \rightarrow CH_2O + O_2$ [1]

Anoxygenic Photosynthesis, for example, *Chromatium* (purple S bacteria)

$2H_2S \rightarrow 2S^0 + 4e^- + 4H^+$

$h\upsilon + Pigment \rightarrow [Pigment^+ + e^-]^a$

$3(ADP+Pi) + [Pigment^+ + e^-]^a \rightarrow 3ATP + Pigment$

$3ATP + CO_2 + 4e^- + 4H^+ \rightarrow CH_2O + HOH + 3 (ADP+Pi)$

Overall: $h\upsilon + CO_2 + 2H_2S \rightarrow CH_2O + 2S^0 + HOH$ [2]

Or Overall: $h\upsilon + 3CO_2 + 2S^0 + 5HOH \rightarrow 2SO_4^{-2} + 3CH_2O + 4H^+$ [3]

[a]*Complex including several components of the photosynthetic system and the respiratory chain.*
(Reproduced from McGill, 1996 with Permission from SBCS and SLCS.)

enough ΔP to produce enough ATP for CO_2 reduction. Cyclic electron transfer can make up the shortfall.

Photolithotrophs use photosynthesis with a reduced mineral such as H_2S as the electron donor to produce S^0 (Table 9.2, Eq. 9.2) or SO_4^{-2} (Table 9.2, Eq. 9.3), which is called anoxygenic photosynthesis. All anoxygenic photosynthetic organisms represented in Fig. 9.5d can use H_2 as H-donor. In addition, the purple nonsulfur bacteria Rhodospirillaceae (Proteobacteria) and the green sulfur bacteria Chloroflexaceae (Chloroflexi) can use organic substrates, the purple-sulfur bacteria Chromatiaceae (Proteobacteria) can use H_2S and organic substrates, and the green sulfur bacteria Chlorobiaceae (Chlorobi) can use H_2S, but not organic substrates. Photon activation of electrons in the PS Rxn generates cyclic transfer of electrons through a series of redox couples, resulting in formation of ATP, and eventually the electron returns to its ground state and to the PS Rxn site. NAD^+ is reduced using reverse electron transfer driven by ΔP generated from photosynthesis. Subsequent transhydrogenation allows NADH to reduce $NADP^+$ to NADPH, which is used to reduce C in the Calvin cycle (Nicholls and Ferguson, 2002). In the Chlorobiaceae, CO_2 is reduced by reversal of the TCA cycle; all others use the Calvin cycle (Gottschalk, 1986). NADH is used to reduce CO_2, so the electron and H transfer to CO_2 in anoxygenic phototrophic bacteria can be summarized by Eq. 2 in Table 9.2.

V   EXAMPLES OF SOIL MICROBIAL TRANSFORMATIONS

The biogeochemical fluxes and transformations of C, N, and other nutrient elements observed at the soil aggregate, pedon, and landscape scales can be reduced to the various metabolic strategies used by soil microorganisms to acquire cellular building blocks, energy, and reducing equivalents. We highlight below some examples of soil microbial transformations that illustrate important metabolic strategies.

A   Organotrophy

Reduced C is the most important energy source for soil microorganisms; a recent global meta-analysis revealed that of numerous soil properties examined, C-availability was the best predictor of phyla-level bacterial abundances (Fierer et al., 2007). It is therefore not surprising that the most dominant microorganisms in soil are those that use reduced organic compounds as their source of C, energy, and reducing equivalents. These heterotrophic chemoorganotrophs are responsible for the decomposition of organic matter, such as plant residues and microbial necromass. For example, lignin degradation is accomplished primarily by fungi in the phyla Basidiomycota and Ascomycota, which can be either saprotrophic (free-living) or root associated (mycorrhizal; Osono, 2007). Cellulose degradation is accomplished mainly by specialized saprotrophic fungi (e.g., *Trichoderma* [Ascomycota]) and common bacterial genera including

Bacillus (Firmicutes), *Pseudomonas* (Proteobacteria), and *Clostridium* (Firmicutes). In its simplest form, the aerobic oxidation of organic matter (frequently represented by $C_6H_{12}O_6$ or CH_2O) to CO_2 is the reverse of the overall oxygenic photosynthesis reaction (Table 9.3). In addition to CO_2, the oxidation of nutrient element-containing organic compounds releases mineral forms of N (NH_3), P (PO_4^{3-}), and S (SO_4^{-2}), which subsequently become available for uptake during primary production. The oxidation of one mole of C in the form of CH_2O generates 4 electrons, whereas the oxidation of one mole of C as L-glutamic acid ($C_5NH_8O_4$) generates only 3.4 electrons (Table 9.4). The amount of ATP synthesized (and therefore of energy gained) from the oxidation of one mole of C from L-glutamate is therefore less than the oxidation of one mole of C from glucose. N-containing energy sources are therefore less energetically favorable than pure carbohydrates or lipids. In excess of the stoichiometric demands for microbial biomass N, the search for energy in N-containing organic molecules in the absence of more energetically favorable alternatives is the basis for much of the mineral N supply to plants from decomposition.

Some soil microorganisms are able to substitute other oxidized inorganic compounds for oxygen as the terminal electron acceptor during anaerobic

---

TABLE 9.3 Oxidation of Reduced C (Represented as CH_2O) Using O_2 as Terminal Electron Acceptor

$CH_2O + HOH \rightarrow CO_2 + 4e^- + 4H^+$

$O_2 + 4e^- + 4H^+ \rightarrow 2HOH + energy$

Overall: $CH_2O + O_2 \rightarrow CO_2 + HOH + energy$

(Reproduced from McGill, 1996 with Permission from SBCS and SLCS.)

TABLE 9.4 Oxidation of N-Containing Organic Molecules to CO_2 as a Way to Obtain Energy. L-Glutamic Acid Oxidation via α-Keto Glutaric Acid is Used as an Example

$2(C_5NH_8O_4) + 2HOH \rightarrow 2NH_3 + 2(C_5H_4O_5) + 6H^+ + 6e^-$

$2(C_5H_4O_5) + 10HOH \rightarrow 10CO_2 + 28H^+ + 28e^-$

$34H^+ + 34e^- + 8.5O_2 \rightarrow 17HOH + energy$

Overall: $2(C_5NH_8O_4) + 8.5O_2 \rightarrow 2NH_3 + 10CO_2 + 5HOH + energy$

(Reproduced from McGill, 1996 with Permission from SBCS and SLCS.)

respiration. Oxygen is the most energetically favorable electron acceptor, but other common alternative electron acceptors in soil are nitrate, sulfate, and carbon dioxide (Ehrlich, 1993). Even under well-aerated conditions, soils are well suited to provide microsites within aggregates where O_2 is excluded or in which it is consumed faster than it can diffuse to microorganisms.

Soil microorganisms, almost all of which are bacteria, that can use nitrate (NO_3^-) as an alternative terminal electron acceptor are facultative anaerobes (e.g., *Pseudomonas denitrificans*), meaning they can perform aerobic respiration in the presence of O_2 or switch to nitrate in its absence (Table 9.5, Eq. 9.3). The reduction of NO_3^- during oxidation of reduced C is called dissimilatory NO_3^- reduction. More specifically, nitrate reduction to N_2 or N_2O is called denitrification, and nitrate reduction to NH_3 is called nitrate respiration. Dissimilatory nitrate reduction is an important biogeochemical process because: (1) it returns biologically available N to the atmosphere in a process that is effectively the reverse of N fixation, (2) it represents a loss of soil fertility, and (3) it is a source of a potent greenhouse gas (e.g., N_2O).

In an analogous process, a diverse group of soil bacteria and archaea can use sulfate (SO_4^{-2}) as an alternative terminal electron acceptor in a process called dissimilatory SO_4^{-2} reduction (Table 9.5, Eq. 9.2). These organisms are strict or obligate anaerobes (e.g., *Desulfovibrio desulfuricans*). The free energy change (ΔG) and ATP/2e$^-$ ratio is greater for NO_3^- respiration than for SO_4^{-2} respiration (Gottschalk, 1986). These metabolic distinctions between NO_3^- respiration and dissimilatory SO_4^{-2} reduction suggest that the latter would not occur in a soil well supplied with NO_3^- or oxygen.

The use of CO_2 as a terminal electron acceptor to produce methane (CH_4) is called methanogenesis. Methanogens, all of which are archaea, are the strictest

---

TABLE 9.5 Carbon Oxidation Using NO_3^- or SO_4^{-2} as Terminal Electron Acceptors

NO_3^- accepts e$^-$

$5CH_2O + 5HOH \rightarrow 5CO_2 + 20H^+ + 20e^-$

$4NO_3^- + 20H^+ + 20e^- + 4H^+ \rightarrow 2\,N_2 + 12HOH + energy$

Overall: $5CH_2O + 4NO_3^- + 4H^+ \rightarrow 5CO_2 + 2\,N_2 + 7HOH + energy$ [1]

SO_4^{-2} accepts e$^-$

$2CH_2O + 2HOH \rightarrow 2CO_2 + 8H^+ + 8e^-$

$SO_4^{-2} + 8H^+ + 8e^- + 2H^+ \rightarrow H_2S + 4HOH + energy$

Overall: $2H_2O + SO_4^{-2} + 2H^+ \rightarrow 2CO_2 + H_2S + 2HOH + energy$ [2]

(Reproduced from McGill, 1996 with Permission from SBCS and SLCS.)

anaerobes normally found in nature and use a limited array of substrates: $H_2 + CO_2$, formate, methanol, methylamines, and acetate. These substrates are formed during fermentation or converted from fermentation products in other anaerobic systems. Two groups of organisms produce methane. One group of organisms that produces methane is chemoorganotrophic. These produce CH_4 from substrates such as methanol, acetate, or methylamines, which contain methyl groups. For methanol fermentation to CH_4, the overall reaction is: $4CH_3OH \rightarrow 3CH_4 + CO_2 + 2HOH$. Acetate is the most common and important, and its conversion to methane is written simply as: $C_2H_4O_2 \rightarrow CH_4 + CO_2$. In the presence of SO_4^{-2}, acetate is oxidized to CO_2 rather than being split to CH_4 and CO_2. Here we see the preference of SO_4^{-2} as a terminal electron acceptor over methane fermentation. The other group is strictly chemolithotrophic organisms that grow on H_2 and CO_2. They are fascinating in their ability to produce all their needs for energy and C from H_2 and CO_2 alone.

This sequence of metabolic strategies, using alternative terminal electron acceptors during the oxidation of reduced organic compounds represents a "redox ladder," where the selection of the terminal electron acceptor is a function of oxygen or redox status and results in decreasing yields of free energy change (Table 9.6, Fig. 9.1).

TABLE 9.6 The "Redox Ladder" of Half-Reactions for Various Terminal Electron Acceptors and Gibbs Free Energy Change Associated with Important Metabolic Reactions Carried out by Organotrophs During Carbon Oxidation

Process	Half Reaction	Heterotrophic Reaction	ΔG^0 (kJ eq^{-1})
Aerobic Respiration	$^1/_4O_2 + H^+ + e^- \rightarrow$ $^1/_2H_2O$	$CH_2O + O_2 \rightarrow CO_2 + H_2O$	-125
Denitrification	$^1/_5NO_3^- + ^6/_5H^+ + e^- \rightarrow$ $^1/_{10}N_2 + ^3/_5H_2O$	$5CH_2O + 4NO_3^- + 4H^+ \rightarrow$ $5CO_2 + 2\,N_2 + 7H_2O$	-119
Iron reduction	$FeOOH + HCO_3^-$ $+ 2H^+ + e^- \rightarrow$ $FeCO_3 + 2H_2O$	$CH_2O + 4FeOOH$ $+ 8H^+ \rightarrow$ $CO_2 + 4Fe^{2+} + 7H_2O$	-42
Sulfate reduction	$^1/_8SO_4^{-2} + ^9/_8H^+ + e^- \rightarrow$ $^1/_8H_2S + ^1/_2H_2O$	$2CH_2O + SO_4 + 2H^+ \rightarrow$ $2CO_2 + H_2S + 2H_2O$	-25
Methanogenesis	$^1/_8CO_2 + H^+ + e^- \rightarrow$ $^1/_8CH_4 + ^1/_4H_2O$	$2CH_2O \rightarrow CO_2 + CH_4$	-23

B Lithotrophy

Chemolithotrophic organisms are relatively restricted compared to chemoorganotrophs because the oxidation of inorganic compounds for energy typically yields much less energy than the oxidation of reduced C. Table 9.7 shows Gibbs free energy for the oxidation of glucose compared to some important inorganic substances used by chemolithotrophs. The oxidation of reduced C provides far more energy than any inorganic alternative. However, balanced against the low energy yields of the oxidation reactions catalyzed by chemolithotrophs is the ability to use energy sources unavailable to other microorganisms. This has the advantage of both reducing resource competition and allowing chemolithotrophs to grow in a range of environments that would be hostile to other organisms. Typically, chemolithotrophs obtain C from CO_2 and are therefore often called chemoautotrophs or chemolithoautotrophs. Most chemolithotrophs are aerobic; however, the diversity of chemolithotrophic strategies is an area of active research, and new chemolithotrophic processes are still being discovered. It was not until the late 1980s that the anaerobic oxidation of ammonia was experimentally demonstrated (Hayatsu et al., 2008). In 2006, it was discovered that a select group of microorganisms gained energy from the anaerobic oxidation of methane, using nitrate and nitrite as electron acceptors (Raghoebarsing et al., 2006). Because any given chemolithotrophic strategy requires highly specialized and unique biochemical machinery, chemolithotrophic organisms generally oxidize only a single compound, which is reflected in the common name of the organism.

TABLE 9.7 Comparison of Gibbs Free Energy Change from Glucose Oxidation versus Other Important Oxidation Reactions Carried out by Chemolithotrophs

Reaction	ΔG^0 (kJ mol^{-1})
$CH_2O + O_2 \rightarrow CO_2 + H_2O$	-2870
$CH_4 + 2O_2 \rightarrow CO_2 + 2H_2O$	-871
$S + \frac{1}{2}O_2 + H_2O \rightarrow SO_4^{-2} + 2H^+$	-587
$NH_4^+ + 1\frac{1}{2}O_2 \rightarrow NO_2^- + 2H^+ + H_2O$	-275
$H_2 + \frac{1}{2}O_2 \rightarrow H_2O$	-237
$HS^- + H^+ + \frac{1}{2}O_2 \rightarrow S + H_2O$	-209
$NO_2^- + \frac{1}{2}O_2 \rightarrow NO_3^-$	-76
$2Fe^{2+} + 2H^+ + \frac{1}{2}O_2 \rightarrow 2Fe^{3+} + 2H_2O$	-31

Methanotrophs are bacterial and archaeal organisms that gain energy from the oxidation of CH_4 to CO_2. Methanotrophy occurs most frequently in soils that are often submerged or water-saturated where a high level of methanogenesis occurs. Methanotrophs have also been found to oxidize atmospheric methane in upland soils. This restricted, but ecologically important, microbial process serves to offset some of the methane emissions produced by methanogens in periodically saturated upland and wetland soils as well as other sources of CH_4 such as ruminants or natural gas leaks.

Nitrifiers are a rare group of chemolithoautotrophs that use energy from oxidation of ammonium or nitrite to fix CO_2. The process of nitrification is accomplished by two separate groups of bacteria: those that oxidize ammonium to nitrite (e.g., *Nitrosomonas* spp., Table 9.8) and those that convert nitrite to nitrate (e.g., *Nitrobacter* spp., Table 9.9). Nitrifiers are agriculturally and environmentally important because the transformation of ammonium to nitrate is an important step in the soil N cycle that counters the processes of N fixation and denitrification.

Current research suggests that the ability to oxidize reduced S is evolutionarily ancient, possibly representing the earliest form of self-sustaining metabolism (Ghosh and Dam, 2009). Various species of reduced S compounds, such as sulfide (H_2S), inorganic S (S_0), and thiosulfate ($S_2O_3^{2-}$), are used as an energy source by S-oxidizing bacteria and archaea (e.g., *Beggiatoa, Paracoccus, Thiobacillus* [(Proteobacteria]). Sulfur oxidation generally occurs in stages by different groups of microorganisms, and the reducing power generated is used to form NADH via reverse electron flow, which is used for CO_2 fixation in the Calvin cycle (Table 9.10).

Ferrous iron (Fe^{2+}) is a soluble form of iron that can be oxidized by several groups of iron-oxidizing bacteria and archaea. The first group (acidophiles,

---

TABLE 9.8 The First Step in Nitrification Where NH_4^+ is Oxidized for Energy and to Reduce CO_2 for Biomass Formation by Autotrophic Chemolithotrophs, such as *Nitrosomonas* spp.

Energy: $NH_4^+ + 2HOH \rightarrow NO_2^- + 6e^- + 8H^+$

$6e^- + 6H^+ + 1.5O_2 \rightarrow 3HOH - 270.7$ kJ

$NH_4^+ + 1.5\ O_2 \rightarrow NO_2^- + HOH + 2H^+ - 270.7$ kJ

C Reduction: $4NH_4^+ + 8HOH \rightarrow 4NO_2^- + 24e^- + 32H^+$

$6CO_2 + 24e^- + 24H^+ \rightarrow 6CH_2O + 6HOH$

$4NH_4^+ + 6CO_2 + 2HOH \rightarrow 4NO_2^- + 6CH_2O + 8H^+$

(Reproduced from McGill, 1996, with Permission from SBCS and SLCS.)

TABLE 9.9 The Second Step in Nitrification Where NO_2^- (nitrite) is Oxidized for Energy and to Reduce CO_2 for Biomass Formation by Autotrophic Chemolithotrophs, such as *Nitrobacter* spp.

Energy: $NO_2^- + HOH \rightarrow NO_3^- + 2e^- + 2H^+$

$2e^- + 2H^+ + 0.5O_2 \rightarrow HOH - 77.4$ kJ

$NO_2^- + 0.5O_2 \rightarrow NO_3^- - 77.4$ kJ

C Reduction: $2NO_2^- + 2HOH \rightarrow 2NO_3^- + 4e^- + 4H^+$

$ATP + CO_2 + 4e^- + 4H^+ \rightarrow CH_2O + HOH$

$ATP + 2NO_2^- + CO_2 + HOH \rightarrow 2NO_3^- + CH_2O$

$3NO_2^- + CO_2 + 0.5\ O_2 + HOH \rightarrow 3NO_2^- + CH_2O$

(Reproduced from McGill, 1996 with Permission from SBCS and SLCS.)

TABLE 9.10 Aerobic Sulfur Oxidation for Energy and to Reduce CO_2 for Biomass Formation by Autotrophic Chemolithotrophs, such as *Thiobacillus* spp.

Energy: $S^0 + 4HOH \rightarrow SO_4^{-2} + 6e^- + 8H^+$

$6e^- + 6H^+ + 1.5O_2 \rightarrow 3HOH - 584.9$ kJ

$S^0 + HOH + 1.5O_2 \rightarrow SO_4^{-2} + 2H^+ - 584.9$ kJ

C Reduction: $4S^0 + 16HOH \rightarrow 4SO_4^{-2} + 24e^- + 32H^+$

$6CO_2 + 24e^- + 24H^+ \rightarrow 6CH_2O + 6HOH$

$4S^0 + 6CO_2 + 10HOH \rightarrow 4SO_4^{-2} + 6CH_2O + 8H^+$

(Reproduced from McGill, 1996 with Permission from SBCS and SLCS.)

e.g., *Acidithiobacillus ferrooxidans*) oxidizes iron in very low pH environments and is important in acid mine drainage. The second group (microaerophiles) oxidizes iron at near-neutral pH and lives at the oxic-anoxic interfaces (e.g., *Leptothrix ochracea*). The third group of iron-oxidizing microbes is the anaerobic photosynthetic bacteria, such as Rhodopseudomonas. Aerobic iron oxidation is an energetically poor process that requires large amounts of iron to be oxidized to generate ATP. Similar to sulfur oxidation, reverse electron flow is used to form NADH used for CO_2 fixation in the Calvin cycle.

C Phototrophy

Although organotrophy and lithotrophy refer specifically to strategies used by organisms to acquire electrons (either reduced C or inorganic minerals), phototrophy refers to a strategy used to acquire energy. In contrast to chemotrophs (which can be chemoorganotrophs or chemolithotrophs, referring to either of the two electron sources discussed earlier), phototrophs use light energy. Most commonly in nature, phototrophs use light energy to fix inorganic CO_2 into reduced C compounds for their own cellular metabolism in a process known as photosynthesis. Photosynthesis represents the ultimate source of the reduced C required by all forms of life for biosynthesis.

All higher plants are photoaquatrophs, meaning they use water as an electron source to drive CO_2 fixation, producing oxygen as a by-product. Prokaryotes, however, employ a much broader range of phototrophic strategies, in some cases using inorganic minerals as an electron source (photolithotrophs), and in other cases using reduced organic compounds as an electron source (photoorganotrophs). In addition, many phototrophic prokaryotes are metabolically flexible and can use a range of organic and inorganic electron sources. The C source employed by phototrophic bacteria is also variable; whereas photoaquatrophs are exclusively autotrophic (i.e., CO_2-fixers); photolithotrophs and photoorganotrophs can either be autotrophic or heterotrophic.

Phototrophy is far more phylogenetically restricted among soil microorganisms than chemotrophic strategies because the light requirement restricts this strategy to the uppermost part of the soil. For example, the phyla Cyanobacteria, also known as blue-green algae, are the only known group of prokaryotic photoaquatrophs. Several prominent groups of anoxyphototrophic bacteria (or photolithotrophs) use a reduced form of S, such as sulfide or thiosulfate, as an electron source. These include the autotrophic purple S bacteria (Chromatiaceae) and green gliding bacteria (Chloroflexaceae), as well as the predominantly heterotrophic green sulfur bacteria (Chlorobiaceae). Photosynthetic members of Chloroflexaceae are noteworthy for exhibiting what may be the earliest form of photosynthetic reaction centers and CO_2 fixation mechanisms. Although cyanobacteria and purple and green phototrophic bacteria are indeed found in soils, they occur primarily in aquatic habitats. By contrast, Heliobacteriaceae (Proteobacteria) are a phylogenetically and ecologically distinct group of photoheterotrophs that have been found only in soils.

Nitrogen fixation is often found in conjunction with phototrophy. The cyanobacteria are considered one of the most important and widespread groups of N-fixers. Members of Rhodospirillaceae, a family of purple non-S bacteria, fix N by coupling light energy with electrons from reduced C or H_2. Some methanogenic archaea are also N-fixers (e.g., Methanococcus (Euryarchaeota)). All described species of heliobacteria are also N-fixers and are thought to be important in the fertility of rice paddy soils, demonstrating that both oxygenic and

TABLE 9.11 Nitrogen Fixation Accompanying Oxygenic and Anoxygenic Photosynthesis

Oxygenic Photosynthesis & N_2 Fixation (e.g., *Nostoc*)

$\upsilon + 6HOH + 19(ADP + Pi) \rightarrow 3O_2 + 12e^- + 12H^+$ 19ATP

$3ATP + CO_2 + 4e^- + 4H^+ \rightarrow CH_2O + HOH + 3 (ADP + Pi)$

$12ATP + N_2 + 6e^- + 6H^+ \rightarrow 2NH_3 + 12 (ADP + Pi)$

$4ATP + 2e^- + 2H^+ \rightarrow H_2 + 4 (ADP + Pi)$

Overall: $\upsilon + 5HOH + CO_2 + N_2 \rightarrow 3O_2 + CH_2O + 2NH_3 + H_2$

Anoxygenic Photosynthesis & N_2 Fixation (e.g., *Chromatium* [purple S bacteria])

$6H_2S \rightarrow 6S^0 + 12e^- + 12H^+$

$\upsilon + Pigment^a \rightarrow [Pigment^+ + e^-]^a$

$19(ADP + Pi) + [Pigment^+ + e^-]^a \rightarrow 10ATP + Pigment^a$

$3ATP + CO_2 + 4e^- + 4H^+ \rightarrow CH_2O + HOH + 3(ADP + Pi)$

$4ATP + 2e^- + 2H^+ \rightarrow H_2 + 4(ADP + Pi)$

$12ATP + N_2 + 6e^- + 6H^+ \rightarrow 2NH_3 + 12(ADP + Pi)$

Overall: $\upsilon + 6H_2S + CO_2 + N_2 \rightarrow 6S^0 + CH_2O + 2NH_3 + H_2 + HOH$

[a]*Complex including several components of the photosynthetic system and the respiratory chain.* (Reproduced from McGill, 1996 with Permission from SBCS and SLCS.)

anoxygenic forms of photosynthesis are associated with the ability to fix N_2 and also to reduce C for release to subsequent organisms along the food chain (Table 9.11).

VI A SIMPLIFIED VIEW OF SOIL MICROBIAL METABOLISM

Energy for soil organisms is obtained by passing electrons (e^-) from donors to acceptors to produce ATP. These e^- donors and acceptors form multiple interconnected oxidation-reduction couples, which lead to cycles through which electrons flow. Because of the central role of oxidation-reduction reactions, O_2 availability is a major control on how these interconnected oxidation-reduction couples operate. One can view most soil biological transformations as a simple framework of interconnected cycles of e^-. Such a conceptualization is a simple and robust way to unite the myriad details about transformations mediated by soil organisms. An electron cycle model accommodates the full spectrum of understanding from detailed presentations of reactions, to global biogeochemical cycles and earth history. It further makes predictions about what we might yet find in nature concerning mineral transformations by soil organisms.

A Model of Interconnected Cycles of Electrons

This system of flowing e^- is like an electrical system with two cycles: an anoxygenic (not O_2-producing) cycle and an oxygenic (O_2-producing) cycle and four circuits—two phototrophic, one chemoorganotrophic, and one chemolithotrophic hooked in parallel (Fig. 9.6). The microbial component consists of four groups of organisms described using the metabolic classification in Fig. 9.3. Three groups of organisms are responsible for C addition to soil and only one (chemoorganotrophs) for its removal. The two cycles are distinguished on the basis of the photosynthetic mechanisms: anoxygenic or oxygenic. The chemoorganotrophic circuit unites the anoxygenic cycle with the oxygenic cycle.

B The Anoxygenic Cycle

Anoxygenic photosynthesis uses energy from sunlight to couple the reduction of C in CO_2 to the anaerobic oxidation of S in S^0 or H_2S. If one is accustomed to thinking of S oxidation in a strictly aerobic sense, then anaerobic S oxidation appears contradictory. Anoxygenic photosynthesis would have been compatible with the anoxic (O_2-free) conditions of Earth's primordial atmosphere. It could have been mediated by anaerobic organisms like present-day photosynthetic S bacteria and is believed to have preceded oxygenic photosynthesis (Blankenship, 2010).

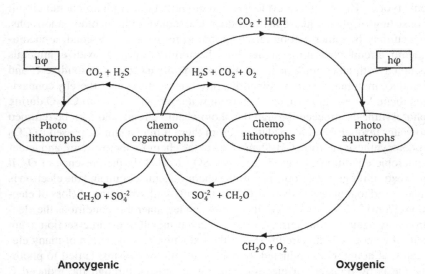

Anoxygenic   **Oxygenic**

FIG. 9.6 Schematic outline of electron flow through cycles starting with anoxygenic (not oxygen-producing) and oxygenic photosynthesis (oxygen-producing). Anoxygenic photosynthesis generates oxidized S species from reduced S under anaerobic conditions, thereby maintaining anaerobic conditions; oxygenic photosynthesis generates oxygen from water, thereby producing aerobic conditions. Both yield CH_2O for biomass production.

Dominance of anoxygenic photosynthesis would have favored anaerobic respiration or fermentative pathways for obtaining energy from the products of photosynthesis. Consequently, one finds a wide representation of anaerobic microorganisms in soil environments. Many elements can cycle under entirely anaerobic conditions due to the syntropic relationships among photolithotrophs and anaerobic chemoorganotrophs. One can thus summarize the anoxygenic cycle as a combination of photosynthesis powered by electromagnetic radiation by photolithotrophs while they reduce C and oxidize elements such as S, combined with decomposition by anaerobic chemoorganotrophs to re-oxidize C and re-reduce S. The cycle can be represented as:

$$4CO_2 + 2H_2S + 4HOH \leftrightarrow 2SO_4^{-2} + 4CH_2O + 4H^+ \tag{9.12}$$

Energy trapped as organic compounds during anoxygenic photosynthesis could be released by oxidation of the reduced C (e$^-$ donor), coupled to reduction of the oxidized minerals (e$^-$ acceptor), and formed during anoxygenic photosynthesis.

C   The Oxygenic Cycle

Oxygenic photosynthesis can be carried out by many eukaryotes, but only by the cyanobacteria among the prokaryotes. Consequently, anoxygenic photosynthesis is dominantly prokaryotic and oxygenic photosynthesis dominated by eukaryotes. The oxygenic cycle has two circuits. First, a photo-chemo circuit consisting of photoaquatrophs to reduce C coupled with chemoorganotrophs, which may be either aerobic or anaerobic to re-oxidize it. Second, a chemo-chemo circuit consisting of aerobic chemolithotrophs to oxidize minerals coupled with anaerobic (or facultative) chemoorganotrophs to oxidize C and rereduce minerals (Fig. 9.6). Electrons flow among the cycles, thereby connecting them. For example, an electron from water may be passed to CH_2O during photosynthesis, proceed through the photo-chemo circuit, and then be returned to water through aerobic oxidation. From there it may again be passed to CO_2 (photosynthesis) to form CH_2O, proceed to the chemo-chemo circuit, and under anaerobic conditions, be used to reduce SO_4^{-2} to H_2S. In the presence of O_2, it proceeds through the aerobic part of the chemo-chemo circuit and the electron is again used to reduce CO_2 to CH_2O concurrently with oxidation of (loss of electrons from) S^{2-}. As it continues its journey, under anaerobic conditions the electron may again be transferred to H_2O, which brings it to the intersection again with the photo-chemo circuit. Hence the oxidation and reduction of many elements, although not conducted by photosynthetic organisms, is tied to photosynthesis by transfers of electrons among organisms through the reduced C and O_2 produced by photosynthesis.

The oxidation of elements is made possible by O_2 from photosynthesis and their reduction by the reduced C from photosynthesis. This alternating oxidation – reduction system involving chemolithotrophs requires that O_2

and CH_2O from photosynthesis travel separately and that there be habitats from which the O_2 is excluded. Soils are uniquely suited to providing such habitats.

The two chemotrophic circuits of Fig. 9.6 can be expanded to distinguish aerobic, facultative, and anaerobic domains based on sensitivity to O_2 (Fig. 9.7). The aerobic domain is set at Eh > 300 mV comprising aerobic chemoorganotrophs and photoaquatrophs as a syntropic system. The facultative domain, between Eh 100 and 300 mV comprises chemoorganotrophs, which may be either aerobic or anaerobic in syntropic associations with chemolithotrophs, which are aerobic. The anaerobic domain consists of strictly anaerobic chemoorganotrophs in association with photoaquatrophs. Anaerobic chemoorganotrophs in the anaerobic domain reduce oxidized minerals generated by aerobic chemolithotrophs in the aerobic domain. Some energy is dissipated through heat loss etc. and must be made up by subsequent photosynthesis.

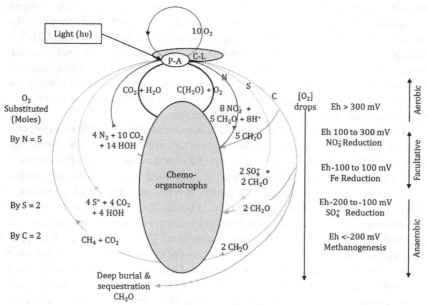

FIG. 9.7 The oxygenic cycles in soil entail cyclic oxidations and reductions of C, N, and S (among other elements), which are driven by solar radiation and controlled by the availability of O_2. Lowering O_2 availability is reflected in lowering Eh values. Oxygenic photoaquatrophs (P-A) produce CH_2O and release O_2; aerobic chemolithotrophs (C-L) oxidize N to NO_3^- or S to SO_4^{-2} using O_2 from photosynthesis and reduce CO_2 to CH_2O autotrophically. Aerobic chemoorganotrophs oxidize CH_2O using O_2 through aerobic respiration (heavy circle), and facultative anaerobic chemoorganotrophs oxidize CH_2O using NO_3^- (nitrate respiration). Anaerobic chemoorganotrophs oxidize CH_2O by reducing SO_4^{-2} or under extremely anaerobic conditions by reducing a portion of the CH_2O itself thereby splitting it into CO_2 (oxidized) and CH_4 (more reduced). For the stoichiometry represented here the moles of O_2 substituted by N, S, or C are designated at the left. Dinitrogen fixation using reduced C ultimately from P-A is required to reduce N_2 to NH_3 prior to its oxidation to NO_3^-. Deep burial of CH_2O opens the oxygenic cycles thereby allowing O_2 to accumulate.

O_2 is an overall control because it inhibits anaerobic processes. Consequently the balance among the three domains in Fig. 9.7 is a function of O_2 availability in local environments or microsites. N_2 fixation is interesting in that it is inhibited by O_2 and mediated by three of the four groups of organisms, with only chemolithotrophs excluded. In addition, the chemoorganotrophs are responsible both for removing N from the pedosphere, by reducing NO_3^- to N_2 and for returning it by further reducing N_2 to NH_3 that is followed by incorporation into organic products.

Figure 9.7 further shows that as the oxidation-reduction potential (Eh) becomes increasingly negative, oxidants become decreasingly effective. Given that the energy available through a redox reaction is directly proportional to the change in oxidation potential between the two couples, less and less energy is released as one moves from aerobic to strictly anaerobic metabolism. Soil organisms have evolved in an energy-limited environment, thus communities of soil organisms use the most energetically favorable energy sources available to them. Such a strategy favors use of O_2 as an electron acceptor followed by NO_3^-, SO_4^{-2} and eventually the use of a portion of the C in CH_2O during methanogenesis. In other words, CH_2O is allocated first to reduction of O_2, then NO_3^-, followed by SO_4^{-2}, and finally methanogenesis. From a practical perspective, then, NO_3^- might be a useful electron acceptor for metabolism of organic contaminants under anaerobic conditions. Indeed, NO_3^- and denitrifying populations have been proposed for removal of organic contaminants under anaerobic conditions.

These coupled cycles of electron flow operate at the metabolic or molecular scale of soil microorganisms, but have important implications in the global biogeochemical cycling of several elements (Falkowski et al., 2008). For instance, the O_2 produced during photosynthesis is consumed stoichiometrically during complete decomposition of photosynthate. It is not possible for O_2 to accumulate in the atmosphere from photosynthesis unless large quantities of CH_2O are not decomposed. We are not awash in nondecomposed plant litter, so how can residual O_2 have accumulated in the atmosphere? Disrupting the cycle of photosynthesis and decomposition would retain O_2 in the atmosphere rather than consuming it in oxidation of photosynthate. Disruption occurs by deep burial and sequestration of CH_2O as represented in Fig. 9.7. Processes yielding soil humus, peat, shale, coal, or petroleum and deep ocean organic sediments, among others over geologic time, all remove CH_2O that microbes would otherwise use to reduce O_2 to water (Logan et al., 1995). Given that C storage could allow O_2 accumulation, what might be the control on maximum O_2 accumulation? Mineral weathering is a sink for O_2 and participates in regulating atmospheric O_2 concentration. In addition, oxidation-reduction reactions, over geological time, control Fe solubility, and due to the reactivity of Fe species with P, control P concentration. Fe and P concentrations in solution, especially in marine environments, often limit photosynthesis and N_2 fixation. Regulation of Fe and P solubility by oxidation state of Fe provides a feedback to atmospheric O_2 concentration (Van Cappellen and Ingall, 1996).

REFERENCES

Blankenship, R.E., 2010. Early evolution of photosynthesis. Plant Physiol. 154, 434–438.

Cleveland, C.C., Liptzin, D., 2007. C:N:P stoichiometry in soil: is there a "Redfield ratio" for the microbial biomass? Biogeochemistry 85, 235–252.

Dick, R.P. (Ed.), 2011. In: Methods in Soil Enzymology. Soil Science Society of America, Madison, WI.

Ehrlich, H.L., 1993. Bacterial mineralization of organic carbon under anaerobic conditions. In: Bollag, J.-M., Stotzky, G. (Eds.), Soil Biochemistry, vol. 8. Dekker, New York, NY, pp. 219–247.

Falkowski, P.G., Fenchel, T., Delong, E.F., 2008. The microbial engines that drive Earth's biogeochemical cycles. Science 320, 1034–1039.

Fierer, N., Bradford, M.A., Jackson, R.B., 2007. Toward an ecological classification of soil bacteria. Ecology 88, 1354–1364.

Fierer, N., Lennon, J.T., 2011. The generation and maintenance of diversity in microbial communities. Am. J. Bot. 98, 439–448.

German, D.P., Weintraub, M.N., Grandy, A.S., Lauber, C.L., Rinkes, Z.L., Allison, S.D., 2011. Optimization of hydrolytic and oxidative enzyme methods for ecosystem studies. Soil Biol. Biochem. 43, 1387–1397.

Ghosh, W., Dam, B., 2009. Biochemistry and molecular biology of lithotrophic sulfur oxidation by taxonomically and ecologically diverse bacteria and archaea. FEMS Microbiol. Rev. 33, 999–1043.

Gottschalk, G., 1986. Bacterial Metabolism. Springer-Verlag, New York.

Hayatsu, M., Tago, K., Saito, M., 2008. Various players in the nitrogen cycle: diversity and functions of the microorganisms involved in nitrification and denitrification. Soil Sci. Plant Nutr. 54, 33–45.

Logan, G.A., Hayes, J.M., Hieshima, G.B., Summons, R.E., 1995. Terminal proterozoic reorganization of biogeochemical cycles. Nature 376, 53–56.

Manzoni, S., Trofymow, J.A., Jackson, R.B., Porporato, A., 2010. Stoichiometric controls dynamics on carbon, nitrogen, and phosphorus in decomposing litter. Ecol. Monogr. 80, 89–106.

McGill, W.B., 1996. Soil sustainability: microorganisms and electrons. In: Solo Suelo 96 Conference 2 CD version, Sociedade Brasileira de Ciência do Solo (SBCS) and Sociedade Latino-Americana de Ciência do Solo (SLCS), Vicosa/MG.

Nannipieri, P., Giagnoni, L., Renella, G., Puglisi, E., Ceccanti, B., Masciandaro, G., Fornasier, F., Moscatelli, M.C., Marinari, S., 2012. Soil enzymology: classical and molecular approaches. Biol. Fertil. Soils 48, 743–762.

Nicholls, D.G., Ferguson, S.J., 2002. Bioenergetics 3. Academic Press, San Diego.

Osono, T., 2007. Ecology of ligninolytic fungi associated with leaf litter decomposition. Ecol. Res. 22, 955–974.

Raghoebarsing, A.A., Pol, A., van de Pas-Schoonen, K.T., Smolders, A.J.P., Ettwig, K.F., Rijpstra, W.I.C., Schouten, S., Damste, J.S.S., Op den Camp, H.J.M., Jetten, M.S.M., Strous, M., 2006. A microbial consortium couples anaerobic methane oxidation to denitrification. Nature 440, 918–921.

Sinsabaugh, R.L., Follstad Shah, J.J., 2012. Ecoenzymatic stoichiometry and ecological theory. In: Futuyma, D.J. (Ed.), Annual Review of Ecology, Evolution, and Systematics, vol. 43. 313–343.

Sinsabaugh, R.L., Lauber, C.L., Weintraub, M.N., Ahmed, B., Allison, S.D., Crenshaw, C., Contosta, A.R., Cusack, D., Frey, S., Gallo, M.E., Gartner, T.B., Hobbie, S.E., Holland, K.,

Keeler, B.L., Powers, J.S., Stursova, M., Takacs-Vesbach, C., Waldrop, M.P., Wallenstein, M.D., Zak, D.R., Zeglin, L.H., 2008. Stoichiometry of soil enzyme activity at global scale. Ecol. Lett. 11, 1252–1264.

Van Cappellen, P., Ingall, E.D., 1996. Redox stabilization of the atmosphere and oceans by phosphorus-limited marine productivity. Science 271, 493–496.

FURTHER READING

The following references are textbooks that provide more in-depth coverage of the basics of microbial physiology and metabolism. While most do not focus exclusively on soil ecosystems and cover a much broader scope of topics, each does contain one or several chapters on microbial energy acquisition, physiology and metabolism from an organismal perspective.

Kim, B.H., Gadd, G.M., 2008. Bacterial Physiology and Metabolism. Cambridge University Press, New York.

White, D., Drummond, J., Fuqua, C., 2011. The Physiology and Biochemistry of Prokaryotes, fourth ed. Oxford University Press, New York.

The following textbook focuses more closely on the connections between bacterial metabolism and environmental biogeochemistry.

Fenchel, T., King, G.M., Blackburn, T.H., 2012. Bacterial Biogeochemistry: The Ecophysiology of Mineral Cycling, third ed. Academic Press, San Diego.

Chapter 10

The Ecology of the Soil Biota and their Function

Sherri J. Morris[1] and Christopher B. Blackwood[2]
[1]Biology Department, Bradley University, Peoria, IL, USA
[2]Department of Biological Sciences, Kent State University, Kent, OH, USA

Chapter Contents

I INTRODUCTION

Ecology is the study of the interactions of organisms with each other and their environment. The name, Oecologie, provided by Haeckel in 1866 is based on the Greek term *oikos*, the family household, which embodied the view of the environment and organisms as a household (Kingsland, 1991). Ecology was initially developed to provide a mechanistic backbone to the science of natural history. Because of this origin, the field of ecology is entwined with evolution, and it is presupposed that the basis of ecological relationships is one of long evolutionary history. Human disturbances to ecosystems can result in interactions among organisms that are not based on evolutionary history and provide opportunities to test this presupposition.

Soil Microbiology, Ecology, and Biochemistry. http://dx.doi.org/10.1016/B978-0-12-415955-6.00010-4
273

Given that the field of ecology focuses on describing patterns of species interactions and distributions, it is odd that ecologists have only recently begun to show interest in the structure and function of microbial communities as fundamental cornerstones of both terrestrial and aquatic ecological systems. The result of this interest has been the development of a wide range of studies that examine soil microbes in the context of ecological research. This interest has resulted in development of the fields of microbial ecology and soil ecology that include the description of the soil biota in an ecological context. This focus retains the fundamental goals of ecologists to understand the mechanisms that determine species distribution and the consequential impact of biotic and abiotic environments.

Ecological research on soil biota differs in several ways from research that focuses on plant and animal systems. For example, the "species concept" adopted by many plant and animal biologists defines a species as an interbreeding group of organisms that is reproductively (and genetically) isolated from other organisms. Bacteria, archaea, and many microbial eukaryotes reproduce asexually. Thus, this part of the biological species concept does not apply to many microorganisms. However, examination of whole genome sequences has revealed the extent of genetic recombination within and among these groups (Fraser et al., 2009). In some lineages, genetic recombination appears to be rampant among closely related strains. Acquisition and loss of genes may be part of the strategy of some bacterial species, allowing for adaptation to a wide range of environmental conditions. For example, the bacterial group *Pseudomonas fluorescens* (class *Gammaproteobacteria*) is widely distributed in soil, including strains that interact with a large diversity of other organisms, including plants, bacteria, archaea, fungi, oomycetes, nematodes, and insects. This phenotypic diversity may be related to a highly variable genome, with only 45-52% of genes in a particular genome also being found in all other *P. fluorescens* strains (Loper et al., 2012). In contrast to frequent recombination among closely related bacterial strains, genetic recombination is less common between distantly related lineages because of their lack of homologous genes and promoters and the limited host ranges of plasmids and viruses that mediate gene transfer (Thomas and Nielsen, 2005). Thus, it is possible that groups of closely related strains with high rates of gene exchange may be analogous to species as described for plants and animals. A strategy is still emerging to adapt microbial taxonomy and ecology to the natural history of microorganisms (Fraser et al., 2009). Methods focusing on genetic markers for microbial phylogeny, coupled with analysis of key genes and physiological traits, are allowing us to make dramatic progress in understanding the ecology of soil organisms despite an evolutionary framework that is still developing.

In this chapter, we focus on the ecology of soil biota, particularly on the processes that drive community structure (number and types of species) and the resultant impacts of their function in ecosystems (processes of energy transformations and nutrient turnover). As soil biota are essential components of every ecosystem, developing an understanding of the ecology of these organisms is essential to understanding terrestrial systems. The need to understand and accurately predict the impacts of global climate change and to develop sustainable practices in agriculture, forestry, ecosystem restoration, and natural resource management provide additional impetus to emphasize the ecology of microorganisms and their roles within soils in the twenty-first century. The integrative approach provided by the field of ecology aids in understanding the complexity of soil systems necessary to manage systems that will not become degraded over the long term. In return, examining soils as complex integrated systems has great potential to influence ecological concepts and the science of ecology in general, and soils can be described as the last ecological frontier.

A glossary of terms used in this chapter is provided in the online Supplemental Material found at http://booksite.elsevier.com/9780124159556.

II MECHANISMS THAT DRIVE COMMUNITY STRUCTURE

A large part of ecology is the study of how organisms become distributed in the environment, which is related to questions such as: Why are some species found in one area and not in others? How stable and repeatable are the groups of species found together? The paradigm that dominated scientific understanding of the distribution of microorganisms throughout the twentieth century was articulated by Lourens G. M. Baas Becking based on the work of Bejerinck: "Everything is everywhere, but the environment selects" (O'Malley, 2007). The implication of this statement is that the distribution of microorganisms is strictly determined by environmental conditions. However, this has been vigorously challenged since molecular methods allowed the study of microbial biogeographic patterns to be unified with general ecological theories (Hanson et al., 2012).

The area an organism lives in is called its *habitat*, and biotic and abiotic conditions of the habitat affect the physiological ability of each organism to survive and reproduce. *Functional traits* are properties of an organism that affect how well the organism performs under a certain set of conditions (McGill et al., 2006). Organisms that live in one habitat make up the *community*, and the numbers and kinds of organisms present is referred to as *community structure*. Communities are composed of populations or subpopulations of various species. A *population* is a collection of all of the organisms belonging to a single species with potential for interaction. This makes the spatial scale of a population dependent on the mobility of the species. A study may encompass only part of a true population (e.g., migrating species) or encompass multiple, isolated populations (e.g., soil bacteria). Given the degree to which species are differentially mobile, it is normal for both situations to arise in the same study.

A Physiological Limitations to Survival

Ecologists often need quantitative information on the suitability of habitats for a particular species to predict population dynamics under changing environmental conditions in new areas or with changes in community structure. A species' functional traits include limitations on the conditions under which individuals, and therefore populations, can grow and reproduce. *Shelford's Law of Tolerance* states that there is a maximum and minimum value for each environmental factor, beyond which a given species cannot survive. This law is usually discussed with respect to environmental characteristics known as *modulators*, such as temperature, pH, or salinity. Modulators impact the physiology of organisms by altering the conformation of proteins and cell membranes and the thermodynamic and kinetic favorability of biochemical reactions (Chapter 9). Species also have an optimal range for each environmental modulator where maximum population growth occurs. Tolerance to modulators can be interactive; for example, in some fungi, tolerance to cold temperature depends on water potential (Hoshino et al., 2009). Normally the geographical range of a species coincides with areas where environmental conditions are within the optimal ranges for the species, with the most optimal conditions at the center of the geographical range. The effects of species being in habitats with modulators outside their tolerance levels are listed in Table 10.1, along with biochemical strategies used by microorganisms that exist under these "extreme" conditions.

Resources are physical components of the environment that are captured by organisms for their use, such as N, energy, or territories. Shelford's law can be applied to most resources, but the responses to different resources are highly interactive. This is partially described in *Liebig's Law of the Minimum*, which states that the resource in lowest supply relative to organismal needs will limit growth. At very low levels of a resource, the organism is unable to accumulate the resource in adequate quantities for metabolism. At very high levels, resources can often be toxic or inhibit growth. The ratio of resources that an organism needs (e.g., the stoichiometry of biomass nutrients) is an important functional trait determining how species performance varies with resource availability, as well as excretion of nutrients that are consumed in excess of need as wastes. Studies have found nutrient availability, specifically environmental C:N:P stoichiometry, can have an impact on microbial biomass C:N:P ratios, with N and P limitations affecting the total amount of soil microbial biomass. However, across soils in general, these ratios really are quite conserved (Cleveland and Liptzin, 2007). The impact of these resource requirements feed back to impact function of the microbial community. Differences in the C:N ratio of fungal and bacterial biomass (as well as other trait differences between fungi and bacteria, such as growth efficiency) have large consequences for soil C and N cycling (Waring et al., 2013).

The response of a species to environmental conditions or resources depends on the genetic makeup of the species. The limits and optima are determined

TABLE 10.1 The Effects of Physical Stresses (Modulators) on Microorganisms and the Biochemical Adaptations They Induce

Modulator	Effect on Cell	Biochemical Adaptation	Organisms that have Required Adaptation
Temperature	Denaturation of enzyme; change in membrane fluidity	Production of proteases and ATP-dependent chaperones (Derré et al., 1999); production of cold-tolerant enzymes by amino acid substitution (Lönn et al., 2002); increases in intracellular trehalose and polyol concentrations and unsaturated membrane lipids, secretion of antifreeze proteins and enzymes active at low temperatures (Robinson, 2001)	Thermophiles; psychrophiles
Water deficit or salt stress	Dehydration and inhibition of enzyme activity	Changes in composition of polysaccharides produced (Coutinho et al., 1999); maintaining salt in cytoplasm, and uptake or synthesis of compatible solutes (Roeßler and Muller, 2001)	Osmophiles; xerophiles; halophiles
pH	Protein denaturation; enzyme inhibition	Increased intrasubunit stability in proteins afforded by increased hydrogen bonds and stronger salt bridges (Settembre et al., 2004); organisms that can secrete a surplus of protons, or block extracellular protons from the cytoplasm by blocking membrane composition (De Jonge et al., 2003); stress regulator genes (de Vries et al., 2001)	Acidophiles

Continued

TABLE 10.1 The Effects of Physical Stresses (Modulators) on Microorganisms and the Biochemical Adaptations They Induce — Cont'd

Modulator	Effect on Cell	Biochemical Adaptation	Organisms that have Required Adaptation
Aeration stress	Oxygen radicals damage membrane lipids, proteins, and DNA	Detoxification of oxygen radicals by catalase, and superoxide dismutase (Wu and Conrad, 2001)	Obligate anaerobes; methanogens; sulfur and N users

Modified from Paul and Clark (1996).

through natural selection and other mechanisms that affect the genome. Some species are adapted to varying conditions through high levels of genomic diversity among strains, as described for *P. fluorescens* earlier. In addition, all organisms have some degree of *phenotypic* or *trait plasticity*, or ability to alter their form or physiology to adjust to the environment. In microorganisms, a change in an environmental condition can induce expression of alternative phenotypes (e.g., proteins, phospholipids) that are adapted to the new conditions, broadening the range of conditions acceptable for the species. For example, presence or absence of plant cell-wall polymers is known to alter expression of enzymes involved in plant cell-wall depolymerization by saprotrophic fungi, some of which carry hundreds of such genes (Aro et al., 2005; Martinez et al., 2009). Transcription of these genes is fine-tuned by modulators such as environmental pH and temperature. This ability for physiological trait adjustment through control over gene transcription allows the fungi to selectively invest in extracellular enzymes resulting in the most efficient utilization of available resources. The cost associated with carrying genes allowing for a large degree of plasticity is extra genetic material that must be duplicated with each cell division, resulting in lower efficiency of resource use. This strategy can be efficient in fluctuating environments, such as the soil surface. In contrast, some symbionts that live within a host and depend on host homeostasis can develop dramatically reduced sets of genes (Gil et al., 2003; Sprockett et al., 2011).

B   Intraspecific Competition

As some organisms grow and reproduce, resources are consumed, reducing their availability for other organisms with the same requirements. Resource limitations in turn reduce activity, growth, and reproduction rates and increase death rates. This reduction in the performance of one organism by another organism that consumes the same resources is called *competition*. One of Darwin's important observations was that because members of the same species have nearly identical resource needs, competition for resources among members of the same species (intraspecific competition) is more fierce than among members of different species (interspecific competition; Kingsland, 1991). The *logistic growth equation* is a mathematical model that describes the effect of intraspecific competition on the change in population size over time (population dynamics). The probability of an individual reproducing minus the probability of death per unit time is equal to the overall amount by which a population grows or shrinks in that period. This is represented in the logistic growth equation as the population specific growth rate, μ, and is a direct manifestation of the suitability of the habitat. The differential equation for logistic growth is

$$\frac{dN}{dt} = \mu \cdot N \tag{10.1}$$

where

$$\mu = r \cdot \left(\frac{K-N}{K}\right) \tag{10.2}$$

and where N is the number of individuals in the population, t is time, and dN/dt is the change in N over time. The first important trait represented in this equation is the intrinsic growth rate of the population (r), which is the value μ approaches when resources are not limiting growth and there is no intraspecific competition. The number of individuals that the resources of a habitat can support (K) is referred to as the carrying capacity and models intraspecific competition with a constant level of resource supply. As a population grows, the resources must be shared among many individuals, decreasing the reproduction rate and increasing the rate of death. In the equation, N approaches K, causing μ to approach zero. If the population is above the carrying capacity, it cannot be supported by the resources present, μ becomes negative, and the population declines. The relationship where population growth rate is sensitive to population size is known as *density-dependent population regulation*.

Lifetime patterns of growth and reproduction, including timing of reproductive and dormant stages, are known as a species' *life history*. Life history strategies have important consequences for population dynamics. The logistic growth equation led to MacArthur and Wilson's (1967) concept of r- and K-selection that is used to generalize species' life histories. K-selected species (with high K values and low r values) have traits that favor the persistence of individuals under conditions of scarce resources and high intraspecific competition. These conditions occur when populations remain near their carrying capacity (K). In contrast, r-selected species have the opposite characteristics with relatively high efficiency in converting resources to offspring. The K-selected strategy is an adaptation to environments in which conditions are relatively stable, resulting in density-dependent mechanisms of population regulation, whereas the r-selected strategy is an adaptation to a variable environment with high resource levels. Pianka (1970) made further predictions about a variety of traits that could be associated with r- versus K-selected species (e.g., r-selected species should have more variable population size, weak competitive interactions, rapid maturation, short life span, and high productivity). However, it was soon realized that selection pressure and reproductive value at different ages can lead to opposite traits (Reznick et al., 2002). Both the logistic growth equation and the r-/K-selection model are now used as conceptual tools. A more detailed understanding of population dynamics is reflected by explicitly modeling resource consumption using the Monod model (Panikov, 1995) or, similarly, using structured demographic models to better understand the effects of life history and age-related variation (Reznick et al., 2002).

The classification of organisms as r- or K-selected is common in soil microbiology. Microbial colonies are typically classified based on the amount of time it takes for them to appear in laboratory isolation media. These designations must be made with reference to a particular environment. Laboratory isolation conditions represent a small range of the conditions encountered in the environment. The environment of a batch culture changes continuously as nutrients are not replenished and wastes are not removed. Hence, r-selected species can be positively identified (with respect to the isolation conditions), but K-selected species cannot. Soil microbiologists have also created other classifications similar to the r-/K-selection dichotomy that are more focused on the species' preferred resources than on intraspecific competition. In 1925, S. Winogradsky used the terms *autochthonous* to describe organisms that grow steadily on resistant organic matter with a constant presence in the environment, and *zymogenous* for organisms that proliferate on fresh organic matter (Panikov, 1995). In another scheme, oligotrophs grow only at low nutrient levels, whereas copiotrophs grow quickly at high nutrient levels.

C Dispersal in Space and Time

As resources become depleted, organisms must move, or disperse, to new areas to avoid the negative effects of competition. Active dispersal involves the expenditure of energy by the organism. Passive dispersal occurs due to the movement of material the organism is attached to or caught in (e.g., wind or water). Passive dispersal can be truly passive with no energy expended, or an organism may prepare morphologically and physiologically by entering a new life stage. Stages for dispersal are typically more resistant, dormant, or mobile than growth stages. The fruiting bodies and spores of fungi and myxococci are examples of elaborate life stages for passive dispersal.

Passive dispersal of bacteria with water flow can occur if the bacteria are not adsorbed onto immobile soil particles or protected by soil structure. Fecal coliforms applied to the soil surface with manure can move several meters and contaminate ground or surface waters in some situations, particularly when preferential flow paths limit interaction of the water and cells with the soil. Cell size limits the passive movement of organisms with water in soil due to the sieving effect of soil particles. Larger-celled microbes, such as yeasts and protozoans, do not experience passive dispersal to as great a degree as bacteria and viruses; however, nondormant protozoans are typically engaged in active dispersal to obtain food. Plant roots, seeds, fungal spores, and chemical substrates found within several centimeters of particular soil bacteria have been shown to induce chemotactic responses (active dispersal) that may be important for responses, such as rhizosphere colonization (Compant et al., 2010).

For hyphal fungi growing in soil, passive dispersal is normally restricted to spores. Spores may be produced in the soil, such as by arbuscular mycorrhizal fungi or in sporocarps (fruiting bodies) above the soil surface. Sporulation is often induced by environmental cues, such as moisture. Spores are also dispersed by animal activity both above- and belowground. Vegetative growth of fungal hyphae can also be considered a form of active dispersal because new areas are being explored. Fungi often have distinct forms of hyphal growth for nutrient acquisition versus dispersal. Hyphae for dispersal, such as rhizomorphs, grow more rapidly, are thicker and tougher, and may be formed by anastomosis (cellular fusion) of multiple smaller hyphae (Rayner et al., 1999). The strategy behind rhizomorphs is to invest little and maintain impermeable surfaces during exploration of a resource-poor environment until a resource-rich patch is encountered. Upon encountering such a patch, there is a proliferation of thinner, more permeable hyphae with a higher surface-to-volume ratio.

The ability of an organism to enter a dormant phase can also be seen as a dispersal mechanism, but through time rather than space. This form of dispersal is one version of the *storage effect*, where reproductive potential is stored across time, resulting in higher reproduction rates under favorable environmental conditions. Entrance of individuals into a dormant life stage may be a developmentally programmed event for some species or may be induced to avoid density-dependent competition. For many organisms, life stages that facilitate passive dispersal in space are also optimal for dispersal in time. This is true of plant seeds and fungal spores.

Species are often distributed in multiple populations that are linked by dispersal and dormancy, which can create significant feedback between populations and contribute to species persistence (Harrison, 1991). This may be particularly important in soil, where there is a large degree of spatial and temporal heterogeneity (Ettema and Wardle, 2002). For example, many bacteria proliferate in spatially and temporally discrete habitats with increased availability of resources, including near roots, particulate organic matter, and worm casts. However, soil microorganisms appear to be generally dormant, as indicated by the increase in numbers and metabolic activity when soil is amended with water or nutrients. Bacterial cells entering a dormant stage are known to undergo a suite of biochemical and morphological changes, including reduction in size. "Dwarf" cells ($<0.07 \, \mu m^3$ biovolume or $<0.3 \, \mu m$ diameter) make up the majority of bacterial cells in soil (Kieft, 2000). These dormant, passively dispersing cells provide the population base for colonization of new patches of resource-rich habitats (Berg and Smalla, 2009; Poll et al., 2010).

D   Interspecific Competition

We have seen in the previous sections how a species' geographic range and occupation of particular habitats is limited by the species' traits resulting in overall adaptation to a set of environmental conditions. The effects of abiotic factors on the survival or growth rate of a population could be plotted with each axis corresponding to one factor. If we imagine many axes, each defining one dimension of an *n*-dimensional space, the region of this space suitable for growth of a species is what Hutchinson (1957) envisioned as the species' *fundamental niche*. The *fundamental niche* of a species describes all combinations of environmental conditions that are acceptable for the persistence of a population. Because of predation and competition with other species, populations of a species will not be present in all habitats that satisfy the species' *fundamental niche*. The reduced hypervolume corresponding to the conditions that a species is actually able to occupy is called its *realized niche*. Interactions between species are fundamental processes in determining the species that will be present in a given location.

Interspecific competition can operate according to the same mechanism as intraspecific competition, except that the individuals competing are from different species. A finite pool of resources is available at any given time, yet if they are consumed faster than they are replenished, growth rates decline. The strongest competitors are able to maintain the highest growth rates despite low levels of resources, thus driving these resources to even lower levels. This form of interspecific competition is known as *resource-based competition* or *exploitative competition*. Given this description of interspecific competition, how do similar species coexist? One common explanation is that spatial and temporal resource variability leads to limited growth in different times and places. Because a species trait can only be optimal under a certain set of conditions, species that are superior competitors for one resource are typically not as competitive for others. Tilman (1982) suggested that the number of similar species that can coexist in a habitat should be equal to the number of potentially limiting resources used in that habitat. It is often assumed that the high heterogeneity of soil conditions supports the enormous microbial diversity.

Species coexistence may also be supported by fluctuating rates of mortality. An event that causes the sudden mortality of an otherwise competitively dominant species or group of species is known as a *disturbance*. Soil tillage is an example of a disturbance that is well known to be detrimental to fungi and favoring bacteria. Species that are *r*-selected may be outcompeted under normal conditions, but flourish when mortality spikes for the dominant species. *Fugitive species* avoid competition by dispersing into habitat patches where the dominant species has become locally extinct. Mortality rates can also be altered through *interference competition*, where one competing species impacts another through direct aggressive action rather than resource use. This type of competition would be likely in systems containing antibiotic-producing organisms. Davelos et al. (2004) confirmed Waksman's much earlier observations when they found a wide variety of antibiotic production and resistant phenotypes present in soil streptomycetes (phylum *Actinobacteria*) at one location, suggesting many organisms are capable of competitive interference. It also suggests organisms have developed mechanisms to avoid this type of interference.

Similar species could also evolve to use different subtypes of the same resource, or their traits and niches can shift in other ways. This is known as *resource partitioning* and was taken as some of the first evidence of competition and natural selection. The concept that no two species having identical niches can coexist is the *competitive exclusion principle*. So, how similar can two species be and still coexist? This has been explored theoretically to a limited extent (beginning with MacArthur and Levins, 1967). Extinctions due to competition have been documented, but evolution can allow an environment to be partitioned into an astonishing array of niches. For example, it has been shown that pure cultures of *Escherichia coli* (class *Gammaproteobacteria*) evolve into distinct, coexisting subtypes because of physiological trait differences (Marx, 2013).

In soil, competition has been exploited as a mechanism for biocontrol, but has also been blamed for the failure of many soil inoculation programs. Fluorescent pseudomonads have been shown to suppress a variety of plant pathogens by secretion of antibiotics (interference competition) and siderophores, which sequester iron (resource competition). Strains of *Fusarium oxysporum* (phylum Ascomycota) that are nonpathogenic can be superior competitors for carbon (C) and root colonization sites (Alabouvette et al., 1996). Organisms introduced to sterilized soil often survive, whereas populations decline rapidly in nonsterile soil. The relatively short half-life of introduced populations has been observed for a variety of groups, including some biocontrol agents, rhizobia, fecal organisms, and genetically modified microbes. This has been attributed to competition, but could also be a result of trophic interactions. In rare cases, inoculated populations have survived (in reduced numbers compared to the inoculum size) if the environment is modified to match their niche requirements or if they are naturally strong competitors.

E Direct Effects of Exploitation

Biological interactions that affect assembly of microbial communities include exploitation and mutualism, as well as intra- and interspecific competition. Exploitation has many forms in the microbial world, including predation, herbivory, parasitism, and pathogenesis. These are trophic interactions, where energy or nutrients are transferred from one organism (prey) to another (the consumer). While studying energy flow through an ecosystem, it may be useful to categorize consumers into *trophic levels* (e.g., herbivores and carnivores). To understand the impact of exploitation on population dynamics, it is important to know if the consumer is a *predator*, causing the immediate death of prey, or a *parasite*, obtaining only a portion of the prey's resources without killing it so that the same organism can be used in the future. Another important distinction is between a *generalist*, which consumes many different prey species, and a *specialist*, which consumes very few. As with all neat categories in ecology, there is a gradient of lifestyles that fall between the extremes. Exploitation in soil biota is widespread. Most soil animals, including protozoa, nematodes, collembola, mites, earthworms, and so on, obtain their resources through exploitation of bacteria, fungi, or plant roots. Invertebrates that ingest plant detritus normally get most of their energy and nutrients from microorganisms residing on the detritus (their "prey") rather than directly from the detritus. Many species of fungi have been shown to attack bacterial colonies and other fungi; there are also fungi that attack soil animals, such as nematodes. *Bdellovibrio* (class Deltaproteobacteria) is a bacterial predator that attacks other bacteria. All organisms also appear to serve as a habitat for an assemblage of smaller organisms, many of which are parasitic. An extreme example of this is the presence of secondary, smaller viral genomes within larger viral genomes, which are dependent on both their bacterial and viral hosts (Swanson et al., 2012).

Predators and parasites will aggregate in patches of high prey density and in areas where prey populations are growing. High numbers of bacteria near roots and decomposing plant litter are thought to increase nematode and protozoan populations, which in turn increases N mineralization from the microbial biomass (Irshad et al., 2011; Koller et al., 2013). High growth rates of bacteria can increase production of viruses in soil (Srinivasiah et al., 2008). Predatory pressure is also a factor in habitat quality for the prey. Predator-free patches serve as refuges for the prey population and can significantly impact metapopulation dynamics. Elliott et al. (1980) found that a finer-textured soil contained more bacteria protected from predation by nematodes compared to a coarse-textured soil. The finer-textured soil contained a larger proportion of pores too small for the nematodes to utilize serving as a refuge for bacteria from nematodes. Amoebae were able to use these pores, consume bacteria, and emerge as food for nematodes. There was a greater growth of nematodes when amoebae were added to a fine- compared to a coarse-textured soil.

The effect of predation on prey population dynamics is an increased death rate. Parasitism is a considerably more complicated phenomenon to model than predation because prey are weakened by parasites, which affects reproduction and death rates. Parasitism can decrease the accumulation of biomass or rate of development. Parasitism also typically increases the death rate, either through prolonged exposure to the parasite, or by making the prey more sensitive to other causes of mortality.

The details of a parasite's transmission route between hosts are critical to understanding how the parasite is spread. Some parasites are able to colonize new hosts from dead tissue. The plant root pathogens in the genera *Gaeumannomyces*, *Rhizoctonia*, and *Pythium* (phyla Ascomycota, Basidiomycota, and Oomycota, respectively) are able to live saprotrophically within plant residue, and then colonize new roots from these habitats. High-quality habitat patches allow pathogenic hyphae to grow further through the soil (to at least 15 cm) to colonize new roots. The probability distribution of colonization of a root from a particular inoculum source would also depend on a variety of other factors, such as the species involved, temperature, moisture, and soil texture. Planting crops at wider distances (i.e., reducing host density) is known to reduce the spread of root diseases. Some parasites are transferred by other species or other components of the environment (*vectors*), and their spread is tightly linked to the dynamics of these factors. Fungal spores, bacteria, and viruses found in soil are transported passively in water and can be transferred by invertebrate vectors, such as plant-parasitic nematodes and mites.

F Indirect Effects of Exploitation

Unlike a nonliving resource, the genetic makeup of prey species will respond to exploitation through evolution resulting in defensive adaptations. Defenses from exploitation can take a variety of forms, including behavioral, morphological, or biochemical. Evolution can also result in the development of new attack strategies in consumers, resulting in a continual coevolutionary arms race between consumers and their prey.

Exploitation can have a large influence on the outcome of competitive interactions between prey species. It can contribute to the coexistence of competing prey species by reducing the population size of the superior competitor. This results in increased resource abundance and ameliorates competition. Most predatory soil fungi, protozoa, nematodes, and collembola will utilize multiple prey species. However, they will either show feeding preferences for or enhanced benefits from particular prey species. It is, therefore, likely that predation regulates community composition (at the species level) by mediating competition between microbes or plant species. The former has been difficult to test because examination of microbial community dynamics *in situ* is complicated; results from laboratory studies do not always provide adequate information about how natural systems operate. Klironomos and Kendrick (1995) found that saprophytic fungi are often the preferred food source for arthropods, yet they will consume mycorrhizal fungi when saprophytic fungi are not available. These relationships can impact competition among plant species by altering the nutrients available from mycorrhizae to plants. Bever (2003) provides a comprehensive review of the mechanisms by which microbes mediate competition in plant communities.

Exploitation pressure at lower trophic levels can be regulated by exploitation at higher trophic levels in a process called a trophic cascade. For example, if the population size of fungal-feeding invertebrate grazers is limited through heavy predation, then the pressure on fungi from the grazers will be low. Although trophic cascades may be common in other environments, a meta-analysis by Sackett et al. (2010) found that they were not widespread in soils when considering entire trophic levels and effects on nutrient cycling. However, exploitation can still regulate species-specific interactions and is often the basis of biocontrol strategies of plant pests. For example, *Trichoderma herzianum* (phylum Ascomycota), a mycoparasitic fungus that attacks the root pathogen *Rhizoctonia solani*, has been introduced to control *Rhizoctonia* density.

The many species-specific trophic relationships between organisms in ecosystems result in a complex web of interactions (a *food web*). When microorganisms are included in soil food webs, the increase in complexity at the species level has been viewed as overwhelming, and these organisms are normally represented by a single trophic level (i.e., an undifferentiated pool of microbial biomass) or are divided into very broad groups (e.g., fungi and bacteria). This is understandable because of the enormous diversity of soil microorganisms, the often unknown role of each taxon in a food web, and the fact that the focus of soil food web studies has typically been on biogeochemical processes, not community structure. However, it also masks unique features of food webs arising when microbial species are included explicitly. There are no "top predators" in food webs containing microorganisms because all organisms are exploited by parasites of varying lethality. Also, the presence of "three-species loops" has been the subject of controversy in food webs of macroscopic organisms and may be possible only when there is differential predation on species due to developmental stage. In microbial systems this food web structure has not been explicitly investigated, but because many predators within the system are generalists it seems likely that such loops can frequently occur due to random encounters. Food webs that include microorganisms must also account for the presence of decomposer organisms and the resource that they utilize, previously dead organisms or their by-products (detritus). This decomposition is critical to the recycling of nutrients that can be used in primary production. Decomposer organisms affect population dynamics of primary producers by supplying nutrients and often by competing with primary producers for the same resources (*immobilization*). In addition, including detritus, nutrients, and decomposers in food webs creates a variety of indirect pathways for organisms to interact (Moore and de Ruiter, 2012; Moore et al., 2004).

G Mutualistic Interactions

Mutualisms, interspecific relationships beneficial to both organisms involved, are also of great ecological significance in soils. Although true mutualisms have been described as mathematically unstable, a diverse array of cross-kingdom partnerships has existed throughout evolutionary history. Soil mutualists have great impact on above- and belowground community dynamics across a wide range of ecosystems. Organisms in soil collaborate with a wide variety of plants to perform nutrient acquisition services in exchange for plant-derived carbohydrates. Although the relationships were originally perceived as bacteria in symbiotic relationships for N acquisition and fungi involved with P acquisition, more recent studies have indicated that fungi are actually involved with the acquisition of almost any limiting nutrient in soil depending on partnering species (Allen, 1991; Smith and Read, 1997).

Mycorrhizae, the relationship between a plant root and fungus, is one of the most ubiquitous soil mutualisms and may be one of the oldest plant relationships (Brundrett, 2002; Stubblefield and Taylor, 1988). There is evidence that this relationship evolved and was lost multiple times in different divisions in the kingdom Fungi and in different groups of plants. Some mycorrhizal fungi have the ability to acquire nutrients directly from decomposing litter (Leake and Read, 1997) and from live animals, such as springtails (Klironomos and Hart, 2001). They can also influence plant water relations (Allen, 1991) and reduce attack on roots by pathogenic fungi (Azcon-Aguilar and Barea, 1992; Gange et al., 1994). These relationships alter the aboveground community both directly, by changing rates of reproduction and death of participant species, and indirectly, by altering competition among plant species. In terms of species specificity, there is not a great deal of consensus on the degree to which partnerships require or benefit from species-specific fungi-plant pairings. There is evidence that there is a wide array of generalists. However, there is concern that important, species-specific relationships may be at risk as a consequence of human disturbances, such as N additions, invasive species, and global climate change, which could limit ecosystem productivity or increase the likelihood of species extinctions (Pickles et al., 2012).

Although mycorrhizae are some of the most terrestrially important mutualists, other soil mutualists are also essential. Much of the N available in soil systems is present as a consequence of N-fixing bacteria, especially through symbiotic relationships in newly establishing communities during primary succession following catastrophic disturbances to soil. Because plants require N for survival, those that are early colonists on highly disturbed sites have often formed symbioses with bacterial N fixers to acquire this necessary nutrient. Early soil formation from rock is impacted by lichen, mutualistic associations of bacteria, or fungi with algae. Lichens stimulate breakdown of rock through hydrolysis and production of organic forms of N through fixation. These associations are also important as macrobiotic crusts necessary for soil stabilization in easily erodible soils.

H Community Impacts on Abiotic Factors

Interactions between organisms have been discussed in terms of resource use. However, environmental modulators (Table 10.1) can also be affected by organisms. The activities of both nitrifying bacteria and plant roots decrease soil pH, and soil temperature is affected by plant and litter cover. This can have positive or negative impacts on the growth of another species, depending on the species' niche requirements.

Some organisms alter the spatial arrangement of components of the environment or serve as new habitat themselves. These organisms are called *ecosystem engineers* and have widespread effects on an ecosystem beyond their own resource use (Jones et al., 1994). Large, competitively dominant organisms, such as trees, are obvious examples of ecosystem engineers. Earthworms are ecosystem engineers because they bury plant litter and create macropores in soil, with large impacts on water infiltration, microbial community structure, and C and nutrient cycling (Eisenhauer et al., 2011; Stromberger et al., 2012).

I Community Variation Among Soil Habitats

Soil conditions are highly variable across a broad range of spatial and temporal scales, including down to the scale of individual soil microaggregates and pores at which single-celled microorganisms typically interact with the soil environment. This variation in conditions has important consequences for microbial community dynamics, ecosystem processes, and interactions with plants. A *landscape,* in ecology, is the particular spatial arrangement of components of the environment that are important in some way to population dynamics of a given species. Landscapes usually include patches of multiple habitats, as well as variability in conditions that affect habitat quality. Unlike some definitions of the term "landscape," this definition does not link landscapes to a particular spatial scale. Instead, it recognizes that landscapes are different for different organisms, depending on the spatial scales over which the organisms interact with the environment (Wiens, 1997).

The habitat that is present in the largest proportion in a landscape, and also has the greatest connectivity, is considered the habitat *matrix*, within which other habitat patches are distributed. In most soils, the habitat matrix is dominated by minerals and nonparticulate, humified organic matter and is not directly exposed to root exudates. We refer to this as "mineral bulk soil." Mineral bulk soil typically harbors a large diversity of microbial species, the majority of microbial biomass, and dominates soil biotic community composition. It is thought that many of the microorganisms in mineral bulk soil are either dormant or nearly dormant, due to a lack of labile organic matter or other resources. However, these dormant microbes rapidly become active if conditions change. Mineral bulk soil can be further divided into pore size classes (micropores, mesopores, and macropores), mineral size classes (clay, silt, and sand fractions), or aggregate size classes (microaggregates and macroaggregates), which may reflect further variability in habitat conditions (Young et al., 2008).

Alternative nonmatrix habitat patches that exist within mineral bulk soil can be created by disturbances or heterogeneity of environmental resources or modulators. After a disturbance, the nonmatrix habitat that is created is defined by an absence of competitively dominant species and is characterized by an abundance of resources due to a lack of competition. Environmental heterogeneity embedded within mineral bulk soil can be caused by many factors, including biological activity, mineralogy, or hydrology. Many soil habitat patches are created through an increased supply of nutrients or labile organic matter and are therefore areas of increased biogeochemical activity. The rhizosphere, fecal matter, and decomposing plant tissue are important examples of this type of habitat. These habitats harbor increased microbial biomass with a differing taxonomic composition (Blackwood and Paul, 2003). Some microorganisms with hyphal growth forms (e.g., many fungi) are able to interact with the environment on a larger spatial scale than are individual rhizospheres or decomposing organic matter particles and therefore integrate over multiple patches of this type. This difference in life-form has important consequences for community and population dynamics and ecosystem processes in the context of spatial variation in soil (Collins et al., 2008; Watkinson et al., 2006).

At larger spatial scales, many types of environmental variation are known to affect the community composition, biomass, and activity of microbes in the soil matrix and other embedded habitats described earlier. Growth of different plant species, and in some cases plant genotypes and developmental stages, cause divergence in the composition of soil microbial communities (e.g., Houlden et al., 2008; Osanai et al., 2013), which can have important consequences for plant health (Berendsen et al., 2012; Berg and Smalla, 2009). Plant species affect microbial communities by releasing different suites of compounds into the rhizosphere and during tissue decomposition, from simple organic acids to complex secondary metabolites. Plants also interact directly with microbial symbionts that may be beneficial or harmful through surface compounds. Microorganisms that proliferate in response to a plant species are typically recruited from the surrounding soil. The microorganisms that proliferate in response to a particular plant species are normally constrained by soil, ecosystem, and land use type (Berg and Smalla, 2009; Lundberg et al., 2012), which can also greatly influence soil microbial community composition. Soil pH is often found to have the best correlation with microbial community composition (Lauber et al., 2009; Tripathi et al., 2012). However, there are many confounded differences among soil types, ecosystem types, and land uses, and it is likely that a complex combination of disturbance and soil factors is involved in differentiating microbial communities at the regional scale.

J   Changes in Community Structure Through Time

Imposed on the habitat mosaic created by environmental heterogeneity and disturbance events, communities also change through time over very short to very long time scales. *Succession* is the replacement of populations in a habitat through time due to ecological interactions. After a habitat patch is created by introduction of environmental heterogeneity or a disturbance event, the initial species that colonize the habitat patch are typically r-selected species with dispersal strategies to increase the chances that they are the first organisms to colonize newly created habitats. These pioneer or "fugitive" species are also capable of making opportunistic use of available resources, or have mechanisms to increase rates of nutrient cycling, such as N fixation. These species are replaced in time by more competitive species; for example, plant species more tolerant of shade or low soil nutrients, or soil microbes capable of producing antibiotics or utilizing more recalcitrant organic matter. Many small-scale habitat patches, such as decomposing plant litter or fecal particles, are defined by a limited pool of labile resources. In addition to colonization by more competitive species with lower dispersal capability, succession in these habitats is driven by a constant change in environmental conditions as resources are used up and the environment is restructured.

Events such as disturbances that create habitat patches are an inherent part of community structure in a large number of systems. Most such events are repeated stochastically at some average rate over large temporal and spatial scales. The constant creation of distinct habitat patches and gradual return to the matrix community creates a *shifting mosaic* of different habitats at different stages of succession (Wu and Loucks, 1995). The proportion of the landscape that is in the matrix community should equal a steady-state value determined by the rate and spatial scale of the disturbance events and the rate of return to the matrix community through succession. This allows fugitive species to depend entirely on the presence of minor habitats with relatively quick turnover rates.

K   Historical and Geographic Contingency Leads to Biogeography

The shifting mosaic picture of the landscape is based on a dynamic equilibrium model. In accordance with the Baas Becking paradigm, quoted at the beginning of this section, with perfect knowledge of organism traits and environmental conditions, one should ideally be able to use the theories described earlier to predict species abundance and community composition in any habitat patch. However, communities that occupy similar habitats in different regions often include different species. It has become clear that there are ecological and evolutionary processes that do not allow populations and communities to reach true stable equilibria predicted by deterministic models. In addition to the niche (or selection-based) processes described earlier, ecological and evolutionary drift (including extinction), mutation (including speciation), and dispersal are broad types of processes that can also affect the distribution of microbial taxa, genotypes, and genes (Hanson et al., 2012). Mutations occur and new types of organisms typically evolve at a particular location; the new gene or species must spread out from this location over time. Physical barriers prevent dispersal to all habitats that could, for example, satisfy a new species' niche requirements. The diversity of barriers is as rich as that of organisms. The environment can also change, resulting in evolutionarily novel conditions and nonequilibrium dynamics, such as those resulting from geological activity, climate change, or human land use. Hence, the global distribution of a gene or species is both a *historical contingency* (i.e., dependent on the particular series of events that occurred in the past) and a *geographic contingency* (dependent on the particular spatial arrangement of elements of a landscape). Although contingencies created by local mutation and drift lead to a biogeography of organisms that cannot be completely explained by deterministic, niche-based processes, dispersal homogenizes the pool of potential colonists and counteracts the effects of mutation and drift.

Examples of historically and geographically contingent biogeographical distributions are particularly apparent when barriers to dispersal break down, such as in the case of exotic species existing outside their native range. Human activity has greatly increased the transport of materials around the globe. Earthworms from Europe were introduced to the Atlantic coast of North America and have been steadily colonizing new soils each year. The root pathogen *Phytophthora infestans* (phylum Oomycota) was introduced from Mexico to the United States, then was transferred to Europe (causing the Irish potato famine), and from there to the rest of the world (Goodwin et al., 1994). Transport of soil is now the subject of international law and regulations. The difference in effects of the introduced species in these two examples is interesting, given the questions raised earlier. *P. infestans* in agroecosystems has a substantial impact because it is involved in aggressive exploitation of the dominant plant (an important ecosystem engineer) and clearly causes system reorganization. On the other hand, earthworms play the role of a detritivore involved in comminution of plant tissue, with large effects on water infiltration, aggregation, and the maintenance of surface plant litter. There is great interest in elucidating the traits that lead some introduced species to become invasive because of the often deleterious effects that invasives have on native species and ecosystems (van der Putten et al., 2007).

Community composition can be structured by ecological, historical, and geographic contingencies, as well as the evolutionary contingencies described above (Belyea and Lancaster, 1999). Ecological drift can result in loss of species through stochastic demographic variation or disturbance. The particular characteristics of a patch's edges, the surrounding habitat patches, and ease of dispersal across other elements of the landscape may all be important determinants of local patch population dynamics. If dispersal from nearby habitats is limited, the order and timing of colonization by different species may be particularly important in determining community composition, rather than fine distinctions among species niche requirements. This emphasis on the potential importance of dispersal is highlighted by the concept of a *metacommunity*, which is a group of communities linked by dispersal across a landscape, resulting in emergent, regionally driven community dynamics. A range of scenarios for the importance of niche characteristics versus stochastic drift, mutation, and dispersal have been described in metacommunity ecology (Leibold et al., 2004).

Broad surveys of soil bacteria indicate that pH is an overriding niche factor structuring community composition (Griffiths et al., 2011; Lauber et al., 2009; Tripathi et al., 2012). However, these broad surveys may mask dispersal limitation and historical contingencies for some groups of organisms (Bissett et al., 2010; Eisenlord et al., 2012) or soil habitats (Hovatter et al., 2011). Likewise, studies of soil fungi indicate that both types of processes are important and often difficult to disentangle (Caruso et al., 2012; Feinstein and Blackwood, 2013; Peay et al., 2010).

III CONSEQUENCES OF MICROBIAL COMMUNITY STRUCTURE FOR ECOSYSTEM FUNCTION

The structure of microbial communities ultimately determines the range of activities that we describe as function. Microbial activities result from the necessity of these organisms to sustain their own life and reproduce in response to a unique set of environmental constraints. The impacts of these activities on the surrounding environment are often discussed as function within a soil or ecosystem context. Soil microbes include organisms that range from strict aerobes to anaerobes, from organisms with high water requirements to low water requirements, from consumers of simple inorganic to complex organic substrates, and from autotroph to heterotroph lifestyles, with some organisms capable of both. As a result of their physiologies and complex life history strategies, these organisms leave a footprint on the systems in which they live. As a result many aspects of the biogeochemical cycles required for terrestrial life are driven by microbial functioning. For example, bacteria in the N cycle consume N as needed for production of body tissues and proteins. Under specific circumstances, bacteria can use N compounds as electron acceptors resulting in increased energy gain. Under conditions of oxygen limitation, other microorganisms can convert nitrate to gaseous products or ammonium; the former process produces N-containing greenhouse gas emissions and may provide a more stable N sink in soils (Rütting et al., 2011). In both cases, these are dissimilatory processes wherein N transformations occur for energy acquisition rather than the acquisition of N for microbial use. The more commonly discussed transformation of N-containing compounds in the N cycle occur in an assimilatory fashion wherein they are used to gain N, and energy is either obtained from that or from another source (Chapter 14).

Many aspects of soil microbial function are related to the role of microbes in determining outcomes in ecosystems. *Ecosystems* are systems defined by a specific composition of interacting populations and the environment within which these organisms interact. They are spatially defined by the interactions of the organisms and their relationship to physical space as an integrated system. Ecosystems are affected by long- and short-term events. For example, trees are still migrating following the last glacial retreat altering the components of ecosystems forcing new species interactions. Even though humans have been on the planet for a very short time period, they have altered ecosystem dynamics through cultivation and burning and, more recently, and across the globe, by increasing soil chemical loads and introducing pesticides and other man-made chemicals. They have also altered the global C cycle by releasing C from ancient organisms from storage pools. The consequence is an altered planetary climate that will impact most species though altered temperature and moisture regimens. All will have impacts on microbial community structure further altering microbial function. The sensitivity of microbes to temperature and moisture make it clear that there will be alterations to belowground processes with climate change; however, ecologists modeling these impacts are not yet certain which biotic processes will increase or decrease. Alterations to the mineralization rate of any nutrient needed by plants will alter ecosystem productivity. Whether process rates increase or decrease depend on the specific changes in temperature and moisture and the interaction between the two consequent impacts on microbial efficiencies, and the degree to which microbes evolve under the new constraints.

The specific components of an ecosystem and the controls over those characteristics are most easily described by the *state factors* originally discussed as soil forming factors by Dokuchaev (Jenny, 1961). The state factors include climate, time, parent material, potential biota, and topography. These factors set bounds on the types and rates of processing and the raw materials available for processing within the ecosystem. Climate is one of the most important factors as it determines rates of processing by controlling moisture availability and temperature. Time is an important factor for evaluating the degree of weathering of soils or vegetative development since a disturbance. The time needed for the development of soils is related interactively with climate as environmentally harsh conditions require a great deal more time for soil development, whereas warmer environments with ample moisture and moderate temperatures require considerably less.

The type of parent material determines the nutrient and water-holding capacity of the medium in which organisms grow and the types of micro- and macroorganisms that can exist. Topography, slope, and aspect characteristics determine access to water, movement of materials, soil depth, degree of weathering of parent material, and total annual energy budget. Topography can alter above- and belowground community structure and consequent microbial activity based on slope location, which has large impacts on moisture and plant productivity and will reflect moisture and the amount of radiant energy received.

Potential biota includes all organisms that can exist or have existed in an area. For example, deep-rooting grasses will differ from other types of plants in their impact on soil development by contributing materials at depth that will be converted into organic matter and turnover slowly. Rooting depth, C:N ratio of materials added to soils, and density and diversity of plants, animals, microbes, and so on, will all contribute differentially to the soil produced. In terms of soil development, microbial community function alters soil chemistry directly through processes that increase nutrient availability (N fixation, P solubilization) and/or alter decomposition rates. Microbial community function alters soil physical structure as a result of their involvement of aggregation, which has a direct bearing on water infiltration, erosion resistance, and nutrient availability.

Microbial function has a direct bearing on ecosystem characteristics. The high diversity of organic compounds available on the terrestrial surface has resulted in a wide range of organisms with a broad spectrum of enzymes. It has been suggested that the microbial community is functionally redundant (i.e., many species have duplicate enzyme systems and roles in decomposition processes so that alterations to microbial community composition do not alter function). Studies over the last several years have challenged this assumption as we have developed more sophisticated approaches that can provide a better understanding of how each species has a unique role in the ecosystem in which they are found. Strickland et al. (2009), by using a microcosm approach, were able to detect differences in community level C mineralization rates using different communities, suggesting that each combination provided a unique set of metabolic physiologies resulting in different rates. These findings are also being supported by metagenomic approaches that evaluate metabolic gene diversity (Röling et al., 2010). Our understanding of the impacts of land use and climate change may remain limited because many current ecosystem models disregard species and community level inputs and constraints on microbial functions. However, there is an increasing number of challenges to this approach (McGuire and Treseder, 2010; Singh et al., 2010).

Communities of micro- and macroorganisms in soil, and their resultant function, are integral to the development and characteristics of all terrestrial ecosystems, and as such, an understanding of soil biology and biochemistry is essential for characterizing ecosystem dynamics. Discussions of ecosystems within an ecological context generally focus on energy flows, elemental cycles, and emergent properties; hence, the focus will be to evaluate the mechanisms by which microbial function impacts these characteristics.

A Energy Flow

The flow of energy in ecosystems is from energy source, to autotroph, to heterotroph. For most systems, the energy source is the sun and the autotrophs are green plants. The amount of energy available within a given system is equivalent to the solar energy that is captured through photosynthesis. The most commonly used term to describe this flow is *net primary productivity* (NPP), which is the total energy uptake by plants in an ecosystem that is available for use by other trophic levels. The amount of NPP in an ecosystem can be predicted most easily by climate or moisture and temperature alone (Fig. 10.1). Biomes are easily plotted along the moisture and temperature gradients showing corresponding

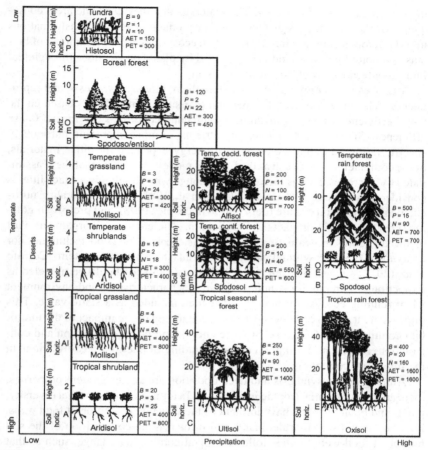

FIG. 10.1 Modified diagram of the distribution of major ecosystem and soil types in relation to precipitation and temperature. Values for elements of structure and function that are used to characterize ecosystems are given for each type. B is the total plant biomass in Mg ha^{-1} year^{-1}, P the total plant production aboveground in Mg ha^{-1} year^{-1}, N the nitrogen uptake by plants in kg ha^{-1} year^{-1}, AET the actual evapotranspiration in millimeters of water per year, PET the potential evapotranspiration in millimeters of water per year. *Reprinted from Aber and Melillo, 2001. Terrestrial Ecosystems, second ed. with permission from Elsevier.*

increases in NPP. At shorter time scales, such as across seasons, NPP is controlled by leaf area, N content, season length, temperature, light, and CO_2 (Chapin et al., 2011). Although autotrophs create a stable energy source using energy from the sun and CO_2 from the atmosphere, they require nutrients from the soil. This requirement makes the contributions by soil microorganisms to processes, such as decomposition and N fixation, every bit as essential to life on this planet as photosynthesis. Most of the true decomposers are heterotrophic osmotrophs. They release enzymes to break down materials and absorb pieces that they can break down for energy. The materials that they consume are a consequence of plant and, to a lesser degree, animal production. Energy transformations in soil microbial communities are reliant on ecosystem NPP. The amount of biomass supported belowground is determined by plant contributions through leaf litter, root turnover, root exudates, and so on.

Microbes use energy gained for metabolism, biomass synthesis, and reproduction. The amount of CO_2 lost per unit of energy gained differs based on the C-use efficiency of the organism. Environmental conditions can impact C-use efficiency (Six et al., 2006). Lack of nutrients, or more important, nutrients in specific ratios, can alter the amount of energy expended to utilize materials. Some computer models, such as the Century model, assume that, of the C assimilated by microbes, 55% is lost as CO_2, although when nonlignin surface litter is considered, a value of 45% is used (Parton et al., 1987). Substrate quality, nutrient availability, temperature (del Giorgio and Cole, 1998), and protozoan grazing (Frey et al., 2001) impact the C utilization efficiency of soil organisms. Part of the difficulty in evaluating efficiency is the lack of good techniques for assaying efficiency without adding materials to a natural system. Cotrufo et al. (2012) suggest that microbial efficiency can and should be modeled as a function of community structure, substrate characteristics, and environment when running ecosystem models rather than included as a static value. This improvement would provide a better understanding of the impact of the microbial community on CO_2 flux, SOM retention, and pool composition and construction, as well as an improved understanding of energy transformations in the microbial community.

Although photosynthesis and decomposition are key ecosystem properties, there are other essential trophic interactions. In terms of soil microbial diversity, plants and plant products have influenced the structure of microbial food webs. First, at the individual scale, plants can influence the composition of the soil microbial food web. Studies following plant community change, such as that which occurs through restoration or the movement of invasive species, have allowed researchers to identify changes in microbial community structure. Bezemer et al. (2010) followed community composition after restoration and found that the individual plant characteristics had the greatest impact on microbial community composition. Their results suggested that it might be more appropriate to examine soil food webs as food webs under individual species rather than a single food web at the level of the plant community. However, they also found that N turnover rates were influenced by the higher-order characteristics of the plant community, such as plant diversity, so plant community structure overall does matter to soil food web dynamics. Other studies have found that individual plant species can alter microbial community composition. Many studies have also found suppression of mycorrhizal fungi following colonization of soils by invasive species, such as garlic mustard (Rout and Callaway, 2012; Wolfe et al., 2008). This alteration to the microbial community will decrease nutrient flow to plants and energy flow to microbial communities through alteration to community structure and, specifically, microbial food web dynamics. In the case of obligate symbionts, trees requiring mycorrhizal fungi for growth and survival will decrease in number impacting diversity, further altering nutrient turnover and ecosystem characteristics.

B Nutrient Cycles

The term *biogeochemical cycles* emphasizes the intertwined roles of biotic and abiotic components for providing necessary molecules for the growth and reproduction of living organisms. The long evolutionary histories of the bacteria and archea have allowed them to develop the machinery necessary to recycle nearly all naturally constructed, complex molecules to the atomic building blocks necessary for the nutrition of higher organisms. Whereas the process of decomposition results in the simultaneous recycling of many different elements, the C, N, and P cycles provide examples of the activities of microbes in these cycles and the consequences of their actions for terrestrial, aquatic, and atmospheric element pools.

Flux in the global C cycle from terrestrial systems is determined by the relative rates of photosynthesis (C uptake) versus autotrophic and heterotrophic respiration as a consequence of decomposition and other microbial processes (C release). Soil organisms and their interactions with each other and with plants significantly impact the rates at which these processes occur. For example, the establishment of a mycorrhizal symbiosis increases photosynthetic rates, especially under conditions of stress, including nutrient or water limitations. These relationships also alter photosynthetic rates indirectly as they mediate competition for resources and alter incidence of root pathogens. The rate at which material is decomposed is impacted, for example, by competition among decomposers for food resources, predation on decomposers by soil animals or other nonsaprophytes, and alterations to abiotic conditions in which the organisms live.

Human-induced alteration to global fluxes in the C cycle is probably the most important ecological experiment of all time. Responses to human alterations will be determined by the biological responses to elevated CO_2, as well as responses to indirect effects, such as temperature and moisture change and climate instability. The global C cycle (Fig. 10.2) has been greatly altered by the large flux of CO_2 to the atmosphere created by fossil fuel use and land use change. The current rate of increase in atmospheric CO_2 concentration (4.1 Pg C per year) is a consequence of imbalance between these anthropogenic sources (7.2 Pg C per year) and CO_2 capture by atmosphere, vegetation, soils, and the ocean (3.1 Pg C per year; Solomon et al., 2007). The relative contribution of microbes to C evolution relative to C storage belowground is of great interest. Net flux to the atmosphere is predicted to increase over time under most scenarios as consumption of organic matter by bacteria and fungi increase with increased temperature and moisture, especially in Arctic systems. There are

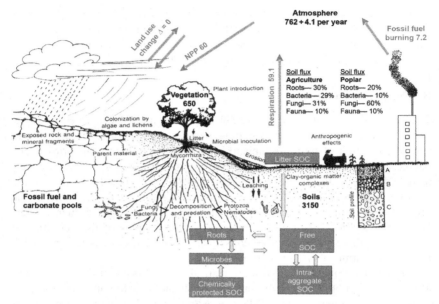

FIG. 10.2   The terrestrial global carbon cycle modified from Paul and Clark (1989, 1996). The pool and flux values, modified from Denman et al. (2007), including preindustrial values for terrestrial flux, are in Pg C for pools and Pg C per year for fluxes. Soil fluxes are driven largely by microbial participants, which researchers are currently working to quantify. Flux values for the global C cycle do not include error values associated with measured values, although they are available with full discussion of measurements in Denman et al. (2007) in the IPCC Fourth Assessment Report. The global C cycle currently includes net flux rates of -2.2 from oceanic systems providing for the balance between surface flux rate (terrestrial+ocean = -3.1 per year), C release to atmosphere from fossil fuels (7.2), and current atmospheric accrual (4.1).

currently a number of studies examining the relative contributions of each belowground participant group to flux (illustrated in Fig. 10.2), C-use efficiency, and the impacts of elevated CO_2 and climate change on these fluxes.

Predicting the impacts of climate change on soil microorganisms is a unique challenge. Soil organisms directly and indirectly control C flux through decomposition. However, they also control the production and consumption of methane and soil N_2O emissions. Soil organisms are not all heterotroph-dependent on plant and animal products for energy and on O_2 as the ultimate electron acceptor in cellular respiration. Soil microorganisms include lithotrophs capable of using materials, such as ammonium and some sulfur compounds, as energy sources. Many soil organisms also use nitrate or sulfate as the ultimate electron acceptors rather than O_2 in O_2 limited areas, allowing them to live anaerobically. As such they have unique roles in C, N, and S cycles and have great impact on global climate change through impacts on CH_4 and N_2O. These unique energy transformations have significant consequences for understanding global climate as they alter rates of ecosystem C flux through photosynthesis and decomposition.

Fluxes within the N cycle are primarily driven by N fixation (conversion from atmospheric N to microbial N); mineralization (ammonification+nitrification) of organic N from sources, such as leaf litter and other plant or animal materials; and gaseous loses, such as through denitrification and ammonia volatilization. The microbial community drives nearly all key processes in this cycle. Mutualistic relationships formed by plants with soil organisms, such as *Rhizobium* (class *Alphaproteobacteria*) and *Frankia* (phylum *Actinobacteria*) are known for their importance in agriculture; however, these relationships are essential in early successional communities where N may not yet be available in soil. In these systems, microbial N fixation provides the initial N source that will allow plants to grow. As plant production increases, the N available for plant growth in these community transitions from atmospheric N fixed by microorganisms to N that is available through decomposition of plant materials.

The rate at which N is returned to the ecosystem following plant uptake is largely determined by plant form, nutrient use efficiency, and ultimately, the rate of mineralization once the organic material is made available to the decomposers. Nitrogen that becomes available for uptake following release from litter or soil organic matter can either be taken up by plants or by soil microorganisms through immobilization, where available N becomes part of microbial tissues rather than plant tissues. When N is limited in soils, plant-microbe competition for N is most often "won" by microbes setting up limitations of N for plants, reducing NPP and litter quality (van der Heijden et al., 2008).

Historically, the amount of N that occurred naturally in ecosystems was a consequence of either N fixation by symbiotic or free-living microbes, or recycling of organic materials by microbes. This is no longer the case as anthropogenic generation and application of fertilizer N and pollutant dispersal has resulted in a doubling of the amount of N currently available for plants in many ecosystems (Chapter 14). This has changed N availability for plant uptake, decomposition rates, plant species germination and competition, and microbial control of N availability in terrestrial ecosystems. Additions of N have been found to increase soil respiration, decrease microbial biomass, and alter enzyme activity across a broad range of soils suggesting that these additions have had significant impacts on the functioning of soil microorganisms (Ramirez et al., 2012).

A great deal of the early literature on the plant-fungal mutualisms focused exclusively on P. The need for P in plants is so great and competition so intense that plants with mutualistic relationships can often grow larger than plants without fungi, even though the plants must provide C for fungal growth. The last several decades of research have expanded our understanding of these mycorrhizal relationships to include exchanges for not only P, but also N, cations, and water, especially in nutrient limited or stressful environments. As was described for C and N, relationships, such as competition, predation, and exploitation, between the mycorrhizal fungi and soil organisms determine the rate at which nutrients and water become available for plants and determine competition within plant communities. The availability of P and other nutrients can also determine the rate of succession. The formation of mutualisms and the arrival of decomposer bacteria and fungi have great potential to alter plant community dynamics. Disturbances, such as fires, that release nutrients to soil can decrease reliance on biogeochemical cycles by hastening the rate at which nutrients become available.

According to Liebig's Law of the Minimum, low quantities of any essential nutrient can cause stress and decrease productivity, so the cycling of all nutrients is important for understanding ecosystem dynamics. In most ecosystems, the nutrients that plants depend on for growth are those that are returned through recycling as a result of microbial action, also called internal cycling or the intrasystem cycle, rather than through fresh nutrient inputs as a result of biotic or abiotic weathering processes. Much of the quantitative work necessary to document the importance of the intrasystem cycle was completed by Likens and others as they documented ecosystem characteristics at the Hubbard Brook Forest (Likens et al., 1977). Further review of the literature suggests that, in general, across systems, decomposition of plant litter by microbes alone can return up to 100% of nutrients required for plant growth (Van der Heijden et al., 2008), suggesting that many systems can rely entirely on local components of nutrient cycles.

C Emergent Properties

Soil dynamics drive elemental cycles, have controlling roles in ecosystem functions, and are largely determinants of *emergent properties* (properties not obvious from the study of processes at finer levels of organization), such as decomposition rates, biodiversity, and system stability. Emergent properties of ecosystems are a consequence of the synergistic effects of environmental components, community composition, and *ecosystem function* (flux of energy and materials through the ecosystem). The idea that the forest is more than the trees conveys the importance of the concept, but it can be applied broadly to every ecosystem. For example, the rate of decomposition is a consequence of any number of interacting factors. Studies of soil organic matter dynamics often evaluate rates as a consequence of substrate composition and availability and decomposer community composition. Christensen (2001) frames organic matter turnover more eloquently as a function of three levels of complexity. The organomineral complexes reflect the inherent characteristics of soil as largely determined by the state factors described earlier and represent the first level of complexity. The consequent aggregation of the primary organomineral complexes are the second level. The third level, soil structure, represents the arrangement of soil aggregates forming the pore network. Soil structure is a complex, often overlooked, consequence of the first two levels and would be considered an emergent property of the first two. The characteristics of the soil determined by soil structure impact all aspects of the soil environment important for biology, such as gas flow, water movement, macro- and micropore spaces, water-holding capacity, and so on. These characteristics interact with the biodiversity, or specifically in this case, the decomposer community structure to determine rates of decomposition; one aspect of ecosystem function within a system.

There have been a number of hypotheses developed to explain the inherent importance of another emergent property, biodiversity within an ecosystem, but few studies have provided convincing support for these hypotheses. Questions such as whether biodiversity impacts characteristics, including ecosystem productivity, nutrient retention, and stability, have permeated the literature over the last several decades. In an era of decreasing global biodiversity and genetic diversity, these questions require immediate attention. Intellectually there is no denying that maintaining biodiversity is essential to maintaining the integrity of systems. There are two main problems with this challenge. The first is that it is difficult to measure the complete biodiversity of a system and determine the degree to which it is in an "undisturbed state." Second, biodiversity is not a term that is indicative of quality, but quantity. Increasing diversity by increasing undesirable species, such as nonnative invasive species, will not maintain the integrity of a system, so biodiversity for biodiversity's sake is not the answer. For many of the discussions of microbial diversity, the arguments have been framed as one of functional redundancy in microbial communities. This argument suggests that there are so many species that have the same function that loss of one will not alter the way that the system operates. Much of the recent research on soil microbial community diversity suggests that this argument and the assertions of Bejerinck discussed earlier are not supported and that the biogeographic distribution and evolution under new environments makes microbial communities unique (Strickland et al., 2009). Further, there is more evidence that process rates are not ultimately controlled by the environment, but by biodiversity. In the past, many studies have focused on plant biodiversity and impacts on ecosystem function. Tilman (1996) demonstrated that ecosystem stability increased with aboveground species diversity. However, belowground systems appear to have more complex dynamics than aboveground systems when it comes to the importance of diversity. Klironomos et al. (2000) found that the presence or absence of AM fungi significantly changed the relationship of plant diversity to aboveground productivity. Without the fungi, productivity increased as plant species were added to a total of 15 species in a linear fashion. Productivity was maximized at 10 plant species in an asymptotic fashion with AM fungi present. A recent review of the literature provides data that suggest that the relationship of microbial diversity and C cycling may be more complex than previously seen for plant or animal studies (Nielsen et al., 2011). At low diversity, there was a significant relationship between diversity and C cycling, such that as diversity increases, C cycling increases. However, when diversity was high, species-specific traits became more important than numbers of species. Dynamics such as these are important for evaluating complex problems such as global climate change, but the lack of easily modeled patterns creates considerable difficulty in implementation. Overall, considerations of the importance of diversity in ecosystem function must include an understanding of the diversity of all participants; otherwise, the importance of diversity among different components may be overlooked.

The amount of stress placed on ecosystems across the globe by anthropogenic influences has changed the way ecosystems operate. Ecosystem stability is an emergent property of community composition and ecosystem function. Novel, frequent, or severe disturbances can be threats to ecosystem stability. There are two main terms used to characterize the way that an ecosystem responds to a disturbance. Whereas a system that does not change appreciably following a disturbance is said to be *resistant*, a system that changes, but returns to its predisturbance state within a reasonable timeframe, is said to be *resilient*. The degree to which ecosystems are resistant or resilient may depend entirely on biodiversity and the amount of stress currently on the system. Shade et al. (2012) examined whether there were principles by which one could predict microbial community resistance and resilience under short-term (pulse) and long-term (press) disturbance. Their conclusion was that soil community composition and function are sensitive to disturbance. Most of the studies examined in the Shade et al. (2012) analysis of the literature reported that microbial communities are neither resistant nor resilient to pulse or push disturbances. However, there may be bias in reporting the response to disturbance as studies that do not find changes in community structure or function may be underreported. As we enter an age of ever increasing anthropogenic disturbance through global climate change, land use change, invasive species, and so on, further studies are needed to evaluate the consequences of the sensitivity of microbial communities to disturbance on ecosystem stability. Unfortunately, as much as this information is needed to protect ecosystems from degradation and against species loss, the stability of ecosystems is an emergent property that cannot easily be measured quantitatively as a single measurement or by quantifying one or more of its component parts.

IV CONCLUSION

Integrating across scales and disciplines is a challenge that defines ecological study. Scientists that study soil microbiology, ecology, and biochemistry can contribute to and benefit from approaching the soil and organisms they examine from an ecological perspective. Society has placed a great burden on scientists by damaging systems before understanding how they operate. Science, especially the field of soil ecology, is now charged with developing an understanding of these systems and finding ways to mitigate the damage. This can only be accomplished by seeking every opportunity to integrate across scientific fields to develop a more comprehensive understanding of the structure and function of soil microbial communities and their influences on ecosystems and the globe.

REFERENCES

Aber, J.D., Melillo, J.M., 2001. Terrestrial Ecosystems, second ed. Academic Press, San Diego.

Alabouvette, C., Lemanceau, P., Steinberg, C., 1996. Biological control of Fusarium wilts: opportunities for developing a commercial product. In: Hall, R. (Ed.), Principles and Practice of Managing Soilborne Plant Pathogens. American Phytopathological Society Press, St. Paul, MN, pp. 192–212.

Allen, M.F., 1991. Ecology of Mycorrhizae. Cambridge University Press, Cambridge.

Aro, N., Pakula, T., Penttila, M., 2005. Transcriptional regulation of plant cell wall degradation by filamentous fungi. FEMS Microbiol. Rev. 29, 719–739.

Azcon-Aguilar, C., Barea, J.M., 1992. Interactions between mycorrhizal fungi and other rhizosphere microorganisms. In: Allen, M.F. (Ed.), Mycorrhizal Functioning. Chapman and Hall, New York, pp. 163–198.

Belyea, L.R., Lancaster, J., 1999. Assembly rules within a contingent ecology. Oikos 86, 402–416.

Berendsen, R.L., Pieterse, C.M.J., Bakker, P.A.H.M., 2012. The rhizosphere microbiome and plant health. Trends Plant Sci. 17, 478–486.

Berg, G., Smalla, K., 2009. Plant species and soil type cooperatively shape the structure and function of microbial communities in the rhizosphere. FEMS Microbiol. Ecol. 68, 1–13.

Bever, J.D., 2003. Soil community feedback and the coexistence of competitors: conceptual frameworks and empirical tests. New Phytol. 157, 465–473.

Bezemer, T.M., Fountain, M.T., Barea, J.M., Christensen, S., Bekker, S.C., Duyts, H., van Hal, R., Harvey, J.A., Hedlund, K., Maraun, M., Mikola, J., Mladenov, A.G., Robin, C., de Ruiter, P.C., Scheu, S., Setälä, H., Smilauer, P., van der Putten, W.H., 2010. Divergent composition but similar function of soil food webs of individual plants: plant species and community effects. Ecology 91, 3027–3036.

Bissett, A., Richardson, A.E., Baker, G., Wakelin, S., Thrall, P.H., 2010. Life history determines biogeographical patterns of soil bacterial communities over multiple spatial scales. Mol. Ecol. 19, 4315–4327.

Blackwood, C.B., Paul, E.A., 2003. Eubacterial community structure and population size within the soil light fraction, rhizosphere, and heavy fraction of several agricultural systems. Soil Biol. Biochem. 35, 1245–1255.

Brundrett, M.C., 2002. Coevolution of roots and mycorrhizas of land plants. New Phytol. 154, 275–304.

Caruso, T., Hempel, S., Powell, J.R., Barto, E.K., Rillig, M.C., 2012. Compositional divergence and convergence in arbuscular mycorrhizal fungal communities. Ecology 93, 1115–1124.

Chapin, F.S., Matson, P.A., Vitousek, P.M., 2011. Principles of Terrestrial Ecosystem Ecology, second ed. Springer, New York.

Christensen, B.T., 2001. Physical fractionation of soil and structural and functional complexity in organic matter turnover. Eur. J. Soil Sci. 52, 345–353.

Cleveland, C.C., Liptzin, D., 2007. C:N:P stoichiometry in soil: Is there a "Redfield ratio" for the microbial biomass? Biogeochemistry 85, 235–252.

Collins, S.L., Sinsabaugh, R.L., Crenshaw, C., Green, L., Porras-Alfaro, A., Stursova, M., Zeglin, L.H., 2008. Pulse dynamics and microbial processes in arid land ecosystems. J. Ecol. 96, 413–420.

Compant, S., Clément, C., Sessitsch, A., 2010. Plant growth-promoting bacteria in the rhizo- and endosphere of plants: their role, colonization, mechanisms involved and prospects for utilization. Soil Biol. Biochem. 42, 669–678.

Cotrufo, M.F., Wallenstein, M.D., Boot, C., Denef, K., Paul, E., 2012. The Microbial Efficiency-Matrix Stabilization (MEMS) framework integrates plant litter decomposition with soil organic matter stabilization: Do labile plant inputs form stable soil organic matter? Glob. Chang. Biol. 19, 988–995.

Coutinho, H.L.C., Kay, H.E., Manfio, G.P., Neves, M.C.P., Ribeiro, J.R.A., Rumjanek, N.G., Beringer, J.E., 1999. Molecular evidence for shifts in polysaccharide composition associated with adaptation of soybean *Bradyrhizobium* strains to the Brazilian Cerrado soils. Environ. Microbiol. 1, 401–408.

Davelos, A.L., Kinkel, L.L., Samac, D.A., 2004. Spatial variation in frequency and intensity of antibiotic interactions among streptomycetes from prairie soil. Appl. Environ. Microbiol. 70, 1051–1058.

de Jonge, R., Takumi, K., Ritmeester, W.S., van Leusden, F.M., 2003. The adaptive response of *Escherichia coli* O157 in an environment with changing pH. J. Appl. Microbiol. 94, 555–560.

de Vries, N., Kuipers, E.J., Kramer, N.E., van Vliet, A.H.M., Bijlsma, J.J.E., Kist, M., Bereswill, S., Vanderbroucke-Grauls, C.M.J.E., Kusters, J.G., 2001. Identification of environmental stress-regulated genes in *Helicobacter pylori* by a lacZ reporter gene fusion system. Helicobacter 6, 300–309.

del Giorgio, P.A., Cole, J.J., 1998. Bacterial growth efficiency in natural aquatic systems. Annu. Rev. Ecol. Syst. 29, 503–541.

Denman, K.L., Brasseur, G., Chidthaisong, A., Ciais, P., Cox, P.M., Dickinson, R.E., Hauglustaine, D., Heinze, C., Holland, E., Jacob, D., Lohmann, U., Ramachandran, S., de Silva Dias, P.L., Wofsy, S.C., Zhang, X., 2007. Couplings between changes in the climate system and biogeochemistry. In: Solomon, S., Qin, D., Manning, M., Chen, Z., Marquis, M., Averyt, K.B., Tignor, M., Miller, H.L. (Eds.), Climate Change 2007: The Physical Science Basis. Contribution of Working Group I to the Fourth Assessment Report of the Intergovernmental Panel on Climate Change. Cambridge University Press, Cambridge, United Kingdom and New York.

Derré, I., Rapoport, G., Msadek, T., 1999. CtsR, a novel regulator of stress and heat shock response, controls clp and molecular chaperone gene expression in Gram-positive bacteria. Mol. Microbiol. 31, 117–131.

Eisenhauer, N., Schlaghamerský, J., Reich, P.B., Frelich, L.E., 2011. The wave towards a new steady state: effects of earthworm invasion on soil microbial function. Biol. Invasions 13, 2191–2196.

Eisenlord, S.D., Zak, D.R., Upchurch, R.A., 2012. Dispersal limitation and the assembly of soil *Actinobacteria* communities in a long-term chronosequence. Ecol. Evol. 2, 538–549.

Elliott, E.T., Anderson, R.V., Coleman, D.C., Cole, C.V., 1980. Habitable pore space and microbial trophic interactions. Oikos 35, 327–335.

Ettema, C.H., Wardle, D.A., 2002. Spatial soil ecology. Trends Ecol. Evol. 17, 177–183.

Feinstein, L.M., Blackwood, C.B., 2013. The spatial scaling of saprotrophic fungal beta diversity in decomposing leaves. Mol. Ecol. 22, 1171–1184.

Fraser, C., Alm, E.J., Polz, M.F., Spratt, B.G., Hanage, W.P., 2009. The bacterial species challenge: making sense of genetic and ecological diversity. Science 323, 741–746.

Frey, S.D., Gupta, V.V.S.R., Elliott, E.T., Paustian, K., 2001. Protozoan grazing affects estimates of carbon utilization efficiency of the soil microbial community. Soil Biol. Biochem. 33, 1759–1768.

Gange, A.C., Brown, V.K., Sinclair, G.S., 1994. Reduction of lack vine weevil larval growth by vesicular-arbuscular mycorrhizal infection. Entomol. Exp. Appl. 70, 115–119.

Gil, R., Silva, F.J., Zientz, E., Delmotte, F., Gonzalez-Candelas, F., Latorre, A., Rausell, C., Kamerbeek, J., Gadau, J., Hölldobler, B., van Ham, R.C.H.J., Gross, R., Moya, A., 2003. The genome sequence of *Blochmannia floridanus*: comparative analysis of reduced genomes. Proc. Natl. Acad. Sci. U. S. A. 100, 9388–9393.

Goodwin, S.B., Cohen, B.A., Fry, W.E., 1994. Panglobal distribution of a single clonal lineage of the Irish potato famine fungus. Proc. Natl. Acad. Sci. U. S. A. 91, 11591–11595.

Griffiths, R.I., Thomson, B.C., James, P., Bell, T., Bailey, M., Whiteley, A.S., 2011. The bacterial biogeography of British soils. Environ. Microbiol. 13, 1642–1654.

Hanson, C.A., Fuhrman, J.A., Horner-Devine, M.C., Martiny, J.B., 2012. Beyond biogeographic patterns: processes shaping the microbial landscape. Nat. Rev. Microbiol. 10, 497–506.

Harrison, S., 1991. Local extinction in a metapopulation context: an empirical evaluation. In: Gilpin, M., Hanski, I. (Eds.), Metapopulation Dynamics: Empirical and Theoretical Investigations. Academic Press, London, pp. 73–88.

Hoshino, T., Xiao, N., Tkachenko, O.B., 2009. Cold adaptation in the phytopathogenic fungi causing snow molds. Mycoscience 50, 26–38.

Houlden, A., Timms-Wilson, T.M., Day, M.J., Bailey, M.J., 2008. Influence of plant developmental stage on microbial community structure and activity in the rhizosphere of three field crops. FEMS Microbiol. Ecol. 65, 193–201.

Hovatter, S.R., Dejelo, C., Case, A.L., Blackwood, C.B., 2011. Metacommunity organization of soil microorganisms depends on habitat type defined by presence of *Lobelia siphilitica* plants. Ecology 92, 57–65.

Hutchinson, G.E., 1957. Concluding remarks. Cold Spring Harb. Symp. 22, 415–427.

Irshad, U., Villenave, C., Brauman, A., Plassard, C., 2011. Grazing by nematodes on rhizosphere bacteria enhances nitrate and phosphorus availability to *Pinus pinaster* seedlings. Soil Biol. Biochem. 43, 2121–2126.

Jenny, H., 1961. Derivation of state factor equations of soils and ecosystems. Soil Sci. Soc. Am. Proc. 25, 385–388.

Jones, C.G., Lawton, J.H., Shachak, M., 1994. Organisms as ecosystem engineers. Oikos 69, 373–386.

Kieft, T.L., 2000. Size matters: dwarf cells in soil and subsurface terrestrial environments. In: Colwell, R.R., Grimes, D.J. (Eds.), Nonculturable Microorganisms in the Environment. American Society for Microbiology Press, Washington, DC, pp. 19–46.

Kingsland, S.E., 1991. Defining ecology as a science. In: Real, L.A., Brown, J.H. (Eds.), Foundations of Ecology: Classic Papers with Commentaries. University of Chicago Press, Chicago.

Klironomos, J.N., Hart, M.M., 2001. Food-web dynamics – animal nitrogen swap for plant carbon. Nature 410, 651–652.

Klironomos, J.N., Kendrick, W.B., 1995. Stimulative effects of arthropods on endomycorrhizas of sugar maple in the presence of decaying litter. Funct. Ecol. 9, 528–536.

Klironomos, J.N., McCune, J., Hart, M., Neville, J., 2000. The influence of arbuscular mycorrhizae on the relationship between plant diversity and productivity. Ecol. Lett. 3, 137–141.

Koller, R., Scheu, S., Bonkowski, M., Robin, C., 2013. Protozoa stimulate N uptake and growth of arbuscular mycorrhizal plants. Soil Biol. Biochem. 65, 204–210.

Lauber, C., Hamady, M., Knight, R., Fierer, N., 2009. Pyrosequencing-based assessment of soil pH as a predictor of soil bacterial community structure at the continental scale. Appl. Environ. Microbiol. 75, 5111–5120.

Leake, J.R., Read, D.J., 1997. Mycorrhizal fungi in terrestrial habitats. In: Wicklow, D.T., Söderström, B. (Eds.), The Mycota IV. Environmental and Microbial Relationships. Springer Verlag, Berlin, pp. 281–301.

Leibold, M.A., Holyoak, M., Mouquet, N., Amarasekare, P., Chase, J.M., Hoopes, M.F., Holt, R.D., Shurin, J.D., Law, R., Tilman, D., Loreau, M., Gonzalez, A., 2004. The metacommunity concept: a framework for multi-scale community ecology. Ecol. Lett. 7, 601–613.

Likens, G.E., Bormann, F.H., Pierce, R.S., Eaton, K.S., Johnson, N.M., 1977. Biogeochemistry of a Forested Ecosystem. Springer Verlag, New York, Heidelberg, and Berlin.

Lönn, A., Gárdonyi, M., van Zyl, W., Hahn-Hägerdal, B., Otero, R.C., 2002. Cold adaptation of xylose isomerase from *Thermus thermophilus* through random PCR mutagenesis gene cloning and protein characterization. Eur. J. Biochem. 269, 157–163.

Loper, J.E., Hassan, K.A., Mavrodi, D.V., Davis II, E.W., Shaffer, B.T., Elbourne, L., Stockwell, V.-O., Hartney, S.L., Breakwell, K., Henkels, M.D., Tetu, S.G., Rangel, L.I., Kidarsa, T.A., Wilson, N.L., van de Mortel, J.E., Song, C., Blumhagen, R., Radune, D., Hostetler, J.B., Brinkac, L.M., Durkin, A.S., Kluepfel, D.A., Wechter, W.P., Anderson, A.J., Kim, Y.C., Pierson III, L.S., Pierson, E.A., Lindow, S.E., Kobayashi, D.Y., Raaijmakers, J.M., Weller, D.-M., Thomashow, L.S., Allen, A.E., Paulsen, I.T., 2012. Comparative genomics of plant-associated *Pseudomonas* spp.: insights into diversity and inheritance of traits involved in multi-trophic interaction. PLoS Genet. 8, e1002784.

Lundberg, D.S., Lebeis, S.L., Paredes, S.H., Yourstone, S., Gehring, J., Malfatti, S., Tremblay, J., Engelbrektson, A., Kunin, V., del Rio, T.G., Edgar, R.C., Eickhorst, T., Ley, R.E., Hugenholtz, P., Tringe, S.G., Dangl, J.L., 2012. Defining the core *Arabidopsis thaliana* root microbiome. Nature 488, 86–90.

MacArthur, R.H., Levins, R., 1967. The limiting similarity, convergence, and divergence of coexisting species. Am. Nat. 101, 377–385.

MacArthur, R.H., Wilson, E.O., 1967. The Theory of Island Biogeography. Princeton University Press, Princeton, NJ, 203 p.

Martinez, D., Challacombe, J., Morgenstern, I., Hibbett, D., Schmoll, M., Kubicek, C.P., Ferreira, P., Ruiz-Duenas, F.J., Martinez, A.T., Kersten, P., Hammel, K.E., Vanden, A., Wymelenberg, V., Gaskell, J., Lindquist, E., Sabat, G., Bondurant, S.S., Larrondo, L.F., Canessa, P., Vicuna, R., Yadav, J., Doddapaneni, H., Subramanian, V., Pisabarro, A.G., Lavin, J.L., Oguiza, J.A., Master, E., Henrissant, B., Coutinho, P.M., Harris, P., Magnuson, J.K., Baker, S.E., Burno, K., Kenealy, W., Joegger, P.J., Kues, U., Ramaiya, P., Lucas, S., Salamov, A., Shapiro, H., Tu, H., Chee, C.L., Misra, M., Xie, G., Teter, S., Yaver, D., James, T., Mokrejs, M., Pospisek, M., Grigoriev, I.V., Brettin, T., Rokhsar, D., Berka, R., Cullen, D., 2009. Genome, transcriptome, and secretome analysis of wood decay fungus *Postia placenta* supports unique mechanisms of lignocellulose conversion. Proc. Natl. Acad. Sci. U. S. A. 106, 1954–1959.

Marx, C.J., 2013. Can you sequence ecology? Metagenomics of adaptive diversification. PLoS Biol. 11, e1001487.

McGill, B.J., Enquist, B.J., Weiher, E., Westoby, M., 2006. Rebuilding community ecology from functional traits. Trends Ecol. Evol. 21, 178–184.

McGuire, K.L., Treseder, K.K., 2010. Microbial communities and their relevance for ecosystem models: decomposition as a case study. Soil Biol. Biochem. 42, 529–535.

Moore, J., de Ruiter, P.C., 2012. Energetic Food Webs. Oxford University Press, UK.

Moore, J.C., Berlow, E.L., Coleman, D.C., de Ruiter, P.C., Dong, Q., Hastings, A., Johnson, N.C., McCann, K.S., Melville, K., Morin, P.J., Nadelhoffer, K., Rosemond, A.D., Post, D.M., Sabo, J.L., Scow, K.M., Vanni, M.J., Wall, D.H., 2004. Detritus, trophic dynamics and biodiversity. Ecol. Lett. 7, 584–600.

Nielsen, U.N., Ayres, E., Wall, D.H., Bardgett, R.D., 2011. Soil biodiversity and carbon cycling: a review and synthesis of studies examining diversity-function relationships. Eur. J. Soil Sci. 62, 105–116.

O'Malley, M.A., 2007. The nineteenth century roots of 'everything is everywhere'. Nat. Rev. Microbiol. 5, 647–651.

Osanai, Y., Bougoure, D.S., Hayden, H.L., Hovenden, M.J., 2013. Co-occurring grass species differ in their associated microbial community composition in a temperate native grassland. Plant Soil 368, 419–431.

Panikov, N.S., 1995. Microbial Growth Kinetics. Chapman & Hall, London, UK, 378 p.

Parton, W.J., Schimel, D.S., Cole, C.V., Ojima, D.S., 1987. Analysis of factors controlling soil organic matter levels in Great Plains grasslands. Soil Sci. Soc. Am. J. 51, 1173–1179.

Paul, E.A., Clark, F.E., 1989. Soil Microbiology and Biochemistry. Academic Press, New York, 273 p.

Paul, E.A., Clark, F.E., 1996. Soil Microbiology and Biochemistry, second ed. Academic Press, New York.

Peay, K.G., Garbelotto, M., Bruns, T.D., 2010. Evidence of dispersal limitation in soil microorganisms: isolation reduces species richness on mycorrhizal tree islands. Ecology 91, 3631–3640.

Pianka, E.R., 1970. On r- and K-selection. Am. Nat. 104, 592–597.

Pickles, B.J., Egger, K.N., Massicotte, H.B., Green, D.S., 2012. Ectomycorrhizas and climate change. Fungal Ecol. 5, 73–84.

Poll, C., Brune, T., Begerow, D., Kandeler, E., 2010. Small-scale diversity and succession of fungi in the detritusphere of rye residues. Microb. Ecol. 59, 130–140.

Ramirez, K.S., Craine, J.M., Fierer, N., 2012. Consistent effects of nitrogen amendments on soil microbial communities and processes across biomes. Glob. Chang. Biol. 18, 1918–1927.

Rayner, A.D.M., Beeching, J.R., Crowe, J.D., Watkins, Z.R., 1999. Defining individual fungal boundaries. In: Worrall, J.J. (Ed.), Structure and Dynamics of Fungal Populations. Kluwer Academic Publishers, Dordrecht, The Netherlands, pp. 19–41.

Reznick, D., Bryant, M.J., Bashey, F., 2002. r- and K-selection revisited: the role of population regulation in life-history evolution. Ecology 83, 1509–1520.

Robinson, C.H., 2001. Cold adaptation in Arctic and Antarctic fungi. New Phytol. 151, 341–353.

Roeßler, M., Müller, V., 2001. Osmoadaptation in bacteria and archaea: common principles and differences. Environ. Microbiol. 3, 743–754.

Röling, W.F., Ferrer, M., Golyshin, P.N., 2010. Systems approaches to microbial communities and their functioning. Curr. Opin. Biotechnol. 21, 532–538.

Rout, M.E., Callaway, R.M., 2012. Interactions between exotic invasive plants and soil microbes in the rhizosphere suggest that 'everything is not everywhere'. Ann. Bot. 110, 213–222.

Rütting, T., Boeckx, P., Muller, C., Klemedtsson, L., 2011. Assessment of the importance of dissimilatory nitrate reduction to ammonium for the terrestrial nitrogen cycle. Biogeosciences 8, 1779–1791.

Sackett, T.E., Classen, A.T., Sanders, N.J., 2010. Linking soil food web structure to above- and belowground ecosystem processes: a meta-analysis. Oikos 119, 1984–1992.

Settembre, E.C., Chittuluru, J.R., Mill, C.P., Kappock, T.J., Ealick, S.E., 2004. Acidophilic adaptations in the structure of *Acetobacter aceti* N5-carboxyaminoimidazole ribonucleotide mutase (PurE). Acta Crystallogr. D 60, 1753–1760.

Shade, A., Peter, H., Allison, S.D., Baho, D.L., Berge, M., Bürgmann, H., Huber, D.H., Langenheder, S., Lennon, J.T., Martiny, J.B.H., Matulich, K.L., Schmidt, T.M., Handelsman, J., 2012. Fundamentals of microbial community resistance and resilience. Front. Microbiol. 3, e417.

Singh, B.K., Bardgett, R.D., Smith, P., Reay, D.S., 2010. Microorganisms and climate change: terrestrial feedbacks and mitigation options. Microbiology 8, 779–790.

Six, J., Frey, S.D., Thiet, R.K., Batten, K.M., 2006. Bacterial and fungal contributions to carbon sequestration in agroecosystems. Soil Sci. Soc. Am. J. 70, 555–569.

Smith, S.E., Read, D., 1997. Mycorrhizal Symbiosis, second ed. Academic Press, London, UK.

Solomon, S., Qin, D., Manning, M., Chen, Z., Marquis, M., Averyt, K.B., Tignor, M., Miller, H.L., 2007. Summary for policymakers. Climate Change 2007: The Physical Science Basis. Cambridge University Press, Cambridge, United Kingdom, 996 p.

Sprockett, D.D., Piontkivska, H., Blackwood, C.B., 2011. Evolutionary analysis of glycosyl hydrolase family 28 (GH28) suggests lineage-specific expansions in necrotrophic fungal pathogens. Gene 479, 29–36.

Srinivasiah, S., Bhavsar, J., Thapar, K., Liles, M., Schoenfeld, T., Wommack, K.E., 2008. Phages across the biosphere: contrasts of viruses in soil and aquatic environments. Res. Microbiol. 159, 349–357.

Strickland, M.S., Lauber, C., Fierer, N., Bradford, M.A., 2009. Testing the functional significance of microbial community composition. Ecology 90, 441–451.

Stromberger, M.E., Keith, A.M., Schmidt, O., 2012. Distinct microbial and faunal communities and translocated carbon in *Lumbricus terrestris* drilospheres. Soil Biol. Biochem. 46, 155–162.

Stubblefield, S.P., Taylor, T.N., 1988. Recent advances in palaeomycology. New Phytol. 108, 3–25.

Swanson, M.M., Reavy, B., Makarova, K.S., Cock, P.J., Hopkins, D.W., Torrance, L., Koonin, E.V., Taliansky, M., 2012. Novel bacteriophages containing a genome of another bacteriophage within their genomes. PLoS ONE 7, e40683.

Thomas, C.M., Nielsen, K.M., 2005. Mechanisms of, and barriers to, horizontal gene transfer between bacteria. Nat. Rev. Microbiol. 3, 711–721.

Tilman, D., 1982. Resource Competition and Community Structure. Princeton University Press, Princeton, 296 p.

Tilman, D., 1996. Biodiversity: population versus ecosystem stability. Ecology 77, 350–363.

Tripathi, B.M., Kim, M., Singh, D., Lee-Cruz, L., Lai-Hoe, A., Ainuddin, A.N., Go, R., Rahim, R.A., Husni, M.H.A., Chun, J., Adams, J.M., 2012. Tropical soil bacterial communities in Malaysia: pH dominates in the equatorial tropics too. Microb. Ecol. 64, 474–484.

van der Heijden, M.G., Bargett, R.D., Van Straalen, N.M., 2008. The unseen majority: soil microbes as drivers of plant diversity and productivity in terrestrial ecosystems. Ecol. Lett. 11, 296–310.

van der Putten, W.H., Klironomos, J.N., Wardle, D.A., 2007. Microbial ecology of biological invasions. ISME J. 1, 28–37.

Waring, B.G., Averill, C., Hawkes, C.V., 2013. Differences in fungal and bacterial physiology alter soil carbon and nitrogen cycling insights from meta-analysis and theoretical models. Ecol. Lett. 16, 887–894.

Watkinson, S., Bebber, D., Darrah, P., Fricker, M., Tlalka, M., Boddy, L., 2006. The role of wood decay fungi in the carbon and nitrogen dynamics of the forest floor. In: Gadd, G.M. (Ed.), Fungi in Biogeochemical Cycles. Cambridge University Press, Cambridge, pp. 151–181.

Wiens, J.A., 1997. Metapopulation dynamics and landscape ecology. In: Hanski, I., Gilpin, M.E. (Eds.), Metapopulation Biology: Ecology, Genetics, and Evolution. Academic Press, San Diego, pp. 43–61.

Wolfe, B.E., Rodgers, V.L., Stinson, K.A., Pringle, A., 2008. The invasive plant *Alliaria petiolata* (garlic mustard) inhibits ectomycorrhizal fungi in its introduced range. J. Ecol. 96, 777–783.

Wu, X.L., Conrad, R., 2001. Functional and structural response of a cellulose-degrading methanogenic microbial community to multiple aeration stress at two different temperature. Environ. Microbiol. 3, 355–362.

Wu, J., Loucks, O.L., 1995. From balance of nature to hierarchical patch dynamics: a paradigm shift in ecology. Q. Rev. Biol. 70, 439–466.

Young, I.M., Crawford, J.W., Nunan, N., Otten, W., Spiers, A., 2008. Microbial distribution in soils: physics and scaling. Adv. Agron. 100, 81–121.

Chapter 11

Plant-Soil Biota Interactions

R. Balestrini, E. Lumini, R. Borriello and V. Bianciotto

Istituto per la Protezione Sostenibile delle Piante, UOS Torino, Viale Mattioli, 10125 Torino, Italy

Chapter Contents

I SOIL BIOTA

Soil functions are essential for the biosphere, and yet soil is perhaps the most difficult and least understood matrix. The main ecological functions of soil include nutrient cycling, C storage and turnover, water maintenance, soil structure arrangement, regulation of aboveground diversity, biotic regulation, buffering, and the transformation of potentially harmful elements and compounds (e.g., heavy metals and pesticides; Haygarth and Ritz, 2009). "Soil biota" is a term that refers to the complete community within a given soil system (Jeffery et al., 2010), although biodiversity can vary from soil to soil (i.e., grassland, arable, forest) and also among different plant species (Wardle et al., 2004). The most dominant group in soil biota, both in terms of number and biomass, is represented by microorganisms (i.e., bacteria, archae, and fungi), which show different nutrient strategies and lifestyles (saprotrophs, pathogens, symbionts). The rhizosphere also contains nematodes, microarthropods (mites and collembola), enchytraeids, and earthworms. Nematodes are considered very important in the soil food web. Only a few nematode species are considered pest organisms, which can cause severe damage to crops (i.e., soybean), whereas others have an applicative interest for the control of insect pests without using pesticides (Neher, 2010).

The soil surface and litter layer also contain numerous macrofauna species (mainly arthropods: beetles, spiders, diplopods, chilopods, snails). Because soil functionality depends on the activity of soil biota, measures of their activity, biomass, and diversity (including the presence and health of root symbionts) have often been proposed as indicators of soil quality.

A wide range of soil microbial diversity and activity remains unexplored despite new strategies and more sophisticated methods for characterizing plant-soil interfaces in terms of variety and structure.

II THE RHIZOSPHERE: ECOLOGICAL NETWORK OF SOIL MICROBIAL COMMUNITIES

Life is patchily distributed in soil, with a tendency for many soil microorganisms to live in aggregates and to form spots of activity (Berg and Smalla, 2009). One of the most important hot spots of activity and diversity in soils is the rhizosphere (Jones and Hinsinger, 2008). The term *rhizosphere* was first introduced by Hiltner (1904) to describe the portion of soil in which microorganism-mediated processes are under the influence of the root system. The rhizosphere only extends a few millimeters from the root surface (Girlanda et al., 2007). However, it can contain up to 10^{11} microbial cells per gram of roots (Berendsen et al., 2012), with the collective microbial community being referred to as the root microbiome. Plants and their microbiomes can be considered as "superorganisms" due, in part, to their reliance on soil microbiota for specific functions and traits (Mendes et al., 2011; van der Heijden et al., 2008).

The release of organic C compounds from plant roots into the surrounding environment is known as rhizodeposition (Koranda et al., 2011). Some plant species allocate 20-50% of their photosynthate to the roots and also release organic C (Bais et al., 2006; Dennis et al., 2010; Jones et al., 2009). This C ranges from 5% to 21% of photosynthetically fixed C (Marschner, 1995) as root exudates, root turnover, and the sloughing off of cells (Dennis et al., 2010). It also feeds microbial communities, provides the resources for root-associated organisms (root herbivores, pathogens, and symbiotic root microbes), and influences their activity and diversity. It is crucial to understand the factors that influence the microbial communities in this habitat due to the importance of plant-organism interactions in the rhizosphere for belowground C flow and allocation, ecosystem functioning, and nutrient cycling (Calderón et al., 2012; Singh et al., 2004). Some negative effects have also been observed (i.e., on soil organic matter (SOM) decomposition) as a consequence of plant (inter specific) competition for nutrients (such as N) and/or low water availability (Koranda et al., 2011; Mol-Dijkstra et al., 2009).

The impact of the root microbiome on plant health can be observed in disease-suppressive soils in which crop plants suffer less from specific soil-borne pathogens than expected. Despite the great interest in this phenomenon, the microbes and mechanisms involved in pathogen control are still largely unknown. Recently, Mendes et al. (2011) identified the key bacterial taxa and genes involved in the suppression of the fungal root pathogen *Rhizoctonia solani* using PhyloChip-based metagenomics of the rhizosphere microbiome with culture-based functional approaches. The authors isolated metagenomic DNA from the sugar beet rhizosphere microbiota; more than 33,000 bacterial and archaeal species were detected, with Proteobacteria, Firmicutes, and Actinobacteria consistently being associated with disease suppression. They concluded that although the Proteobacteria produced an antifungal compound, the presence of the complex community also played a crucial role in protecting roots.

Bulgarelli et al. (2012) and Lundberg et al. (2012) have provided a comprehensive picture of the *Arabidopsis* root microbiome and its relationship with the soil microbial community. The rhizosphere and the root endophytic microbiomes, including the species that infiltrate the root cortex and live as endophytes until their release back into the soil on root senescence, were considered using 454 sequencing (Roche) of 16S rRNA gene amplicons to compare soil, the rhizosphere, and endophytic bacterial communities. Rhizosphere communities are less diverse than those of bulk soil, suggesting that plant roots, either directly or indirectly, select for specific organisms for growth. In both these studies, the endophytes were a smaller, distinct group (50% fewer species identified than in the rhizosphere), dominated by Actinobacteria, followed by Proteobacteria. Firmicutes, Bacteroidetes, and Cyanobacteria. Even though the soil type is an important force that shapes microbial communities, plant species/genotype can select similar communities in different soils (Berendsen et al., 2012).

It has recently been reported that diverse soil microbiomes have a different impact on plant growth and leaf metabolome composition in *Arabidopsis thaliana* (Badri et al., 2013). Moreover, plant inoculation has been shown to lead to a significant inhibition of herbivore insect larval feeding in comparison with control plants using several tested soil microbiomes.

A great deal of effort has also been dedicated to characterizing the other microbial communities (i.e., fungi) that are involved in the rhizosphere. Many rhizospheric fungi establish diverse interactions with plants, ranging from antagonistic to mutualistic. Some of the fungi residents in the rhizosphere (i.e., root endophytes) colonize the plant root without forming typical mycorrhizal associations (see Section IV) and can also colonize the host cells without visible disease symptoms. They can also interact with host-root cells in several ways along the saprotrophy-biotrophy continuum, dependent on the plant species and environmental conditions (Peterson et al., 2008). The molecular mechanisms involved in these plant-fungus interactions are not well known, and the question about their role is still open. Dark septate, endophytic fungi (DSE; e.g., *Phialocephala* genus) are frequent colonizers of the roots of forest trees, shrubs, orchids, and a broad range of plants in alpine and subalpine habitats in northern temperate regions (Grünig et al., 2002).

Fungi belonging to the Sebacinales order (Basidiomycota) are known to be involved in a variety of mycorrhizal symbioses (Weiß et al., 2011). A few members of this fungal order do not form typical mycorrhizal structures but have raised considerable interest because of their ability to enhance plant growth and increase the resistance of their host plants to abiotic stress factors and fungal pathogens. For this reason, they are commonly referred to as "endophytes." These fungal endophytes have the ability to colonize the roots of a variety of host plants: bryophytes, pteridophytes, and several families of herbaceous angiosperms, including liverworts, wheat, maize, and the model plant *A. thaliana* (Brassicaceae). The model organism of this group of fungi is *Piriformospora indica*, which was originally isolated from the soil in the Indian Thar Desert (Weiß et al., 2011). Plants colonized by *Piriformospora* display beneficial effects, including enhanced host growth and resistance to biotic/abiotic stresses and improved nitrate and phosphate assimilation (Zuccaro et al., 2011). A beneficial effect of the filter-sterilized culture filtrate of *P. indica* has been reported to increase the productivity of lignans in hairy root cultures of *Linum album* (Kumar et al., 2012). The genome of this fungus (25 Mb) has recently been reported in association with transcriptional responses during the colonization of live and dead barley roots (Zuccaro et al., 2011). A comparative analysis of this genome with those of other Basidiomycota and Ascomycota fungi with different lifestyle strategies highlighted features of both biotrophism and saprotrophism.

Some saprotrophic fungal species can also play a beneficial role on plants and soil by functioning as natural biocontrol agents. *Trichoderma* spp., the most widely applied biocontrol fungi, has been extensively studied using a variety of research approaches (Lorito et al., 2010). Together with other beneficial microbes (bacteria), they help to maintain the general suppressiveness of the disease and the fertility of soils. Biocontrol, *Trichoderma* species are active ingredients in many commercial biofungicides used to control a range of economically important fungal plant pathogens (Harman, 2000). The antagonistic activity of *Trichoderma* strains is attributable to complex mechanisms, including the activity of cell wall-degrading enzymes, direct competition for nutrients, antibiosis, and the induction of plant defense responses at both local and systemic levels (Harman et al., 2004). The species *Fusarium oxysporum* is well represented among the rhizosphere, soil-borne fungal communities throughout the world. All strains of *F. oxysporum* are able to grow and survive for long periods of time on SOM and in the rhizosphere of many plant species (Garrett, 1970). Although all strains exist saprophytically, some are also well known for inducing wilt or root rot on plants, whereas others are considered as nonpathogenic. The interactions between pathogenic and nonpathogenic strains result in the control of the disease in suppressive soils. They are able to compete for nutrients in the soil and for infection sites on the root and can trigger plant-defense reactions, thus inducing systemic resistance (Fravel et al., 2003).

The development of "-omics" tools, and a greater understanding of the biology of rhizosphere microbes, have helped to demonstrate the contribution of the root microbiome to both disease resistance and nutrient cycling. The knowledge generated by these studies provides new opportunities for the application of these beneficial agents to optimize plant health, nutrition, and yields in sustainable agriculture and forestry (Chaparro et al., 2012).

III   NEW INSIGHTS INTO ROOT/SOIL MICROBES THROUGH METAGENOMIC/METATRANSCRIPTOMIC APPROACHES

The extraction and direct sequencing of DNA from environmental samples, known as "environmental genomics" or "metagenomics" (Hugenholtz and Tyson, 2008), have provided information on microbial diversity in several environments. This approach, coupled with the introduction of "high-throughput" sequencing techniques, such as 454 sequencing platforms, has greatly changed the analysis of environmental samples in terms of yields, cost, and speed and has resulted in thousands of sequences being obtainable in a short time. These techniques were first used to characterize bacterial communities in different environmental compartments (Simon and Daniel, 2011), and only during the last few years has this approach been applied to the description of fungal biodiversity. Soil is a huge repository of biodiversity, with several billion prokaryotic and eukaryotic microorganisms. In terms of biomass in the soil, fungi are the dominant eukaryotics among microorganisms and are involved in: (1) decomposition and mineralization processes, (2) pathogenic and symbiotic interactions with plant roots, (3) controlling soil structure, and (4) regulating aboveground biodiversity. The soil biodiversity in an ecosystem is closely related to the biological functioning of the soil, and the investigation of the environmental factors that act on fungal communities is essential for the ecological characterization of a site (Orgiazzi et al., 2012). The first studies on fungal diversity using a metagenomic approach in combination with high-throughput technology appeared in 2009 (Buée et al., 2009; Jumpponen and Jones, 2009). During the past several years, additional investigations have focused their attention on soil fungal communities from different biomes (Table 11.1). As a result, an impressive unexpected fungal biodiversity has been discovered, and among the huge number of fungal DNA sequences that have been obtained, many have not yet been identified (environmental DNA sequences).

Even though molecular approaches can provide information on new species, traditional taxonomy methods, based on morphology and culture isolation, cannot be abandoned. For example, among the different fungal lineages only derived from environmental DNA sequences, a group of ubiquitous soil fungi that often dominate belowground fungal assemblages has uncertain assignment in the fungal tree of life. Rosling et al. (2011) have been able to cultivate, for the first time, fungi belonging to the aforementioned enigmatic group. Archaeorhizomycetes were previously only known from soil environmental DNA sequences, suggesting that they play a role in decomposition. The exact ecological niches and the complete life cycle of this group still remain unknown, but

TABLE 11.1 Overview of papers published on soil fungal assemblages using Next Generation Sequencing (updated July 2013)

Reference	Ecosystem/ Environment	Target Region	Major Findings
Buée et al. (2009)	Six different forest soils/France	ITS1	This study has validated the effectiveness of the NGS 454 platform for the surveying of soil fungal diversity.
Öpik et al. (2009)	Boreal forest roots/ Estonia	SSU	An increase in the root AM fungi identified in a forest from those previously observed using cloning and sequencing approaches.
Jumpponen et al. (2010a)	Different strata/ Tallgrass prairie soils	ITS2	Fungal community richness and diversity decline with soil depth and the communities are distinct among the strata.
Jumpponen et al. (2010b)	*Quercus* spp. ECM/urban and rural sites	ITS1	Identification of seasonal trends within fungal communities.
Lim et al. (2010)	Island soils/ Yellow Sea, Korea	SSU	High diversity of soil fungi has been reported. The dominant Ascomycota species have been described as lichen-forming fungi litter/wood decomposer plant parasites, endophytes, and saprotrophs while most of the Basidiomycota sequences belonged to ECM and wood rotting fungi.
Lumini et al. (2010)	AM symbiotic fungi, Mediterranean soils/Italy	SSU	This study reports a large AMF-soil-based sequence dataset for the first time and describes the differences in AM fungal communities along a land-used gradient.
Ovaskainen et al. (2010)	Saw dust from spruce logs/ Finland	ITS1	A modeling approach for the correct identification of a given taxonomic level has been developed which focuses on dead wood-inhabiting fungi.

TABLE 11.1 Overview of papers published on soil fungal assemblages using Next Generation Sequencing (updated July 2013)—Cont'd

Reference	Ecosystem/ Environment	Target Region	Major Findings
Rousk et al. (2010)	Different pH spots/arable soil	SSU	The fungal community composition was less affected by pH in relation to the bacterial community suggesting that fungi exhibit wider pH ranges for growth.
Sugiyama et al. (2010)	Organic and conventional potato farms/ Colorado Soils	SSU	Slightly higher diversity and evenness have been observed inside the microbial community in organic farms compared to conventional farms. Quantitative-PCR has been performed on three potato pathogens to validate the 454 data.
Tedersoo et al. (2010)	Tropical mycorrhizal fungi; ECM/ Cameroon	ITS1	Comparison of 454 pyrosequencing and Sanger sequencing techniques for the identification of a tropical fungal community. Technical biases have also been discussed.
Dumbrell et al. (2011)	Temperate grassland/Roots	SSU	This study reveals distinct root AM fungal communities in winter and summer. A seasonally changing supply of host-plant carbon mirroring changes in temperature and sunshine hours may be the driving force behind changes in AM fungal assemblages.
Lentendu et al. (2011)	Alpine tundra habitats/Soils	ITS1	This study shows that fungal beta-diversity patterns can be explained by environmental conditions demonstrating that diversity estimation based on 454 FLX sequencing is robust to support ecological studies. The ecological significance of rare/unidentified molecular taxa has been reported confirming the existence of a fungal rare biosphere.

Continued

TABLE 11.1 Overview of papers published on soil fungal assemblages using Next Generation Sequencing (updated July 2013)—Cont'd

Reference	Ecosystem/ Environment	Target Region	Major Findings
Mello et al. (2011)	Truffle-grounds/ Soil	ITS1/ ITS2	This is the first report on a comparison between two ITS regions on soil fungal populations.
Arfi et al. (2012)	Mangrove trees	ITS1/ ITS2/ SSU	Host specificity has been reported to be a key factor in the distribution of the fungal communities and their diversity in both aerial and intertidal parts of the trees.
Becklin et al. (2012)	Rhizosphere from alpine plant species/Roots	SSU	Plant host identity has a stronger effect on rhizosphere fungi than on habitat with even a stronger effect for the symbiotic AM fungal community.
Blaalid et al. (2012)	ECM from herb *Bistorta vivipara* along a primary succession gradient in front of a glacier/Norway	ITS1	High fungal diversity dominated by basidiomycetes has been found in *B. vivipara* root systems. The total fungal diversity increased slightly initially, with a significant increase towards the climax of vegetation.
Daghino et al. (2012)	Serpentine soils	ITS1/ ITS2	First meta-genomic description of saprotrophic fungal diversity in serpentine substrates. The results indicate the absence of correlation between the substrate and the fungal community assemblages.
Dai et al. (2012)	Wheat fields/ Canadian prairie soils	SSU	Although some AM fungal taxa were evenly distributed in all examined soils, different AM fungi showed varied distribution patterns in several Canadian Chernozem soils.

TABLE 11.1 Overview of papers published on soil fungal assemblages using Next Generation Sequencing (updated July 2013)—Cont'd

Reference	Ecosystem/ Environment	Target Region	Major Findings
Danielsen et al. (2012)	Young transgenic poplar plantation/ Soil and roots	ITS1	This study has revealed that the predominant groups in soil are saprophytic pathogenic and endophytic fungi. AM diversity is higher in the soil than in the roots, while poplar roots were dominated by ECM fungi. The results suggest that roots and soil constitute distinct ecological fungal biomes.
Davison et al. (2012)	Natural forest ecosystem/ Estonian soils	SSU	Heterogeneous vegetation of the natural forest study system has an impact on the spatial structure in soil AM fungal communities. The results suggest that natural ecosystems contain a relatively constant pool of AMF taxa in the soil from which plants select species to form mycorrhizae during the year.
Ihrmark et al. (2012)	Artificial and natural communities/ Soils and roots	ITS2	The authors have presented three new primers to amplify the fungal ITS2 region for 454-sequencing.
Kubartová et al. (2012)	Decaying spruce logs/ Norway	ITS2	The results indicate that the highly abundant fungal fruiting species may respond to environmental characteristics in different ways from other fungal species.
Lin et al. (2012)	Arable soils/North China	SSU	Shifts in AM fungal assemblage in soil mirrored long-term fertilization. Long-term P and N fertilization significantly decrease AM fungal community richness and diversity.

Continued

TABLE 11.1 Overview of papers published on soil fungal assemblages using Next Generation Sequencing (updated July 2013)—Cont'd

Reference	Ecosystem/ Environment	Target Region	Major Findings
Orgiazzi et al. (2012)	Land-use background soils/ Mediterranean, Italy	ITS1/ ITS2	The data suggest that investigation of the belowground fungal community may provide useful elements on aboveground features, such as vegetation coverage and agronomic procedures.
Verbruggen et al. (2012)	Bt maize plants/ Soils	SSU	Consistent changes have not been found in AM fungal communities associated to transgenic and non-transgenic maize cultivars.
Xu et al. (2012a)	Pea fields/Roots, rhizosphere and soil	ITS1	Both plant health status and field have significant effects on the fungal community structure in roots, while only field shapes the rhizosphere and soil fungal communities.
Xu et al. (2012b)	Pea field with a soil health gradient	ITS1	The study shows that fungal communities in soils with diseased plants are significantly different from soils with healthy plants.
Yu et al. (2012)	*Pisum sativum* roots	ITS1	Succession patterns and high diversity have been shown for pea-root associated fungal communities during the plant growth cycle.
Blaalid et al. (2013)	ECM from herb *Bistorta vivipara*	ITS1/ ITS2	Evaluation of the usability of ITS1 and ITS2 regions as DNA metabarcoding markers for fungi.
Clemmensen et al. (2013)	Boreal forest islands/ Sweden	ITS2	This study, using a 454 sequencing approach together with stable isotope analyses, highlights the central role of root-associated fungi in soil C sequestration in boreal forests (see the Perspective by Treseder and Holden, 2013).

TABLE 11.1 Overview of papers published on soil fungal assemblages using Next Generation Sequencing (updated July 2013)—Cont'd

Reference	Ecosystem/ Environment	Target Region	Major Findings
Öpik et al. (2013)	Root samples/six continents and five climatic zones	SSU	This study improves the information about AM fungi distribution and reports a new global dataset targeting previously unstudied geographical areas.
Orgiazzi et al. (2013)	Heterogeneous ecosystems/ Europe	ITS1/ ITS2	The largest attempt so far to comprehensively identify soil fungal beta-diversity patterns at multiple scales over long distances using NGS methods. Comparative 454 data analysis has evidenced that spatial distance acts on the soil fungal assemblages. It is reported that the nature of plant cover is one of the main environmental factors that separate most of the examined soil samples.

AM, arbuscular mycorrhizae/al; ECM, ectomycorrhizae/al.

their isolation, *in vitro* cultures, and the possibility of having genome sequencing, will help elucidate their role in the ecosystems in which they are present.

Metagenomics is a DNA-based approach and therefore cannot differentiate between expressed and nonexpressed genes, thereby failing to reflect the actual metabolic activity (Sorek and Cossart, 2010). For this reason, metagenomics studies have recently been supplemented by metatranscriptomics, which is defined as the study and exploration of the transcriptomes that are present in an environmental sample (Bailly et al., 2007; Damon et al., 2011, 2012). Increased information on the genes involved in specific pathways, and the greater number of available fungal genome sequences (Grigoriev et al., 2011), have allowed for a faster progression in this area of research. This progression could also be attributable to the possibility of directly processing RNA extracted from the soil. These studies have increasingly examined the richness and composition of expressed functional gene-coding for key enzymes in C and N cycling (e.g., cellobiohydrolase, laccase, nitrate reductase) in forest and agricultural environments (Baldrian et al., 2012; Edwards et al., 2011; Gorfer et al., 2011; Kellner and Vandenbol, 2010; Kellner et al., 2011; Weber et al., 2011).

Several large-scale metatranscriptomics projects have the aim of exploring the interaction between plants and soil fungi communities, including mycorrhizal symbionts and saprotrophic soil fungi. The increasing availability of reference sequences for different fungal species and genes should allow the functional fungal components inside ecosystems to be identified in addition to the ways fungal communities change in response to various environmental perturbations to be highlighted.

IV   AN IMPORTANT RHIZOSPHERIC COMPONENT: THE MYCORRHIZAL FUNGI

Mycorrhizal fungi are specialized root symbionts that engage in intimate associations with a great diversity of plants (Balestrini et al., 2012; Smith and Read, 2008; Fig. 11.1). Mycorrhizal associations are also found between fungi and the underground gametophytes of many bryophytes and pteridophytes, as well as the sporophytes of most pteridophytes and the roots of seed plants (Bonfante and Genre, 2008; Smith and Read, 2008).

The associations between soil mycorrhizal fungi and roots, referred to as mycorrhizae, are an essential feature of the biology of most terrestrial plants. The success of mycorrhizal fungi is linked to the nutritional benefits that they confer to their plant hosts; they take up phosphate (Pi), N, and other macronutrients, in addition to microelements and water from the soil, and then deliver them to the plant. The fungus receives photosynthetic carbohydrates. The extra-radical, mycorrhizal mycelium that grows outside the roots captures water and nutrients from the soil and represents an important sink for host C. Carbohydrates and mineral nutrients are exchanged across the interface between the plant and the fungus inside the roots (Smith and Smith, 1990). Mycorrhizal fungi can also perform other significant roles, including protection of the plant

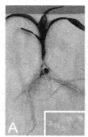

FIG. 11.1   Mycorrhizae features in different plant/fungus interactions. (A) Maize roots colonized with arbuscular endomycorrhizal (AM) fungi, which form intracellular structures. Inset: (A) AM fungal spores. (B) *Orchis purpurea* plant colonized by endomycorrhizal fungi from natural grassland. (C) *Corylus avellana/Tuber melanosporum* ectomycorrhizae (ECM) with a typical clavate aspect. (*Image B courtesy of Enrico Ercole.*)

from biotic and abiotic stress (i.e., altering host environmental tolerances to water deficit or pollutants, or by reducing susceptibility to pathogens; Aroca et al., 2007; Smith and Read, 2008). They are essential for plant performance in both natural and agricultural ecosystems and have been shown to have a beneficial effect on growth and productivity under salt stress conditions (Estrada et al., 2012).

The variety of mycorrhizal associations established between plants and fungi can be divided into two main types (endo- and ectomycorrhizae), primarily on the basis of the structures formed inside or outside the plant roots and the taxonomic position of the symbiotic partners (Smith and Read, 2008). Approximately two-thirds of the mycorrhizal plants form endosymbiosis with arbuscular mycorrhizal (AM) fungi and represent the most common mycorrhizal association. Ericoid mycorrhizal fungi form endomycorrhizae with plants of the genus *Ericales* and play an important ecological role as efficient SOM degraders, but are mainly restricted to heathlands (Read and Perez-Moreno, 2003). Ectomycorrhizae (ECM) are made up of a relatively small number of plants, about 8000, mostly woody plants, yet have an extensive occupancy of biomes (Plett and Martin, 2011). Through mutualistic symbioses with ECM fungi, these tree species have acquired metabolic capabilities that have allowed otherwise unavailable ecological niches to be utilized.

On the fungal side, the phylogenetic classification of this group of mycorrhizal fungi (Arbuscular, Ericoid, Ecto-) has revealed that they are formed by a different range of fungi. Arbuscular mycorrhizal fungi belong to Glomeromycota on the basis of nuclear ribosomal RNA phylogeny (Schüßler et al., 2001) with about 220 species. The AM fungal taxonomy is in continual evolution (Oehl et al., 2011; Schüßler and Walker, 2010). These fungi are obligate biotrophs and represent 5-10% of the soil microbial biomass of appropriate soils (Fitter et al., 2011). Even though AM fungal spores can germinate in the absence of host plants (presymbiotic phase), they depend on the establishment of intracellular symbiosis (Fig. 11.2) to complete the fungal life cycle and

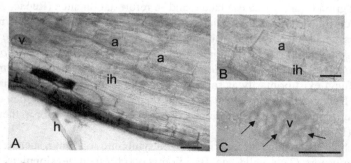

FIG. 11.2   Grapevine roots colonized by AM fungi (cotton blue staining). a, arbuscules; h, extraradical hypha; ih, intraradical hyphae; v, vesicles. Arrows point to lipid droplets inside a vesicle. Bars = (A) 40 μm; (B and C) 15 μm.

produce the next generation of spores. In agricultural systems, AM fungi form associations with almost all important crops (i.e., maize, wheat, soybean, rice, tomato), with applicative potential in low-input agriculture.

The AM fungi quickly receive plant-fixed C and then gradually release it to the rest of the microbial community present in the (myco-) rhizosphere, thus demonstrating its prominent role in the transfer of C between plants and soil (Drigo et al., 2010; Fitter et al., 2005). The host plant allocates C to the cooperative fungal partners, which in turn improves nutrient (phosphate) transfer to the roots (Kiers et al., 2011). The AM fungi have also been associated with soil aggregation through both the extensive, extraradical mycelium and an AM fungal glycoprotein called glomalin-related soil protein (GRSP) (Rillig, 2004). Glomalin, the authentic gene product, has been characterized as a putative homolog of heat shock protein (HSP) 60 (Hammer and Rillig, 2011). The GRSP represents a potentially important soil organic C pool (Rillig et al., 2003) and has been reported to have an impact on soil structure by increasing the stability of soil aggregates (Maček et al., 2012; Rillig and Mummey, 2006).

The ECM fungi, which are best known for their fruiting structures (e.g., truffles, bolets, amanitas, chantarelles), often grow next to tree trunks in woodlands and colonize their roots. However, ECM roots often contain fungi that cannot be linked to epigeous fruiting bodies (Anderson and Cairney, 2007). These fungi involve at least 6000 species, including a large number of Basidiomycota and some Ascomycota (Smith and Read, 2008). Most ECM fungi and plants can associate with multiple partners, although specialists occur in ECM symbiosis in both fungi and plants (Kennedy et al., 2012; Molina et al., 1992). During the symbiotic phase, ECM fungi form a fungal sheath (the mantle) that develops outside the root (Fig. 11.3). This mycelium is linked to extramatrical hyphae that are involved in substrate exploration, mineral nutrition, and water uptake. Some hyphae penetrate between the epidermal and outer cortical cells to form an intercellular hyphal network inside the root tissues (the Hartig net). The early growth and survival of forest plantations can be improved by inoculating seedlings with selected ECM fungal strains in soils deficient in ECM fungi (polluted areas, agricultural, or treeless lands) and in reforestation sites (Repáč, 2011).

Mycorrhizal fungi that associate with members of the Ericaceae and related families are all Ascomycota (Smith and Read, 2008), such as *Oiodendron maius* and fungi belonging to the Helotiales (i.e., *Meliniomyces bicolor*, *Meliniomyces variabilis*, and *Rhizoscyphus ericeae*). It is not certain whether ericoid mycorrhizal fungi exist primarily as soil saprotrophs or as mycorrhizal associates of plants. If they are less dependent on plants than AM or ECM fungi, their capacity to form mycorrhizal associations would not be a driving factor in their evolution (Brundrett, 2002).

Orchids also require an association with symbiotic soil fungi, such as *Tulasnella calospora* (Tulasnellaceae, Basidiomycota) for their development, which support seed germination and can assist the adult plant growth (Selosse and Roy, 2009; Waterman and Bidartondo, 2008).

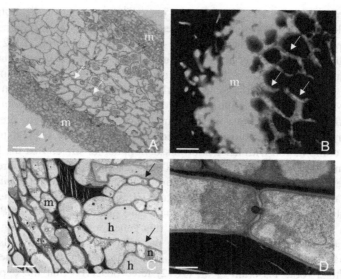

FIG. 11.3 Truffle ectomycorrhizae. (A) Semithin section of an ectomycorrhiza (ECM) of *Corylus avellana* and *Tuber melanosporum* under a light microscope after staining with Toluidine blue, showing the mantle (m) developed around the root and the Hartig net compartment among host cells (arrows). Arrowheads point to the external mycelium. (B) Section of a *Tuber* spp. ECM under a confocal microscope after treatment with Wheat Germ Agglutinin (WGA)-FITC to localize chitin in the fungal cell wall (green signal). The loose external and the compact inner mantle (m) ensheathing the root, and the Hartig net mycelium progressing among host cells (arrows) are seen. (C) Ultrastructural view of a *C. avellana/T. melanosporum* ECM. Arrows point to the Hartig net mycelium. H, host cell; m, mantle; n, fungal nucleus. (D) Magnification of the mantle region. A typical septum of Ascomycota is visible. Bars = (A) 12 μm; (B) 10 μm; (C) 3.2 μm; (D) 1.3 μm.

Physiological and molecular targeted approaches, and more recently "-omics" investigations, have been useful for unraveling the mechanisms at the base of nutrient acquisition in mycorrhizal symbioses (Balestrini and Lanfranco, 2006; Sawers et al., 2008). Due to the agronomic potential of these associations in low-impact agriculture, several studies have focused on the nutritional exchanges that occur between symbionts in AM symbiosis with improved phosphate (Pi) uptake considered the main benefit for plants (Bucher, 2007; Javot et al., 2007). The fungal phosphate transporters expressed in the extraradical mycelium (Benedetto et al., 2005; Harrison and Buuren, 1995; Maldonado-Mendoza et al., 2001; Requena et al., 2003) are involved in the uptake of Pi from the substrate. Phosphate, taken up by the extraradical mycelium from soil solutions, is translocated through the AM fungal hyphae as polyphosphate (PolyP).

The fast synthesis of PolyP is crucial for efficient hyphal Pi uptake (Tani et al., 2009). PolyP is then released as free Pi into the periarbuscular space, and the subsequent import of Pi across the periarbuscular membrane into the plant cell is mediated by plant phosphate transporters (PTs).

Mycorrhiza-specific PTs, which are responsible for plant Pi uptake in arbuscule-containing cells, have been characterized in several plants (Glassop et al., 2005; Guether et al., 2009a; Harrison et al., 2002; Nagy et al., 2005; Paszkowski et al., 2002; Rausch et al., 2001). The best characterized symbiotic plant PT is the *Medicago truncatula* low-affinity transporter MtPT4, which is located in the periarbuscular membrane (Harrison et al., 2002). MtPT4 activity is required for AM symbiosis; in mtPT4 mutants, which lack MtPT4 function, arbuscules accumulate PolyP and run into premature senescence; thus the symbiosis is blocked (Javot et al., 2007). Gene expression data, in combination with laser-microdissection, demonstrated that the five, so far, identified *LePT* genes in tomato (Nagy et al., 2005) are differentially expressed in cortical cell populations during the interaction with *Glomus mosseae* (Balestrini et al., 2007). The tomato PT transcripts are present inside the arbuscule-containing cells, suggesting that plants guarantee maximum P uptake through a functional redundancy inside this gene family. On the fungal side, the presence of *GmosPT* transcripts in the arbuscule-containing cells offers confirmation of the previous results of Benedetto et al. (2005) on the whole mycorrhizal root. Tisserant et al. (2012) have observed that the expression pattern of *GintPT* in *Glomus intraradices* DAOM 197198 is in agreement with the results obtained for *GmosPT* in *G. mosseae* (Balestrini et al., 2007; Benedetto et al., 2005). These results suggest that the efflux of phosphate probably occurs in competition with its uptake and that the fungus may reabsorb part of the Pi released into the interface space (Balestrini et al., 2007; Tisserant et al., 2012). In addition to playing a role in P nutrition, studies on fungal physiology (Allen and Shachar-Hill, 2009; Govindarajulu et al., 2005; Leigh et al., 2009) have shown that AM fungi have the capacity to transfer substantial amounts of N and S from the soil to the host plant.

Sulfate, ammonium, nitrate, and other N-transporters, which may be involved in S and N uptake by the plant from the fungus, have been identified and characterized (Casieri et al., 2012; Guether et al., 2009a, b). A laser microdissection approach showed that transcripts corresponding to both plant and fungal ammonium transporter (AMT) genes, which are involved in the uptake of NH_4 (considered the most likely N form acquired by the plant in the AM interaction) are present in the arbuscule-containing cells. This suggests that competition might exist between the symbionts for the N present in the interfacial apoplast, as already suggested for Pi transport (Gomez et al., 2009; Guether et al., 2009a; Pérez-Tienda et al., 2011). It is well known that ECM fungi, in particular, play a pivotal role in the uptake of N by increasing the N nutrition of their host plants (Smith and Read, 2008). The ECM fungi, which are localized in organically enriched soil horizons, have been shown to have organic N uptake systems (Martin, 2007). In addition, AM and ECM mycelia are effective scavengers of either ammonium or nitrate with ECM mycelia particularly effective in ammonium uptake. Transporters, as well as assimilating enzymes involved in nitrate and ammonium uptake, have been found and characterized in ECM fungi, such as *Pisolithus, Laccaria,* and *Tuber* (Jargeat et al., 2003; Javelle et al., 2003; Montanini et al., 2003; 2006) and in the AM fungus *G. intraradices* DAOM 197198 (Pérez-Tienda et al., 2011; Tisserant et al., 2012).

Sucrose, the primary transported carbohydrate in plants, can only be used as a C source by fungi if hydrolyzed by plant apoplastic invertases (Schaarschmidt et al., 2006; Wright et al., 1998). The plant cells involved in symbiosis have to acquire sugar either directly from the phloem or from the apoplast and then must release it at the symbiotic interface, where the fungus utilizes it. Examples of plant candidates for sugar transport to sink tissues are the AM-inducible hexose transporter MTST1 in *M. truncatula* (Harrison, 1996) and some sucrose transporter genes (SUT; Doidy et al., 2012), which have been suggested to be involved in sugar fluxes toward the fungal symbiont. The AM fungus can take up hexoses in the colonized cells (Pfeffer et al., 1999). A hexose transporter, MST2, from an AM fungus has only recently been characterized, suggesting an involvement in fungal nutrition during symbiosis (Helber et al., 2011). On the other hand, fungal hexose transporters, putatively involved in the C uptake at the symbiotic interface, have already been identified in ECM fungi, such as the model fungi *Amanita muscaria* and *Laccaria bicolor* (Nehls et al., 2010 and references therein). Elevated gene expression for these transporters has been observed in the fungal intraradical structure (Hartig net) during the interaction with the plant (Nehls et al., 2001).

The activation of several glucose transporters in the Hartig net compartment (e.g., the plant/fungus interface) has been demonstrated using laser-microdissection to dissect the two fungal compartments (i.e., mantle and Hartig net) in *Tuber melanosporum* ECMs (Hacquard et al., 2013). The *T. melanosporum* genome has been shown to also contain a gene encoding an invertase; the corresponding transcripts have been found in the Hartig net compartment (Hacquard et al., 2013; Martin et al., 2010). These results support the hypothesis that this Ascomycota ECM fungus has the ability to hydrolyze plant-derived sucrose, providing the substrate for the glucose transporters at the symbiotic interface (Hacquard et al., 2013). In contrast, most Basidiomycota ECM fungi have been shown to be reliant on host invertase to hydrolyze sucrose (Martin et al., 2008; Nehls et al., 2010).

The symbiotic fungus, in addition to the conversion of glycogen, trehalose, and/or mannitol (Nehls et al., 2010; Shachar-Hill et al., 1995), converts a high percentage of plant-derived C in lipids (Dalpé et al., 2012). These have been used as biochemical tool for mycorrhizal fungi quantification (Bååth et al., 2004; Ngosong et al., 2012) and turnover (Calderón et al., 2012). The metabolism of AM fungi, where lipids represent the main C form (mainly as triacylglycerol, TAG), has been investigated to verify the C allocation during symbiotic interactions (Dalpé et al., 2012). The intraradical mycelium acquires hexoses from the plant, which are then metabolized to lipids, mainly neutral lipids (such as TAG).

Tisserant et al. (2012) have recently identified transcripts that code for the α and β subunits of the fatty acid (FA) synthase complex in the intracellular mycelium of an AM fungus, in addition to the acetyl-CoA carboxylase transcript, confirming that AM fungi possess fatty acid synthetic capacities. Fatty acid 16:1ω5 is dominant in both membrane and storage lipids (Dalpé et al., 2012), and several studies have proposed it as a tool to quantify AM fungal biomass and to determine the distribution of AM fungi in soil (Ngosong et al., 2012; van Aarle and Olsson, 2003). The pathways related to FAs biosynthesis and degradation in ECM fungi have been reconstructed in *L. bicolor* using lipid composition, gene annotation, and transcriptional analysis (Reich et al., 2009). Nineteen different FAs have been detected in *Laccaria* mycelium, one of which was the FA enriched in AM fungi, even though at low concentrations.

Consistent with their extensive genetic variability, mycorrhizal associations show different functional abilities and offer distinct advantages to the host plant. Identification of the events that lead to the formation of the several mycorrhizal interactions, including the mechanisms involved in nutrient transport, is a necessary step for the exploitation of these associations in agro-forest programs.

V   THE CONTRIBUTION OF FUNGAL GENOME PROJECTS

Genome projects of several mycorrhizal fungi, in comparison with the genomes of fungi with different lifestyles, can reveal how evolution has played a crucial role in fashioning fungal genomes to suit their own specialization. The first sequenced genomes of the ECM *L. bicolor*, belonging to Basidiomycota, and the Ascomycota *T. melanosporum* have provided new insights into the mutually symbiotic relationship between the host plant and the fungus (Martin et al., 2008, 2010), but have revealed few features that are common to both symbiotic partners. This suggests that mutualistic symbiosis has evolved in different ways in Ascomycota and Basidiomycota (Martin et al., 2008, 2010; Plett and Martin, 2011). The Joint Genome Institute (JGI) established the Mycorrhizal Genomics Initiative, which targets ecologically diverse ranges of mycorrhizal fungi from different orders (Grigoriev et al., 2011; Plett and Martin, 2011). Analyses of these genomes, a few of which have already been released (http://genome.jgi.doe.gov), will offer a great opportunity to conduct a deeper investigation of the diversity of mechanisms involved in the different mycorrhizal symbiosis, including ecto- and endomycorrhizal associations (Marmeisse et al., 2013). The endomycorrhizal genome sequences of the ericoid fungus *O. maius* and the orchid *T. calospora* species are available.

In the absence of a complete genome sequence for members of the phylum Glomeromycota, new information has been derived from the availability of a genome-wide catalog of gene expression in *R. irregularis*, obtained from different fungal structures (germinated spores, extra- and intraradical mycelium, arbuscules; Tisserant et al., 2012). These data show that the AM fungus shares features with ECM symbionts and obligate biotrophic pathogens, such as up-regulation of the membrane transporters and small protein genes secreted during symbiosis, as well as a lack of expression of PCW polysaccharides-degrading enzymes. These data have been improved thanks to the release of the *R. irregularis* genome sequence (Tisserant et al., 2013; Lin et al., 2014) and the mitochondrial complete sequences of several AM fungal species (Formey et al., 2012; Lee and Young, 2009; Nadimi et al., 2012; Pelin et al., 2012). One of the objectives for conducting comparative analyses of several fungal genomes (Floudas et al., 2012) is to reconstruct the evolution of lignin degradation mechanisms. The results suggest that peroxidases, acting on lignin, expanded in the lineage of white rot species (leading to the Agaromycetes ancestor) and contracted in parallel lineages leading to brown rot and mycorrhizal species (Floudas et al., 2012).

VI SUMMARY

Recent studies have highlighted the importance of the relationship between the plant and soil/root microbiome and how this interaction impacts plant health and productivity (Chaparro et al., 2012; van der Heijden et al., 2008). It is known that there is a complex dialog between soil microbiome and plants, which can be mediated by root exudates. A role for volatile organic compounds (VOCs) has also recently been investigated (Insam and Seewald, 2010; Müller et al., 2013). However, the actors involved in this conversation, and the ways in which they interact with each other, must still be elucidated. Thanks to the development of "-omics" tools (genomics, transcriptomics, proteomics, metabolomics, volatilomics), which are continuously in evolution, the microbiome "universe" will one day be understood. A better understanding of the rules for soil microbe/plant interactions will open the way for the manipulation of specific microbiomes which will result in increased soil health and plant fertility (Chaparro et al., 2012).

ACKNOWLEDGMENTS

The research conducted by the authors of this chapter was funded by CNR (Commessa AG. P04.025, Premio DAA2009, Progetto Premiale 2011), BIOBITs Project (Regione Piemonte), FIORIBIO and PURE (FP7-265865) EU projects. This chapter was completed during the first part of 2013. It was not possible to further update the state of the art, with minor exceptions.

REFERENCES

Allen, J.W., Shachar-Hill, Y., 2009. Sulfur transfer through an arbuscular mycorrhiza. Plant Physiol. 149, 549–560.

Anderson, I.C., Cairney, J.W.G., 2007. Ectomycorrhizal fungi: exploring the mycelial frontier. FEMS Microbiol. Rev. 31, 388–406.

Arfi, Y., Buée, M., Marchand, C., Levasseur, A., Record, E., 2012. Multiple markers pyrosequencing reveals highly diverse and host-specific fungal communities on the mangrove trees *Avicennia marina* and *Rhizophora stylosa*. FEMS Microbiol. Ecol. 79, 433–444.

Aroca, R., Porcel, R., Ruiz-Lozano, J.M., 2007. How does arbuscular mycorrhizal symbiosis regulate root hydraulic properties and plasma membrane aquaporins in *Phaseolus vulgaris* under drought, cold, or salinity stresses? New Phytol. 173, 808–816.

Bååth, E., Nilsson, L.O., Göransson, H., Wallander, H., 2004. Can the extent of degradation of soil fungal mycelium during soil incubation be used to estimate ectomycorrhizal biomass in soil? Soil Biol. Biochem. 36, 2105–2109.

Badri, D.V., Zolla, G., Bakker, M.G., Manter, D.K., Vivanco, J.M., 2013. Potential impact of soil microbiomes on the leaf metabolome and on herbivore feeding behavior. New Phytol. 198 (1), 264–273. http://dx.doi.org/10.1111/nph.12124.

Bailly, J., Fraissinet-Tachet, L., Verner, M.-C., Debaud, J.-C., Lemaire, M., Wésolowski-Louvel, M., Marmeisse, R., 2007. Soil eukaryotic functional diversity, a metatranscriptomic approach. Int. Soc. Microb. Ecol. J. 1, 632–642.

Bais, H.P., Weir, T.L., Perry, L.G., Gilroy, S., Vivanco, J.M., 2006. The role of root exudates in rhizosphere interactions with plants and other organisms. Annu. Rev. Plant Biol. 57, 233–266.

Baldrian, P., Kolařík, M., Štursová, M., Kopecký, J., Valášková, V., Větrovský, T., Žifčáková, L., Šnajdr, J., Rídl, J., Vlček, Č., et al., 2012. Active and total microbial communities in forest soil are largely different and highly stratified during decomposition. Int. Soc. Microb. Ecol. J. 6, 248–258.

Balestrini, R., Lanfranco, L., 2006. Fungal and plant gene expression in arbuscular mycorrhizal symbiosis. Mycorrhiza 16, 509–524.

Balestrini, R., Gómez-Ariza, J., Lanfranco, L., Bonfante, P., 2007. Laser micro-dissection reveals that transcripts for five plant and one fungal phosphate transporter genes are contemporaneously present in arbusculated cells. Mol. Plant Microbe Interact. 20, 1055–1062.

Balestrini, R., Sillo, F., Kohler, A., Schneider, G., Faccio, A., Tisserant, E., Martin, F., Bonfante, P., 2012. Genome-wide analysis of cell wall-related genes in *Tuber melanosporum*. Curr. Genet. 58, 165–177.

Becklin, K.M., Hertweck, K.L., Jumpponen, A., 2012. Host identity impacts rhizosphere fungal communities associated with three alpine plant species. Microb. Ecol. 63, 682–693.

Benedetto, A., Magurno, F., Bonfante, P., Lanfranco, L., 2005. Expression profiles of a phosphate transporter gene (*GmosPT*) from the endomycorrhizal fungus *Glomus mosseae*. Mycorrhiza 15, 620–627.

Berendsen, R.L., Pieterse, C.M., Bakker, P.A., 2012. The rhizosphere microbiome and plant health. Trends Plant Sci. 17, 478–486.

Berg, G., Smalla, K., 2009. Plant species and soil type cooperatively shape the structure and function of microbial communities in the rhizosphere. FEMS Microbiol. Ecol. 68, 1–13.

Blaalid, R., Carlsen, T., Kumar, S., Halvorsen, R., Ugland, K.I., Fontana, G., Kauserud, H., 2012. Changes in the root-associated fungal communities along a primary succession gradient analysed by 454 pyrosequencing. Mol. Ecol. 21, 1897–1908.

Blaalid, R., Kumar, S., Nilsson, R.H., Abarenkov, K., Kirk, P.M., Kauserud, H., 2013. ITS1 versus ITS2 as DNA metabarcodes for fungi. Mol. Ecol. Resour. 13 (2), 218–224. http://dx.doi.org/10.1111/1755-0998.12065.

Bonfante, P., Genre, A., 2008. Plants and arbuscular mycorrhizal fungi: an evolutionary-developmental perspective. Trends Plant Sci. 13, 492–498.

Brundrett, M.C., 2002. Coevolution of roots and mycorrhizas of land plants. New Phytol. 154, 275–304.

Bucher, M., 2007. Functional biology of plant phosphate uptake at root and mycorrhiza interfaces. New Phytol. 173, 11–26.

Buée, M., Reich, M., Murat, C., Morin, E., Nilsson, R.H., Uroz, S., Martin, F., 2009. 454 Pyrosequencing analyses of forest soils reveal an unexpectedly high fungal diversity. New Phytol. 184, 449–456.

Bulgarelli, D., Rott, M., Schlaeppi, K., van Themaat, E.V.L., Ahmadinejad, N., Assenza, F., Rauf, P., Huettel, B., Reinhardt, R., Schmelzer, E., et al., 2012. Revealing structure and assembly cues for *Arabidopsis* root-inhabiting bacterial microbiota. Nature 488, 91–95.

Calderón, F.J., Schultz, D.J., Paul, E.A., 2012. Carbon allocation, belowground transfers, and lipid turnover in a plant-microbial association. Soil Sci. Soc. Am. J. 76, 1614.

Casieri, L., Gallardo, K., Wipf, D., 2012. Transcriptional response of *Medicago truncatula* sulphate transporters to arbuscular mycorrhizal symbiosis with and without sulphur stress. Planta 235, 1431–1447.

Chaparro, J.M., Sheflin, A.M., Manter, D.K., Vivanco, J.M., 2012. Manipulating the soil microbiome to increase soil health and plant fertility. Biol. Fertil. Soils 48, 489–499.

Clemmensen, K.E., Bahr, A., Ovaskainen, O., Dahlberg, A., Ekblad, A., Wallander, H., Stenlid, J., Finlay, R.D., Wardle, D.A., Lindahl, B.D., 2013. Roots and associated fungi drive long-term carbon sequestration in boreal forest. Science 339, 1615–1618.

Daghino, S., Murat, C., Sizzano, E., Girlanda, M., Perotto, S., 2012. Fungal diversity is not determined by mineral and chemical differences in serpentine substrates. PLoS One 7, e44233.

Dai, M., Hamel, C., St. Arnaud, M., He, Y., Grant, C., Lupwayi, N., Janzen, H., Malhi, S.S., Yang, X., Zhou, Z., 2012. Arbuscular mycorrhizal fungi assemblages in Chernozem great groups revealed by massively parallel pyrosequencing. Can. J. Microbiol. 58, 81–92.

Dalpé, Y., Trépanier, M., Sahraoui, A.L.-H., Fontaine, J., Sancholle, M., 2012. Lipids of Mycorrhizas. In: Hock, B. (Ed.), Fungal Associations. Springer, Berlin Heidelberg, pp. 137–169.

Damon, C., Vallon, L., Zimmermann, S., Haider, M.Z., Galeote, V., Dequin, S., Luis, P., Fraissinet-Tachet, L., Marmeisse, R., 2011. A novel fungal family of oligopeptide transporters identified by functional metatranscriptomics of soil eukaryotes. Int. Soc. Microb. Ecol. J. 5, 1871–1880.

Damon, C., Lehembre, F., Oger-Desfeux, C., Luis, P., Ranger, J., Fraissinet-Tachet, L., Marmeisse, R., 2012. Metatranscriptomics reveals the diversity of genes expressed by eukaryotes in forest soils. PLoS One 7, e28967.

Danielsen, L., Thürmer, A., Meinicke, P., Buée, M., Morin, E., Martin, F., Pilate, G., Daniel, R., Polle, A., Reich, M., 2012. Fungal soil communities in a young transgenic poplar plantation form a rich reservoir for fungal root communities. Ecol. Evol. 2, 1935–1948.

Davison, J., Öpik, M., Zobel, M., Vasar, M., Metsis, M., Moora, M., 2012. Communities of arbuscular mycorrhizal fungi detected in forest soil are spatially heterogeneous but do not vary throughout the growing season. PLoS One 7, e41938.

Dennis, P.G., Miller, A.J., Hirsch, P.R., 2010. Are root exudates more important than other sources of rhizodeposits in structuring rhizosphere bacterial communities? FEMS Microbiol. Ecol. 72, 313–327.

Doidy, J., van Tuinen, D., Lamotte, O., Corneillat, M., Alcaraz, G., Wipf, D., 2012. The *Medicago truncatula* sucrose transporter family: characterization and implication of key members in carbon partitioning towards arbuscular mycorrhizal Fungi. Mol. Plant 5, 1346–1358.

Drigo, B., Pijl, A.S., Duyts, H., Kielak, A.M., Gamper, H.A., Houtekamer, M.J., Boschker, H.T.S., Bodelier, P.L.E., Whiteley, A.S., Van Veen, J.A., et al., 2010. Shifting carbon flow from roots into associated microbial communities in response to elevated atmospheric CO_2. Proc. Natl. Acad. Sci. USA 107, 10938–10942.

Dumbrell, A.J., Ashton, P.D., Aziz, N., Feng, G., Nelson, M., Dytham, C., Fitter, A.H., Helgason, T., 2011. Distinct seasonal assemblages of arbuscular mycorrhizal fungi revealed by massively parallel pyrosequencing. New Phytol. 190, 794–804.

Edwards, I.P., Zak, D.R., Kellner, H., Eisenlord, S.D., Pregitzer, K.S., 2011. Simulated atmospheric N deposition alters fungal community composition and suppresses ligninolytic gene expression in a Northern hardwood forest. PLoS One 6, e20421.

Estrada, B., Barea, J. M., Aroca, R., and Ruiz-Lozano, J. M. (2012). A native *Glomus intraradices* strain from a Mediterranean saline area exhibits salt tolerance and enhanced symbiotic efficiency with maize plants under salt stress conditions. Plant Soil 1-17.

Fitter, A.H., Gilligan, C.A., Hollingworth, K., Kleczkowski, A., Twyman, R.M., Pitchford, J.W., Programme T. M. of the N.S.B., 2005. Biodiversity and ecosystem function in soil. Funct. Ecol. 19, 369–377.

Fitter, A.H., Helgason, T., Hodge, A., 2011. Nutritional exchanges in the arbuscular mycorrhizal symbiosis: implications for sustainable agriculture. Fungal Biol. Rev. 25, 68–72.

Floudas, D., Binder, M., Riley, R., Barry, K., Blanchette, R.A., Henrissat, B., Martínez, A.T., Otillar, R., Spatafora, J.W., Yadav, J.S., et al., 2012. The Paleozoic origin of enzymatic lignin decomposition reconstructed from 31 fungal genomes. Science 336, 1715–1719.

Formey, D., Molès, M., Haouy, A., Savelli, B., Bouchez, O., Bécard, G., Roux, C., 2012. Comparative analysis of mitochondrial genomes of *Rhizophagus irregularis*—syn. *Glomus irregulare*—reveals a polymorphism induced by variability generating elements. New Phytol. 196, 1217–1227.

Fravel, D., Olivain, C., Alabouvette, C., 2003. *Fusarium oxysporum* and its biocontrol. New Phytol. 157, 493–502.

Garrett, S.D., 1970. Pathogenic Root-infecting Fungi. Cambridge University Press, Cambridge, London.

Girlanda, M., Perotto, S., Bonfante, P., 2007. Mycorrhizal Fungi: their habitats and nutritional strategies biology of the fungal cell. In: Howard, R.J., Gow, N.A.R. (Eds.), The Mycota IV, second ed. Springer-Verlag, Berlin, pp. 223–249.

Glassop, D., Smith, S.E., Smith, F.W., 2005. Cereal phosphate transporters associated with the mycorrhizal pathway of phosphate uptake into roots. Planta 222, 688–698.

Gomez, S.K., Javot, H., Deewatthanawong, P., Torres-Jerez, I., Tang, Y., Blancaflor, E.B., Udvardi, M.K., Harrison, M.J., 2009. *Medicago truncatula* and *Glomus intraradices* gene expression in cortical cells harboring arbuscules in the arbuscular mycorrhizal symbiosis. BMC Plant Biol. 9, 10.

Gorfer, M., Blumhoff, M., Klaubauf, S., Urban, A., Inselsbacher, E., Bandian, D., Mitter, B., Sessitsch, A., Wanek, W., Strauss, J., 2011. Community profiling and gene expression of fungal assimilatory nitrate reductases in agricultural soil. Int. Soc. Microb. Ecol. J. 5, 1771–1783.

Govindarajulu, M., Pfeffer, P.E., Jin, H., Abubaker, J., Douds, D.D., Allen, J.W., Bücking, H., Lammers, P.J., Shachar-Hill, Y., 2005. Nitrogen transfer in the arbuscular mycorrhizal symbiosis. Nature 435, 819–823.

Grigoriev, I.V., Cullen, D., Goodwin, S.B., Hibbett, D., Jeffries, T.W., Kubicek, C.P., Kuske, C., Magnuson, J.K., Martin, F., Spatafora, J.W., et al., 2011. Fueling the future with fungal genomics. Mycology 2, 192–209.

Grünig, C.R., Sieber, T.N., Rogers, S.O., Holdenrieder, O., 2002. Genetic variability among strains of *Phialocephala fortinii* and phylogenetic analysis of the genus *Phialocephala* based on rDNA ITS sequence comparisons. Can. J. Bot. 80, 1239–1249.

Guether, M., Balestrini, R., Hannah, M., He, J., Udvardi, M.K., Bonfante, P., 2009a. Genome-wide reprogramming of regulatory networks, transport, cell wall and membrane biogenesis during arbuscular mycorrhizal symbiosis in *Lotus japonicus*. New Phytol. 182, 200–212.

Guether, M., Neuhauser, B., Balestrini, R., Dynowski, M., Ludewig, U., Bonfante, P., 2009b. A mycorrhizal-specific ammonium transporter from *Lotus japonicus* acquires nitrogen released by arbuscular mycorrhizal fungi. Plant Physiol. 150, 73–83.

Hacquard, S., Tisserant, E., Brun, A., Legué, V., Martin, F., Kohler, A., 2013. Laser microdissection and microarray analysis of *Tuber melanosporum* ectomycorrhizas reveal functional heterogeneity between mantle and Hartig net compartments. Environ. Microbiol. 15 (6), 1853–1869. http://dx.doi.org/10.1111/1462-2920.12080.

Hammer, E., Rillig, M.C., 2011. The influence of different stresses on glomalin levels in an arbuscular mycorrhizal fungus—salinity increases glomalin content. PLoS One 6, e28426.

Harman, G.E., 2000. Myths and dogmas of biocontrol changes in perceptions derived from research on *Trichoderma harzianum* T-22. Plant Dis. 84, 377–393.

Harman, G.E., Howell, C.R., Viterbo, A., Chet, I., Lorito, M., 2004. *Trichoderma* species—opportunistic, avirulent plant symbionts. Nat. Rev. Microbiol. 2, 43–56.

Harrison, M.J., 1996. A sugar transporter from *Medicago truncatula*: altered expression pattern in roots during vesicular-arbuscular (VA) mycorrhizal associations. Plant J. 9, 491–503.

Harrison, M.J., van Buuren, M.L., 1995. A phosphate transporter from the mycorrhizal fungus *Glomus versiforme*. Nature 378, 626–629.

Harrison, M.J., Dewbre, G.R., Liu, J., 2002. A phosphate transporter from *Medicago truncatula* involved in the acquisition of phosphate released by arbuscular mycorrhizal fungi. Plant Cell 14, 2413–2429.

Haygarth, P.M., Ritz, K., 2009. The future of soils and land use in the UK: soil systems for the provision of land-based ecosystem services. Land Use Policy 26, S187–S197.

Helber, N., Wippel, K., Sauer, N., Schaarschmidt, S., Hause, B., Requena, N., 2011. A versatile monosaccharide transporter that operates in the arbuscular mycorrhizal fungus *Glomus* sp is crucial for the symbiotic relationship with plants. Plant Cell 23, 3812–3823.

Hiltner, L., 1904. Über neuere Erfahrungen und Probleme auf dem Gebiete der Bodenbakteriologie unter besonderer Berücksichtigung der Gründüngung und Brache. Arb DLG 98, 59–78.

Hugenholtz, P., Tyson, G.W., 2008. Microbiology: metagenomics. Nature 455, 481–483.

Ihrmark, K., Bödeker, I.T.M., Cruz-Martinez, K., Friberg, H., Kubartova, A., Schenck, J., Strid, Y., Stenlid, J., Brandström-Durling, M., Clemmensen, K.E., et al., 2012. New primers to amplify the fungal ITS2 region—evaluation by 454 sequencing of artificial and natural communities. FEMS Microbiol. Ecol. 82, 666–677.

Insam, H., Seewald, M.S.A., 2010. Volatile organic compounds (VOCs) in soils. Biol. Fertil. Soils 46, 199–213.

Jargeat, P., Rekangalt, D., Verner, M.-C., Gay, G., Debaud, J.-C., Marmeisse, R., Fraissinet-Tachet, L., 2003. Characterization and expression analysis of a nitrate transporter and nitrite reductase genes, two members of a gene cluster for nitrate assimilation from the symbiotic basidiomycete *Hebeloma cylindrosporum*. Curr. Genet. 43, 199–205.

Javelle, A., André, B., Marini, A.-M., Chalot, M., 2003. High-affinity ammonium transporters and nitrogen sensing in mycorrhizas. Trends Microbiol. 11, 53–55.

Javot, H., Pumplin, N., Harrison, M.J., 2007. Phosphate in the arbuscular mycorrhizal symbiosis: transport properties and regulatory roles. Plant Cell Environ. 30, 310–322.

Jeffery, S., Gardi, C., Jones, A., Montanarella, L., Marmo, L., Miko, L., Ritz, K., Peres, G., Römbke, J., W, Van der Putten, 2010. European Atlas of Soil Biodiversity. European Commission. Publications Office of the European Union, Luxembourg.

Jones, D.L., Hinsinger, P., 2008. The rhizosphere: complex by design. Plant Soil 312, 1–6.

Jones, D.L., Nguyen, C., Finlay, R.D., 2009. Carbon flow in the rhizosphere: carbon trading at the soil-root interface. Plant Soil 321, 5–33.

Jumpponen, A., Jones, K.L., 2009. Massively parallel 454 sequencing indicates hyperdiverse fungal communities in temperate *Quercus macrocarpa* phyllosphere. New Phytol. 184, 438–448.

Jumpponen, A., Jones, K.L., Blair, J., 2010a. Vertical distribution of fungal communities in tallgrass prairie soil. Mycologia 102, 1027–1041.

Jumpponen, A., Jones, K.L., Mattox, J.D., Yeage, C., 2010b. Massively parallel 454-sequencing of fungal communities in *Quercus* spp. ectomycorrhizas indicates seasonal dynamics in urban and rural sites. Mol. Ecol. 19, 41–53.

Kellner, H., Vandenbol, M., 2010. Fungi Unearthed: transcripts encoding lignocellulolytic and chitinolytic enzymes in forest soil. PLoS One 5, e10971.

Kellner, H., Luis, P., Portetelle, D., Vandenbol, M., 2011. Screening of a soil metatranscriptomic library by functional complementation of *Saccharomyces cerevisiae* mutants. Microbiol. Res. 166, 360–368.

Kennedy, P.G., Matheny, P.B., Ryberg, K.M., Henkel, T.W., Uehling, J.K., Smith, M.E., 2012. Scaling up: examining the macroecology of ectomycorrhizal fungi. Mol. Ecol. 21, 4151–4154.

Kiers, E.T., Duhamel, M., Beesetty, Y., Mensah, J.A., Franken, O., Verbruggen, E., Fellbaum, C.R., Kowalchuk, G.A., Hart, M.M., Bago, A., et al., 2011. Reciprocal rewards stabilize cooperation in the mycorrhizal symbiosis. Science 333, 880–882.

Koranda, M., Schnecker, J., Kaiser, C., Fuchslueger, L., Kitzler, B., Stange, C.F., Sessitsch, A., Zechmeister-Boltenstern, S., Richter, A., 2011. Microbial processes and community composition in the rhizosphere of European beech—the influence of plant C exudates. Soil Biol. Biochem. 43, 551–558.

Kubartová, A., Ottosson, E., Dahlberg, A., Stenlid, J., 2012. Patterns of fungal communities among and within decaying logs, revealed by 454 sequencing. Mol. Ecol. 21, 4514–4532.

Kumar, V., Rajauria, G., Sahai, V., Bisaria, V.S., 2012. Culture filtrate of root endophytic fungus *Piriformospora indica* promotes the growth and lignan production of *Linum album* hairy root cultures. Process Biochem. 47, 901–907.

Lee, J., Young, J.P.W., 2009. The mitochondrial genome sequence of the arbuscular mycorrhizal fungus *Glomus intraradices* isolate 494 and implications for the phylogenetic placement of *Glomus*. New Phytol. 183, 200–211.

Leigh, J., Hodge, A., Fitter, A.H., 2009. Arbuscular mycorrhizal fungi can transfer substantial amounts of nitrogen to their host plant from organic material. New Phytol. 181, 199–207.

Lentendu, G., Zinger, L., Manel, S., Coissac, E., Choler, P., Geremia, R.A., Melodelima, C., 2011. Assessment of soil fungal diversity in different alpine tundra habitats by means of pyrosequencing. Fungal Divers. 49, 113–123.

Lim, Y.W., Kim, B.K., Kim, C., Jung, H.S., Kim, B.-S., Lee, J.-H., Chun, J., 2010. Assessment of soil fungal communities using pyrosequencing. J. Microbiol. 48, 284–289.

Lin, X., Feng, Y., Zhang, H., Chen, R., Wang, J., Zhang, J., Chu, H., 2012. Long-term balanced fertilization decreases arbuscular mycorrhizal fungal diversity in an arable soil in north china revealed by 454 pyrosequencing. Environ. Sci. Technol. 46, 5764–5771.

Lin, K., Limpens, E., Zhang, Z., Ivanov, S., Saunders, D.G.O., et al., 2014. Single nucleus genome sequencing reveals high similarity among nuclei of an endomycorrhizal fungus. PLoS Genet. 10, e1004078. http://dx.doi.org/10.1371/journal.pgen.1004078.

Lorito, M., Woo, S.L., Harman, G.E., Monte, E., 2010. Translational research on *Trichoderma*: from 'omics to the field. Annu. Rev. Phytopathol. 48, 395–417.

Lumini, E., Orgiazzi, A., Borriello, R., Bonfante, P., Bianciotto, V., 2010. Disclosing arbuscular mycorrhizal fungal biodiversity in soil through a land use gradient using a pyrosequencing approach. Environ. Microbiol. 12, 2165–2179.

Lundberg, D.S., Lebeis, S.L., Paredes, S.H., Yourstone, S., Gehring, J., Malfatti, S., Tremblay, J., Engelbrektson, A., Kunin, V., del Rio, T.G., et al., 2012. Defining the core *Arabidopsis thaliana* root microbiome. Nature 488, 86–90.

Maček, I., Kastelec, D., Vodnik, D., 2012. Root colonization with arbuscular mycorrhizal fungi and glomalin-related soil protein (GRSP) concentration in hypoxic soils from natural CO_2 springs. Agric. Food Sci. 21, 62–71.

Maldonado-Mendoza, I.E., Dewbre, G.R., Harrison, M.J., 2001. A phosphate transporter gene from the extra-radical mycelium of an arbuscular mycorrhizal fungus *Glomus intraradices* is regulated in response to phosphate in the environment. Mol. Plant Microbe Interact. 14, 1140–1148.

Marmeisse, R., Nehls, U., Öpik, M., Selosse, M.A., Pringle, A., 2013. Bridging mycorrhizal genomics, metagenomics and forest ecology. New Phytology 198, 343–346.

Marschner, H., 1995. Mineral Nutrition of Higher Plants. Academic Press, UK.

Martin, F., 2007. Fair trade in the underworld: the ectomycorrhizal symbiosis. In: Howard, R.J., Gow, N.A. R. (Eds.), The Mycota VIII Biology of the Fungal Cell. Springer-Verlag, Berlin, pp. 289–306.

Martin, F., Aerts, A., Ahrén, D., Brun, A., Danchin, E.G.J., Duchaussoy, F., Gibon, J., Kohler, A., Lindquist, E., Pereda, V., et al., 2008. The genome of *Laccaria bicolor* provides insights into mycorrhizal symbiosis. Nature 452, 88–92.

Martin, F., Kohler, A., Murat, C., Balestrini, R., Coutinho, P.M., Jaillon, O., Montanini, B., Morin, E., Noel, B., Percudani, R., et al., 2010. Périgord black truffle genome uncovers evolutionary origins and mechanisms of symbiosis. Nature 464, 1033–1038.

Mello, A., Napoli, C., Murat, C., Morin, E., Marceddu, G., Bonfante, P., 2011. ITS-1 versus ITS-2 pyrosequencing: a comparison of fungal populations in truffle grounds. Mycologia 103, 1184–1193.

Mendes, R., Kruijt, M., de Bruijn, I., Dekkers, E., van der Voort, M., Schneider, J.H.M., Piceno, Y.M., DeSantis, T.Z., Andersen, G.L., Bakker, P.A., et al., 2011. Deciphering the rhizosphere microbiome for disease-suppressive bacteria. Science 332, 1097–1100.

Mol-Dijkstra, J.P., Reinds, G.J., Kros, H., Berg, B., De Vries, W., 2009. Modelling soil carbon sequestration of intensively monitored forest plots in Europe by three different approaches. For. Ecol. Manage. 258, 1780–1793.

Molina, R., Massicotte, H., Trappe, J.M., 1992. Specificity phenomena in mycorrhizal symbiosis: community ecological consequences, practical implications. In: Allen, M.F. (Ed.), Mycorrhizal Functioning: An Integrated Plant Fungal Process. Chapman and Hall, London, pp. 357–423.

Montanini, B., Betti, M., Márquez, A.J., Balestrini, R., Bonfante, P., Ottonello, S., 2003. Distinctive properties and expression profiles of glutamine synthetase from a plant symbiotic fungus. Biochem. J. 373, 357–368.

Montanini, B., Viscomi, A.R., Bolchi, A., Martin, Y., Siverio, J.M., Balestrini, R., Bonfante, P., Ottonello, S., 2006. Functional properties and differential mode of regulation of the nitrate transporter from a plant symbiotic ascomycete. Biochem. J. 394, 125–134.

Müller, A., Faubert, P., Hagen, M., Zu Castell, W., Polle, A., Schnitzler, J.P., Rosenkranz, M., 2013. Volatile profiles of fungi-chemotyping of species and ecological functions. Fungal Genet. Biol. 54, 25–33.

Nadimi, M., Beaudet, D., Forget, L., Hijri, M., Lang, B.F., 2012. Group I intron-mediated trans-splicing in mitochondria of *Gigaspora rosea* and a robust phylogenetic affiliation of arbuscular mycorrhizal fungi with *Mortierellales*. Mol. Biol. Evol. 29 (9), 2199–2210. http://dx.doi.org/10.1093/molbev/mss088.

Nagy, R., Karandashov, V., Chague, V., Kalinkevich, K., Tamasloukht, M., Xu, G., Jakobsen, I., Levy, A.A., Amrhein, N., Bucher, M., 2005. The characterization of novel mycorrhiza-specific phosphate transporters from *Lycopersicon esculentum* and *Solanum tuberosum* uncovers functional redundancy in symbiotic phosphate transport in solanaceous species. Plant J. 42, 236–250.

Neher, D.A., 2010. Ecology of plant and free-living nematodes in natural and agricultural soil. Annu. Rev. Phytopathol. 48, 371–394.

Nehls, U., Bock, A., Einig, W., Hampp, R., 2001. Excretion of two proteases by the ectomycorrhizal fungus *Amanita muscaria*. Plant Cell Environ. 24, 741–747.

Nehls, U., Göhringer, F., Wittulsky, S., Dietz, S., 2010. Fungal carbohydrate support in the ectomycorrhizal symbiosis: a review. Plant Biol. 12, 292–301.

Ngosong, C., Gabriel, E., Ruess, L., 2012. Use of the signature fatty acid 16:1ω5 as a tool to determine the distribution of arbuscular mycorrhizal fungi in soil. J. Lipids 2012, Article ID 236807.

Oehl, F., Sieverding, E., Palenzuela, J., Ineichen, K., Da Silva, G.A., 2011. Advances in Glomeromycota taxonomy and classification. Int. Mycol. Assoc. FUNGUS 2, 191–199.

Öpik, M., Metsis, M., Daniell, T.J., Zobel, M., Moora, M., 2009. Large-scale parallel 454 sequencing reveals host ecological group specificity of arbuscular mycorrhizal fungi in a boreonemoral forest. New Phytol. 184, 424–437.

Öpik, M., Zobel, M., Cantero, J.J., Davison, J., Facelli, J.M., Hiiesalu, J., Jairus, T., Kalwij, J.M., Koorem, K., Leal, M.E., et al., 2013. Global sampling of plant roots expands the described molecular diversity of arbuscular mycorrhizal fungi. Mycorrhiza 23, 411–430. http://dx.doi.org/10.1007/s00572-013-0482-2.

Orgiazzi, A., Lumini, E., Nilsson, R.H., Girlanda, M., Vizzini, A., Bonfante, P., Bianciotto, V., 2012. Unraveling soil fungal communities from different Mediterranean land-use backgrounds. PLoS One 7, e34847.

Orgiazzi, A., Bianciotto, V., Bonfante, P., Daghino, S., Ghignone, S., Lazzari, A., Lumini, E., Mello, A., Napoli, C., Perotto, S., et al., 2013. 454 Pyrosequencing analysis of fungal assemblages from geographically distant, disparate soils reveals spatial patterning and a core mycobiome. Diversity 5, 73–98.

Ovaskainen, O., Nokso-Koivisto, J., Hottola, J., Rajala, T., Pennanen, T., Ali-Kovero, H., Miettinen, O., Oinonen, P., Auvinen, P., Paulin, L., et al., 2010. Identifying wood-inhabiting fungi with 454 sequencing—what is the probability that BLAST gives the correct species? Fungal Ecol. 3, 274–283.

Paszkowski, U., Kroken, S., Roux, C., Briggs, S.P., 2002. Rice phosphate transporters include an evolutionarily divergent gene specifically activated in arbuscular mycorrhizal symbiosis. Proc. Natl. Acad. Sci. U.S.A. 99, 13324–13329.

Pelin, A., Pombert, J.-F., Salvioli, A., Bonen, L., Bonfante, P., Corradi, N., 2012. The mitochondrial genome of the arbuscular mycorrhizal fungus *Gigaspora margarita* reveals two unsuspected trans-splicing events of group I introns. New Phytol. 194, 836–845.

Pérez-Tienda, J., Testillano, P.S., Balestrini, R., Fiorilli, V., Azcón-Aguilar, C., Ferrol, N., 2011. GintAMT2, a new member of the ammonium transporter family in the arbuscular mycorrhizal fungus *Glomus intraradices*. Fungal Genet. Biol. 48, 1044–1055.

Peterson, R.L., Wagg, C., Pautler, M., 2008. Associations between microfungal endophytes and roots: do structural features indicate function? Botany 86, 445–456.

Pfeffer, P.E., Douds, D.D., Bécard, G., Shachar-Hill, Y., 1999. Carbon uptake and the metabolism and transport of lipids in an arbuscular mycorrhiza. Plant Physiol. 120, 587–598.

Plett, J.M., Martin, F., 2011. Blurred boundaries: lifestyle lessons from ectomycorrhizal fungal genomes. Trends Genet. 27, 14–22.

Rausch, C., Daram, P., Brunner, S., Jansa, J., Laloi, M., Leggewie, G., Amrhein, N., Bucher, M., 2001. A phosphate transporter expressed in arbuscule-containing cells in potato. Nature 414, 462–470.

Read, D.J., Perez-Moreno, J., 2003. Mycorrhizas and nutrient cycling in ecosystems—a journey towards relevance? New Phytol. 157, 475–492.

Reich, M., Göbel, C., Kohler, A., Buée, M., Martin, F., Feussner, I., Polle, A., 2009. Fatty acid metabolism in the ectomycorrhizal fungus *Laccaria bicolor*. New Phytol. 182, 950–964.

Repáč, I., 2011. Ectomycorrhizal inoculum and inoculation techniques. In: Rai, M., Varma, A. (Eds.), Diversity and Biotechnology of Ectomycorrhizae. In: Soil Biology, 25. Springer-Verlag, Berlin, pp. 43–63.

Requena, N., Breuninger, M., Franken, P., Ocón, A., 2003. Symbiotic status, phosphate, and sucrose regulate the expression of two plasma membrane H+-ATPase genes from the mycorrhizal fungus *Glomus mosseae*. Plant Physiol. 132, 1540–1549.

Rillig, M.C., 2004. Arbuscular mycorrhizae, glomalin and soil quality. Can. J. Soil Sci. 84, 355–363.

Rillig, M.C., Mummey, D.L., 2006. Mycorrhizas and soil structure. New Phytol. 171, 41–53.

Rillig, M.C., Ramsey, P.W., Morris, S., Paul, E.A., 2003. Glomalin, an arbuscular-mycorrhizal fungal soil protein, responds to land-use change. Plant Soil 253, 293–299.

Rosling, A., Cox, F., Cruz-Martinez, K., Ihrmark, K., Grelet, G.-A., Lindahl, B.D., Menkis, A., James, T.Y., 2011. Archaeorhizomycetes: unearthing an ancient class of ubiquitous soil fungi. Science 333, 876–879.

Rousk, J., Bååth, E., Brookes, P.C., Lauber, C.L., Lozupone, C., Caporaso, J.G., Knight, R., Fierer, N., 2010. Soil bacterial and fungal communities across a pH gradient in an arable soil. Int. Soc. Microb. Ecol. J. 4, 1340–1351.

Sawers, R.J.H., Yang, S.-Y., Gutjahr, C., Paszkowski, U., 2008. The molecular components of nutrient exchange in arbuscular mycorrhizal interactions. In: Siddiqui, Z.A., Akhtar, M.S., Futai, K. (Eds.), Mycorrhizae: Sustainable Agriculture and Forestry. Springer, The Netherlands, pp. 37–59.

Schaarschmidt, S., Roitsch, T., Hause, B., 2006. Arbuscular mycorrhiza induces gene expression of the apoplastic invertase LIN6 in tomato (*Lycopersicon esculentum*) roots. J. Exp. Bot. 57, 4015–4023.

Schüßler, A., and Walker, C. (2010). The Glomeromycota. A species list with new families and new genera. (Edinburgh & Kew, UK: The Royal Botanic Garden; Munich, Germany: Botanische Staatssammlung Munich; Oregon, USA: Oregon State University). URL: http://www.amf-phylogeny.com.

Schüßler, A., Gehrig, H., Schwarzott, D., Walker, C., 2001. Analysis of partial Glomales SSU rRNA gene sequences: implications for primer design and phylogeny. Mycol. Res. 105, 5–15.

Selosse, M.-A., Roy, M., 2009. Green plants that feed on fungi: facts and questions about mixotrophy. Trends Plant Sci. 14, 64–70.

Shachar-Hill, Y., Pfeffer, P.E., Douds, D., Osman, S.F., Doner, L.W., Ratcliffe, R.G., 1995. Partitioning of intermediate carbon metabolism in VAM colonized leek. Plant Physiol. 108, 7–15.

Simon, C., Daniel, R., 2011. Metagenomic analyses: past and future. Appl. Environ. Microbiol. 77, 1153–1161.

Singh, B.K., Millard, P., Whiteley, A.S., Murrell, J.C., 2004. Unravelling rhizosphere-microbial interactions: opportunities and limitations. Trends Microbiol. 12, 386–393.

Smith, S.E., Read, D.J., 2008. Mycorrhizal Symbiosis, third ed. Academic Press, London, UK.

Smith, S.E., Smith, F.A., 1990. Structure and function of the interfaces in biotrophic symbioses as they relate to nutrient transport. New Phytol. 114, 1–38.

Sorek, R., Cossart, P., 2010. Prokaryotic transcriptomics: a new view on regulation, physiology and pathogenicity. Nat. Rev. Genet. 11, 9–16.

Sugiyama, A., Vivanco, J.M., Jayanty, S.S., Manter, D.K., 2010. Pyrosequencing assessment of soil microbial communities in organic and conventional potato farms. Plant Disease 94, 1329–1335.

Tani, C., Ohtomo, R., Osaki, M., Kuga, Y., Ezawa, T., 2009. ATP-Dependent but proton gradient-independent polyphosphate-synthesizing activity in extraradical hyphae of an arbuscular mycorrhizal fungus. Appl. Environ. Microbiol. 75, 7044–7050.

Tedersoo, L., Nilsson, R.H., Abarenkov, K., Jairus, T., Sadam, A., Saar, I., Bahram, M., Bechem, E., Chuyong, G., Kõljalg, U., 2010. 454 Pyrosequencing and Sanger sequencing of tropical mycorrhizal fungi provide similar results but reveal substantial methodological biases. New Phytol. 188, 291–301.

Tisserant, E., Kohler, A., Dozolme-Seddas, P., Balestrini, R., Benabdellah, K., Colard, A., Croll, D., Da Silva, C., Gomez, S.K., Koul, R., et al., 2012. The transcriptome of the arbuscular mycorrhizal fungus *Glomus intraradices* (DAOM 197198) reveals functional tradeoffs in an obligate symbiont. New Phytol. 193, 755–769.

Tisserant, E., Malbreil, M., Kuo, A., Kohler, A., Symeonidi, A., Balestrini, R., Charron, P., Duensing, N., et al., 2013. Genome of an arbuscular mycorrhizal fungus provides insight into the oldest plant symbiosis. Proc. Natl. Acad. Sci. U.S.A. 110, 20117–20122.

Treseder, K.K., Holden, S.R., 2013. Fungal carbon sequestration. Science 339, 1528–1529.

van Aarle, I.M., Olsson, P.A., 2003. Fungal lipid accumulation and development of mycelial structures by two arbuscular mycorrhizal fungi. Appl. Environ. Microbiol. 69, 6762–6767.

van der Heijden, M.G.A., Bardgett, R.D., Van Straalen, N.M., 2008. The unseen majority: soil microbes as drivers of plant diversity and productivity in terrestrial ecosystems. Ecol. Lett. 11, 296–310.

Verbruggen, E., Kuramae, E.E., Hillekens, R., De Hollander, M., Kiers, E.T., Röling, W.F.M., Kowalchuk, G.A., van der Heijden, M.G.A., 2012. Testing potential effects of maize expressing the *Bacillus thuringiensis* Cry1Ab endotoxin (Bt maize) on mycorrhizal fungal communities via DNA- and RNA-based pyrosequencing and molecular fingerprinting. Appl. Environ. Microbiol. 78, 7384–7392.

Wardle, D.A., Bardgett, R.D., Klironomos, J.N., Setälä, H., van der Putten, W.H., Wall, D.H., 2004. Ecological linkages between aboveground and belowground biota. Science 304, 1629–1633.

Waterman, R.J., Bidartondo, M.I., 2008. Deception above, deception below: linking pollination and mycorrhizal biology of orchids. J. Exp. Bot. 59, 1085–1096.

Weber, C.F., Zak, D.R., Hungate, B.A., Jackson, R.B., Vilgalys, R., Evans, R.D., Schadt, C.W., Megonigal, J.P., Kuske, C.R., 2011. Responses of soil cellulolytic fungal communities to elevated atmospheric CO_2 are complex and variable across five ecosystems. Environ. Microbiol. 13, 2778–2793.

Weiß, M., Sýkorová, Z., Garnica, S., Riess, K., Martos, F., Krause, C., Oberwinkler, F., Bauer, R., Redecker, D., 2011. Sebacinales everywhere: previously overlooked ubiquitous fungal endophytes. PLoS One 6, e16793.

Wright, D.P., Read, D.J., Scholes, J.D., 1998. Mycorrhizal sink strength influences whole plant carbon balance of *Trifolium repens* L. Plant Cell Environ. 21, 881–891.

Xu, L., Ravnskov, S., Larsen, J., Nicolaisen, M., 2012a. Linking fungal communities in roots, rhizosphere, and soil to the health status of *Pisum sativum*. FEMS Microbiol. Ecol. 82, 736–745.

Xu, L., Ravnskov, S., Larsen, J., Nilsson, R.H., Nicolaisen, M., 2012b. Soil fungal community structure along a soil health gradient in pea fields examined using deep amplicon sequencing. Soil Biol. Biochem. 46, 26–32.

Yu, L., Nicolaisen, M., Larsen, J., Ravnskov, S., 2012. Molecular characterization of root-associated fungal communities in relation to health status of Pisum sativum using barcoded pyrosequencing. Plant Soil 357, 395–405.

Zuccaro, A., Lahrmann, U., Güldener, U., Langen, G., Pfiffi, S., Biedenkopf, D., Wong, P., Samans, B., Grimm, C., Basiewicz, M., et al., 2011. Endophytic life strategies decoded by genome and transcriptome analyses of the mutualistic root symbiont *Piriformospora indica*. PLoS Pathog. 7, e1002290.

Chapter 12

Carbon Cycling: The Dynamics and Formation of Organic Matter

William Horwath

Department of Land, Air, and Water Resources, University of CA, Davis, CA, USA

Chapter Contents

I INTRODUCTION

Carbon (C) is the 15th most abundant element in the Earth's crust and the 4th most abundant element in the universe. It is nonmetallic and tetravalent, allowing four electrons to form covalent chemical bonds. The range of organic compounds formed is immense—from complex compounds to simple molecules, such as carbon dioxide (CO_2) and methane (CH_4), which are the primary controllers of climate since the origin of the Earth's atmosphere. The early Earth contained very little C. It was deposited on Earth's primordial surface from the bombardment of C-rich asteroids and meteorites (Anders, 1989). In addition to fertilizing the Earth with C, these extraterrestrial objects were the source of the vast amounts of water found in today's oceans (Morbidelli et al., 2000). The extraterrestrial C consisted of organic compounds, such as hydrocarbons, organic acids, and amino compounds, which are hypothesized as contributions to the evolution of life. This resulted in complex processes leading to the transfer of C among Earth's mantle, atmosphere, oceans, land, and life resulting in what is called today the global C cycle. This cycle is composed of both a long-term geological cycle and a short-term biological cycle.

II GEOLOGICAL CARBON CYCLE

Throughout Earth's history, the geological C cycle has dominated the fate of C. The accumulation of CO_2 and water in Earth's primordial atmosphere resulted in the first major process of the geological cycle; the weathering of exposed rocks that produced dissolved carbonates (Fig. 12.1). Carbonic acid (HCO_3^-) found in precipitation and water began dissolving rocks containing calcium (Ca)-magnesium (Mg) and silicate (Si).

$$2CO_2 + 2H_2O + CaSiO_3 \rightarrow Ca^{2+} + 2HCO_3^- + H_4SiO_4 \qquad (12.1)$$

The Ca and Mg carbonates are transported in surface runoff and subsurface flow to the ocean, where the following reaction occurs:

$$Ca^{2+} + 2HCO_3^- \rightarrow CaCO_3 \downarrow + CO_2 + H_2O \qquad (12.2)$$

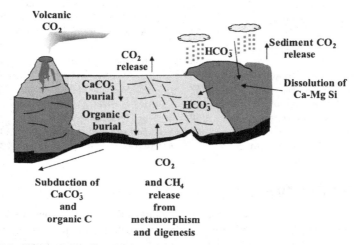

FIG. 12.1   The long-term C cycle.

The overall reaction consumes atmospheric CO_2. The process is dependent on temperature, where warmer temperatures increase the consumption of CO_2 through increased rates of weathering. Some of the carbonates formed during weathering are deposited in soils as $CaCO_3$, termed caliche during pedogenesis, whereas the remainder is lost equally through outgassing from overland flow and carbonate deposition in the marine environment. The process of creating sedimentary rock and its subduction into the Earth's mantle is a major sink for C and represents a near-permanent geologic reservoir (Berner, 2004). The subducted carbonate-bearing rocks undergo thermal weathering deep in the Earth's crust, and a fraction of the C is released back into the atmosphere as CO_2 primarily through volcanic activity and crust outgassing (Fig. 12.1). This process represents a critical step in controlling the Earth's atmospheric CO_2 content. Sedimentary rocks have sequestered most of Earth's C over its 4.5-billion-year history, with a small fraction existing in the atmosphere, oceans, fossil C deposits, vegetation, and soils.

The evolution of photosynthetic organisms altered the geologic C cycle by providing another sink for atmospheric CO_2. Approximately 3-3.5 billion years ago, cyanobacteria began the process of accelerating the removal of atmospheric CO_2 and introduced oxygen (O_2) into the atmosphere. The cyanobacteria also fixed dinitrogen and introduced organic and reactive inorganic nitrogen (N) to the ocean. Eventually, the green alga (Trebouxia) or cyanobacteria (Nostoc) formed a symbiosis with fungi (particularly Ascomycota), termed lichens, that allowed both photosynthesis and biological N fixation to occur in the terrestrial environment beginning in the Devonian period over 400 million years ago (see Supplemental Fig. S12.1 in the online Supplemental Material at http://booksite.elsevier.com/9780124159556). Lichens accelerated rock weathering through the production of organic acids and introduced organic matter (OM) and organic N into the terrestrial environment. More advanced, vascular plant life evolved later in the Devonian period. This perturbed the long-term carbonate sedimentation cycle to a greater extent by accelerating the weathering of rocks through hydrolysis processes associated with root exudates and OM decomposition and by the removal of CO_2 from the atmosphere through deposition of OM in soils and sediments.

The proliferation of vascular plants in the low-lying coastal areas of the supercontinent of Gondwana through the Carboniferous period led to the deposition of vast amounts of OM. This deposited OM is the source of the fossil fuels so vital to humanity today (see more information in the online Supplemental Material at http://booksite.elsevier.com/9780124159556). The evolution of lignin, a component of the secondary cell wall of vascular plants, is surmised to have contributed to the large quantities of plant debris that led to the formation of large coal reserves during the Carboniferous and Permian periods. It has been hypothesized that the initial burial of large amounts of lignaceous plant material occurred as a result of the absence of lignin degrading organisms, which evolved later in the Paleozoic period (Robinson, 1990). This may be why the

Carboniferous and Permian coal reserves outweigh those from any other period and will likely never again be a significant sink for Earth's C.

The long-term geological weathering process and subduction of carbonate-bearing rocks act as a global thermostat. As CO_2 in the atmosphere is consumed through the weathering process, the greenhouse gas (GHG) effect is diminished causing the climate to cool. The evolution of photosynthetic organisms provided another consumption process for CO_2 through the deposition of C in soil and sediments. During cold periods when fewer plants exist, the outgassing of CO_2 overcomes consumption processes leading to warming. Other geological processes can also influence the geologic C cycle. The uplift of continental plates through tectonic activity can enhance Ca-Mg, Si rock weathering through exposure of large areas of unweathered rocks. The rapid uplift, in geological terms, of the Himalaya Mountains is thought to have increased Ca-Mg, Si rock weathering, leading to the consumption of large amounts of atmospheric CO_2. This has been hypothesized to result in the cooling during the late Cenozoic period due to depletion of atmospheric CO_2 (Raymo, 1991).

III BIOLOGICAL C CYCLE

The biological C cycle is dominated by the interplay of terrestrial and marine photosynthesis and decomposition (Fig. 12.2). The processes are closely balanced, each being sinks of about 2 Gt C annually (Le Quéré et al., 2009). Two C-based gases, CO_2 and CH_4, dominate the biological C cycle. Perturbation events of the biological C cycle, such as ice ages, cause changes in the concentration of these GHGs on a timescale of hundreds to thousands of years. The GHGs absorb reflected infrared radiation from the Earth's surface, trapping heat in the lower atmosphere (see more information in the online Supplemental Material at http://booksite.elsevier.com/9780124159556). Earth's high CO_2 concentration through the Cretaceous period kept the Earth's temperature fairly consistent, ideal for promotion of the evolution of life. Other GHGs produced by microbes, such as nitrous oxide (N_2O; see Chapter 14), also play an important role in regulating the Earth's temperature. Variations in the sun's energy output and changes in the Earth's orbital inclination have also contributed to climate change over the Quaternary. Imbrie and Imbrie (1979) suggest that changes in Earth's orbital inclination that affect the intensity and duration of sunlight near the poles cause the rhythm of the ice ages.

In 1850, the atmosphere contained about 285 ppmv CO_2 (McCarroll and Loader, 2004) compared to over 400 ppmv or 790 Pg C (also Gt) today (Houghton, 2007). Fossil fuel use, forest burning, and the conversion of extensive areas of virgin land to agriculture have led to a net transfer of terrestrial C to the atmosphere (Falkowski et al., 2000). Today, humans are emitting CO_2 at a rate of about 10 Gt year^{-1} (Le Quéré et al., 2009; see online Supplemental Material at http://booksite.elsevier.com/9780124159556; Units of the global C cycle). The emission of fossil CO_2-C constitutes about 8.5 Gt C year^{-1} of

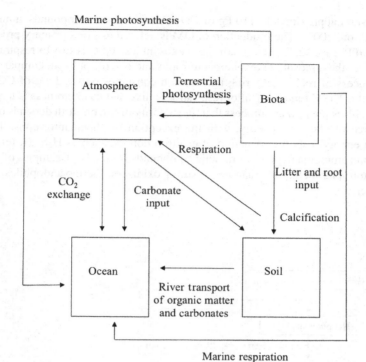

FIG. 12.2   The short-term C cycle.

the total, with the remainder associated with deforestation and land use change. This has been partially offset by the annual net uptake of about 2 Gt of C in the oceans and 1 Gt on land (Normile, 2009). It has been suggested that terrestrial and ocean C sequestration are underestimated. For example, the expansion of South American tropical forests on former grassland savannas has been occurring for 4000 years leading to increases in ecosystem C up to 10-fold (Silva et al., 2008). Nitrogen deposition from anthropogenic activities has increased plant production in some areas causing increases in CO_2 uptake through increased net primary production (NPP). These biological processes, combined with considerable uncertainty in the gross rates of Ca and Mg dissolution from terrestrial sources, likely explain discrepancies in the global C budget. The following discussion elaborates on the two most important components of the biological cycle; photosynthesis and decomposition.

A   Photosynthesis

Photosynthesis is responsible for cycling significant amounts of CO_2-C between the atmosphere and the biosphere (see more information in the online Supplemental Material at http://booksite.elsevier.com/9780124159556).

It converts approximately 110 Pg of CO_2-C into organic compounds annually (Houghton, 2007). The production of OM is defined as gross primary production (GPP; Fig. 12.3). About half the photosynthesized C is lost to respiration back into the atmosphere by photoautotrophs and heterotrophs. In comparison, the process of rock weathering and erosion consumes only 1.1 Pg of CO_2-C annually. The C remaining after respiration in live and dead biomass is termed NPP. This is the primary process that the majority of life on earth depends on as a source of chemical energy, with the exception of chemoautotrophs. They obtain energy from the oxidation of electron donors, such as H_2S, H, ferrous iron, and ammonia to produce organic compounds from CO_2. Examples of chemoautotrophs include methanogens, sulfur oxidizers, thermoacidophiles, and nitrifiers.

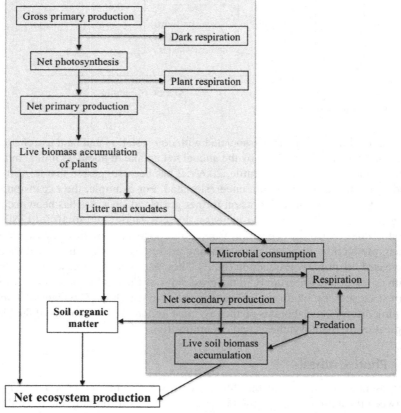

FIG. 12.3 Various components of gross primary production and net ecosystem production.

B   Decomposition of Plant and Microbial Products

A consortium of organisms capable of degrading and utilizing specific chemical structures and substrates perform the decomposition of plant litter and microbial necromass and other inputs, including leachates and exudates from various sources. Free-living microorganisms and fauna consume (decompose) the majority of NPP both in the ocean and in terrestrial environments (Fig. 12.3). The turnover of NPP provides an energy source for heterotrophs to build new biomass termed net secondary production (NSP). Live biomass can also be consumed through animal, insect, and microbial grazing. The process of decomposition occurs on the order of days-to-decades and is dependent on temperature and moisture and the quality of the live and senesced biomass. A small fraction of plant and heterotrophic decomposer constituents are transformed into soil organic matter (SOM), which can persist for thousands of years and is an important stable C pool making up a significant proportion of terrestrial C stocks. The C remaining after decomposition of NPP and NSP is termed net ecosystem production. The decomposition of NSP, including microorganisms and to a smaller extent fauna, represents a crucial step that contributes to maintenance of soil and sediment C. All of the NPP and NSP processes are often expressed in terms of g C m^{-2} year^{-1} or similar units.

The various components of NPP added to soil vary greatly as a source of energy and nutrients. In terrestrial environments, up to 50% of NPP inputs are C polymers, including hemicellulose, pectins, and cellulose (Table 12.1). These major cell wall structural components contain limited essential nutrients, particularly N and P. Cytoplasmic constituents, such as sugars, amino compounds, organic acids, and proteins, comprise up to 10% of plant residue dry

TABLE 12.1 Percentage of Cytoplasmic and Cell Wall Components in Plants (adapted from Horwath, 2002)

Plant Component	% of Total
Waxes and pigment	1
Amino acids, sugars, nucleotides etc.	5
Starch	2-20
Protein	5-7
Hemicellulose	15-20
Cellulose	4-50
Lignin	8-20
Secondary compounds	2-30

weight providing labile energy and essential nutrients to complete decomposition. The following describes substrate inputs to soils and sediments and their decomposition.

C   Cytoplasmic Constituents of Plants and Microorganisms

The various cytoplasmic constituents of plants and microorganisms are biochemically very diverse (Tables 12.1 and 12.2). Cytoplasmic proteins contain the majority of N in plant and microbial tissue. The protein content of plant tissues ranges from 1% in cell walls to 22% in meristematic regions and seeds. In leaves, the photosynthetic enzyme ribulose bisphosphatase can comprise up to

TABLE 12.2 Biochemical and Molecular Composition of Prokaryotic Cells[a]

Molecule	Dry Weight (%)[b]	Molecules per Cell	Different Kinds of Molecules
Total macromolecules	96	24,610,000	2500
Protein	55	2,350,000	1850
Polysaccharide	5	4,300	2
Lipid	9.1	22,000,000	4
Lipopolysaccharide	3.4	1,430,000	1
DNA	3.1	2	1
RNA	20.5	255,500	660
Murein	2.5	1	1
Glycogen	2.5	4,360	1
Total Monomers	3.0	_[c]	350
Amino acids	0.5	–	100
Sugars	3.0	–	50
Nucleotides	2.0	–	200
Inorganic	1.0	–	18
Total	100%		

[a]Data from Neidhardt et al. (1996).
[b]Dry weight calculated on basis of 70% cell water content.
[c]Reliable estimates are lacking.

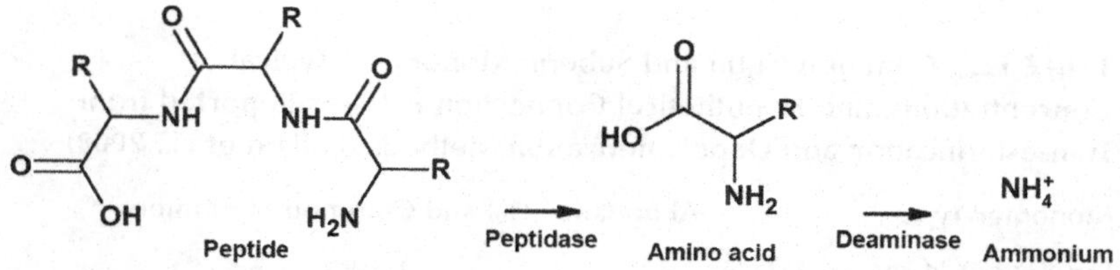

FIG. 12.4 Breakdown of protein/polypeptides into amino acids and further mineralization to ammonium is a source of N for decomposition and microbial growth.

70% of the total, leaf-tissue N. Microorganisms contain up to 55% protein (Table 12.2). Proteins also contain significant amounts of sulfur (S) in the form of the amino acids cysteine and methionine. Proteins and peptides are hydrolyzed by proteases and peptidases to peptides and individual amino acids during decomposition (Fig. 12.4). The RNA content represents the next largest N component up to 20.5% of dry weight of microorganisms with plants being significantly lower due to the substantial cell wall component. The majority, up to 80%, is ribosomal RNA (23S, 16S, and 5S RNA). Fungi have similar structures with a slightly larger size (e.g., 18S and 25S RNA). These labile N sources, such as proteins, nucleic acids, and other C containing compounds, provide the initial nutrients and energy source for decomposition and are a major area of modern studies, such as in proteomics and metabolomics.

D   Plant and Microbial Lipids

Lipids are a diverse class of compounds ranging from simple fatty acids to complex sterols, phospholipids, cutins, and suberins (Table 12.3). They are determined by sequential extraction with nonpolar solvents, such as hexane and chloroform. The average lipid content of most plants is about 5% of dry weight with leaves containing the greatest amount. The lipid content of high-cutin containing plants, such as conifers and succulents, may reach 10% or more of the dry weight. The durability of lipids depends on their chemical complexity. Long-chain aliphatic fatty acids and phospholipids, components of membranes, are degraded relatively quickly depending on the degree of saturation or double bond content. More complex, resin-like material, such as cutin and suberin, can be recalcitrant and can form decomposition-resistant substances in soil and sediment. The hydrophobic character of both lipids and resins allows them to sorb into hydrophobic domains of OM, shielding them from decomposers.

Microbial lipids are similar in function to those of plants and animals, but differ structurally. Appreciable amounts of lipids are present in fungal spores and hyphae. The lipid content of fungal mycelium averages 17% by weight and ranges from 1% to 55% depending on the species. Various phospholipids

TABLE 12.3 Common Cutin and Suberin Monomers, Typical Concentrations, and Hypothetical Connection Patterns Reported from Transesterification and Depolymerization Methods (Pollard et al., 2008)

Monomer Type	Abundance (%) and Common Monomers	
Unsubstituted fatty acids	1-25% C16:0, C18:0, C18:1, C18:2	1-10% C18:0 to C24:0
ω-Hydroxy fatty acids	1-32% C16:0, C18:1, C18:2	11-43% C18:1, C16:0 to C26:0
α, ω-Dicarboxylic acids	Usually <5% C16:0, C18:0, C18:1, C18:2	24-45% C18:1, C18:2, C16:0 to C26:0
Mid-chain functionalized monomers		
Epoxy-fatty acids	0-34% C18:0 (9, 10-epoxy) C18:1 (9, 10 epoxy)	0-30% C18:1 (9, 10-epoxy-hydroxyl) C18:(9, 10-epoxy-1, 18—diacid)
Polyhydroxy-fatty acids	16-92% C16:0 (10, 16-dihydroxy) C18:0 (9, 10, 18-trihydroxy)	0-2% C10:0 (9, 10, 18-trihydroxy)
Polyhydroxy α, ω-dicarboxylic acid	Trace	0-8% C18:0 (9, 10-dihydroxy)
Fatty alcohols		
Alkan-1-ols and alken-1-ols	0-8% C16:0, C:18:1	1-6% C18:0 to C22:0
α, ω-Alkanediols and α, ω-alkenediols	0-5% C18:1	0-3% C22:0
Glycerol	1-14%	14-26%
Phenols	0-1% Ferulate	0-10% Ferulate, courmarate, sinapate, caffeate

are unique and are useful for identifying microorganisms (Chapter 7). Assays of specific fungal lipids, such as ergosterol, have proven useful for quantifying fungal biomass and for qualitative evidence of their diversity. The bacterial cell is typically about 9% lipid (Table 12.2). Microbial lipids degrade readily in their unprotected state, but like plant lipids, once sorbed into the hydrophobic domains of OM, they become protected from decomposition.

Lipids can accumulate in both acidic and high OM soils where they can constitute up to 30% of the total SOC (Stevenson, 1994). High OM soils contain the greatest amount of lipids. Clay soils have higher lipid content than coarse textured or sandy soils. Soil disturbance, such as tillage, erosion, or fire, can release these compounds into the decomposition cycle. Lipases, a class of lipid-degrading enzymes, in oleaginous microorganisms, particularly molds and yeast, degrade plant-derived lipids. *Pseudomonas aeruginosa* and *Burkholderia* sp., both Proteobacteria and *Bacillus subtilis* (Firmicutes), are representative microorganisms that show enhanced degradation of lipids in wastewater treatment systems. The decomposition of more complex lipids, such as sterols and cutins, require multifaceted enzymatic systems or groups of enzymes. Cutin is a polymer network of oxygenated C-16 and C-18 fatty acids cross-linked by ester bonds (Table 12.3). The accumulation, in soil, of polyaromatic hydrocarbons released from fossil fuel combustion is gaining attention because they behave similarly to soil lipids.

E   Starch

Starch is an energy-storage compound used for cellular maintenance in plants, especially perennials. It is a plant polymer synthesized from glucose and stored in plastids as grains of 1-100 nm in diameter (see Fig. S12.2 in the online Supplemental Material at http://booksite.elsevier.com/9780124159556 for glucose structure). It consists of two glucose polymers, amylose and amylopectin. Amylose contains long unbranched chains of (1-4)-glucose units (Fig. 12.5a). For most plants, amylose can account for up to 30% of the total starch. Amylopectin has a similar structure linked to every 20-30 glucose residues by (1-6)-glucose bonds (Fig. 12.5b). A class of enzymes known as amylases degrades starch readily. It is commonly used in the food and beverage industry to convert starch to sugar. α-Amylase catalyzes the cleavage of glucosyl bonds producing small glucans called dextrins along with glucose and maltose. β-Amylose cleaves maltose residues from the nonreducing end of the starch molecule. The final degradation of maltose, short-chained glucans and dextrins to glucose, is achieved by β-glycosidase. Many bacteria and fungi produce these enzymes to convert insoluble starch into soluble, easily metabolized glucose. Bacteria include *Bacillus* (Firmicutes) and the Proteobacteria *Pseudomonas*, *Escherichia coli*, and *Serratia marscens*. Fungi include *Trichoderma* and *Aspergillus* (both in the phylum Ascomycota) and *Rhizopus* (Zygomycota).

F   Hemicelluloses, Pectins, and Cellulose

Hemicellulose, pectins, and cellulose, as a group of compounds, are the most abundant biosynthesized polymers on earth. The majority of these carbohydrates are found as part of the cytoskeleton in the primary and secondary plant cell walls. The structure of the plant cell wall is composed of various layers with enrichment in certain lignin monomers (Fig. 12.6a). Intimately associated with the cell wall carbohydrate and lignin framework are protein networks thought to contribute to its structural integrity (Fig. 12.6b). These structural proteins are

Amylose

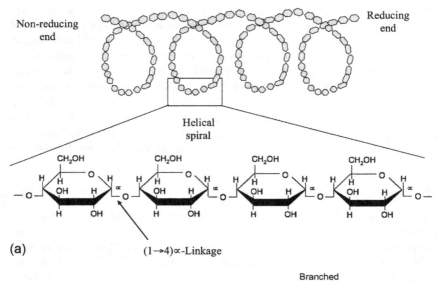

(a)

(1→4)∝-Linkage

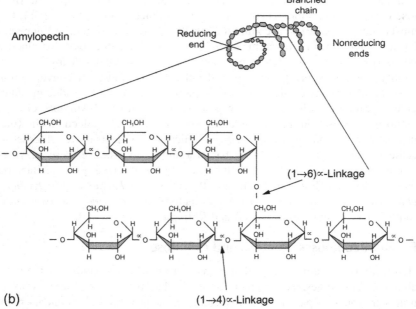

(b)

(1→4)∝-Linkage

FIG. 12.5 The structure of starch showing the (a) ∝(1-4) linkage of amylose and (b) ∝(1-6) branched linkage of amylopectin.

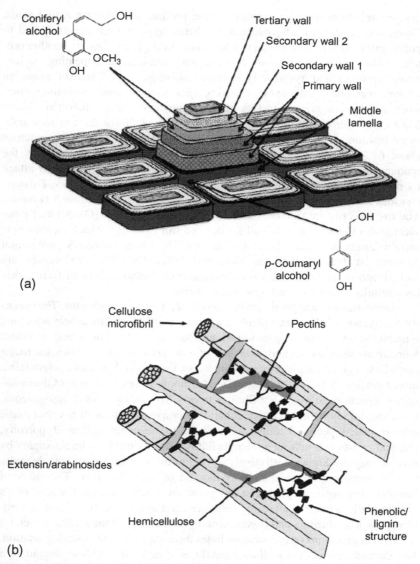

FIG. 12.6 The structure of the plant cell wall is composed of various layers and enrichment in certain lignin monomers (a). The framework of pectins, hemicellulose, and lignin is joined together by a protein called extension (b).

glycoproteins enriched in hydroxyproline, proline, and glycine amino acids. Glycoproteins contain oligosaccharide chains (glycans) covalently linked to polypeptide side chains through a process known as glycosylation. Another protein called extensin, similar to collagen, consists of repeating serine-hydroxyproline and tyrosine-lysine-tyrosine sequences. Extensin plays an important architectural role cross-linking lignin to the pectin/hemicellulose network and may comprise up to 15% of the primary cell wall (Lodish et al., 2000). Most of its hydroxyprolines are glycosylated with chains of three or four arabinose residues. The serine residues are linked to galactose making extensin about 65% carbohydrate. In addition to providing structural integrity, the unique amino acid and glycosation sequence may prevent microbial attack by presenting unrecognizable cleavage sites to proteolytic enzymes. Polysaccharides are long chains of sugars that are covalently linked through H bonds. The vast majority are aldoses with the empirical formula (CH_2O). All the monosaccharides in the plant cell wall are derived from glucose, which on alteration form a variety of 5 and 6 C sugars (see Fig. S12.2 in the online Supplemental Material at http://booksite.elsevier.com/9780124159556). The sugars are locked into polymerized chains by pyranose or furanose rings to form either hemicellulose/pectic or cellulose microfibrils.

Hemicelluloses and pectin are composed of 5 C sugars or glycans. The majority of glycans of flowering plants consist of D-xylose, D-glucuronic acid, and D-arabinose. A common cross-linked structure is (1-4)-D-glucan and -D-xylose found in all dicotyledons and about half of monocots. In many grasses, the major cross-linked glycan is glucuronoarabinoxylan. Cereals and grasses also contain a mixed linkage of 1-3(1-4)-D-glucans as a distinguishing component of the cross-linked glycan-cellulose microfibril network. The mixture of heterogeneous, branched, highly hydrated polysaccharides composed mainly of D-galacturonic acid is called pectin. Pectins are thought to influence cell wall porosity, pH, and cell-to-cell adhesion. Hemicelluloses are reduced to simple sugars by several enzymes known collectively as pectinases. For this reason, the degradation of pectin and hemicellulose is thought to occur together. The study of hemicellulose/pectin degradation has received much attention because of its importance as a point of attack by pathogens and as a means for microbial symbionts, such as rhizobia and mycorrhiza, to gain access to the middle lamella.

The pectinase enzyme system includes three stages of substrate degradation. The degradation of hemicellulose/pectin is similar to cellulose degradation except that more enzymes are involved. The decoupling of the hemicellulose and pectin cross-linked glycan structure is thought to provide access for other cell wall degrading enzymes. The release of oligogalacturonides and simple sugars repress the enzyme system and is thought to control microbial succession during litter decomposition. Pectinolytic soil bacteria include *Erwinia* and *Pseudomonas*, both Proteobacteria, *Arthrobacter* (Actinobacteria), and *Bacillus* (Firmicutes). Common pectinolytic fungi in the Phylum Ascomycota, include *Aspergillus*, *Sclerotina*, *Penicillium* and *Fusarium*, and *Rhizopus*

(Zygomycota). Yeast in the phylum Ascomycota, such as Conidia and *Kluyver-omyces*, also exhibits pectinase activity.

Cellulose, the most abundant plant polysaccharide, accounts for 15-30% of the primary cell wall and a greater percentage of the secondary cell wall, especially in woody species. It consists of glucose units linked by β(1-4) bonds to form D-glucan chains (Fig. 12.7). The chains are cross-linked by H bonds to form paracrystilline assemblages called microfibrils. The average microfibril is composed of 36 individual glucan chains, with several thousand individual glucan molecules, reaching a length of 2-3 mm. Cellulose microfibrils are cross-linked into a network or scaffold with glycans or hemicellulose. In addition, plants can also synthesize callose by linking glucans in the β(1-3) configuration similar to that found in yeast and fungi. Callose is found in phloem sieve plates, pollen tubes, cotton fibers, and other specialized cells. Callose is produced in response to wounding, such as fungal hyphae penetration of the primary cell wall.

Cellulose microfibrils are decomposed by a group of enzymes called cellulases, consisting of endoglucanase, exoglucanase, and β-glucosidase, also known collectively as cellobiases. The cellulase enzymes have a distinct role in cleaving specific bonds within the microfibril network, which causes disruption of the crystalline structure followed by depolymerization into short glucose chains. Endoglucanases act randomly on both soluble and insoluble glucose chains by cleaving the β(1-4) linkages, yielding glucose and cello-oligosaccharides. Subsequently, exoglucanases, including glucanhydrolase, act on the nonreducing ends of the cellulose chains, yielding glucose and cellobiose (glucose dimmers) and cellotriose (glucose trimers). In the final step of decomposition, β-glucosidase hydrolyzes the glucose chain fragments to glucose. Breakdown of the glucose subunits occurs rapidly, and the products can inhibit the activity of the cellulase system.

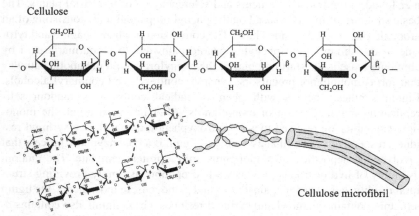

FIG. 12.7 Cellulose structure showing β(1-4) linkages combining glucose residues into a single chain polymer. Chains are cross-linked by hydrogen bonds to form cellulose microfibrils.

A wide range of fungi can degrade cellulose, but only a few have demonstrated the complete depolymerization and hydrolyzation of the crystalline microfibril structure *in vitro*. The cellulase system of the fungus *Trichoderma* (Ascomycota) has been extensively studied and shows a large production of *endo*-β-glucanases and *exo*-β-glucanases, but low levels of β-glucosidases. This fungus gained notoriety in the Second World War, Pacific Theatre, where it destroyed the canvas tents of U.S. serviceman. The genus *Aspergillus* (Ascomycota) produces large amounts of *endo*-β-glucanase and β-glucosidases, but low levels of *exo*-β-glucanases. Chaetonium (Ascomycota) is found on a wide variety of cellulose materials from paper to composts, especially in moist environments, and produces thermotolerant cellulases that may be commercially useful for converting cellulose to simple sugars. Other ascomycetous fungi with extensively studied cellulase systems include *Acremonium celluloyticus*, *Penicillium*, *Fusarium*, and *Agarica*.

Bacteria have less cellulytic capacity than fungi. Bacterial cellulases are organized into globular, scaffolding proteins called cellusomes, bound to their cell walls. These structures coordinate the cellulase system to attack the crystalline microfibrils, increasing the activity or efficiency of the individual enzymes. Common aerobic soil bacteria that depolymerize cellulose are *Cellulomas* (Actinobacteria), *Cellovibrio* and *Pseudomonas* (Proteobacteria), and *Bacillus* (Firmicutes). Anaerobic bacteria include *Acetobacter* (Proteobacteria), *Bacteriodes* (Bacteroidetes), *Clostridium* and *Ruminococcus* (Firmicutes), and *Fibrobacter* (Fibrobacteres).

G Lignin

Lignin is a unique hydrocarbon comprising 8-20% of the secondary cell wall of terrestrial plants. It is a complex, dense, amorphous, secondary cell wall polymer found in the trachea elements and sclerenchyma of terrestrial plants. The basic structure of lignin is based on the phenyl propanoid unit, consisting of an aromatic ring and a 3-C side chain. The conversion of phenylalanine and tyrosine in the shikimic and phenylpropanoid pathways to *p*-courmaric acid by ammonia lyase is the starting point of the phenylpropanoid metabolic pathway that forms monolignol precursors, synapyl, coniferyl, and coumaryl alcohols. Lignin synthesis begins with phenoxy radical coupling, a random self-replicating polymerization of monolignols. After polymerization, the monolignol subunits are referred to as *p*-hydroxyphenol, guaiacyl, and syringyl residues, respectively. The final assembly of lignin is a nonenzymatic process that involves the deposition of monolignols onto a protein template. The random assembly of hydrocarbons produces a hydrophobic assembly providing structural rigidity and a barrier against pests and pathogens. In most dicots, the lignin structure contains guaiacyl and syringyl residues. Grass lignin also contains *p*-hydroxyphenol in small amounts. Lignin is cross-linked to hemicellulose via a cell wall protein called extensin.

The dense nature, hydrophobicity, and nonspecific structure of lignin make it difficult for enzymes to attack. Fungi are the most efficient lignin degraders, playing a key role in terrestrial C cycling and in pathogenesis. Lignin is not a component of marine plants. Fungal species that degrade lignin are grouped into soft rot, brown rot, and white rot fungi based on the color of the decaying substrate. Various fungi represented by Imperfecti (Deuteromycota) and Ascomycetes (Ascomycota) cause soft rot. These fungi prefer polysaccharides, but have the ability to remove methylated side chains (R—O—CH$_3$) and cleave aromatic rings, yet cannot completely degrade the lignin structure. The soft-rot fungi are important in mesic environments and appear to degrade hardwood lignin more effectively than softwoods, with Chaetomium and Preussia being representative organisms (both in the phylum Ascomycotathe).

Both brown and white rot fungi (from the phylum Basidiomycota) perform the majority of wood decay. Brown rots lack ring-cleaving enzymes, but readily degrade hemicellulose and cellulose of intact wood. They modify and degrade lignin through demethylation and removal of methylated side chains producing hydroxylated phenols. The oxidation of the lignin aromatic structures causes the characteristic brown color. The separation of polysaccharides from lignin is thought to occur through nonenzymatic oxidation via the production of hydroxyl radicals ($^\bullet$OH). It is theorized that hydroxyl radicals enter the lignin structure by creating pores to allow access to enzymes (Watanabe, 2003). The $^\bullet$OH may be produced from the reaction of Fe(II) with hydrogen peroxide (H$_2$O$_2$) via the Fenton reaction:

$$Fe(II) + H_2O_2 \rightarrow Fe(III) + OH^- + {}^\bullet OH \tag{12.3}$$

Other transition metals, such as Cu and Mn, may also be used in this process. The disruption of the lignin structure allows for access to the underlying sugar polymer structures. This gives the brown rots the ability to degrade intact wood without completely disrupting the lignin structure. Representative organisms include Poria and Gloeophyllum.

White rot fungi are the most effective lignin degraders. Several thousand species of white rots are known from the Basidiomycetes and Ascomycetes. The Basidiomycota most studied are *Phanerochaete chrysoporum* and *Coriolus versicolor*. *Pleurotus ostreatus*, the oyster mushroom, and *Lintinula edodes*, the shiitake mushroom, are grown commercially for food. Ascomycetes include Xylaria, Libertella, and Hypoxylon. They produce lignolytic enzymes that oxidatively cleave phenylpropane units, and demethylate, convert aldehyde groups (R-CHO) to carboxyl groups (R-COOH), and cleave aromatic rings, resulting in the complete destruction of lignin to CO$_2$ and water. Lignin degradation is repressed by more easily degraded substrates, and very little lignin C is used for microbial growth. White rots use three classes of extracellular-lignin-degrading enzymes: the phenol oxidase laccase, lignin peroxidase, and manganese oxidase. Laccase and manganese peroxidase cannot directly oxidize nonphenolic structures, which make up 70-90% of lignin. Lignin peroxidase can oxidize both phenolic and nonphenolic lignin structures.

The degree of lignin phenolic decomposition can be described by the relative distribution of acidic and aldehydic phenolic units within the vanillyl and syringyl phenol families (Kögel-Knabner, 1986). As lignin is degraded, carboxylic acid units are formed from the lignin polymer during cleavage of phenylpropanoid Cα—Cβ bonds. This leads to an increase in carboxylic acid containing phenolic units with respect to phenolic units with an aldehyde side chain, which significantly increases the O and N content of decomposed lignin (Horwath and Elliott, 1996). The change in the acid to aldehyde ratio for vanillyl and syringyl units reflects the degree of lignin degradation. Kögel-Knabner (1986) showed that the degree of lignin decomposition increased with soil depth. This approach provides for a quantitative measure of the degree of lignin phenol degradation in soil, but because most of the lignin is decomposed to liberate sugar polymers, it cannot be used to estimate the turnover of the original starting plant material.

Bacteria may attack parts of the lignin structure to gain access to the energy-rich cellulose and hemicellulose. Gram-negative, aerobic bacteria in the family of Pseudomonadaceae, the Azotobacter and Neisseriaceae, both Proteobacteria, and common Actinobacteria, such as *Nocardia* and *Streptomyces*, have a limited ability to degrade lignin. Bacteria found in the guts of ruminants and some arthropods have some limited ability to degrade lignin. Lignin is also solubilized in termites by gut-inhabiting *Streptomyces* to free energy-rich cellulose and hemicelluloses for further digestion. The effect of the high pH (9-11) in termite foreguts and the proportion of true lignase activity relative to depolymerization and by-product formation are not completely understood.

H   Plant Secondary Compounds

Plants produce an array of organic compounds that are not associated with growth and development. Extractable phenols and tannins are significant components of some terrestrial plants, especially in forest systems, and can comprise up to 30% of their dry weight. These secondary metabolites or plant secondary compounds are divided into three major groups: terpenoids, alkaloids, and phenylpropanoids. The polyphenolic compounds range in molecular weight from 500 to 3000 Da. Many play a significant role in protection against herbivory and microbial infection as attractants for pollinators and seed dispersers and as alleleopathic agents. They readily precipitate proteins through tannin reactions. The tanning of leather is an example of where a natural product is protected from microbial attack. Two major classes of tannins, termed condensed and hydrolysable, are found in higher plants (Fig. 12.8). Condensed tannins, also called proanthocyanidins, are polymers of three-ring flavanols joined by C—C bonds. Hydrolysable tannins are grouped into gallotannins and ellagitannins composed of gallic acid or hexahydroxydiphenic acid esters,

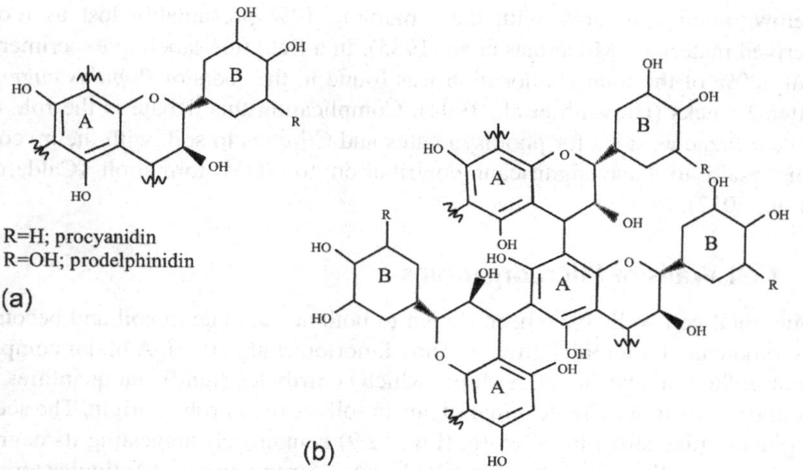

FIG. 12.8 Structures of tannins: (a) a condensed tannin trans monomer unit, and (b) a condensed tannin trimer showing different intermonomer linkages (C4–C8 and C4–C6).

respectively, linked to a sugar moiety. The rate of decomposition can be directly related to the content of plant secondary compounds. The ability to utilize precipitated proteins has been hypothesized to be a competitive advantage for some decomposers (Kraus et al., 2003).

I   Roots and Root-Derived Materials

The allocation of photosynthate belowground can vary from 10% to 70% of total NPP (Finlay and Soderstrom, 1992). Root biomass can range as a proportion of the aboveground from 90% in the tundra to 9% in agricultural crops (Jobbágy and Jackson, 2000). The proportion of root biomass in forest systems is about 20%, with up to 80% in grasslands. Standing biomass does not reflect root exudation and turnover (growth and death), and roots can be a greater C input to soils than the aboveground shoot portion. Root exudates include the secretion of ions, enzymes, mucilage, and a diverse array of primary and secondary metabolites. Root exudates contain carbohydrates (sugars) and amino compounds, providing a significant energy and nutrient source in the rhizosphere. The amount of C entering soil as root exudates has been fiercely debated, mainly due to the complexity of studying roots *in situ*. About half of the photosynthate allocated belowground is respired within 2 weeks by the roots, mycorrhiza, and rhizosphere organisms (Harris et al., 1985; Horwath et al., 1994). Root-derived materials are important substrates that contribute to rhizosphere processes, but their role in C cycling is not as well understood due to the aforementioned methodological constraints. Natural grasslands labeled with $^{14}CO_2$ retained 52% of the assimilated C aboveground and 36% in belowground structures, with the remaining 12% presumably lost as root-derived materials (Milchunas et al., 1985). In a field tree-labeling experiment, only 20% of the total C allocation was found in the roots of *Populus eugenii* after 2 weeks (Horwath et al., 1994). Complicating this debate is the role of mycorrhizae as sinks for photosynthates and C inputs to soil, with the mycorrhiza said to have significant contribution to SOM formation (Calderon et al., 2012).

J Cell Walls of Microorganisms

Microbial cell walls have been shown to both accumulate in soil and become by-products of microbial growth (Throckmorton et al., 2012). A major component of the fungal cell wall is chitin, which contributes significant quantities of amino sugars to soil. Most amino sugars in soil are of microbial origin. The acetylglucosamine structure of chitin (Fig. 12.9) contains N, suggesting its degradation is unlikely to be limited by N. *Streptomyces* (Actinobacteria), *Pseudomonas* (Proteobacteria), and *Bacillus* and *Clostridium*, both from the phylum Firmicutes, degrade fungal cell walls. The cell walls of *Phytophthera* (Heterokontophyta) contain fibers of cellulose and $\beta(1\text{-}4)$ glucans that may interact with other cell components to affect their decomposition. Polymers of $\beta(1\text{-}3)$ and $\beta(1\text{-}6)$ glucan are found in organisms such as *Fusarium* (Ascomycota). Hyphae containing both chitin and glucans require both chitinase and glucanase to be degraded. Dark-colored pigments, often referred to as melanins, provide protection and contribute to lower turnover rates of fungal cell wall constituents. Other substances resembling glycoproteins of fungal origin, identified as glomalin, can accumulate in soil. This fairly stable fungal product is produced by arbuscular mycorrhizae and likely contributes to aggregate formation (Rillig, 2004).

Bacterial cell walls are composed of a rigid layer of two sugar derivatives, *N*-acetylglucosamine and *N*-acetylmuramic acid chains. Gram-negative

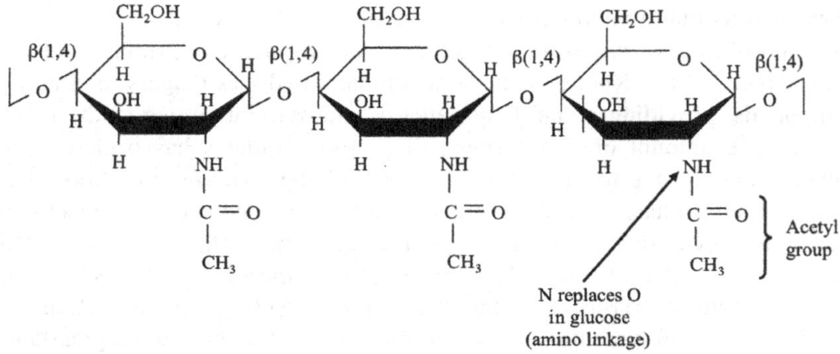

FIG. 12.9   The structure of chitin showing the $\beta(1\text{-}4)$ linking acetylglucosamine residues.

bacteria contain additional layers outside the ridged layer creating more complex cell walls. These chains are linked by a limited number of amino acids through peptide bonds (Fig. 12.10). The cross-linkage peptidoglycan (murein) structures are composed of repeating units of L-alanine, D-alanine, D-glutamic acid, and either lysine or diaminopimelic acid. These components are connected to form a repeating structure called the glycan tetrapeptide. In Gr⁻ bacteria, about 10% of the cell wall is peptidoglycan in contrast to Gr⁺ bacteria, in which peptidoglycan can make up to 90% of the cell wall. The thick wall of Gr⁺ bacteria contains peptidoglycans linked to other wall constituents that include a variety of polysaccharide and polyphosphate molecules joined through phosphoid-ester linkages (Kögel-Knabner, 2002). Most contain D-alanine. More than 100 different types of peptidoglycans have been identified. The biochemical diversity of microbial cell walls is thought to affect their decomposition, but more studies are needed to address this hypothesis.

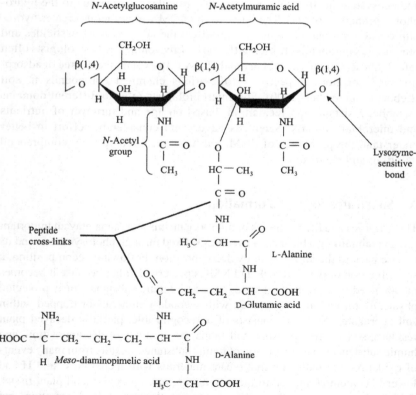

FIG. 12.10 The structure of one of the repeating units of peptidoglycan found in gram-negative bacteria.

IV ORGANIC MATTER

OM in soils and sediments represents residual compounds and the organic structures remaining following the decomposition of plant, fauna, and microbial inputs. The formation and decay of OM is an essential ecosystem process contributing to the regulation of atmospheric trace gases, particularly CO_2, N_2O, and CH_4. During the early nineteenth century, De Saussure recognized that the level of OM in soils directly contributed to plant growth by supplying essential nutrients such as N, P, and S. The chemical structure of OM has an approximate stoichiometry of C/N/P/S of 100/10/1/1 and is the primary available source of these nutrients in soil and sediments. The chemical characterization of SOM started in the eighteenth century as scientists recognized that various soils had different levels of extractable C (Stevenson, 1994).

Soils and sediments continue to be valued natural resources for producing food and fiber in our modern society. Today they are impacted by human activity through disturbance that causes erosion, land use change that releases C and other GHGs to the atmosphere, and pollutants that threaten both human and ecosystem health. Pollutants are of particular concern due to the hydrophobic properties of OM. The interaction of OM with minerals creates hydrophilic and hydrophobic domains that affect the adsorption of pesticides and other toxic compounds (Karickhoff, 1981). Environmental toxicologists often refer to these domains as rubbery and glassy C to explain the degree of adsorption and sorption, respectively, of various chemical compounds in soil (Leboeuf and Weber, 2000). Soil microbiologists refer to different domains as labile, resistant, and recalcitrant based on age and turnover of nutrients and microbial biomass. Scientists have spent considerable effort to better understand the properties of SOM, but have yet to completely comprehend its origin and chemistry.

A Substrates for OM Formation

The activities and life strategies of microorganisms and fauna play an important role in maintaining a balance between CO_2 fixed through photosynthesis and its release back to the atmosphere by decomposition. Following decomposition, a small fraction of the C in NPP and NSP is preserved either because it becomes metabolized to a recalcitrant state, such as a humic substance, or is protected physically through the association with secondary minerals or trapped within soil aggregates. Soil OM consists of unrecognizable, partially decayed plant residues, soil microorganisms, soil fauna, by-products of decomposition, and humic substances. The formation of humic substances results from many events of oxidation and reduction that create materials with increased C and H and lower O content compared to the original animal, microbial, and plant tissue. During decomposition, N compounds react through radical coupling and increase the N content of humic substances (Fig. 12.11). Humic substances

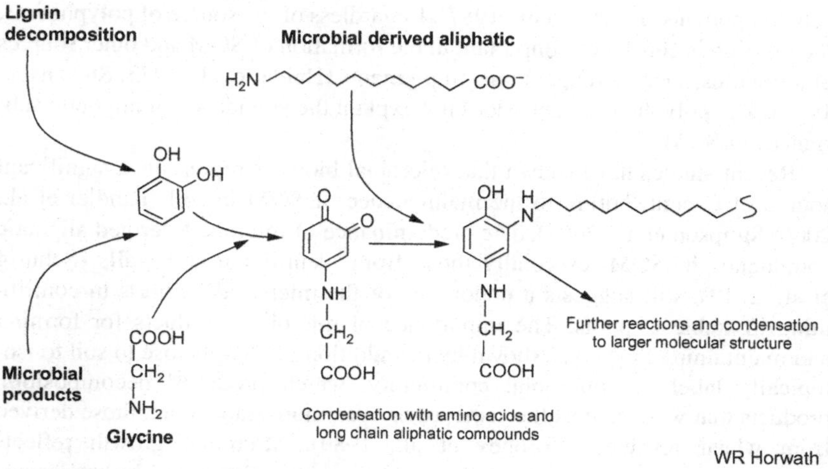

FIG. 12.11 Hypothesized mechanism for the formation and protection of humic substances showing the importance of microbial products and altered lignin components in condensation reactions and formation of soil aggregates.

consist of approximately 50-55% C, 5% H, 33% O, 4.5% N, 1% S, and 1% P. Metals and micronutrients, such as Al, Ca, Zn, and Cu, are present in much smaller amounts.

Maillard (Stevenson, 1994) suggested that the accumulation of N in humic substances occurred through the interaction of reducing sugars and amino acids to form dark brown polymers similar to humic acids termed the "Browning" reaction. The Browning reaction gained recognition because microorganisms readily produce sugar and amino compounds in addition to plant residue sources. This mechanism could also explain the formation of humic substances in aquatic environments, such as the ocean, where no lignin is formed. The mechanism involves rearrangement and fragmentation of compounds to form three intermediates: (1) aldehydes and ketones, (2) reductones, and (3) furfurals. All react readily with amino compounds to form dark-colored end products through the formation of double bonds. Jokic et al. (2004) suggest that the action of manganese (IV) oxide on sugar-amino acid condensation under ambient conditions results in the formation of heterocyclic N compounds, showing that typical soil redox reactions may be sufficient to promote this reaction. Other theories explaining the accumulation of N in OM include the "Lignin Theory" proposed by Waksman (1936) and the "Polyphenol Theory" promoted by Flaig et al. (1975). These theories suggest that quinone structures from lignin decomposition polymerize with amino acids to form high-molecular-weight nitrogenous substances. However, these theories do not explain the absence of the unique depleted ^{13}C signature of lignin phenols in stabilized OM (Wedin et al., 1995). Lignin is depleted in ^{13}C by -2‰ to -6‰ compared to other plant cell components (Benner et al., 1987). Regardless of the source of polyphenols, the reaction is considered important in the formation of SOM and other sources of aromatics, such as fungal melanin pigments (Haider et al., 1975; Stevenson, 1994). The polyphenol theory does not explain the abundance of aliphatic substances in SOM.

Recent studies have shown that microbial biomass represents a significant source of C contributing to the maintenance of SOM in soil (Kindler et al., 2006; Simpson et al., 2007). The predominance of microbial-derived aliphatic compounds in SOM, especially those from identifiable cell walls (Schurig et al., 2013), still suggests a major role of the microbial biomass in contributing to stable C pools. The importance of microbial products for forming and maintaining SOM was shown by the addition of ^{14}C-glucose to soil to isotopically label the microbial community, which produced decomposition products that were at least the same as, or even more stable than, those derived from wheat residues (Voroney et al., 1989). Microbial growth reflects the dominant groups of organisms present in a soil. Fungi and bacteria comprise approximately 90% of total soil biomass (Rinnan and Bååth, 2009), and the turnover of their nonliving biomass (necromass) is estimated to contribute as much as 80% to the maintenance and accumulation of SOM (Liang and Balser, 2010).

It has been hypothesized that cellular biochemistry from diverse microbial groups determines the form and amount of C destined to form stable SOM (Throckmorton et al., 2012). Martin and Haider (1979), using laboratory incubations, showed that fungi contributed more C to SOM than other microbial groups and suggested it was due to their complex cell walls and pigments. This observation was supported by the observation that fungi have higher C use efficiency (CUE) compared to bacteria that could potentially lead to greater contribution of fungal C to stable SOM (Rinnan and Bååth, 2009). However, the overlapping ranges of CUEs for soil fungi and bacteria bring this theory into question (Six et al., 2006).

Diverse bacterial cell wall structure (e.g., Gram-stain classification) likely influences decomposition (i.e., mineralization) because Gr^+ bacteria contain a greater amount of peptidoglycan in their cell walls than Gr^- bacteria. Greater peptidoglycan content is often associated with slower decomposition, as these compounds have been shown to accumulate in soil. However, the macromolecular structure of peptidoglycan varies across species and growth conditions making predictions of decomposability challenging (Vollmer et al., 2008). Limited field studies that compare fungal and bacterial turnover have not produced definitive results of whether or not diverse microbial groups contribute differentially to stable soil C (Strickland and Rousk, 2010). Throckmorton et al. (2012) showed that the contribution of diverse biochemistry of major microbial groups contributes equally to maintaining OM. The results suggest that the input of various microbial groups is more dependent on their abundance rather than their unique cellular biochemistries.

B   Theories of SOM Stabilization

The mechanisms leading to the persistence of SOM in soils and sediments are highly debated (Schmidt et al., 2011). A persistent view of SOM formation and maintenance has centered on the molecular components of inputs, such as lignin content, and/or bulk chemical composition, such as C to N ratio, to determine the dynamics of SOM processes. This approach has been useful in describing the kinetics of SOM turnover, but does not address the structure or protection mechanisms responsible for preserving C fractions that are greater than 1000 years or their function. Hassink (1997) proposed that the capacity of soils to preserve organic C and N was related to the content of clay and silt or soil texture. The stabilization of SOM with secondary minerals, such as clay and amorphous oxides, notably iron oxides, to form stable organomineral complexes was observed decades earlier and continues to be used as an explanation of stabilization (Allison et al., 1949; Hayes and Clapp, 2001; Theng, 1982).

Hassink (1995) showed that the heavy, mineral-dominated and small aggregate (<20 μm) soil fractions have turnover rates of 0.5-0.7×10^{-4} day^{-1}, independent of soil textural class. These organomineral fractions typically contain the majority of soil C, particularly in temperate areas. It follows that understanding the mechanisms leading to the stabilization of soil and sediment C relies on knowledge of mechanisms in play at the surface of minerals. Organomineral interactions are hypothesized to occur through physical bonds (dipole interactions and van der Waals forces), ionic bonds, hydrogen bonds, hydrophobic interactions, and ligand exchange (Fig. 12.12). These mechanisms include the binding mechanisms described earlier and their interaction with colloids, such as iron oxides that lead to micelle formation and flocs. Other metals and micronutrients, such as Al, Ca, Zn, and Cu, present in much smaller amounts, can also fulfill this role.

Other theories of SOM stabilization suggest that chemical recalcitrance explains its stabilization and persistence. During the 1960s and 1980s, exhaustive chemical oxidation and reduction to degrade humic substances found that amphiphilic and alkane compounds were dominant features of SOM (Hatcher et al., 1981; Schnitzer, 1978). Preston and Schnitzer (1984) identified chemical structures and side chains using NMR techniques confirming these observations. Later, Haider and Schulten (1985), using advanced, analytical mass spectrometry, verified the significant aliphatic content of SOM. The results suggested that aromatic structures were cross-linked by longer chain aliphatic compounds, carboxyl, and amine groups. Schulten et al. (1997), using pyrolysis-methylation gas chromatograph/mass spectrometry (Py-GC/MS), confirmed that alkyl-aryl compounds consisted of aromatic rings covalently bonded to aliphatic chains. They also showed that the aromatic fraction contained significant heterocyclic N (pyrroles and pyridines), N-derivatives of benzene, and long-chain nitriles. Whether these structures impart inherent recalcitrance versus stabilization by the mineral phase is still debated.

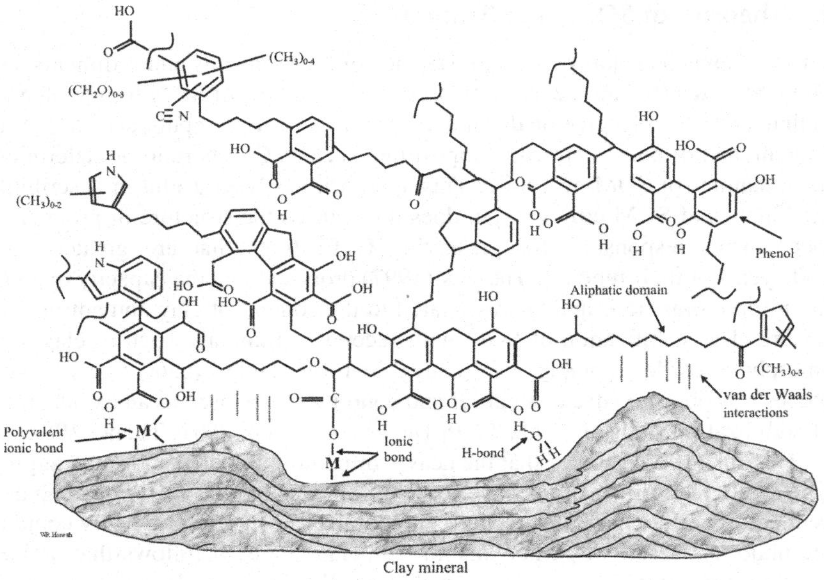

FIG. 12.12 Idealized structure of humic acid showing high aliphatic content showing physiochemical interactions with a clay mineral. Organomineral interactions M denote various cations, such as iron and calcium.

Swift (1999) postulated that the macromolecular structure of humic substances occurred through the self-aggregation of humic substances on minerals and oxides. Piccolo (2001) expanded on this theory by hypothesizing the supramolecular structure of humic substances were stabilized predominantly by weak disruptive forces, such as van der Waals, instead of covalent linkages. Sutton and Sposito (2005) furthered the aggregation theory by suggesting humic substances form micellar structures in aqueous environments, supporting Swift's hypothesis. The self-aggregating micellar theory explains the hydrophobic properties of OM and the inability of degradative enzymes to function in hydrophobic domains. However, these studies were mainly done on extracted humic substances, providing scant information on their original structure or composition and their interaction with minerals. The self-aggregating micellar theory explains the hydrophobic properties of OM and the inability of degradative enzymes to function in hydrophobic domains.

Von Lützow et al. (2006) summarized important mechanisms controlling the stability of SOM in soil as: (1) formation of structures such as micelles or biochar, (2) metal bridging to oxides and clay minerals, and (3) physical isolation within aggregates. It is also clear that even though some soil C is >1000 years old, the stability of the majority of soil C is relative rather than absolute suggesting that without some sort of protection mechanism soil C is likely to be consumed by the decomposer community (Dungait et al., 2012).

This concept partly confirms the notion of "C saturation" proposed by Six et al. (2002); however, it does not explain why similar textured soils in different climates contain different amounts of SOM. It is likely that the interplay between mineral surface attraction and micelle formation provide the continuum where soil C resides. Soils with a longer duration of moisture seasonally likely maximize and maintain micelle formation and maintenance processes. Dryer soils or soils with frequent wetting and drying likely disrupt micelle formation resulting in lower SOM contents. The preceding research illustrates the complexity in understanding the structure and composition of SOM.

Building on the notion of micellar features and structures, a zonal model for the organization of compounds with varying properties and functional groups, was proposed by Kleber et al. (2007). The zonal model assumes SOM is stabilized through a self-organization mechanism characterized by a heterogeneous mixture of compounds that display amphiphilic and hydrophobic properties in an aqueous environment (Fig. 12.13). The interactions in the zonal or layered model are likely dominated by polar functional groups that bind to the mineral

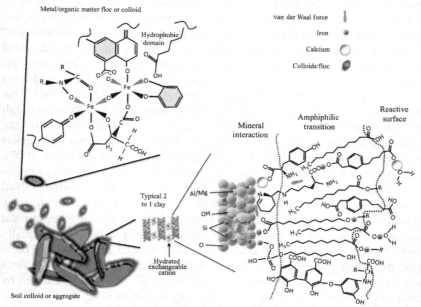

FIG. 12.13 The figure depicts proposed flocculation processes of dissolved organic matter (OM) with metals, such as Fe, as on mechanisms leading to stable soil C. The flocs eventually interact with OM on minerals and aggregates. The process is dynamic with flocs and OM in a state of equilibrium. The OM associated with mineral interacts by forming layers of different chemical properties: (1) mineral surface interacts where van der Waals and chemical bonding occurs; (2) an amphiphilic layer where covalent bonding and hydrophobic interactions occur together; and (3) an outer boundary or surface where the charged portion of compounds react with the water environment. The outer boundary area is the location where floc likely interact.

surface through hydrogen or covalent bonding (Chien et al., 1997; Kaiser and Guggenberger, 2003). The organomineral layer would be further stabilized via ligand exchange coordinating with hydroxyls to form inner-sphere complexes. The near-mineral zone likely also contains proteins both in the folded and unfolded state adhering to the mineral surface displaying hydrostatic and hydrophobic interactions, such as van der Waals interactions.

The presence of proteins near the mineral surface explains the low C:N ratio of isolated organomineral fractions (Hassink, 1997; Rasmussen et al., 2008). The area away from the mineral surface, where the amphiphilic portions of the compounds adhering to the mineral surface reside together with other covalent interactions is hypothesized to exhibit strong hydrophobic properties similar to the lipid bilayer of cell membranes (Kleber et al., 2007; Sutton and Sposito, 2005). The presence of hydrophobic domains containing aliphatic compounds supports the partitioning of nonpolar compounds observed in soils (Chiou et al., 1983). The partitioning is analogous to a hexane/water mixture where hydrophobic compounds, such as some pollutants and pesticides, readily sorb from the water phase. The entropic heat of a hydrophobic or amphiphilic type compound partitioning into a hydrophobic domain is less than staying in the solution, thus driving the phase partitioning process. At the boundary or surface of the hydrophobic layer, the conglomerated structure is posited to extend into the polar water environment. In this outer edge region, it is hypothesized that broken edges of micellar structures, proteins, and possibly other N-containing compounds, interact loosely through cation bridging, hydrogen bonding, and other interactions (Kleber et al., 2007; Stevenson, 1994). This active, outer boundary of the organomineral assemblage is thought to have high exchange rates with soil solution. However, the requirement to stabilize the zonal model with hydrophobic compounds, such as lipids, does not support the observations of low C to N of isolated mineral fractions (Kleber et al., 2007).

The formation of a micellar type feature likely occurs as a result of multivalent ions, such as Fe^{3+} and Al^{3+} and likely Ca^{2+}, combining with dissolved OM to form flocs that exhibit both micellar and covalent features (Henneberry et al., 2012). The primary functional group adhering to these metal oxides are carboxyls. According to spectral data, the compounds that initially form flocs are aromatic in character with aliphatic and simple soluble compounds adhering following floc formation (Henneberry et al., 2010). The flocs continue to grow in size until their density is greater than water followed by their precipitation from solution with the only limitation being the abundance of metal oxides to promote flocking. As these floc/micellar formations precipitate, they likely interact with the SOM coatings on the organomineral particles as described earlier.

Toosi et al. (2012) showed that soil solution dissolved SOM is at equilibrium with the organomineral phase in the absence of fresh OM inputs. In the presence of fresh OM inputs, an exchange between new input C and older organomineral C occurs. This process is highly dependent on soil moisture, which influences the stability of both micellar and organomineral assemblages. The frequency of wetting and drying cycles influences the stability of the micellar features exposing unstable C containing structures to enzymatic attack as mentioned earlier. Even though the properties of OM, such as phase partitioning, are well characterized, debate on the structure and formation of SOM will continue. The limitations in determining the structure of SOM lie in the vast chemical complexity of SOM/mineral interactions and in methodological and analytical instrumentation capabilities.

C Aggregate Protection of OM Fractions in Soil

Another C protection mechanism in soils is physical isolation within aggregates. Early studies attributed soil aggregation to the cohesive properties of polysaccharides, fine roots, fungal hyphae, and organometal interactions with clays (Fig. 12.14; Tisdall and Oades, 1982; Swift, 1991). In a review of soil structure, Bronik and Lal (2005) suggested these and other processes, including phosphate precipitation, carbonate formation, and flocculation with metals (i.e., silica, Ca, and Fe), promote aggregation. Soil aggregates range in size from >2 mm through <53 μm to silt-sized aggregates (2-53 μm). It is hypothesized that smaller-sized aggregates lead to the formation of larger aggregates in agriculture soils (Six et al., 2000). The individual organomineral complexes described in the previous section likely promote aggregate formation through the interaction of the micellar regions to create an additive hydrophobic domain that stabilizes a continuum of organomineral colloids.

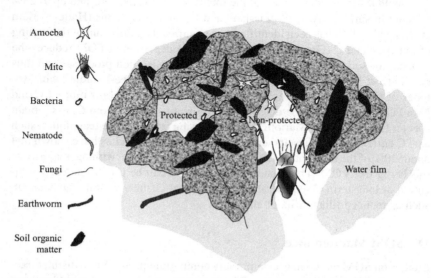

FIG. 12.14 Soil aggregate showing OM adsorbing to and binding minerals. Protected areas and water films are shown depicting habitats for soil microorganisms.

Soil C can be minimally defined as being unbound or bound to minerals throughout the aggregate structure. The unbound fraction or particulate organic matter (POM) is physically protected or occluded and is defined as no longer being recognizable as plant residues, but not yet completely degraded to individual compounds. POM is isolated by flotation in dense liquids, such as cesium chloride, bromoform, sodium polytungstate, or silica gel with specific gravities ranging from 1.4 to 2.2 g cm^{-3}. It can exist both outside and inside aggregates. Disrupting soil aggregates by sonication or chemical dispersion releases occluded POM. Occluded POM is protected against decomposition within aggregates by physical isolation. The formation and stability of occluded POM depends on the frequency of soil disturbance (i.e., less in tilled soils), faunal activity (i.e., worms increase aggregation), and type of plant community (grasslands have larger quantities and size of aggregates from higher root inputs and microbial and faunal activity).

Studies have shown that cultivation depletes C associated with both POM and mineral-associated fractions (Cambardella and Elliott, 1992; Grandy and Robertson, 2006; Tiessen and Stewart, 1983). Separating soil particles by sieving or ultrasonification into specific aggregate-size fractions followed by density separation yields POM of varying turnover rates and capacity to influence nutrient cycling. Unbound or free POM exists outside aggregates and as a result has residence times of 1-7 years (Buyanovsky et al., 1994). Occluded POM in aggregates dominated by silt has residence times of up to 400 years, whereas clay aggregate-protected POM can reach 1000 years. In general, the residence time of aggregate C increases with decreasing aggregate size.

The POM represents 5-15% of the total C of cultivated soils and up to 25% or more in surface layers of grassland and forest soil horizons (Haile-Mariam et al., 2008). POM has been identified as a significant determinant of the cycling of nutrients and C (Gregorich et al., 2006). The depletion of POM reduces the capacity of soil to supply nutrients through mineralization processes and thus could affect NPP adversely. The dynamics of POM are best studied with isotopes of C and N. For example, the transformation and stabilization of C and N in POM fractions can be determined following a switch from $C_4 \leftrightarrow C_3$ plant species (Balesdent and Mariotti, 1987). By using other isotope techniques, such as ^{14}C enrichment studies and carbon-14 dating, many studies have shown that about 20% of the total soil C turns over in less than 5 years, with the majority of this being free POM. The accumulation of POM can be used as an early indication of long-term C sequestration as a result of changes in soil management, such as reduced tillage (Six et al., 2002).

D SOM Maintenance

Studies on SOM maintenance commonly compartmentalize it into distinct fractions with different turnover rates. The compartmentalization of SOM has been based on physical fractionation, chemical extraction, analytical and mass spectroscopy, and conceptualization. The compartmentalization of SOM fractions is an effort to understand the significance of specific SOM fractions in influencing the cycling and stabilization of C and nutrients. All of these efforts have resulted in defining "operational fractions" with scant information on the structure and function. What is evident is that SOM is a mixture of fractions defined by chemical and physical properties. Future research should be aimed at determining the nature of these complex fractions with an emphasis on function, particularly their roles in governing and influencing biogeochemical processes.

V QUANTITY, DISTRIBUTION, AND TURNOVER OF CARBON IN SOIL AND SEDIMENTS

The C in soils and sediments plays a major role in sustaining ecosystems by influencing nutrient availability and cycling and through impacting physical properties, such as water-holding capacity and porosity. It also assumes a critical role influencing the amount of atmospheric CO_2 and other GHGs, such as CH_4 and N_2O. As mentioned earlier, a large portion of the increase in atmospheric CO_2 is attributed to land use change that increases the turnover of C in soils and sediments. The quantity of soil and sediment C is dependent on the balance between NPP and decomposition processes. These processes are closely matched and have maintained CO_2 levels below 300 ppm for millions of years. The increasing use of fossil C will continue to increase atmospheric CO_2 levels, now exceeding 400 ppm, in the foreseeable future (Solomon et al., 2009).

The total global NPP is estimated to be 61.8 Pg C year^{-1} (Table 12.4; Fig. 12.15). On a geological timescale, photosynthesis has been slightly larger than decomposition leading to estimated reserves of fossil fuels that represent four times the C stored in the soil. Oceanic photosynthesis and decay are nearly equal to terrestrial photosynthesis, yet are carried out by biota of only 45 Pg. A petogram is equivalent to 10^{15} g or 10^9 metric tons. The oceans have large reserves of dissolved C, both inorganic and organic, especially in the thermocline and in the deep sea representing a C pool of 38,000 Pg, substantially larger than soil C. Woody components of the land biota account for 75% of stored terrestrial plant C. Subtracting the values for respiration from GPP gives NPP values of 50-60 Pg, indicating that it takes approximately 10 years for the terrestrial plants to recycle their C. This can be attributed to fast turnover rates for leachates and easily decomposable residues relative to the tens, hundreds, and even thousands of years necessary for the turnover of woody components and humified material. The deep ocean and fossil C have even slower turnover rates. Tropical forests represent about a third of global NPP at 20.1 Pg C year^{-1}. They are characterized by high production and decomposition rates where nutrients are actively cycling through plant litter and new growth. Optimal temperature and

TABLE 12.4 The Area, Mean Annual Precipitation and Temperature, NPP, Inputs, and Global Stocks of Soil C for Biomes[a]

Biome	Area (×10^12 m^-2)	Area (% of Total)	MAP (mm)	MAT C	Mean Soil C (kg m^-2)	NPP (PgC/y)	C Inputs (kg m^-2 y^-1)	Terrestrial C Stock (PgC) Plant	Soil (0-3 m)	Total
Tropical forests	15.4	12.2	1400-4500	23-28.5	12.0	20.1	2.03	340	692	1032
Temperate forests	12	9.5	750-2500	9-14.5	8.7	7.4	0.85	139	262	401
Boreal forests	11.1	8.8	600-1800	4.5	16.4	2.4	0.50	57	150	207
Tropical savannas	24	19.0	500-1350	23.5	5.4	13.7	0.48	79	345	424
Temperate grasslands	9	7.1	450-1400	9	13.3	5.1	0.30	23	172	195
Mediterranean shrubland	3.9	3.1	400-600	14	7.6	1.3	0.46	17	124	141
Deserts	18.2	14.4	125-500	4.5-25	3.4	3.2	0.08	10	208	218
Tundra	8.8	7.0	250-1500	2.3	19.6	0.5	0.10	2	818[b]	820
Croplands	21.2	16.8	+	9-28.5	7.9	3.8	0.48	4	248	252
Wetlands g	2.8	2.2	+	4.5-28.5	72.3	4.3	0.17	15	450	465
Total	126.4	100			16.7	61.8	0.05	686	3051	3737

[a]Amundson (2001) and Jobbágy and Jackson (2000).
[b]Tarnocai et al. (2009).

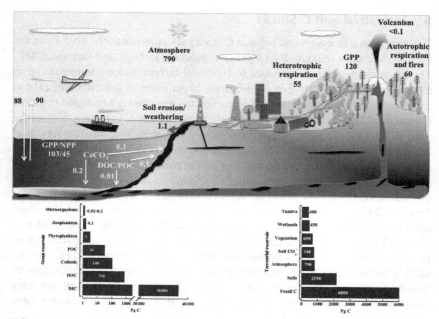

FIG. 12.15   The global C cycle showing major ocean and terrestrial pools and annual transfers of C. The values are presented in petograms, which are equivalent to 10^{15} g or 10^9 metric tons. *(Adapted from IPCC, 2007.)*

moisture conditions for decomposition prevail in moist tropical environments, leading to relatively closed nutrient cycles in the absence of disturbance. Temperate forests represent about 12% of global NPP at 7.4 Pg C year^{-1}. Nonforest biomes represent about 45% of global NPP, with savannas (14%) and grasslands (19%) accounting for the majority. In northern latitude biomes, boreal forests and tundra represent 8% of global NPP at 2.4 and 2 Pg C, respectively. In this colder climate, lower temperatures lead to slower decomposition of plant litter. The accumulation of plant matter in the forest floor litter layers can immobilize essential nutrients, particularly N and P, leading to reduced NPP. Disturbance, such as fire, is required to release nutrients to sustain NPP. Desert biome NPP is limited by moisture and large temperature swings of more than 20 °C daily, and some of the decomposition is abiotic in nature, such as UV-induced oxidation. These xeric environments have a global NPP of 3.2 Pg C year^{-1}, and abiotic decomposition forces appear to be significant. Future changes in global NPP will occur as climate change brings altered precipitation and temperature regimes. Increasing CO_2 levels may increase NPP through C fertilization and increased water use efficiency. The direction of NPP has been predicted to increase; however, uncertainties in these estimates make the magnitude of increase unpredictable.

A Terrestrial Soil C Stocks

The total global belowground organic C stocks are approximately 3195 Pg in the upper 3 m of soil excluding litter layers and biochar (Jobbágy and Jackson, 2000). Though rarely mentioned, inorganic soil C or soil carbonates are estimated to be up to 748 Pg to a depth of 100 cm. Significantly larger stores of soil carbonate exist at greater soil depths (Batjes, 1996). Soil carbonates are often not included in global C estimates. Total nonforest biome, belowground organic C content is 65% of the total at 2091 Pg. Tropical forests have 692 Pg followed by temperate forests at 262 Pg and boreal at 150 Pg. The accumulation of soil C in the soil surface is from aboveground litter accumulation and root inputs, particularly fine root turnover. In most soils, the highest concentration of C is found in the 0-20 cm depth with the exceptions being entisols, histisols, and soils with buried A horizons (Jobbágy and Jackson, 2000; Fig. 12.16). Basing C turnover on surface versus whole solum amounts provides a different perspective of actively cycling versus older C (Table 12.5). Soil C in the surface is influenced by new C inputs compared to deeper soils where older C accumulates (Rumpel and Kögel-Knabner, 2011). Approximately half the soil C turns over very slowly, with mean residence times often of thousands of years (Paul et al., 2001). Table 12.5 shows various turnover rates based on the amount of soil C found at different soil depths. (Turnover rates

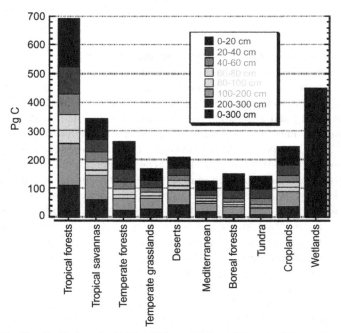

FIG. 12.16 The distribution of soil C by different depths to 300 cm in different biomes of the world. Wetland soils are only depicted from 0 to 300 cm.

TABLE 12.5 Turnover Rates of Soil C Inputs in Various Soil Depths

	Soil Depth (cm)					
	0-20 cm		0-40 cm		0-300 cm	
	K	Turnover	K	Turnover	K	Turnover
Biome	(y^{-1})	(y)	(y^{-1})	(y)	(y^{-1})	(y)
Tropical forests	0.187	5.3	0.119	8.4	0.045	22.2
Tropical savannas	0.162	6.2	0.098	10.2	0.033	29.9
Temperate forests	0.105	9.5	0.073	13.7	0.039	25.9
Temperate grasslands	0.063	15.9	0.040	24.8	0.016	63.3
Deserts	0.039	25.4	0.023	42.6	0.007	144.2
Mediterranean	0.060	16.7	0.038	26.2	0.014	69.7
Boreal forests	0.099	10.1	0.066	15.1	0.037	27.4
Tundra	0.019	52.3	0.011	89.8	0.006	164.8
Croplands	0.159	6.3	0.102	9.8	0.041	24.4
Wetlands					0.001	945.4

are discussed in detail in Chapter 17.) Turnover rates shown for the 0-20 and 0-40 cm soil depths are generally two to four times faster than in the 0-300 cm soils for all biomes. Land use change, such as conversion to agriculture or disturbance from deforestation, affects the interpretation of the actively cycling soil C more in the surface rather than the deeper soil C.

The highest accumulation of C is not always associated with high NPP. Wetlands have high soil C ($33 \, kg \, m^{-2}$) but relatively low C inputs ($0.17 \, kg \, m^{-2} \, year^{-1}$) combined with very slow decomposition rates (Tables 12.4 and 12.5). Turnover of NPP is slowed by inundation causing low redox potentials and anoxic conditions. In cold tundra and boreal forest soils, turnover of NPP is limited by inputs and low temperature causing C to accumulate as litter and POM.

B   Ocean C Pools

The largest pool of C in the ocean is dissolved inorganic C or carbonates at 38,000 Pg C, the majority of which is in the form of carbonates derived primarily from solution-dissolution reactions with the atmosphere. There is a net influx of 2 Pg C into the ocean annually. The next largest pool of ocean C is dissolved organic C (DOC) and other colloids at 800 Pg C. The DOC is produced via phytoplankton exudation and their grazing by zooplankton. Particulate organic C, such as dead organisms and fecal pellets, represents 30 Pg C. Phytoplankton and zooplankton contain 3 and 0.1 Pg C, respectively. Bacterial biomass is the small C pool ranging from 0.02 to 0.2 Pg C.

VI ROLE OF CLIMATE CHANGE ON THE GLOBAL C CYCLE

Rising levels of GHG, specifically CO_2, CH_4, and nitrous oxide (Chapter 14), will impact both the biological and geological C cycles.

A Results from Elevated CO_2 Studies

Higher atmospheric CO_2 levels should lead to higher NPP. Large-scale ecosystem level experiments (free atmospheric carbon exchange [FACE]) have revealed that terrestrial plants, particularly tree species, sometimes respond to elevated CO_2 by increasing photosynthesis while reducing transpiration (Ainsworth and Long, 2005; Huang et al., 2007). The results confirm small-scale chamber experiments showing light-saturated C uptake, diurnal assimilation, growth, and above- and belowground biomass production increase, while stomatal conductance decreases under elevated CO_2 conditions (Ainsworth and Long, 2005). Van Kessel et al. (2000) showed that grasses and legumes respond differently to elevated CO_2. Nitrogen-fixing clovers dramatically increased yield under elevated CO_2, whereas ryegrass responded only to increased N inputs. Yield increases in crop FACE experiments were less than anticipated compared to controlled chamber studies (Ainsworth and Long, 2005). Although no reasons have been specifically identified for the lack of a CO_2 fertilization effect, changes in soil resource availability (such as C-enriched litters that negatively influence the cycling of nutrients) and environment (increased temperature/changes in moisture status) have been implicated in the explanation of the lack of growth stimulus under elevated CO_2. Tree FACE results are particularly noteworthy because forest biomes dominate global NPP, suggesting that global climate models that include feedback between terrestrial biosphere and the atmosphere are poorly constrained (Norby et al., 2005). Silva and Horwath (2013) reviewed 66 studies of changes in tree growth over the last century and concluded that only six studies showed evidence of increased NPP. The majority of studies showed a growth decline. These results indicate that other factors, such as progressive nutrient limitation or drought stress, may offset any positive influence of CO_2 fertilization.

B Soil Processes in a Warmer Climate

The impact of climate change, specifically warming, will likely change both NPP and decomposition rates and the production and consumption of GHG.

Significantly more C is stored in soils than is found in the atmosphere, which suggests that warming could release soil C to the atmosphere through increased decomposition. It is estimated that the C stored in the litter layer above the mineral soil will decrease by 15% by 2100 (Davidson and Janssens, 2006). Additional losses of the same magnitude are expected to occur in the surface, mineral soil horizons. More important are potentially larger losses of up to 25% in C-rich soils in wetlands, peatlands, and permafrost. In addition to losses in soil C, it is expected that rising CO_2 will increase CH_4 and N_2O production from increased root growth and reduced soil water losses (van Groenigen et al., 2011). However, when depth of soils is taken into consideration, tropical soils store as much C as their temperate counterparts.

C Methane in the C Cycle

Methane is the most abundant hydrocarbon in the atmosphere, yet it comprises less than 1% of the global C budget. The current global abundance of CH_4 is 1775 ppb, giving a total atmospheric burden of approximately 5000 Tg (Solomon et al., 2007). Greater amounts of CH_4 are found as natural gas in fossil fuel deposits, as hydrates or clathrates compounds in ice (e.g., permafrost), and in the ocean floor. As a GHG, CH_4 is responsible for about 21% of the total radiative forcing attributed to the major GHGs or 0.48 W m^{-2}. The total annual global sources of CH_4 are between 503 and 610 Tg with sinks about 492-556 Tg, resulting in an annual average increase of 22 Tg (Table 12.6). The major removal mechanism of about 95% of the CH_4 from the atmosphere involves radical chemistry, when it reacts with the hydroxyl radical (•OH), formed from water vapor that is broken down by ozone in combination with ultraviolet radiation. The reaction occurs in the troposphere resulting in a methane lifetime of 8-10 years. The remaining 5% is removed by methanotrophs in the upland soil environment.

Methane emissions into the atmosphere come from a variety of biogenic and industrial processes. Microbial production of methane results from the decomposition of organic materials in the absence of oxygen. Biogenic CH_4 sources account for the majority of global emissions. Carbon dioxide is used as an electron acceptor, and a reduced organic compound is used as the donor. The reduction of CO_2 occurs in under reduced conditions, such as in wetlands, or where oxygen diffusion is severely limited, such as within soil aggregates. Waterlogged soils, such as rice paddies and wetlands, waste disposal sites, and enteric fermentation in ruminants, are typical examples of methanogenic habitats. Rice cropping systems, a highly productive constructed wetland, account for about 15% of total global CH_4 annual emissions. Approximately 1% of global NPP is converted to CH_4, half of which is oxidized to CO_2 by methanotrophs in the soil (Reeburgh, 2003). The production and distribution of fossil fuels contributes about 7% to total CH_4 emissions. The recent increase in hydraulic fracturing, commonly called "fracking," to remove CH_4 or natural gas from rock

TABLE 12.6 Global Sources and Sinks for Methane (Adapted from Bousquet et al. (2006) and Forster et al. (2007).)

Sources	Tg CH_4 y^{-1}
Natural sources	
Wetlands	100-231 (166)[a]
Termites	20-29 (24.5)
Oceans	4-15 (9.5)
Anthropogenic sources	
Ruminants	76-92 (84)
Rice production	31-112 (71.5)
Biomass burning	14-88 (51)
Waste (landfills and human and animal sewage)	35-69 (22)
Fossil fuel production/distribution	74-110 (92)
Total sources	**503-610 (557)**
Sinks	
Atmospheric removal	458-556 (507)
Soil microbial oxidation	21-34 (28.5)
Total sinks	**492-581 (535)**
Atmospheric increase (y^{-1})[b]	**11-29 (22)**

[a]*Average values in parentheses.*
[b]*Calculated using average values.*

formations composed of shale, is increasing emissions by about 1.0 Tg. Recent evidence suggests there is a renewed growth in atmospheric CH_4 following almost a decade of decline (Dlugokencky et al., 1998; Rigby et al., 2008). Methane is a more potent GHG compared to CO_2, and its dynamics have garnered much interest from the scientific community.

D   Elevated CO_2 and Ocean Processes

The impact of rising CO_2 on ocean NPP is less certain for a number of reasons. The NPP of oceans (35-70 pg C year^{-1}) is comparable to terrestrial NPP (61.8 Pg C year^{-1}); however, its productivity is limited by light and nutrients (phosphorus, nitrate, silicic acid, and micronutrients, especially iron; Bopp and Le Quéré, 2009). Increasing atmospheric CO_2 causes a decline in pH due to a shift in seawater carbonate chemistry. The process of ocean acidification alters seawater chemical speciation and biogeochemistry of many elements and compounds. The most pronounced effect is the lowering of $CaCO_3$ saturation that will impact calcifying organisms from plankton to corals (Doney et al., 2009). Other effects of changing calcification include changes in the Mg to Ca ratio (Ries, 2005). Under the scenario of increasing atmospheric CO_2 levels of 1% $year^{-1}$, ocean pH may decrease by an estimated 0.3 pH units from present day value of 8.1 (Orr et al., 2005). A decrease in the calcification rate will potentially lead to a decrease in the burial rate of carbonates.

VII FUTURE CONSIDERATIONS

The C cycle is complex having both a geological and a short-term biological cycle. Because the geological C cycle is on the order of millions of years, no immediate solutions to affect this cycle could be implemented to mitigate anthropogenically caused climate change. At the current rate of emissions, it will take thousands of years to consume the CO_2 already emitted into the atmosphere from fossil C combustion (Solomon et al., 2009). As for the climate change issues imminently facing humanity, the biological cycle could be manipulated by recycling existing C with biofuels, sequestering active C by ocean iron fertilization, reforestation, and geologic capture, in addition to a number of other emerging technologies. Ecosystem management solutions to maximize soil and sediment C sequestration provide only short-term solutions and are limited by the need for perpetual management to ensure the stability of the sequestered C. Engineered solutions, such as solar energy, ocean tidal and wave power, nuclear energy, and geothermal energy, will need to be implemented to reduce our reliance on C-rich fossil fuels. More energy efficient devices and machinery would also decrease energy use. However, as technology improves to extract fossil fuels, particularly natural gas, the bridge to alternative energy will remain long. Longer-term solutions must involve reducing the emissions of other primary GHGs, namely CH_4 and N_2O. Pressure to grow food crops to feed an estimated 9 billion population will continue to act negatively on the potential to use ecosystem management to offset increasing GHG emissions. In particular, soil disturbance effects on C sequestration, N_2O emission from N fertilizers, and CH_4 emission from ruminants and rice production will need to be addressed to reduce their negative influence on the climate.

The major production and turnover components of the C cycle are fairly well characterized at the process level. However, on a global scale, the interaction of organismal and metabolic diversity, sources and sinks for C, and anthropogenic influences have yet to be fully appreciated. These interactions must be better understood to adequately predict ecosystem response to perturbations such as climate change.

REFERENCES

Ainsworth, E.A., Long, S.P., 2005. What have we learned from 15 years of free-air CO_2 enrichment (FACE)? A meta-analytic review of the responses of photosynthesis, canopy. New Phytol. 165, 351–371.

Allison, F.E., Sherman, M.S., Pinck, L.A., 1949. Maintenance of soil organic matter: I. Inorganic soil colloid as a factor in retention of carbon during formation of humus. Soil Sci. 68, 463–478.

Anders, E., 1989. Prebiotic organic matter from comets and asteroids. Nature 342, 255–257.

Balesdent, J., Mariotti, A., 1987. Natural ^{13}C abundance as a tracer for studies of soil organic matter dynamics. Soil Biol. Biochem. 19, 25–30.

Batjes, N.H., 1996. Total carbon and nitrogen in the soils of the world. Eur. J. Soil Sci. 47, 151–163.

Benner, R., Fogel, M.L., Sprague, E.K., Hodson, R.E., 1987. Depletion of ^{13}C in lignin and its implications for stable carbon isotope studies. Nature 329, 708–710.

Berner, R.A., 2004. The Phanerozoic Carbon Cycle: CO_2 and O_2. Oxford University Press, New York.

Bopp, L., Le Quéré, C., 2009. Ocean carbon cycle. Am. Geophys. Union. http://dx.doi.org/10.1029/2008GM000780.

Bronik, C.J., Lal, R., 2005. Soil structure and management: a review. Geoderma 124, 3–22.

Buyanovsky, G.A., Aslam, M., Wagner, G.H., 1994. Carbon turnover in soil physical fractions. Soil Sci. Soc. Am. J. 58, 1167–1173.

Calderon, F.J., Schultz, D.J., Paul, E.A., 2012. Carbon allocation, belowground transfers, and lipid turnover in a plant-microbial association. Soil Sci. Soc. Am. J. 76, 1614–1623.

Cambardella, C.A., Elliott, E.T., 1992. Particulate soil organic-matter changes across a grassland cultivation sequence. Soil Sci. Soc. Am. J. 56, 777–783.

Chien, Y.Y., Kim, E.G., Bleam, W.F., 1997. Paramagnetic relaxation of atrazine solubilized by humic micellar solutions. Environ. Sci. Technol. 31, 3204–3208.

Chiou, C.T., Porter, P.E., Schmedding, D.W., 1983. Partition equilibria of nonionic organic compounds between soil organic matter and water. Environ. Sci. Technol. 17, 227–231.

Davidson, E.A., Janssens, I.A., 2006. Temperature sensitivity of soil carbon decomposition and feedbacks to climate change. Nature 440, 165–173.

Dlugokencky, E.J., Masarie, K.A., Lang, P.M., Tans, P.P., 1998. Continuing decline in the growth rate of the atmospheric methane burden. Nature 393, 447–451.

Doney, S.C., Fabry, V.J., Feely, R.A., Kleypas, J.A., 2009. Ocean acidification: the other CO_2 problem. Ann. Rev. Mar. Sci. 1, 169–192.

Dungait, J.J., Hopkins, D.W., Gregory, A.S., Whitmore, A.P., 2012. Soil organic matter turnover is governed by accessibility not recalcitrance. Glob. Chang. Biol. 18, 1781–1796.

Falkowski, P.R., Scholes, J., Boyle, E., Canadell, J., Caneld, D., Elser, J., Gruber, N., Hibbard, K., Högberg, P., Linder, S., Mackenzie, F.T., Moore III, B., Pedersen, T., Rosenthal, Y., Seitzinger, S., Smetacek, V., Steffen, W., 2000. The global carbon cycle: a test of our knowledge of Earth as a system. Science 290, 291–296.

Finlay, A.H., Soderstrom, B., 1992. Mycorrhiza and carbon flow to the soil. In: Allen, M. (Ed.), Mycorrhiza Functioning. Chapman & Hall, London, UK, pp. 134–160.

Flaig, W., Beutelspacher, H., Rietz, E., 1975. Chemical composition and physical properties of humic substances. In: Gieseking, J.E. (Ed.), Soil Components. Springer Verlag, Berlin, pp. 1–211.

Grandy, A.S., Robertson, G.P., 2006. Aggregation and organic matter protection following cultivation of an undisturbed soil profile. Soil Sci. Soc. Am. J. 270, 1398–1406.

Gregorich, E.G., Beare, M.H., McKim, U.F., Skjemstad, O.J., 2006. Chemical and biological characteristics of physically uncomplexed organic matter. Soil Sci. Soc. Am. J. 70, 975–985.

Haider, K., Schulten, H.R., 1985. Pyrolysis field ionization mass spectrometry of lignins, soil humic compounds and whole soil. J. Anal. Appl. Pyrolysis 8, 317–331.

Haider, K., Martin, J.P., Filip, Z., 1975. Humus biochemistry. In: Paul, E.A., McLaren, A.D. (Eds.), Soil Biochemistry, vol. 4. Marcel Dekker, New York, pp. 195–244.

Haile-Mariam, S., Collins, H.P., Wright, S., Paul, E.A., 2008. Fractionation and long-term laboratory incubation to measure soil organic matter dynamics. Soil Sci. Soc. Am. J. 72, 370–378.

Harris, D., Pacovsky, R.S., Paul, E.A., 1985. Carbon economy of soybean-mycorrhiza-rhizobium-glomus associations. New Phytol. 101, 427–440.

Hassink, J., 1995. Decomposition rate constants of size and density fractions of soil organic matter. Soil Sci. Soc. Am. J. 59, 1631–1635.

Hassink, J., 1997. The capacity of soils to preserve organic C and N by their association with clay and silt particles. Plant Soil 191, 77–87.

Hatcher, P.G., Maciel, G.E., Dennis, L.W., 1981. Aliphatic structure of humic acids; a clue to their origin. Org. Geochem. 3, 43–48.

Hayes, M.H.B., Clapp, C.E., 2001. Humic substances: considerations of compositions, aspects of structure, and environmental influences. Soil Sci. 166, 723–737.

Henneberry, Y.K., Kraus, T.E.C., Fleck, J.A., Krabbenhoft, D.P., Bachand, P.M., Horwath, W.R., 2010. Removal of inorganic mercury and methylmercury from surface waters following coagulation of dissolved organic matter with metal-based salts. Sci. Total Environ. 409, 631–637.

Henneberry, Y.K., Kraus, T.E.C., Peter, N., Horwath, W.R., 2012. Structural stability of coprecipitated natural organic matter and ferric iron under reducing conditions. Org. Geochem. 48, 81–89.

Horwath, W.R., Elliott, L.F., 1996. Ryegrass straw component decomposition during mesophilic and thermophilic incubations. Biol. Fertil. Soils 21, 227–232.

Horwath, W.R., Pregitzer, K.S., Paul, E.A., 1994. ^{14}C allocation in tree-soil systems. Tree Physiol. 14, 1163–1176.

Houghton, R.R., 2007. Balancing the global carbon budget. Annu. Rev. Earth Planet. Sci. 35, 313–347.

Huang, J.G., Bergeron, Y., Denneler, B., Berninger, F., Tardif, J., 2007. Response of forest trees to increased atmospheric CO_2. Crit. Rev. Plant Sci. 26, 265–283.

Imbrie, J., Imbrie, K., 1979. Ice Ages: Solving the Mystery. Enslow Publishers, Short Hills, NJ.

IPCC, 2007. Climate change 2007: the physical science basis. In: Solomon, S., Qin, D., Manning, M., Chen, Z., Marquis, M., Averyt, K.B., Tignor, M., Miller, H.L. (Eds.), Contribution of Working Group I to the Fourth Assessment Report of the Intergovernmental Panel on Climate Change. Cambridge University Press, Cambridge, United Kingdom and New York, USA.

Jobbágy, E.G., Jackson, R.B., 2000. The vertical distribution of soil organic carbon and its relation to climate and vegetation. Ecol. Appl. 10, 423–436.

Jokic, A., Schulten, H.R., Cutler, J.N., Schnitzer, M., Huang, P.M., 2004. A significant abiotic pathway for the formation of unknown nitrogen in nature. Geophys. Res. Lett. 31, L05502. http://dx.doi.org/10.1029/2003GL018520.

Kaiser, K., Guggenberger, G., 2003. Mineral surfaces and soil organic matter. Eur. J. Soil Sci. 54, 1–18.

Karickhoff, S.W., 1981. Semi-empirical estimation of hydrophobic pollutants on natural sediments and soils. Chemosphere 10, 833–846.

Kindler, R., Miltner, A., Richnow, H., Kastner, M., 2006. Fate of gramnegative bacterial biomass in soil – mineralization and contribution to SOM. Soil Biol. Biochem. 38, 2860–2870.

Kleber, M., Sollins, P., Sutton, R., 2007. A conceptual model of organo-mineral interactions in soils: self-assembly of organic molecular fragments into zonal structures on mineral surfaces. Biogeochemistry 85, 9–24.

Kögel-Knabner, I., 1986. Estimation and decomposition pattern of the lignin component in forest humus layers. Soil Biol. Biochem. 18, 589–594.

Kögel-Knabner, I., 2002. The macromolecular organic composition of plant and microbial residues as inputs to soil organic matter. Soil Biol. Biochem. 34, 139–162.

Kraus, T.E.C., Dahlgren, R.A., Zasoski, R.J., 2003. Tannins in nutrient dynamics of forest ecosystems – a review. Plant Soil 25, 41–66.

Le Quéré, C., Raupach, M.R., Canadel, J.G., Marland, G., Bopp, L., Ciais, P., Conway, T.J., Doney, S.C., Feely, R.A., Foster, P., Friedlingstein, P., Gurney, K., Houghton, R.A., House, J.I., Huntingford, C., Levy, P.E., Lomas, M.R., Majkut, J., Metzl, N., Ometto, J.P., Peters, G.P., Prentice, I.C., Randerson, J.T., Running, S.W., Sarmiento, J.L., Schuster, U., Sitch, S., Takahashi, T., Viovy, N., van der Werf, G.R., Woodward, F.I., 2009. Trends in the sources and sinks of carbon dioxide. Nat. Geosci. 2, 831–836.

Leboeuf, E.J., Weber, W.J., 2000. Macromolecular characteristics of natural organic matter. 1. Insights from glass transition and enthalpic relaxation behavior. Environ. Sci. Technol. 34, 3623–3631.

Liang, C., Balser, T.C., 2010. Microbial production of recalcitrant organic matter in global soils: implications for productivity and climate policy. Nat. Rev. Microbiol. http://dx.doi.org/10.1038/nrmicro2386 cl.

Lodish, H., Berk, A., Zipursky, L.S., Matsudaira, P., Baltimore, D., Darnell, J., 2000. Molecular Cell Biology, fourth ed. W. H. Freeman & Co., New York.

Martin, J.P., Haider, K., 1979. Biodegradation of ^{14}C-labeled model and cornstalk lignins, phenols, model phenolase humic polymers, and fungal melanins as influenced by a readily available carbon source and soil. Appl. Environ. Microbiol. 38, 91–110.

McCarroll, D., Loader, N.J., 2004. Stable isotopes in tree rings. Q. Sci. Rev. 23, 771–801.

Milchunas, D.G., Lauenroth, W.K., Singh, J.S., Cole, C.V., Hunt, H.W., 1985. Root turnover and production by 14C dilution – implications for C partitioning in plants. Plant Soil 88, 353–365.

Morbidelli, A., Chambers, J., Lunine, J.I., Petit, J.M., Robert, F., Valsecchi, G.B., Cyr, K.E., 2000. Source regions and timescales for the delivery of water to the Earth. Meteorit. Planet. Sci. 35, 1309–1320.

Norby, R.J., DeLucia, E.H., Gielen, B., Calfapietra, C., Giardina, C.P., King, J.S., Ledford, J., McCarthy, H.R., Moore, D.J.P., Ceulemans, R., De Angelis, P., Finzi, A.C., Karnosky, D.F., Kubiske, M.E., Lukac, M., Pregitzer, K.S., Scarascia-Mugnozza, G.E., Schlesinger, W.H., Oren, R., 2005. Forest response to elevated CO^2 is conserved across a broad range of productivity. Proc. Natl. Acad. Sci. U. S. A. 102, 18052–18056.

Normile, D., 2009. Round and round: a guide to the carbon cycle. Science 325, 1642–1643.

Orr, J.C., Fabry, V.J., Aumont, O., Bopp, L., Doney, S.C., 2005. Anthropogenic ocean acidification over the twenty-first century and its impact on calcifying organisms. Nature 437, 681–686.

Paul, E.A., Collins, H.P., Leavitt, S.W., 2001. Dynamics of resistant soil carbon of midwestern agricultural soils measured by naturally occurring ^{14}C abundance. Geoderma 104, 239–256.

Piccolo, A., 2001. The supramolecular structure of humic substances. Soil Sci. 166, 810–832.

Preston, C.M., Schnitzer, M., 1984. Effects of chemical modifications and, extractants on the carbon-13 NMR spectra of humic materials. Soil Sci. Soc. Am. J. 48, 305–311.

Rasmussen, C., Southard, R.J., Horwath, W.R., 2008. Litter type and soil minerals control temperate forest soil carbon response to climate change. Glob. Chang. Biol. 14, 2064–2080.

Raymo, D.N., 1991. Geochemical evidence supporting T. C. Chamberlin's theory of glaciation. Geology 19, 344–347.

Reeburgh, W.S., 2003. Global methane biogeochemistry. In: Keeling, R.F., Holland, H.D., Turekian, K.K. (Eds.), Treatise on Geochemistry: The Atmosphere, vol. 4. Elsevier-Pergamon, Oxford, UK, pp. 65–89.

Ries, J.B., 2005. Aragonite production in calcite seas: effect of seawater Mg/Ca ratio on the calcification and growth of the calcareous alga *Penicillus capitatus*. Paleobiology 31, 445–458.

Rigby, M., Prinn, R.G., Fraser, P.J., Simmonds, P.G., Langenfelds, R.L., Huang, J., Cunnold, D.M., Steele, L.P., Krummel, P.B., Weiss, R.F., O'Doherty, S., Salameh, P.K., Wang, H.J., Harth, C.M., Muhle, J., Porter, L.W., 2008. Renewed growth of atmospheric methane. Geophys. Res. Lett. 35, http://dx.doi.org/10.1029/2008GL036037, L22805.

Rillig, M.C., 2004. Arbuscular mycorrhizae, glomalin, and soil aggregation. Can. J. Soil Sci. 84, 355–363.

Rinnan, R., Bååth, E., 2009. Differential utilization of carbon substrates by bacteria and fungi in tundra soil. Appl. Environ. Microbiol. 75, 3611–3620.

Robinson, J.M., 1990. Lignin, land plants and fungi: biological evolution affecting Phanerozoic oxygen balance. Geology 15, 607–610.

Rumpel, C., Kögel-Knabner, I., 2011. Deep soil organic matter—a key but poorly understood component of terrestrial C cycle. Plant Soil 338, 143–158.

Schmidt, M.W.I., Torn, M.S., Abiven, S., Dittmar, T., Guggenberger, G., Janssens, I.A., Kleber, M., Kögel-Knabner, I., Lehmann, J., Manning, D.A.C., Nannipieri, P., Rasse, D.P., Weiner, S., Trumbore, S.E., 2011. Persistence of soil organic matter as an ecosystem property. Nature 478, 49–56.

Schnitzer, M., 1978. Humic substances: chemistry and reactions. In: Schnitzer, M., Khan, S.U. (Eds.), Soil Organic Matter, vol. 8. Elsevier Scientific Publication Company, New York, pp. 1–64.

Schulten, H.-R., Schnitzer, M., 1993. A state-of-the-art structural concept for humic substances. Naturwissenschaften 80, 29–30.

Schulten, H.-R., Sorge-Lewin, C., Schnitzer, M., 1997. Structure of "unknown" soil nitrogen investigated by analytical pyrolysis. Biol. Fertil. Soils 24, 249–254.

Schurig, C., Smittenberg, R.R., Berger, G., Kraft, J., Woche, S., Goebel, M.O., Heipieper, H.J., Miltner, E., Kaestner, M., 2013. Microbial cell envelope fragments and the formation of soil organic matter: a case study from a glacier forefield. Biogeochemistry 113, 595–612.

Silva, L.C.R., Horwath, W.R., 2013. Explaining global increases in water use efficiency: why have we overestimated responses to rising atmospheric CO_2 in natural forest ecosystems? PLoS One 8, e53089. http://dx.doi.org/10.1371/journal.pone.0053089.

Silva, L.C.R., Sternberg, L.S.L., Haridasan, M., Hoffman, W.A., Miralles-Wilhelm, F., Franco, A.C., 2008. Expansion of gallery forests into central Brazilian savannas. Glob. Chang. Biol. 14, 1–11.

Simpson, A.J., Simpson, M.S., Smith, E., Kelleher, B.P., 2007. Microbially derived inputs to soil organic matter: are current estimates too low? Environ. Sci. Technol. 41, 8070–8076.

Six, J., Elliott, E.T., Paustian, K., 2000. Soil macroaggregate turnover and microaggregate formation: a mechanism for C sequestration under no-tillage agriculture. Soil Biol. Biochem. 32, 2099–2103.

Six, J., Conant, R.T., Paul, E.A., Paustian, K., 2002. Stabilization mechanisms of soil organic matter: implications for C-saturation of soils. Plant Soil 241, 155–176.

Six, J., Frey, S.D., Thiet, R.K., Batten, K.M., 2006. Bacterial and fungal contributions to carbon sequestration in agroecosystems. Soil Sci. Soc. Am. J. 70, 555–569.

Solomon, S., Qin, D., Manning, M., Chen, Z., Marquis, M., Averyt, K.B., Tignor, M., Miller, H.L., 2007. Contribution of Working Group I to the Fourth Assessment Report of the Intergovernmental Panel on Climate Change. Cambridge University Press, Cambridge.

Solomon, S., Plattnerb, G.K., Knuttic, R., Friedlingsteind, P., 2009. Irreversible climate change due to carbon dioxide emissions. Proc. Natl. Acad. Sci. U. S. A. 106, 1704–1709.

Stevenson, F.J., 1994. Humus Chemistry: Genesis, Composition, Reactions. John Wiley & Sons, Inc., New York.

Strickland, M.S., Rousk, J., 2010. Considering fungal:bacterial dominance in soil – methods, controls, and ecosystem implications. Soil Biol. Biochem. 42, 1385–1395.

Sutton, R., Sposito, G., 2005. Molecular structure in soil humic substances: the new view. Environ. Sci. Technol. 39, 9009–9015.

Swift, R.S., 1991. Effects of humic substances and polysaccharides on soil aggregation. In: Wilson, M.S. (Ed.), Advances in Soil Organic Matter Research: The Impact on Agriculture and the Environment. Redwood Press Ltd., Wiltshire, England, pp. 45–60.

Swift, R.S., 1999. Macromolecular properties of soil humic substances: fact, fiction and opinion. Soil Sci. 164, 790–802.

Theng, B.K.G., 1982. Clay-polymer interactions: summary and perspectives. Clays Clay Miner. 30, 1–10.

Throckmorton, H.M., Bird, J.A., Dane, L., Firestone, M.K., Horwath, W.R., 2012. The source of microbial C has little impact on soil organic matter stabilization in forest ecosystems. Ecol. Lett. 15, 1257–1265.

Tiessen, H., Stewart, J.W.B., 1983. Particle-size fractions and their use in studies of soil organic matter: II. Cultivation effects on organic matter composition in size. Soil Sci. Soc. Am. J. 47, 509–514.

Tisdall, J.M., Oades, J.M., 1982. Organic matter and water-stable aggregates in soils. J. Soil Sci. 33, 141–163.

Toosi, E.R., Doane, T.A., Horwath, W.R., 2012. Abiotic solubilization of soil organic matter, a less-seen aspect of dissolved organic matter production. Soil Biol. Biochem. 5, 12–21.

van Groenigen, K.J., Osenberg, C.W., Hungate, B.A., 2011. Increased soil emissions of potent greenhouse gases under increased atmospheric CO_2. Nature 475, 214–216.

Van Kessel, C., Horwath, W.R., Hartwig, U., Harris, D., Lüscher, A., 2000. Net soil carbon input under ambient and elevated CO_2 concentrations: isotopic evidence after four years. Glob. Chang. Biol. 6, 123–135.

Vollmer, W., Blanot, D., de Pedro, M.A., 2008. Peptidoglycan structure and architecture. FEMS Microbiol. Rev. 32, 149–167.

Von Lützow, M., Kögel-Knabner, I., Ekschmitt, K., Matzner, E., Guggenberger, G., Marschner, B., Flessa, H., 2006. Stabilization of organic matter in temperate soils: mechanisms and their relevance under different soil conditions – a review. Eur. J. Soil Sci. 57, 426–445.

Voroney, R.P., Paul, E.A., Anderson, D.W., 1989. Decomposition of wheat straw and stabilization of microbial products. Can. J. Soil Sci. 69, 63–77.

Waksman, S.A., 1936. Humus, Origin, Chemical Composition, and Importance in Nature. Balliere, Tindall and Cox, London.

Watanabe, T., 2003. Microbial degradation of lignin-carbohydrate complexes. In: Koshijima, T., Watanabe, T. (Eds.), Association between Lignin and Carbohydrates in Wood and Other Plant Tissues. Springer-Verlag, New York, pp. 237–287.

Wedin, D.A., Tieszen, L.L., Dewey, B., Pastor, J., 1995. Carbon isotope dynamics during grass decomposition and soil organic matter formation. Ecology 76, 1383–1392.

FURTHER READING

Amundson, R., 2001. The carbon budget in soils. Annu. Rev. Earth Planet. Sci. 2, 535–562.

Tarnocai, C., Canadell, J.G., Schuur, E.A.G., Kuhry, P., Mazhitova, G., Zimov, S., 2009. Soil organic carbon pools in the northern circumpolar permafrost region. Global Biogeochem. Cycles 23, GB2023. http://dx.doi.org/10.1029/2008GB003327.

Chapter 13

Methods for Studying Soil Organic Matter: Nature, Dynamics, Spatial Accessibility, and Interactions with Minerals

Claire Chenu[1], Cornelia Rumpel[2] and Johannes Lehmann[3]
[1]Agro Paris Tech, UMR Bioemco, Thiverval Grignon, France
[2]CNRS, UMR Bioemco, Thiverval Grignon, France
[3]Department of Crop and Soil Sciences, Cornell University, Ithaca, NY, USA

Chapter Contents

Soil Microbiology, Ecology, and Biochemistry. http://dx.doi.org/10.1016/B978-0-12-415955-6.00013-X
383

I INTRODUCTION

Soil organic matter (SOM) can be defined as all organic materials found in soil that are part of or have been part of living organisms. It is a continuum of materials at various stages of transformation due to both abiotic and biotic processes. Despite its low concentrations in soil, organic matter is of major qualitative importance. It contributes to soil chemical fertility, being a reserve of nutrients released with mineralization (N, P, and S) and retains nutrient cations on its negative charges (Ca^{++}, K^+, Mg^{++}). It also is the basis of soil biological activity, being the source of C and energy for heterotrophs, from microorganisms through all food webs. SOM plays a major role in soil physical fertility, increasing water retention, aggregating mineral particles, and thus reducing soil erosion. Moreover, SOM has a major role in the environment: (a) it retains organic pollutants, heavy metals, and radionucleides, thus protecting water quality and (b) it produces nitrates and phosphates by mineralization, thereby reducing water quality. The dynamics of SOM lead to mineralization or storage of CO_2, methane, and nitrous oxide, which are major greenhouse gasses (see Chapter 12).

Characterization of SOM fulfills different aims (Table 13.1): (a) understanding its origin and formation in relation to pedogenesis; (b) understanding and predicting SOM contents and soil organic C (SOC) stocks as a function of land use and of cropping and forestry practices in different climatic and pedogenic contexts; (c) understanding and predicting soil properties and functions influenced by SOM; (d) identifying critical SOM contents relative to these functions and identifying indicators of soil quality; and (e) predicting ecosystem response to perturbations.

Given the variety of the inputs (i.e., different plants and plant tissues, microorganisms, and animals) and their different stages of decomposition, SOM is an extremely complex mixture. The complexity of SOM, which intrigued early chemists and soil scientists, has led to attempts to subdivide it into several fractions by various approaches. Each approach has provided a different picture of SOM, adding to the difficulty of its understanding.

Soils are extremely heterogeneous materials and environments due to: (a) the diversity of their mineral and organic constituents and that of their living inhabitants, (b) their size spectrum from the nanometer to the meter (Fig. 13.1), and (c) the complex 3-D spatial arrangements of these constituents, which define a network of voids of various sizes, more or less filled with water and air. Soils are the juxtaposition of many microenvironments, in which the constituents and the

TABLE 13.1 Rationale for Methods to Study Soil Organic Matter

Questions	Methods Category	Examples of Methods
What is SOM made of?	Characterization methods	Elemental analysis, biochemistry methods, FTIR, NMR, pyrolysis, NEXAFS
What are SOM chemical and physical properties?	Fractionation methods (physical and chemical)	Particle size fractionation, solubility in acids and bases
	Characterization methods	Measurement of charge, complexation properties, hydrophobicity
Where is SOM located in soil?	Physical fractionation methods	Aggregate size fractionation
	Visualization methods	Optical microscopy, scanning electron microscopy, nanoSIMS, STXM and NEXAFS
Which processes control SOM dynamics and stabilization in soil?	Fractionation methods (physical and chemical)	Aggregate size fractionation, HF hydrolysis
	Visualization methods	Optical microscopy, scanning electron microscopy, nanoSIMS, STXM and NEXAFS
Is it possible to isolate-characterize SOM fractions corresponding to kinetic pools?	Fractionation methods (physical and chemical)	Particle size fractionation, chemical and thermal oxidation, long-term incubation
What is the turnover rate of SOM?	Isotopic methods, incubation	^{13}C natural labeling, ^{14}C dating, ^{14}C bomb labeling, tracer addition experiments

physical and physicochemical conditions (e.g., redox potential) may differ considerably between the microenvironments and from the average soil characteristics. As a consequence, the traditional and pragmatic approach that consists of predicting soil properties and functions from its average composition and characteristics is strongly limited. This applies in particular to SOM matter dynamics and functions. Two categories of methods try to account for the structural heterogeneity of soil, even at nanometer spatial scales: fractionation methods, which

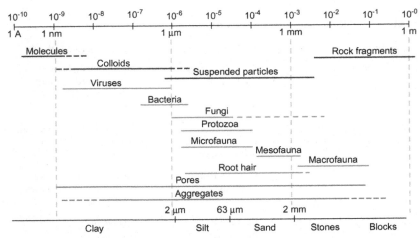

FIG. 13.1   Size of soil constituents and structural features. *Adapted from Totsche and Kögel-Knabner (2004).*

attempt to separate SOM into meaningful subsets, and visualization methods, which intend to observe SOM in its environment. An integrated understanding of organic matter functions within a specific pedological context is best attained by the combination of fractionation methods with detailed molecular analysis of SOM composition and measurement of its dynamics.

II   QUANTIFYING SOIL ORGANIC MATTER

SOM quantification is generally performed on soil that has passed a 2-mm sieve, thus excluding plant roots, litter, coarse plant debris, or large fauna. It is rarely, and with great difficulty, measured directly. Most methods measure its most abundant element, C, and multiply the SOC content by a factor of 1.72-2 to obtain SOM contents. These conversion factors correspond to the fact that the C content of SOM usually ranges from 50% to 58%. Organic C contents can be determined by wet digestion, using dichromate to oxidize organic molecules, as in the classical Walkley-Black method (Nelson and Sommers, 1996), or by dry combustion. Dry combustion converts soil C to CO_2, which is quantified by various methods, such as infrared detection or thermoconductivity. If the soil contains carbonates, either combustion is performed after acid pretreatment to remove them (following adjustment of pH in acid solution or performing an acid fumigation) or the total C needs to be corrected for the inorganic C content measured separately.

III   FRACTIONATION METHODS

The complexity of SOM has historically led to numerous attempts to separate SOM into subsets, or fractions, each being homogeneous in their characteristics or in their dynamics. The aims of fractionation methods are to subdivide the complex continuum of SOM to identify and characterize its constituents and to isolate "functional pools." These correspond to:

(1) kinetic pools (i.e., subsets of OM homogeneous in their turnover rate within the pool and distinct on that characteristic from other pools, each pool being characterized by specific turnover rates and by a pool size; Christensen, 2001);

(2) pools of OM that are controlled by a specific stabilization mechanism (e.g., recalcitrance, spatial inaccessibility, and organomineral interactions; von Lutzow et al., 2007); and

(3) pools of OM that have a major implication in a given soil function, these not being restricted to biological functions, such as the recycling of C and N by biodegradation and mineralization, but also to physical and chemical functions, such as cation exchange capacity or pollutant sorption (Feller et al., 2001).

The dynamics of SOM are usually described in models by several kinetic pools, three pools being the most frequent: (1) one pool, named either labile or active, is assigned turnover rates of weeks to months; (2) an intermediate or slow pool with turnover rates of years to decades; and (3) a passive to inert pool with turnover rates of centuries to millennia (Falloon et al., 1998). Ongoing research is being conducted to identify SOM fractions matching these conceptual pools, under the well-known headline of "modeling the measurable or measuring the modelable" (Christensen, 1996; Elliott et al., 1996). Achieving this would increase the efficiency for the calibration and testing of models.

Fractionation methods separate the SOM continuum into fractions by either biological, chemical, or physical approaches. The separated fractions can be characterized by (a) their size (soil mass, % of soil C, N, S, or P), (b) elemental and biochemical composition, as well as their physicochemical or biological characteristics, and (c) their C and N turnover rates.

In addition to being meaningful (i.e., able to separate functional pools or distinct organic components), fractionation methods should be (a) adaptable to soils with different textures, mineralogy, pH, and water logging conditions; (b) sensitive to changes in management and environmental conditions; and (c) repeatable and standardizable.

A Chemical and Thermal Fractionation

1 Extraction Methods

The objective of extraction methods is to bring as much as possible of the target SOM in solution for analysis (see Table S13.1 in online Supplemental Material at http://booksite.elsevier.com/9780124159556).

1.1 Dissolved Organic Matter

Dissolved organic matter (DOM) is typically defined as SOM smaller than 0.45 μm in soil science. It is either obtained with different extractants, from cold water to saline solutions with different ionic strengths, after mixing a given mass of soil with the extractant and centrifuging the slurry to recover DOM, or it may be recovered directly *in situ* by using suction cups or lysimeters (Haynes, 2005; Zsolnay, 2003). DOM generally represents less than 1% of organic C in agricultural soils and a small percentage of TOC in forest soils. It is made of diverse molecules, from low-molecular weight to colloidal moieties, and is heterogeneous in turnover rates, associating labile components with OM that has decadal turnover times (Haynes, 2005; Kalbitz et al., 2000; Zsolnay, 2003).

1.2 Solubility in Bases and Acids: Humic Substances and Extracts

The oldest fractionation of SOM was introduced in the late eighteenth century by F. Achard (1786). Alkali solutions, such as NaOH or pyrophosphate ($Na_4P_2O_7$), solubilize large amounts of SOM (up to 80% of TOC with NaOH; Stevenson, 1982). Oxygen-containing functional groups (carboxyl, alcohols, phenols) are ionized at high pH, and thus molecules rich in such functional groups are rendered soluble. In contrast, nonpolar, alkyl-C molecules are not extracted by alkali solutions. The enzymatic and oxidative depolymerisation of plant polymers increases the abundance of carboxylic functional groups and thus the solubility of plant material in alkali increases with decomposition. The organic molecules solubilized by alkali are named humic substances, but are strictly speaking an operational definition of the organic matter extracted by the method (Kleber and Johnson, 2010; Schmidt et al., 2011). Lowering the pH below 2, by using HCl, precipitates organic molecules named humic acid extracts, whereas others, named fulvic acids, remain soluble (Fig. 13.2). The material insoluble in alkali is named humin. Extensive studies have been

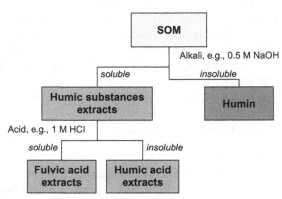

FIG. 13.2 Humic extracts separation.

performed since the nineteenth century to characterize SOM using this approach. The relative abundance of humic and fulvic acid extracts varies according to soil type and was used to decipher and characterize pedogenesis (Duchaufour, 1976). Although these methods allow for the extraction and analysis of a large proportion of SOM, they suffer drawbacks: (a) formation of artifacts: the alkaline extraction products do not represent OM as it is present *in situ*, as extracted organic molecules present distinct characteristics from those existing in soil in an adsorbed state (Baldock and Nelson, 2000) or directly found using spectromicroscopy (Lehmann et al., 2008) and (b) although it was thought that humic substance extracts consist of pedogenic macromolecules, it appears that both humic and fulvic acid fractions can be described as a mixture of biomolecules (Kelleher and Simpson, 2006), including inherited macromolecules and supramolecular associations of small organic entities (Piccolo, 2002). This separation procedure is extensively used to study the interaction between SOM and pollutants because the chemical reactivity of SOM constituents can be directly investigated.

2 Hydrolysis Methods

Hot water (80-100 °C) is used to isolate a fraction of SOM that is hypothesized to be readily available to microorganisms (Haynes, 2005; see Table S13.2 in online Supplementary Materials at http://booksite.elsevier.com/9780124159556). Hot water extracts typically correspond to 1-5% of SOC and contains mainly carbohydrates and N-rich compounds. This fraction is very sensitive to land use and agricultural practices and is thus used as an early indicator of changes in management and as a proxy for organic compounds promoting soil aggregation. Hot water soluble compounds are mainly composed of organic compounds with rapid turnover (Balesdent, 1996).

Concentrated solutions of acids, such as 6 N HCl or 1 M H_2SO_4 have been extensively used to extract and characterize soil N and C based on their reactivity (Paul et al., 2006). In the presence of acids and usually at high temperature, organic bonds are cleaved by dissociated water molecules. Fatty acids, proteins, and polysaccharides (including hemicellulose and cellulose) are susceptible to acid hydrolysis treatment, whereas long-chain alkyls, waxes, lignin, and other aromatics, including pyrogenic C, are resistant. The composition of the HCl insoluble fraction could be affected by the formation of artifacts, the so-called melanoidins, by recombination of carbohydrates and proteins following hot acid treatment. Reflux acid hydrolysis with 6 N HCl at 116 °C typically solubilizes 65-80% of soil N (Bremner, 1967) and 30-87% of SOC (Paul et al., 2006). On average, acid hydrolysis-resistant SOC is 1200-1400 years older than bulk SOC (Paul et al., 2006). It was proposed as a proxy for the passive pool of SOC that would allow the identification of chemically recalcitrant organic C. The size of the acid-resistant organic fraction varies with soil management, such as afforestation or cultivation, showing that the amount of SOC remaining after acid hydrolysis is not only due to its intrinsic chemical properties, but also due to its interaction with minerals.

3 Oxidation Methods

As biodegradation of SOM is an oxidative process, mild oxidants, such as $KMnO_4$ (0.33 or 0.03 M) have been proposed to mimic it (Blair et al., 1995). Potassium permanganate oxidizes 13-28% of SOC, with the abundance of this fraction being very sensitive to land management. However, the turnover rate of organic C resistant to $KMnO_4$ has not been measured to date, and it is not clear whether this method offers a proxy for a SOC kinetic pool.

Strong oxidants are used classically to eliminate SOM prior to textural analysis or mineralogical analysis. As not all SOM can be oxidized, these methods have been proposed to isolate a kinetically stable organic C pool. Hydrogen peroxide (H_2O_2, 10-50%), disodium peroxodisulfate ($Na_2S_2O_8$ 8%, buffered with $NaHCO_3$), and sodium persulfate (NaOCl, 6%) oxidize most SOM (see Table S13.2 in online Supplementary Material at http://booksite.elsevier.com/9780124159556), with variations depending on soil constituents and pH. Disodium peroxodisulfate and sodium persulfate are considered more efficient because of the additional dispersing action of sodium, which increases the accessibility of SOM to the oxidant in the soil matrix (Mikutta et al., 2005). Resistant compounds are essentially long chain, alkyl, and pyrogenic C. Resistance to wet oxidation also seems to depend on soil microaggregation and the presence of sorbents, such as clay minerals and oxides. Organic matter resistant to these oxidants was found to be older than average SOM by 75-3000 years (Eusterhues et al., 2005; Kleber et al., 2005; Fig. 13.3). Strong oxidants therefore have the potential to isolate a kinetically stable fraction.

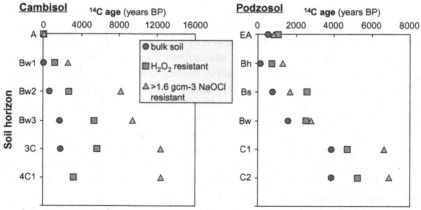

FIG. 13.3 Age of bulk soil organic carbon and organic fractions resistant to oxidation in two temperate soils; a dystric Cambisol (left) and a Podzosol (right). *Using data from Kögel-Knabner et al. (2008).*

4   Destruction of the Mineral Phase

Hydrofluoric (HF) acid hydrolyzes silicates, carbonates, and most oxides, but does not affect apatite, spinel, and zircon. Treatment of soil with HF is assumed to free the OM adsorbed to the hydrolyzed minerals or coprecipitated with them. If the previously adsorbed or coprecipitated OM is solubilized, it may be lost with the HF solution. Loss of SOC after HF treatment can thus be used to quantify nonparticulate SOM adsorbed to or coprecipitated with minerals. Application of HF treatment to subsoil horizons dissolved up to 92% of SOC, showing that most of the SOM in these horizons is adsorbed to minerals (Eusterhues et al., 2003). Moderate losses of SOC (10-30%) and small alterations of the chemical composition are found in litter layer samples after HF treatment, showing that SOM can be affected by the treatment as well (von Lutzow et al., 2007). The HF soluble fraction is not always older than HF-resistant SOM, hence HF treatments do not isolate kinetic pools. This chemical fractionation allows quantification of organic matter that is stabilized by adsorption to minerals (Eusterhues et al., 2003). However, as the HF reagent is extremely toxic, this fractionation procedure may be adopted only with adequate safety considerations.

Sodium dithionite solutions are known to solubilize all iron and aluminum oxides, whereas sodium oxalate is used to solubilize poorly crystallized iron and aluminum oxides (Shang and Zelazny, 2008). These reactants thus break the bonds between oxide minerals and the adsorbed or coprecipitated organic compounds. Although these methods solubilize a fair amount of SOM in different soil types, the use of a bicarbonate buffer or of the oxalate itself adds C in solution and impedes measurement of the turnover of the extracted or the resistant SOC.

5   Calorimetric Methods

Thermal methods typically use a temperature program to scan the reactivity of the sample over a wide temperature range and may provide a detailed picture of the thermal properties of the material, which is dependent on its chemistry and surface properties (Plante et al., 2009). These methods have been applied to various organic materials and were applied in soil science to distinguish labile and stable fractions of SOM (Plante et al., 2009). There is a significant relationship between thermal parameters and the biological stability of organic matter in soil (Plante et al., 2011).

B   Physical Fractionation

1   Introduction and Definitions

Physical fractionation separates subsets of soil according to physical criteria, such as size and/or density or magnetic susceptibility. This is done after various degrees of dispersion are applied to soil to break bonds between elements of the soil structure. Physical fractionation procedures seek to minimize chemical alterations and aim to separate primary soil particles, primary organomineral or secondary organomineral associations.

Primary organomineral associations relate to the "primary structure of soils as defined by soil texture" (Christensen, 1992). They result from the association of SOM with primary mineral particles and are isolated after complete dispersion of soils. *Secondary organomineral associations* are aggregates, separated after limited dispersion of soil. In many soils the complete dispersion of particles, especially clay-sized particles, is not possible if organic matter is not oxidized; hence, true primary organomineral complexes cannot be isolated (Chenu and Plante, 2006). However, these terms will be used here for clarity and their limits discussed when needed.

Physical fractionation can pursue different objectives. One is to separate soil organic constituents in their original state (i.e., in the state they have in soils) without any chemical alteration. This is one of the reasons behind the separation of particulate organic matter (POM), which is essentially composed of weakly decomposed plant debris, root fragments, fungal hyphae, spores, seeds, fecal pellets, and faunal bodies. Another objective is to separate SOM in its physical surroundings with the assumption that the association of organic matter with minerals and their location within the 3-D soil architecture controls their dynamics (physicochemical and physical protection). Physical fractionation methods have brought a considerable amount of information on SOM and have led to major advances in the understanding of soil C dynamics and stabilization processes.

2 Soil Dispersion

Physical fractionation schemes represented in Fig. 13.4 imply some degree of soil dispersion prior to application of the separation criteria. A variety of dispersion techniques and protocols exist, complicated by a lack of standardization. Indeed standardization is seldom possible using energy units (e.g., how to express the energy provided by agitating a certain mass of soil in water?). Furthermore, a given objective of dispersion (e.g., to disperse all particles >50 μm diameter) will require different energy inputs depending on the soil type and land use.

Weak dispersion of soil is achieved by immersing it in water with no or some agitation. The least dispersion is obtained using field capacity moist soil as done to measure aggregate stability (Angers and Mehuys, 1993; Arias et al., 1997). Strong dispersion of soil can be achieved by mechanical agitation of soil in water with the addition of glass or agate beads (Balesdent et al., 1998) or by the use of ultrasonic vibrations. The energy released by ultrasonic probes depends on the probes themselves, the diameter of the probe, its depth of immersion, and the volume of the vessel. Calorimetric measurements can be used to calibrate and standardize ultrasonic dispersion (Schmidt et al., 1999). Excessive energy might cause breakdown of small plant remnants, such as POM, or mica-type mineral particles. For a range of acidic to mildly acidic temperate soils,

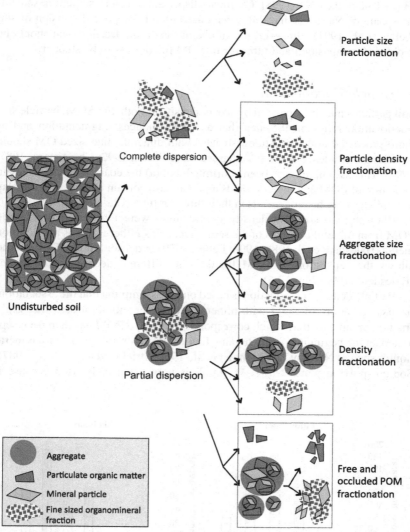

FIG. 13.4 Schematic presentation of physical fractionation methods. Fine sized organomineral fraction can be clay+silt fraction or clay-size fraction, depending on protocols.

sonication energies between 450 and 500 J ml^{-1} were shown to completely disperse the soil without abrasion artifacts (i.e. to yield the same particle size distribution as the textural analysis; Balesdent et al., 1991; Schmidt et al., 1999). To avoid POM abrasion, soil dispersion can also be performed in two steps: (1) agitation with glass beads to allow for the separation between POM and a fraction <50 μm and (2) an ultrasonic dispersion of the <50 μm fraction

(Balesdent et al., 1991, 1998). Chemical dispersants, such as sodium hexameta-phosphate or Na resins, have also been used to aid complete dispersion of soil (Feller et al., 1991). However, use of chemicals during fractionation should be avoided if composition and turnover of SOM fractions is to be studied.

3 Particle Fractionation

Soil particles may be separated by size or density or both. For SOM, particle size fractionation relies on the idea that as decay induces fragmentation and as more processed organic matter associates with minerals, fine-sized OM should have longer residence times (Cambardella and Elliott, 1992; Christensen, 1987, 1992). This assumption has been confirmed, but (a) the coarser fractions remain a mixture of POM and mineral sands and may also contain char and (b) the clay size fractions are heterogeneous in their turnover time (Christensen, 1992).

Both particle size and density fractionation were introduced to separate POM from mineral particles of the same size, using either water as the flotation media (Balesdent et al. 1988, 1998; Feller, 1979) or denser liquids. This method allows the separation of POM fractions in different decomposition stages (Balesdent, 1996; Fig. 13.5).

Particle density fractionation is based on the assumption that the association, or disassociation, of SOM with minerals determines its turnover time. Light fraction organic matter, which corresponds roughly to POM has then been sep-arated, at the beginning, using organic liquids (bromoform), then dense mineral liquids such as NaI, $ZnCL_2$ (Monnier et al., 1962; Strickland and Sollins, 1987). Sodium polytungstate ($Na_6[H_2W_{12}O_{40}]$) is now extensively used because it

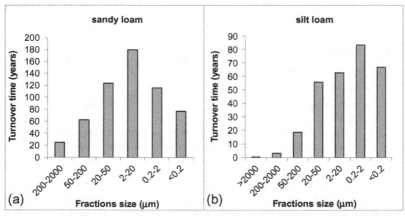

FIG. 13.5 Turnover time of particle fractions in temperate cultivated soils. Particles were separated by size (fractions ≤50 μm) or by both size and density (POM, i.e., fractions >50 μm). Turnover time was estimated from ^{13}C natural abundance. *Data from Balesdent et al. (1987), Balesdent (1996).*

spans a wide range of densities (from 1.0 to 3.1 g cm⁻³), with low viscosity and little toxicity. Moreover, it can be recycled (Six et al., 1999). However, some solubilisation of organic matter takes place, which is negligible when separating POM, but may amount to 20% of SOC when using it to fractionate fine-sized OM (Virto et al., 2008).

The nature of minerals affects their capacity to protect OM from decomposition. Therefore, a second rationale for performing density fractionation is the separation of distinct mineral species with their associated organic matter. Poorly crystallized aluminosilicates, feldspars, and iron oxides have been separated using solutions of sodium polytungstate based on their different densities (Basile-Doelsch et al., 2007). A limit to this approach is the incomplete dispersion of particles.

4 Aggregate Fractionation

Aggregate size fractionation is based on the premises that: (1) organic matter binds mineral particles together. Aggregate size fractionation can then be used to separate the organic matter responsible for aggregation (Puget et al., 1995) and (2) aggregates of different sizes correspond to different microbial habitats and provide contrasting conditions for decomposition (Beare et al., 1994; Elliott, 1986). Aggregates can be separated after immersion in water of initially dry or wet soil and by using different intensities of agitation or slaking. Increasing the applied energies will allow the separation of differently sized aggregates, as the smaller the aggregates, the higher the energy needed to break the aggregates. According to this concept, macroaggregates >250 µm (Elliott, 1986), sand-sized microaggregates (Besnard et al., 1996), or silt-sized microaggregates (Virto et al., 2008) can be separated. A two-step procedure has been introduced to separate sand-sized microaggregates located within macroaggregates (Six et al., 2000; Fig. 13.6). One limitation of the aggregate-size fractionation is that the size fractions contain true aggregates

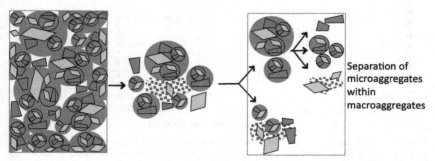

FIG. 13.6 Schematic presentation of separation of microaggregates within macroaggregates. *Method developed by Six et al. (2000).*

as well as free organic and mineral particles. The use of aggregate fractionation has allowed the demonstration of aggregate hierarchy in soils and the investigation of the complex interactions between soil structure and SOM dynamics (see review by Six et al., 2004). Although aggregates of different sizes are characterized by different turnover rates of their OM (Six and Jastrow, 2002), this approach does not allow to separate kinetic pools of SOC. The SOM located within the aggregate can also be stabilized because of its chemical recalcitrance and adsorption to minerals.

Organomineral fractions have a density intermediate between that of pure organic matter and pure mineral particles (Chenu and Plante, 2006; Fig. 13.7). Dense mineral solutions can thus be used to separate true aggregate fractions with intermediate densities from dispersed organic fractions (POM) and mineral fractions (Fig. 13.5). This approach has shown that occluded POM is physically protected from biodegradation by virtue of its location within aggregates (Besnard et al., 1996; Six et al., 2000).

Several methods have been developed to peel aggregates with the idea that organic matter should be more accessible in the outer volume of aggregates and consequently have faster turnover rates (Bol et al., 2004; Wilcke et al., 1996). However, these methods show gradients only in soils where aggregates turn over very slowly.

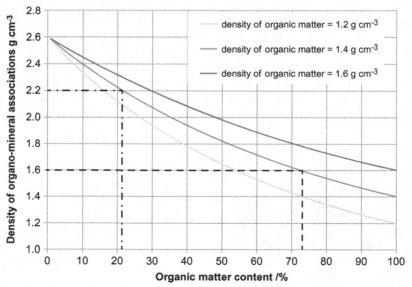

FIG. 13.7 Calculated density of organomineral associations assuming a density of either 1.2, 1.4, or 1.6 g cm⁻³ for the organic matter and a density of 2.6 g cm⁻³ for the mineral phase. Dashed drop-lines allow visualization of the OM content at the boundary when separations are performed at densities of 1.6 and 2.2 g cm⁻³. *Redrawn from Chenu and Plante (2006).*

IV   CHARACTERIZATION METHODS

Molecular characterization of SOM composition is essential because most pedological processes operate at the molecular scale. The objective of the analysis of biochemical constituents, as well as molecular composition of SOM, is to gain a better understanding of their origin and transformation processes. Solubilization of SOM is necessary because of the technical limitations for the characterization of SOM in its solid state. The advent of technical developments allowing for the characterization of the composition of functional groups using solid-state nuclear magnetic resonance (NMR) spectroscopy and molecular composition using analytical pyrolysis has led to less use of NaOH fractionation. In general, a combination of different methods is needed to obtain a good understanding of the origin and chemical composition of SOM. All methods presented in this section can be applied to both bulk soils and to soil fractions, provided the sample amount is sufficient (generally several hundred milligrams; Fig. 13.8).

A   Wet Chemical Methods for Analyses of Biochemical SOM Constituents

Wet chemical methods include solvent extraction for free lipids as well as acid and base hydrolysis or chemical oxidation to break macromolecules into their monomer constituents before analyses. Although the biochemical composition of pure organic matter in plant litter or organic soil horizons can be analyzed easily with

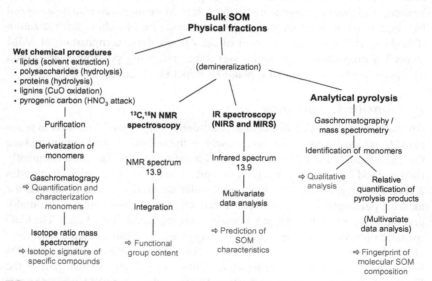

FIG. 13.8 Analysis scheme and information to be obtained with biogeochemical methods most commonly used for SOM analyses.

wet chemical methods, the presence of mineral interactions hampers the wet chemical analysis of SOM compounds in mineral soil horizons. Here, less than 50% of SOM can be accounted for when using these methods (Hedges et al., 2000).

1 Carbohydrates

Carbohydrates are abundant biochemical constituents of SOM. They account for approximately 50% of the plant litter entering the soil system (Kögel-Knabner, 2002). Although carbohydrates are labile compounds, which are usually rapidly metabolized by soil microbial biomass, their degradation may be reduced by adsorption to the mineral matrix (Miltner and Zech, 1998), as well as by their incorporation into soil aggregates (Puget et al., 1999). To understand the nature and functioning of carbohydrates in soil, it is important to know whether they are plant or microbial derived. Their monomers can be of specific origin (e.g., xylose and arabinose are of plant origin, whereas mannose and galactose are predominantly microbial; Cheshire, 1979). Determination of carbohydrates in soil at a molecular level involves acid hydrolysis, purification, and derivatization of the monosaccharides, which are subsequently analyzed. Cellulose, because of its stable crystalline structure, requires harsh hydrolysis conditions involving strong acids and high temperature, such as hot H_2SO_4 (Cheshire, 1979). Cold H_2SO_4 (Cheshire, 1979), cold HCl (Kiem et al., 2000), or hot TFA (Amelung et al., 1996) are used for the hydrolysis of hemicelluloses and other amorphous polysaccharides.

Determination of total monomer concentration after hydrolysis can be achieved by colorimetric methods (e.g., by using MBTH (3-methyl-2-benzothiazolinone hydrazone hydrochloride)). Monomers can also be analyzed by liquid chromatography or gas chromatography after derivatization (Cheshire, 1979). Characterization of the carbohydrate composition of SOM showed a predominant microbial signature, indicating possible stabilization by interaction with the mineral phase (Rumpel et al., 2010).

2 Aromatic Compounds

Aromatic compounds in soil include plant-derived lignin and fire-derived pyrogenic C, which are degraded more slowly by the soil microbial biomass and are therefore thought to be preserved in soil for the long term. The most frequently used method for soil lignin analysis is cupric oxide (CuO) oxidation. It includes a chemical breakdown of its macromolecular structure by CuO in an alkaline medium purification using solid phase extraction and subsequent gas chromatographic analyses of the released monomers (Hedges and Ertel, 1982). The CuO oxidation products include vanillyl, syringyl, and p-coumaryl units with their aldehyde, ketone, and acid side chains. The sum of CuO oxidation products is used as an indicator of soil lignin quantity, even if only a small part of the lignin is accessible with this procedure. The ratios between CuO products characterize the lignin composition and its state of degradation (Kögel-Knabner,

2000). CuO oxidation can be combined with stable isotope analyses, which allows for assessment of its dynamics (Dignac et al., 2005). The composition and fate of lignins was widely studied using this method, and it is now generally accepted that soil lignin turnover occurs at a decadal timescale (Thévenot et al., 2010).

Several methods can be used to quantify pyrogenic C (i.e., condensed fire derived C; Hammes et al., 2007). Wet-chemical analyses of pyrogenic C quantity and composition can be achieved by biomarker approaches (e.g., the analyses of benzene carboxylic acids (BPCA) after digestion with TFA, followed by nitric acid oxidation and silylation of the monomers before analysis by gas chromatography; Brodowski et al., 2005). Pyrogenic C consists of a continuum ranging from partly burned OM to completely graphitized materials. Therefore, different methods largely target different sections of the entire pyrogenic C spectrum but also have specific strengths and weaknesses that have to be evaluated before choosing one particular approach.

3 Aliphatic Compounds

Aliphatic compounds present in soil comprise free and ester-bound lipids. Free lipids are analyzed after solvent extraction by methanol/chloroform or methanol/dichloromethane mixtures and by using an accelerated solvent extraction system at high temperature and pressure (Bligh and Dyer, 1959 ; Wiesenberg et al., 2004). Total free lipid content can be determined gravimetrically after solvent evaporation. Characterization of lipid composition is usually carried out by gas chromatography-mass spectrometry (GC-MS) after derivatization using silylation or transmethylation. GC-MS analyses identify alkanes, alkenes, alcohols, sterols, fatty acids, and aromatic hydrocarbons. Lipids are commonly used as biomarkers and have successfully been applied to assess SOM composition following land use changes (Bull et al., 2000).

Phospholipid fatty acids can be used as biomarkers for all living microorganisms in soil except archea. They are fractionated after solvent extraction using silicic acid chromatography and then analyzed by capillary gas chromatography (Zelles et al., 1992). Branched glycerol dialkyl glycerol tetraethers (GDGT) may be used as indicators of past temperature and pH effects. They are separated by solid phase extraction and analyzed by high-pressure liquid chromatography/mass spectrometry (HPLC/MS; Hopmans et al., 2000).

Ester-bound lipids, which include cutin and suberin, are released by either saponification or transesterification, the former being more efficient (Mendez-Millan et al., 2010). Their relative abundance allows for the differentiation of root and shoot origin of SOM (Mendez-Millan et al., 2010).

4 Organic Nitrogen

Organic N compounds make up most of the N in soils and are important contributors to stabilized SOM. Moreover, N-containing compounds are, by order of decreasing abundance, proteins, free amino acids, amino sugars, and DNA. Proteins and amino acids may be analyzed after acid hydrolysis using 6 M HCl at high temperature. Hydrolysable proteins can be quantified calorimetrically by detection of the amino acids after their reaction with ninhydrine (Stevenson and Cheng, 1970). Quantification may also be carried out by gas chromatography after purification and derivatization. Amino acid enantiomers in soils may be used to assess the residence time of amino acids (Amelung and Zhang, 2001). The common hydrolysis procedure with 6 M HCl leaves a part of organic N unextracted, but the use of methanesulfonic acid (MSA) hydrolysis and nonderivatized amino acid and amino sugar quantification by ion chromatography with pulsed amperometric detection greatly increases the recovered N (Martens and Loeffelmann, 2003).

B Spectroscopic Methods to Characterize Functional Groups of SOM

Spectroscopic methods may give a more complete picture of bulk SOM composition than wet-chemical methods. However, due to their low sensitivity and interference of their signals by the mineral phase, or paramagnetic compounds, demineralization is often recommended as a pretreatment. This is usually carried out with HF acid, which was found to only slightly alter SOM composition, while greatly improving spectral quality (Rumpel et al., 2006).

1 Infrared Spectroscopy

Fourier transformed infrared analyses are especially useful for elucidating degradation pathways of litter compounds (Wershaw et al., 1996). Near infrared spectroscopy (NIRS) uses the near-infrared region of the electromagnetic spectrum (from about 0.75 to 2.5 μm wavelengths), whereas midinfrared spectroscopy (MIRS) exploits the region between 3 and 8 μm. The interpretation of the spectra is difficult because inorganic soil components produce signals that overlap with SOM absorption bands. This can be overcome by applying IR only to SOM in chemical extracts, demineralization prior to measurement (Rumpel et al., 2006; Fig. 13.9), or by subtracting IR spectra recorded before and after SOM removal from the samples (Skjemstad et al., 1996). Infrared peaks of soils and OM are often too complex to provide visual quantitative data. Multivariate data analysis is then associated with regression modeling (i.e., the modeling of relationships between two sets of measurements to predict certain parameters from IR spectra (Nduwamungu et al., 2009)). Such predictions can reduce the time and costs of the analyses of the predicted parameter. Midinfrared spectroscopy in combination with prediction modeling has been used to predict lignite C contribution in mine soils (Rumpel et al., 2001), to characterize soil fractions (Calderón et al., 2011), as well as bulk SOM composition (Terhoeven-Urselmans et al., 2006).

2   Solid-State NMR Spectroscopy

Solid-state NMR spectroscopy provides information on an atomic level that depends on the chemical environment of the observed nucleus. The technique gives an overview of the functional group composition of C- and N-containing compounds in SOM (Kögel-Knabner, 2000). Two NMR pulse programs may be used for solid samples: cross-polarization magic angle spinning (CPMAS), which detects C via its interaction with H atoms, and single-pulse or Bloch decay experiments, which detects C directly. As NMR spectroscopy is an expensive, time-consuming method due to the low abundance of C compared to H, CPMAS is generally preferred because it requires less measurement time. However, at high magnetic fields and low spinning speeds, spinning side bands can overlap with relevant signals and obscure the intensity distribution (Knicker, 2011). Aromatic C with low content of H-atoms, close to C, may not be seen by this method. Despite these drawbacks, ^{13}C and ^{15}N CPMAS NMR spectroscopy is widely used in soil science assuming that quantitative information about alkyl, O-alkyl, aromatic, and carboxylic groups present in SOM may be obtained with the right NMR parameters (Knicker, 2011; Fig. 13.9). ^{13}C CPMAS NMR analyses are of great use for characterizing biodegradation of plant litter (Kögel-Knabner, 2002), for showing the state of SOM biodegradation (Baldock et al., 1997), and to analyze the contribution of pyrogenic C in soil (Knicker, 2011). It can be used in combination with ^{13}C-labeled substances to

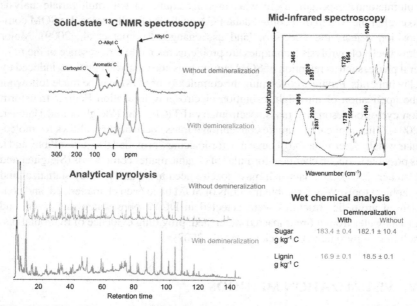

FIG. 13.9   Characterization of SOM in topsoil (A horizon) of a Dystric Cambisol under forest with four methods before and after demineralisation. *Data from Rumpel et al. (2006).*

study the fate of plant material in terrestrial environments (Kelleher et al., 2006). ^{15}N CPMAS NMR spectroscopy is less widely used due to the fact that most organic N seen by this method occurs in soils in the form of amide. NMR spectroscopy is a nonsensitive method due to the fact that NMR spectra consist of the accumulation of several hundred or thousands of signals.

C   Molecular Analysis of SOM

Molecular analyses of SOM in solid samples can be achieved by analytical pyrolysis or thermochemolysis. These two techniques are based on the heating of SOM with or without tetramethylammonium hydroxide (TMAH) under N_2 atmosphere to thermally break down SOM. The pyrolysis products are analyzed by GC-MS or pyrolysis molecular beam mass spectrometry py-MBMS. Carbohydrate-derived compounds can be distinguished from lignin-derived compounds, N-containing compounds, aliphatic compounds, and unspecific compounds. The interpretation of pyrolysis data requires a detailed knowledge of the pyrolysis behavior of the compounds under study as many pyrolysis products can originate from an SOM component, and thermal secondary reactions exist. Choice of polarity of the chromatographic column is decisive in GC-MS: polar compounds (i.e., sugars, lignins, and proteins) should be analyzed on polar columns, whereas for poly-aromatic compounds and aliphatic compounds, an apolar column is necessary (Dignac et al., 2006). Py-MBMS is a more rapid analysis that does not result in losses on the GC columns and is especially useful when used in conjunction with multivariate analysis to exploit the wealth of recorded data. PCA analysis of bulk molecular SOM composition detects the effects of land use change (Rumpel et al., 2009). Major drawbacks of pyrolysis techniques are problems related to the presence of the mineral phase, such as retention of SOM compounds and catalytic reactions induced by clay minerals. Progress concerning the complexity of SOM may be made following the application of ultrahigh resolution electrospray ionization Fourier transform ion cyclotron resonance mass spectrometry (FTICR-MS) (Sleighter and Hatcher, 2007). This application, unlike conventional mass spectrometry, allows for molecular weight determination of large macromolecules, provided they are soluble, and it is possible to obtain chemical formulas of organic matter compounds (Sleighter and Hatcher, 2007). The data-rich mass spectra need to be analyzed by statistical and graphical tools. Water-soluble fractions of SOM have been characterized, and condensed aromatic structures were detected in DOM, pore water, and river- and groundwater from a fire impacted watershed, providing evidence of microbial dissolution of pyrogenic C (Hockaday et al., 2007).

V   VISUALIZATION METHODS

The objective of visualization methods are twofold: (1) to gain morphological information on SOM to gain a better understanding of its dynamics or roles (e.g., the state of decomposition of plant debris) and (2) to locate SOM within its natural environment and study its relationship with mineral soil constituents, its accessibility to decomposers, or its contribution to soil structure. Visualization methods allows for the interpretation of the spatial heterogeneity and complexity of soils. They can be distinguished on the basis of:

- their spatial resolution that determines which constituents of SOM can be observed;
- how SOM is distinguished from minerals or voids; and
- the information given on the elemental or chemical nature of SOM.

Methods for the visualization of SOM, their principles, resolution and associated preparation methods are presented in Table S13.2 (see online Supplementary Material at http://booksite.elsevier.com/9780124159556).

A Light Microscopy

It is possible to observe the soil fabric using transmitted light after embedding soil in a resin and preparing thin sections. This allows identification of several organic matter features based on their shape and color: plant remnants, fecal pellets, and organic matter coatings on mineral grains. The origin of SOM (e.g., plant vs. leachates), its processing by soil fauna or flora, and its degree of association with minerals can be identified at scales of several micrometers to millimeters (Fig. 13.10a). Incident light at selected wavelengths can excite fluorescence within the sample, either due to autofluorescence of organic molecules (such as chitin) or to stains bound to organic molecules. Epifluorescence microscopy has a better spatial resolution (0.2 μm) than conventional light microscopy and can be used to visualize and enumerate bacteria and fungi within soil fabrics (Postma and Altemüller, 1990; Nunan et al., 2001; Fig. 13.10b). It is not used to locate dead organic matter because of the weak fluorescence recorded and the autofluorescence of minerals. The same limitation applies to confocal laser microscopy, which is seldom applied to soils, although it can be used on fresh soils to locate bacteria previously stained and inoculated (Gilbert et al., 1998; Minyard et al., 2012).

B Scanning Electron Microscopy

Through the use of scanning electron microscopy (SEM), the surface of the sample is scanned by an electron beam, and secondary emitted electrons are recorded (Table S13.2, see online Supplementary Material at http://booksite.elsevier.com/9780124159556). This allows for reconstruction of a 3-D image of the topography of the sample, with a resolution <1 μm. Specific preparation methods (low temperature SEM or cryoSEM) or specific SEMs (environmental SEM) are necessary to preserve the fragile hydrated structures of organic matter, microorganisms, or clay minerals (Chenu and Tessier, 1995). In SEM, organic matter is identified by its shape (Fig. 13.10c). SEM can be combined with energy dispersive X-ray fluorescence, which gives access to the elemental composition of samples.

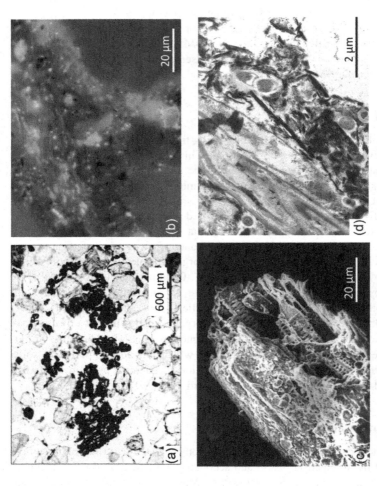

FIG. 13.10 Visualization of organic matter in soil fabrics. (a) Partly decomposed root debris in a Podzol observed with bright field light microscopy (Buurman et al., 2008) ; (b) bacteria located in a soil thin section using epifluorescence microscopy after the sample had been stained with Calcofluor white (Nunan et al., 2003); (c) plant vessel (POM) occluded by clay particles, SEM (Chenu, unpublished); and (d) bacteria in their soil habitat as visualized with TEM (UPb staining). Six bacteria can be observed in more or less close vicinity to the plant cell wall remnant (Chenu, unpublished).

In that case, the presence of organic matter can be deduced from C and N signals. SEM has been very useful in studies of soil aggregation by organic matter (Chenu, 1993; Dorioz et al., 1993), identification of POM in soil fractions (Besnard et al., 1996; Golchin, 1994), and localization of bacteria in their microhabitat (Chenu et al., 2001; Gaillard et al., 1999).

C Transmission Electron Microscopy

When transmission electron microscopes send an electron beam through very thin samples, the transmitted electrons form an image in which thicker regions of the sample, or regions with a higher atomic number, appear darker. Diffraction of part of the incident electrons provides additional contrast and information, such as the visualization of clay lattices. With the resolution of TEM being 0.2 nm, it is well adapted to the study of fine soil particles (Table S13.2, see online Supplementary Material at http://booksite.elsevier.com/9780124159556). Sample preparation involves the deposition of a suspension of soil particles on a grid or the preparation of an ultrathin section of the sample, after chemical fixation, to preserve organic matter features, dehydration, and embedding in a resin (Elsass et al., 2007). Organic matter appears white or pale gray due to its low mean atomic numbers. It is thus usually stained with heavy metals to enhance its contrast, either prior to embedment or directly on the thin section. Stains are contrasting agents with broad or more narrow (e.g., polysaccharides) specificity. Excellent visualization of decaying plant material, microorganisms, and their association with minerals can be obtained (Fig. 13.10d). Energy dispersive X-ray spectroscopy (EDS) additionally informs on the atomic composition of the sample. Another method, electron energy loss spectroscopy (EELS), analyzes the energy distribution of electrons that have crossed the sample. This carries information on the atomic composition and chemical bonding (Watteau et al., 2002). One advantage of EELS over EDS X-ray spectroscopy is the low sensitivity of the latter to low atomic number elements (i.e., those characteristic of organic matter) and the ability to obtain information about chemical bonds, not just atomic identification. TEM provided early evidence of the heterogeneity of soils at the microscale, including the description of diverse bacterial habitats and of sites where SOM could be inaccessible to microorganisms (Foster, 1988; Foster and Rovira, 1976; Kilbertus and Proth, 1979). TEM also showed the patchy distribution of organic matter among clay-sized particles and the occurrence of clay organic matter aggregates at submicrometric scales (Chenu and Plante, 2006).

D Atomic Force Microscopy

In atomic force microscopy (AFM), the surface of the sample is scanned with a physical probe attached to a cantilever that records the forces arising between the probe and the surface of the sample. This process gives an image of the topography of the sample with unmatched lateral and vertical resolution and can be applied to wet samples, but they need to be nearly flat. Thus AFM is seldom used in soil science, presumably because of the roughness of soil particles. It has, however, aided in the visualization of soil mineral and organic colloids and the identification of the patchy distribution of organic matter at the surface of clay minerals (Cheng et al., 2009; Citeau et al., 2006).

E X-Ray Spectro-Microscopy

Scanning transmission X-ray microscopes (STXM) are synchrotron-based, soft X-ray instruments that permit visualization and chemical mapping of thin specimens. An image of the transmitted X-rays is formed with the same principles as with TEM with a spatial resolution of a few tens of nanometers. Element distribution is obtained using X-rays absorbed by the sample at energy levels characteristic of atoms or bonds of atoms (Lehmann et al., 2009; Thieme et al., 2010). STXM can be applied to hydrated samples, provided that they are very thin. It has been used on fine particle deposits, ultrathin sections of soil embedded in sulfur, and ultrathin sections of shock-frozen microaggregates (Kinyangi et al., 2006; Solomon et al., 2012; Wan et al., 2007). STXM can be coupled with near-edge X-ray fine structure (NEXAFS) spectroscopy (for higher-z elements). This method is also called X-ray absorption, fine structure (XANES) and gives information on the functional groups within the organic matter at the same time as spatial resolution. This extremely powerful set of methods and instruments has allowed visualization of the spatial distribution of organic matter at a submicrometric scale as discrete particles, as well as coatings among soil mineral particles. A spatially distinct distribution of different chemical qualities of organic matter has been found in several studies (e.g., aromatic vs. carboxylic; Kinyangi et al., 2006; Lehmann et al., 2008; Solomon et al., 2012; Wan et al., 2007; Fig. 13.11). Provided the energy of the X-ray beam is strong enough, elements such as Si, Ca, Fe, Al, and K can also be mapped (Solomon et al., 2012; Wan et al., 2007). This shows that the heterogeneity of the distribution of OM is also dependent on the adsorbent mineral phase. Spatial distribution of S and P can be determined, albeit with slightly lower spatial resolution of about 80 nm (Brandes et al., 2007; Lehmann et al., 2009).

F NanoSIMS

Secondary ion mass spectrometry is based on sputtering a few atomic layers from the surface of a sample using a primary ion beam and analyzing the emitted secondary ions, distinguished by their mass-to-charge ratio, and ejected from a sample with a mass spectrometer. Primary ions of cesium (Cs^+) or oxygen (O^-) are used to eject ions of opposite charge (e.g., $^{12}C^-$, $^{13}C^-$, $^{12}C^{14}N^-$, $^{12}C^{15}N^-$, and $^{28}Si^-$). Maps of elements and isotopes can be recorded with a spatial resolution of 1μm with NanoSIMS (nano scale secondary ion mass spectrometry). The operation must be performed under high vacuum, so samples need to be dry and should be flat

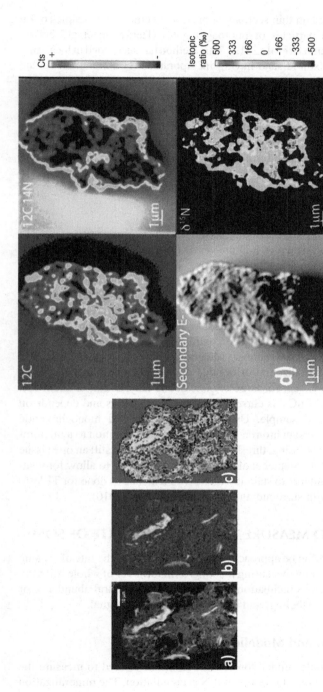

FIG. 13.11 Visualization of organic matter in soil fabric. (a), (b), and (c) Distribution of organic matter in a microaggregate from an Oxisol using NEXAFS combined with STXM. (a) Map of total C concentration using a given portion of the NEXAFS spectra (subtraction of energy region at 280–282 eV from 290–292 eV). The brighter the zone, the higher the C concentration. (b) Map of aromatic C, using another spectral region. Again, the brighter colors are the richer in aromatic compounds. (c) A cluster map of different C forms (takes also into account aliphatic C, carboxylic C, and phenolic C, where different colors correspond to different types of spectra recorded in NEXAFS (Lehmann et al., 2008). (d) NanoSIMS images of a microaggregate from a soil to which ^{15}N labeled litter had been added showing the secondary electron image for morphology and the ^{12}C, [12]C-^{14}N, and δ^{15}N spatial distributions. Points a, b, and c have very high δ^{15}N, showing sites of high assimilation of the labeled litter (Remusat et al., 2012).

(nm rugosity). It is used on thin sections of previously embedded samples or on deposits of fine-sized particles of microaggregates (Herrmann et al., 2007a; Mueller et al., 2012; Remusat et al., 2012). This method has succeeded in localizing bacteria in a coarse-textured sand having incorporated ^{15}N ammonium sulfate (Herrmann et al., 2007b) and analyzing the spatial and temporal distribution of assimilated [13]C-^{15}N labeled amino acids or ^{15}N labeled litter at the scale of individual soil particles or microaggregates (Fig. 13.11d; Mueller et al., 2012; Remusat et al., 2012). NanoSIMS has a very high mass resolution, but adequate standards are necessary to obtain quantitative analysis. To date, it is the only visualization method that provides dynamic information on biogeochemical processes affecting SOM, given the possibility to locate isotopes of C and N.

G   X-Ray Computed Microtomography

Another methodology rapidly expanding in soil science is X-ray microcomputed tomography (μCT) that generates 3-D images of fresh undisturbed soil samples. The sample is subjected to a beam of X-rays and X-ray radiographs are recorded at different angles during stepwise rotation of the sample around its vertical axis. The transmission of the X-rays is attenuated though the sample, depending on the density and composition (average atomic number) of the object. After image processing, cross-sectional images of the attenuation coefficients allow for reconstruction of a 3-D image of the sample. Contrast in the images comes from the differences in average atomic number. Mineral particles and voids can easily be distinguished, but visualization of organic matter poses problems that have not yet been resolved, except for large organic debris (Elyeznasni et al., 2012). The spatial resolution of μCT is currently several micrometers and depends on the size of the scanned sample. Using small samples and monochromatic X-ray beams issued from synchrotron lines improves the resolution to a micrometer. Accurate image processing, thresholding, and analysis are still an open issue in μCT (Peth et al., 2008; Sleutel et al., 2008). One possibility to allow for visualization of organic matter is to stain it with heavy elements, as done for TEM to enhance its contrast with surrounding minerals (Peth et al., 2010).

VI   METHODS TO MEASURE THE TURNOVER RATE OF SOM

The dynamics of SOM can be approached by either following the fate of organic material added to soil or by measuring the long-term turnover of whole soil C by different methods, such as incubation and modeling, ^{13}C natural abundance or labeling, ^{14}C dating or labeling, and by following the ^{15}N signal.

A   Soil Incubation and Modeling of CO_2 Evolution

Incubation of soil samples under controlled conditions is used to measure the mineralization of SOM by CO_2 or mineral N accumulation. The mineralization curve is modeled, using first-order kinetics (Chapter 17) to derive turnover rate constants (Paul et al., 2006). This approach can also be used to biologically fractionate SOM (i.e., separate kinetic pools of C within SOM on the basis of the shape of the mineralization curve). Paul et al. (1999) used the results of long-term incubation, acid hydrolysis, and carbon dating to establish the kinetically derived pools and MRTs necessary to model field CO_2 evolution rates. The active pool representing 2-3% of the SOC of three different cultivated field sites was found to have a field-equivalent MRT of 36-66 days. The slow SOC pools of these sites accounted for 40-42% of the SOC with MRTs of 9-13 years. The slightly greater than 50% of the SOC resistant to acid hydrolysis, and thus calculated as the resistant pools, had ^{14}C MRTs of 1350-1450 years. Utilization of these values in a Century model, together with measured field moisture and temperature measurements, produced daily field CO_2 predictions very similar to those actually measured.

One of the limits of the incubation approach is that it concentrates on organic materials that have relatively short residence times in soils. The use of long-term bare fallows (i.e., experiments in which the soil has been left bare of any vegetation for years to decades) is very useful as the organic matter initially present in the soil progressively decomposes over time under field conditions (Barré et al., 2010).

B Decomposition of Organic Materials Added to Soil

The decomposition and mineralization of plant residues, organic wastes, or organic molecules in soil is classically measured by adding them to soil in ^{14}C, ^{13}C, or ^{15}N labeled forms and incubating the whole. The evolved CO_2 from the added organic matter (^{14}C-CO_2, ^{13}C-CO_2) or mineral N (^{15}N) is then measured (Chapter 7), or the remaining ^{14}C or ^{13}C in soil is monitored over time. Because of the high cost of OM enriched in isotopes, such studies are typically performed at laboratory scale or on small plots at field scale. An exception is Free Air CO_2 Enrichment (FACE) experiments, in which the elevated CO_2 is depleted in ^{13}C. Mineralization of the added material can also be estimated by measuring the difference of evolved CO_2 between the sample with the OM addition and a reference soil, provided the amounts of added C are sufficient and no priming effect takes place (priming effect is the enhanced mineralization of SOM observed after addition of fresh SOM to soil).

The decomposition of litter is currently measured *in situ* using litterbags (i.e., small nylon or fiberglass mesh bags with a known amount of litter that are exposed *in situ* on top of or within soil). Decomposition rates, using first-order kinetics, are determined from the mass loss or the C loss within the litterbags. Litterbag experiments account for litter decomposition due to microbial decomposition and leaching, but depending on the mesh size, they may not account for litter fragmentation by soil fauna and just tend to overestimate decomposition rates (Cotrufo et al., 2010). These methods typically quantify the short-term decomposition rate of the added organic matter (days-to-decades).

C ^{13}C Natural Abundance

The ^{13}C natural abundance method utilizes known vegetation changes from C_3 plants (Calvin cycle) to C_4 (Hatch-Slack cycle) plants with contrasting photosynthetic pathways. The plant compounds of these photosynthetic types have contrasting ^{13}C to ^{12}C ratios (typically $\delta^{13}C \approx -27‰$ for C_3 type vegetation and $\delta^{13}C \approx -12‰$ for C_4 type vegetation). Although microbial decomposition of these materials causes some isotopic fractionation, the resulting SOM still bears the isotopic signature of the parent vegetation.

Hence, where a change of vegetation has occurred at some known date, measurement of the change in the ^{13}C natural abundance of SOC allows the determination of the rate of loss of the C derived from the initial vegetation and the rate of incorporation of the C from the new vegetation using first-order kinetics (Fig. 13.12; Balesdent and Mariotti, 1996). Despite a narrow range of situations where it can be applied, this method has allowed major progress in measuring and understanding the dynamics of C in soil because the turnover rate of C is measured *in situ* and over decades and can be applied to bulk soil C as well as to any physical and chemical fraction of SOM. Its coupling with gas chromatography-based separation methods permits the measurement of the turnover rate of molecular entities *in situ* (Amelung et al., 2008). However, this

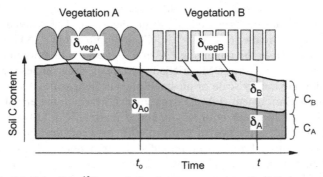

FIG. 13.12 Principle of the ^{13}C natural abundance measurement of soil C turnover. At a given time t, the total C content of soil is:

$$C = C_A + C_B \tag{1}$$

The isotopic composition of SOM at this time is given by the mass balance equation:

$$\delta(C_A + C_B) = \delta_A C_A + \delta_B C_B \tag{2}$$

From which the fraction of new carbon in the sample, F, can be derived:

$$F = C_B / C = (\delta - \delta_A)/(\delta_B - \delta_A) \tag{3}$$

Under steady state conditions, F is a direct expression of the turnover of soil C. *Reprinted from Balesdent and Mariotti (1996).*

approach has several requirements and limitations. The isotopic signature of SOM is slightly different from that of the parent vegetation (δ_A and δ_B cannot be measured directly, Fig. 13.12). A reference situation with no vegetation change is therefore needed, as its isotopic composition provides an estimate of δ_A. It also helps to account for the variability of ^{13}C content between plant organs or among organic molecules; isotopic measurements of any considered fraction will be performed on both the chronosequence samples as well as on the reference situation. The ^{13}C natural abundance method is suited to the estimation of the turnover of C within years to decades, and data from such measurements should specify the length of the tracer exchange.

D ^{14}C Dating

After the death of organisms, their ^{14}C content, initially in equilibrium with that of the atmosphere, decreases with first-order kinetics by radioactive decay, with a half-life of 5568 years. Soil C pools constantly receive new C inputs from plants and lose C through decomposition. The $^{14}C/^{12}C$ ratio of a given organic matter fraction reflects both the radioactive decay (time elapsed since the C was fixed by photosynthesis) and the rate of decomposition of the organic matter. Measuring the ^{14}C content in soil provides an estimate of the age of SOM within a time frame of 200-40,000 years; samples with an age less than 200 years are considered modern. This method is suited to measure turnover rates of centuries to millennia (Paul et al., 1997; Scharpenseel et al. 1989; Torn et al., 1997).

Aboveground nuclear bomb testing released very large amounts of ^{14}C into the atmosphere between 1950 and 1965, causing a sharp increase and then a slow decrease in the atmospheric ^{14}C concentration. This peak of bomb ^{14}C entered the vegetation and is progressively incorporated into SOM, creating a large-scale *in situ* labeling of SOM that allows us to measure turnover rates from ≈ 4 to 100 years (Trumbore, 2009).

E Compound-Specific Assessment of SOM Turnover

Compound-specific stable isotope analyses may be carried out at natural abundance or after labeling. Isotopic mass ratio spectrometers were first coupled to gas chromatographs, and methods for stable isotope measurements of lignins, sugars, and lipids were established. Recent developments include stable isotope measurements of molecules separated by liquid chromatography (Bai et al., 2013). The advantage of this technique is that no derivatization, and therefore no correction for C additions, is required. Measurement of ^{14}C at the molecular level was performed using a preparative gas chromatography to analyze the ^{14}C age of single molecules (Kramer et al., 2005).

The different methods available to measure turnover rates of bulk soil C or specific compounds give access to different turnover ranges and thus to different C pools and are complementary. The duration of laboratory incubations is insufficient to mobilize C having turnover rates of centuries to millennia. Laboratory incubations, measurements of the decomposition of added organic matter, hence address C pools with turnover times of days to a few decades. Available vegetation changes for ^{13}C natural abundance studies usually go on for a few decades, yet very seldom for a century. Therefore the tracer (the ^{13}C signature of the new vegetation) cannot label SOM having turnover rates of millennia. [13]C natural abundance and ^{14}C bomb spiking are appropriate methods for measuring turnover rates of years-to-decades. In turn, ^{14}C dating gives access to the organic matter that has the slowest turnover rate (i.e., centuries-to-millennia). Appropriate modeling is then necessary to account for the heterogeneous character of SOM turnover in soils (Chapter 17).

REFERENCES

Amelung, W., Zhang, X.D., 2001. Determination of amino acid enantiomers in soils. Soil Biol. Biochem. 33, 553–562.

Amelung, W., Cheshire, M.V., Guggenberger, G., 1996. Determination of neutral and acidic sugars in soil by capillary gas-liquid chromatography after trifluoroacetic acid hydrolysis. Soil Biol. Biochem. 28, 1631–1639.

Amelung, W., Brodowski, S., Sandhage-Hofmann, A., Bol, R., 2008. Combining biomarker with stable isotope analyses for assessing the transformation and turnover of soil organic matter. In: Sparks, D.L. (Ed.), Advances in Agronomy, vol. 100. Academic Press, Burlington, pp. 155–250.

Angers, D.A., Mehuys, G.R., 1993. Aggregate stability to water. In: Carter, M.R. (Ed.), Soil Sampling and Methods of Analysis. CRC Press, Boca Raton, pp. 651–657.

Arias, M., Barral, M.T., Diaz Fierros, F., 1997. Influence of dispersion procedures on measurements of particle size distribution and composition of the granular fraction of pasture and arable soils. Agrochimica 41, 63–77.

Bai, Z., Bode, S., Huygens, D., Zhang, X.D., Boeckx, P., 2013. Kinetics of amino sugar formation from organic residues of different quality. Soil Biol. Biochem. 57, 814–821.

Baldock, J.A., Nelson, P.N., 2000. Soil organic matter. In: Sumner, M.E. (Ed.), Handbook of Soil Science. CRC Press, Boca Raton, pp. B25–B84.

Baldock, J.A., Oades, J.M., Nelson, P.N., Skene, T.M., Golchin, A., Clarke, P., 1997. Assessing the extent of decomposition of natural organic materials using solid-state 13C NMR spectroscopy. Aust. J. Soil Res. 35, 1061–1083.

Balesdent, J., 1996. The significance of organic separates to carbon dynamics and its modelling in some cultivated soils. Eur. J. Soil Sci. 47, 485–494.

Balesdent, J., Mariotti, A., 1996. Measurement of soil organic matter turnover using 13C natural abundance. In: Boutton, J.W., Yamakasi, S.I. (Eds.), Mass Spectrometry of Soils. Marcel Dekker, New York, pp. 83–111.

Balesdent, J., Mariotti, A., Guillet, B., 1987. Natural 13C abundance as a tracer for studies of soil organic matter dynamics. Soil Biol. Biochem. 19, 25–30.

Balesdent, J., Wagner, G.H., Mariotti, A., 1988. Soil organic matter turnover in long term field experiments as revealed by carbon-13 natural abundance. Soil Sci. Soc. Am. J. 52, 118–124.

Balesdent, J., Pétraud, J.P., Feller, C., 1991. Effet des ultrasons sur la distribution granulométrique des matières organiques des sols. Science du Sol 29, 95–106.

Balesdent, J., Besnard, E., Arrouays, D., Chenu, C., 1998. The dynamics of carbon in particle-size fractions of soil in a forest-cultivation sequence. Plant Soil 201, 49–57.

Barré, P., Eglin, T., Christensen, B.T., Ciais, P., Houot, S., Kätterer, T., van Oort, F., Peylin, P., Poulton, P.R., Romanenkov, V., Chenu, C., 2010. Quantifying and isolating stable soil organic carbon through long-term bare fallow experiments. Biogeosciences 7, 3839–3850.

Basile-Doelsch, I., Amundson, R., Stone, W.E.E., Borschneck, D., Bottero, J.Y., Moustier, S., Masin, F., Colin, F., 2007. Mineral control of carbon pools in a volcanic soil horizon. Geoderma 137, 477.

Beare, M.H., Cabrera, M.L., Hendrix, P.F., Coleman, D.C., 1994. Aggregate-protected and unprotected organic matter pools in conventional-tillage and no-tillage soils. Soil Sci. Soc. Am. J. 58, 787–795.

Besnard, E., Chenu, C., Balesdent, J., Puget, P., Arrouays, D., 1996. Fate of particulate organic matter in soil aggregates during cultivation. Eur. J. Soil Sci. 47, 495–503.

Blair, G.J., Lefroy, R.D.B., Lise, L., 1995. Soil carbon fractions based on their degree of oxidation and the development of carbon management index for agricultural systems. Aust. J. Agric. Res. 46, 1459–1466.

Bligh, E.G., Dyer, W.J., 1959. A rapid method of total lipid extraction and purification. Can. J. Biochem. Physiol. 54, 911–917.

Bol, R., Amelung, W., Friedrich, C., 2004. Role of aggregate surface and core fraction in the sequestration of carbon from dung in a temperate grassland soil. Eur. J. Soil Sci. 55, 71–77.

Brandes, J.A., Ingall, E., Paterson, D., 2007. Characterization of minerals and organic phosphorus species in marine sediments using soft X-ray fluorescence spectromicroscopy. Mar. Chem. 103, 250–265.

Bremner, J.M., 1967. Nitrogenous compounds in soils. In: McLaren, A.D., Peterson, G.H. (Eds.), Soil Biochemistry. Dekker, New York, N.Y., pp. 19–66.

Brodowski, S., Amelung, W., Haumaier, L., Abetz, C., Zech, W., 2005. Morphological and chemical properties of black carbon in physical soil fractions as revealed by scanning electron microscopy and energy-dispersive X-ray spectroscopy. Geoderma 128, 116.

Bull, I.D., Nott, C.J., vanBergen, P.F., Poulton, P.R., Evershed, R.P., 2000. Organic geochemical studies of soils from the Rothamsted classical experiments—VI. The occurrence and source of organic acids in an experimental grassland soil. Soil Biol. Biochem. 32, 1367–1376.

Buurman, P., Jongmans, A.G., Nierop, K.G.J., 2008. Comparison of Michigan and Dutch podzolized soils: organic matter characterization by micromorphology and pyrolysis-GC/MS. Soil Sci. Soc. Am. J. 72, 1344–1356.

Calderón, F.J., Reeves III, J.B., Collins, H., Paul, E.A., 2011. Chemical differences in soil organic matter fractions determined by diffuse-reflectance mid-infrared spectroscopy. Soil Sci. Am. J. 75, 568–579.

Cambardella, C.A., Elliott, E.T., 1992. Particulate organic matter across a grassland cultivation sequence. Soil Sci. Soc. Am. J. 56, 777–783.

Cheng, S.Y., Bryant, R., Doerr, S.H., Wright, C.J., Williams, P.R., 2009. Investigation of surface properties of soil particles and model materials with contrasting hydrophobicity using atomic force microscopy. Environ. Sci. Technol. 43, 6500–6506.

Chenu, C., 1993. Clay- or sand-polysaccharides associations as models for the interface between microorganisms and soil: water-related properties and microstructure. Geoderma 56, 143–156.

Chenu, C., Plante, A.F., 2006. Clay-sized organo-mineral complexes in a cultivation chronosequence: revisiting the concept of the "primary organo-mineral complex." Eur. J. Soil Sci. 56, 596–607.

Chenu, C., Tessier, D., 1995. Low temperature scanning electron microscopy of clay and organic constituents and their relevance to soil microstructures. Scanning Microsc. 9, 989–1010.

Chenu, C., Hassink, J., Bloem, J., 2001. Short term changes in the spatial distribution of microorganisms in soil aggregates as affected by glucose addition. Biol. Fertil. Soils 34, 349–356.

Cheshire, M.V., 1979. Nature and Origin of Carbohydrates in Soils. Academic Press, London.

Christensen, B.T., 1987. Decomposability of organic matter in particle size fractions from fielP soils with straw incorporation. Soil Biol. Biochem. 19, 429–436.

Christensen, B.T., 1992. Physical fractionation of soil and organic matter in primary particle size and density separates. Adv. Soil Sci. 20, 1–90.

Christensen, B.T., 1996. Matching measurable soil organic matter fractions with conceptual pools in simulation models of carbon turnover: revision of model structures. In: Powlson, D., Smith, P., Smith, J. (Eds.), Evaluation of Soil Organic Matter Models. Springer Verlag, Berlin, p. 38.

Christensen, B.T., 2001. Physical fractionation of soil and structural and functional complexity in organic matter turnover. Eur. J. Soil Sci. 52, 345–353.

Citeau, L., Gaboriaud, F., Elsass, F., Thomas, F., Lamy, I., 2006. Investigation of physico-chemical features of soil colloidal suspensions. Colloids Surf., A: Physicochem Eng Aspects 287, 94–105.

Cotrufo, M.F., Ngao, J., Marzaioli, F., Piermatteo, D., 2010. Inter-comparison of methods for quantifying above-ground leaf litter decomposition rates. Plant Soil 334, 365–376.

Dignac, M.-F., Bahri, H., Rumpel, C., Rasse, D.P., Bardoux, G., Balesdent, J., Girardin, C., Chenu, C., Mariotti, A., 2005. Carbon-13 natural abundance as a tool to study the dynamics of lignin monomers in soil: an appraisal at the Closeaux experimental field (France). Geoderma 128, 1–17.

Dignac, M.F., Houot, S., Derenne, S., 2006. How the polarity of the separation column may influence the characterization of compost organic matter by pyrolysis-GC/MS. J. Anal. Appl. Pyrolysis 75, 128–139.

Dorioz, J.M., Robert, M., Chenu, C., 1993. The role of roots, fungi and bacteria on clay particle organization. An experimental approach. Geoderma 56, 179–194.

Duchaufour, P., 1976. Dynamics of organic matter in soils of temperate regions Its action on pedogenesis. Geoderma 15, 31–40.

Elliott, E.T., 1986. Aggregate structure and carbon, nitrogen and phosphorus in native and cultivated soils. Soil Sci. Soc. Am. J. 50, 627–633.

Elliott, E.T., Paustian, K., Frey, S.D., 1996. Modeling the measurable or measuring the modelable: a hierarchical approach to isolating meaningful soil organic matter fractionations. In: Powlson, D., Smith, P., Smith, J. (Eds.), Evaluation of Soil Organic Matter Models, 38, Springer Verlag, Berlin, pp. 161–180.

Elsass, F., Chenu, C., Tessier, D., 2007. Transmission electron microscopy for soil samples: preparation methods and use. In: Drees, R., Kingery, B. (Eds.), Methods in Soil Analysis. Soil Science Society of America, Madison, pp. 235–268, Part 5, 9.

Elyeznasni, N., Sellami, F., Pot, V., Benoit, P., Vieuble-Gonod, L., Young, I., Peth, S., 2012. Exploration of soil micromorphology to identify coarse-sized OM assemblages in X-ray CT images of undisturbed cultivated soil cores. Geoderma 179, 38–45.

Eusterhues, K., Rumpel, C., Kleber, M., Kogel-Knabner, I., 2003. Stabilisation of soil organic matter by interactions with minerals as revealed by mineral dissolution and oxidative degradation. Org. Geochem. 34, 1591–1600.

Eusterhues, K., Rumpel, C., Kogel-Knabner, I., 2005. Organo-mineral associations in sandy acid forest soils: importance of specific surface area, iron oxides and micropores. Eur. J. Soil Sci. 56, 753–763.

Falloon, P., Smith, P., Coleman, K., Marshall, S., 1998. Estimating the size of the inert organic matter pool from total soil organic carbon content for use in the Rothamsted carbon model. Soil Biol. Biochem. 30, 1207–1211.

Feller, C., 1979. Une méthode de fractionnement granulométrique de la matière organique des sols. Application aux sols tropicaux à textures grossières, très pauvres en humus. Cah. ORSTOM, sér. Pédologie XVII, 39–345.

Feller, C., Burtin, B., Gérard, B., Balesdent, J., 1991. Utilisation des résines sodiques et des ultrasons dans le fractionnement granulométrique de la matière organique des sols. Intérêts et limites. Science du Sol 29, 77–93.

Feller, C., Balesdent, J., Nicolardot, B., Cerri, C., 2001. Approaching "functional" soil organic matter pools through particle-size fractionation: examples for tropical soils. In: Lal, R., Kimble, J.M., Stewart, B.A. (Eds.), Assessment Methods for Soil Carbon. CRC Press, Boca Raton, pp. 53–67.

Foster, R.C., 1988. Microenvironments of soil microorganisms. Biol. Fertil. Soils 6, 189–203.

Foster, R.C., Rovira, A.D., 1976. Ultrastructure of wheat rhizosphere. New Phytol. 76, 343–352.

Gaillard, V., Chenu, C., Recous, S., Richard, G., 1999. Carbon, nitrogen and microbial gradients induced by plant residues decomposing in soil. Eur. J. Soil Sci. 50, 567–578.

Gilbert, B., Assmus, B., Hartmann, A., Frenzel, P., 1998. In situ localization of two methanotrophic strains in the rhizosphere of rice plants. FEMS Microbiol. Ecol. 25, 117–128.

Golchin, A., 1994. Soil structure and carbon cycling. Aust. J. Soil Res. 32, 1043–1068.

Hammes, K., Schmidt, M.W.I., Smernik, R.J., Currie, L.A., Ball, W.P., Nguyen, T.H., Louchouam, P., Houel, S., Gustafsson, O., Elmquist, M., Cornelissen, G., Skjemstad, J.O., Masiello, C.A., Song, J., Peng, P., Mitra, S., Dunn, J.C., Hatcher, P.G., Hockaday, W.C., Smith, D.M., Hart-kopf-Froeder, C., Boehmer, A., Luer, B., Huebert, B.J., Amelung, W., Brodowski, S., Huang, L., Zhang, W., Gschwend, P.M., Flores-Cervantes, D.X., Largeau, C., Rouzaud, J.N., Rumpel, C., Guggenberger, G., Kaiser, K., Rodionov, A., Gonzalez-Vila, F.J., Gonzalez-Perez, J.A., de la Rosa, J.M., Manning, D.A.C., Lopez-Capel, E., Ding, L., 2007. Comparison of quan-tification methods to measure fire-derived (black/elemental) carbon in soils and sediments using reference materials from soil, water, sediment and the atmosphere. Glob. Biogeochem. Cycles 21(3). http://dx.doi.org/10.1029/2006GB002914.

Haynes, R.J., 2005. Labile organic Matter fractions as central components of the quality of agricul-tural soils: an overview. In: Sparks, D.L. (Ed.), Advances in Agronomy, vol. 85. Academic Press, San Diego, CA, p. 221.

Hedges, J.I., Ertel, J.R., 1982. Characterization of lignin by capillary chromatography of cupric oxide oxidation products. Anal. Chem. 54, 174–178.

Hedges, J.I., Eglinton, G., Hatcher, P.G., Kirchman, D.L., Arnosti, C., Derenne, S., Evershed, R.P., Kögel-Knabner, I., de Leeuw, J.W., Littke, R., Michaelis, W., Rullkötter, J., 2000. The molecularly-uncharacterized component of nonliving organic matter in natural environments. Org. Geochem. 31, 945–958.

Herrmann, A.M., Clode, P.L., Fletcher, I.R., Nunan, N., Stockdale, E.A., O'Donnel, A.G., Murphy, D.V., 2007a. A novel method for the study of the biophysical interface in soils using nano-scale secondary ion mass spectrometry. Rapid Commun. Mass Spectrom. 21, 29–34.

Herrmann, A.M., Ritz, K., Nunan, N., Clode, P.L., Pett-Ridge, J., Kilburn, M.R., Murphy, D.V., O'Donnell, A.G., Stockdale, E.A., 2007b. Nano-scale secondary ion mass spectrometry—a new ana-lytical tool in biogeochemistry and soil ecology: a review article. Soil Biol. Biochem. 39, 1835–1850.

Hockaday, W.C., Grannas, A.M., Kim, S., Hatcher, P.G., 2007. The transformation and mobility of charcoal in a fire-impacted watershed. Geochim. Cosmochim. Acta 71, 3432–3445.

Hopmans, E.C., Schouten, S., Pancost, R.D., van der Meer, M.T.J., Damste, J.S.S., 2000. Analysis of intact tetraether lipids in archaeal cell material and sediments by high performance liquid chro-matography/atmospheric pressure chemical ionization mass spectrometry. Rapid Commun. Mass Spectrom. 14, 585–589.

Kalbitz, K., Solinger, S., Park, J.H., Michalzik, B., Matzner, E., 2000. Controls on the dynamics of dissolved organic matter in soils: a review. Soil Sci. 165, 277–304.

Kelleher, B.P., Simpson, A.J., 2006. Humic substances in soils: are they really chemically distinct? Environ. Sci. Technol. 40, 4605–4611.

Kelleher, B.P., Simpson, M.J., Simpson, A.J., 2006. Assessing the fate and transformation of plant residues in the terrestrial environment using HR-MAS NMR spectroscopy. Geochim. Cosmochim. Acta 70, 4080–4094.

Kiem, R., Knicker, H., Korschens, M., Kogel-Knabner, I., 2000. Refractory organic carbon in C-depleted arable soils as studied by 13C NMR spectroscopy and carbohydrate analysis. Org. Geochem. 31, 655–668.

Kilbertus, G., Proth, J., 1979. Observation d'un sol forestier (rendzine) en microscopie électronique. Can. J. Microbiol. 25, 943–946.

Kinyangi, J., Solomon, D., Liang, B.I., Lerotic, M., Wirick, S., Lehmann, J., 2006. Nanoscale biogeocomplexity of the organomineral assemblage in soil: application of STXM microscopy and C 1s-NEXAFS spectroscopy. Soil Sci. Soc. Am. J. 70, 1708–1718.

Kleber, M., Johnson, M.G., 2010. Advances in understanding the molecular structure of soil organic matter: implications for interactions in the environment. Adv. Agron. 106, 78–143.

Kleber, M., Mikutta, R., Torn, M.S., Jahn, R., 2005. Poorly crystalline mineral phases protect organic matter in acid subsoil horizons. Eur. J. Soil Sci. 56, 717–725.

Knicker, H., 2011. Solid state CPMAS 13C and 15N NMR spectroscopy in organic geochemistry and how spin dynamics can either aggravate or improve spectra interpretation. Org. Geochem. 42, 867–889.

Kögel-Knabner, I., 2000. Analytical approaches for characterizing soil organic matter. Org. Geochem. 31, 609–625.

Kögel-Knabner, I., 2002. The macromolecular organic composition of plant and microbial residues as inputs to soil organic matter. Soil Biol. Biochem. 34, 139–162.

Kögel-Knabner, I., Guggenberger, G., Kleber, M., Kandeler, E., Kalbitz, K., Scheu, S., Eusterhues, K., Leinweber, P., 2008. Organo-mineral associations in temperate soils: integrating biology, mineralogy, and organic matter chemistry. J. Plant Nutr. Soil Sci. 171, 61–82.

Kramer, C., Gleixner, G., 2006. Variable use of plant- and soil-derived carbon by microorganisms in agricultural soils. Soil Biol. Biochem. 38, 3267–3278.

Lehmann, J., Solomon, D., Kinyangi, J., Dathe, L., Wirick, S., Jacobsen, C., 2008. Spatial complexity of soil organic matter forms at nanometre scales. Nat. Geosci. 1, 238–242.

Lehmann, J., Brandes, J., Fleckenstein, H., Jacobsen, C., Solomon, D., Thieme, J., 2009. Synchrotron-based near-edge X-ray Spectroscopy of natural organic matter in soils and sediments. In: Senesi, N., Xing, P., Huang, P.M. (Eds.), Biophysico-Chemical Processes Involving Natural Nonliving Organic Matter in Environmental Systems. Wiley, Hoboken, NJ, pp. 729–781.

Martens, D.A., Loeffelmann, K.L., 2003. Soil amino acid composition quantified by acid hydrolysis and anion chromatography-pulsed amperometry. J. Agric. Food Chem. 51, 6521–6529.

Mendez-Millan, M., Dignac, M.F., Rumpel, C., Rasse, D.P., Derenne, S., 2010. Molecular dynamics of shoot vs. root biomarkers in an agricultural soil estimated by natural abundance C-13 labelling. Soil Biol. Biochem. 42, 169–177.

Mikutta, R., Kleber, M., Kaiser, K., Jahn, R., 2005. Review: organic matter removal from soils using hydrogen peroxide, sodium hypochlorite, and disodium peroxodisulfate. Soil Sci. Soc. Am. J. 69, 120–135.

Miltner, A., Zech, W., 1998. Carbohydrate decomposition in beech litter as influenced by aluminium, iron and manganese oxides. Soil Biol. Biochem. 30, 1–7.

Minyard, M.L., Bruns, M.A., Liermann, L.J., Buss, H.L., Brantley, S.L., 2012. Bacterial associations with weathering minerals at the regolith-bedrock interface, Luquillo Experimental Forest, Puerto Rico. Geomicrobiol. J. 29, 792–803.

Monnier, G., Turc, L., Jeanson-Lusignang, C., 1962. Une méthode de fractionnement densimétrique par centrifugation des matières organiques des sols. Ann. Agron. 13, 55.

Mueller, C.W., Koelbl, A., Hoeschen, C., Hillion, F., Heister, K., Herrmann, A.M., Koegel-Knabner, I., 2012. Submicron scale imaging of soil organic matter dynamics using NanoSIMS - From single particles to intact aggregates. Org. Geochem. 42 (12), 1476–1488.

Nduwamungu, C., Ziadi, N., Parent, L.E., Tremblay, G.F., Thuries, L., 2009. Opportunities for, and limitations of, near infrared reflectance spectroscopy applications in soil analysis: a review. Can. J. Soil Sci. 89, 531–541.

Nelson, D.W., Sommers, L.E., 1996. Total carbon, organic carbon and organic matter. In: Sparks, D.L., Page, A.L., Helmke, P.A., Loeppert, R.H., Soltanpour, P.N., Tabatabai, M.A., Johnston, C.T., Sumner, M.E. (Eds.), Methods of Soil Analysis. In: Chemical MethodsSoil Science Society of America, Madison, WI, pp. 961–1010, Part III.

Nunan, N., Ritz, K., Crabb, D., Harris, K., Wu, K., Crawford, J.W., Young, I.M., 2001. Quantification of the in situ distribution of soil bacteria by large-scale imaging of thin sections of undisturbed soil. FEMS Microbiol. Ecol. 37, 67.

Nunan, N., Wu, K., Young, I.M., Crawford, J.W., Ritz, K., 2003. Spatial distribution of bacterial communities and their relationships with the micro-architecture of soil. FEMS Microbiol. Ecol. 44, 203.

Paul, E.A., Follett, R.F., Leavitt, S.W., Halvorson, A., Peterson, G.A., Lyon, D.J., 1997. Radiocarbon dating for determination of soil organic matter pool sizes and dynamics. Soil Sci. Soc. Am. J. 61, 1058–1067.

Paul, E.A., Harris, D., Collins, H.P., Schulthess, U., Robertson, G.P., 1999. Evolution of CO_2 and soil carbon dynamics in biologically managed, row crop agroecosystems. Appl. Soil Ecol. 11, 53–65.

Paul, E.A., Morris, S.J., Conant, R.T., Plante, A.F., 2006. Does the acid hydrolysis-incubation method measure meaningful soil organic carbon pools? Soil Sci. Soc. Am. J. 70, 1023–1035.

Peth, S., Horn, R., Beckmann, F., Donath, T., Fischer, J., Smucker, A.J.M., 2008. Three dimensional quantification of intra aggregate pore space features using synchrotron radiation based microtomography. Soil Sci. Soc. Am. J. 72, 897–907.

Peth, S., Chenu, C., Garnier, P., Nunan, N., Pot, V., Beckmann, F., Ogurreck, M., Horn, R., 2010. Non-invasive localization of soil organic matter by Osmium staining using SR-µCT. Annual report of the Hamburger Synchrotron Strahlungslabor HASYLAB am Deutschen Elektronen-Synchrotron DESY in der Helmholtz-Gesellschaft HGF.

Piccolo, A., 2002. The supramolecular structure of humic substances: a novel understanding of humus chemistry and implications in soil science. Adv. Agron. 75, 57–134, Academic Press.

Plante, A.F., Fernandez, J.M., Leifeld, J., 2009. Application of thermal analysis techniques in soil science. Geoderma 153, 1–10.

Plante, A.F., Fernandez, J.M., Haddix, M.L., Steinweg, J.M., Conant, R.T., 2011. Biological, chemical and thermal indices of soil organic matter stability in four grassland soils. Soil Biol. Biochem. 43, 1051–1058.

Postma, J., Altemüller, H.J., 1990. Bacteria in thin soil sections stained with the fluorescent brightener calcofluor white M2R. Soil Biol. Biochem. 22, 89–96.

Puget, P., Chenu, C., Balesdent, J., 1995. Total and young organic carbon distributions in aggregates of silty cultivated soils. Eur. J. Soil Sci. 46, 449–459.

Puget, P., Angers, D.A., Chenu, C., 1999. Nature of carbohydrates associated with water-stable aggregates of two cultivated soils. Soil Biol. Biochem. 31, 55–63.

Remusat, L., Hatton, P.J., Nico, P.S., Zeller, B., Kleber, M., Derrien, D., 2012. NanoSIMS study of organic matter associated with soil aggregates: advantages, limitations, and combination with STXM. Environ. Sci. Technol. 46, 3943–3949.

Rumpel, C., Janik, L.J., Skjemstad, J.O., Kogel-Knabner, I., 2001. Quantification of carbon derived from lignite in soils using mid-infrared spectroscopy and partial least squares. Org. Geochem. 32, 831–839.

Rumpel, C., Rabia, N., Derenne, S., Quenea, K., Eusterhues, K., Kogel-Knabner, I., Mariotti, A., 2006. Alteration of soil organic matter following treatment with hydrofluoric acid (HF). Org. Geochem. 37, 1437–1451.

Rumpel, C., Chabbi, A., Nunan, N., Dignac, M.F., 2009. Impact of landuse change on the molecular composition of soil organic matter. J. Anal. Appl. Pyrolysis 85, 431–434.

Rumpel, C., Eusterhues, K., Kögel-Knabner, I., 2010. Non-cellulosic neutral sugar contribution to mineral associated organic matter in top-and subsoil horizons of two acid forest soils. Soil Biol. Biochem. 42, 379–382.

Scharpenseel, H.W., Becker-Heidmann, P., Neue, H.U., Tsutsuki, K., 1989. Bomb-carbon, 14C dating and 13C measurements as tracers of organic matter dynamics as well as of morphogenetic and turbation processes. Sci. Total Environ. 81, 99–110.

Schmidt, M.W., Rumpel, C., Kögel-Knabner, I., 1999. Evaluation of an ultrasonic dispersion procedure to isolate primary organomineral complexes from soils. Eur. J. Soil Sci. 50, 87–94.

Schmidt, M.W.I., Torn, M.S., Abiven, S., Dittmar, T., Guggenberger, G., Janssens, I.A., Kleber, M., Kogel-Knabner, I., Lehmann, J., Manning, D.A.C., Nannipieri, P., Rasse, D.P., Weiner, S., Trumbore, S.E., 2011. Persistence of soil organic matter as an ecosystem property. Nature 478, 49–56.

Shang, C., Zelazny, L.W., 2008. Selective dissolution techniques of mineral analysis of soils and sediments. In: Ulery, A.L., Drees, L.R. (Eds.), Methods of Soil Analysis. Part 5. Mineralogical Methods. Soil Science Society of America, Madison, pp. 33–80.

Six, J., Jastrow, J.D., 2002. Organic matter turnover. In: Lal, R. (Ed.), Encyclopedia of Soil Science. Marcel Dekker, NY, pp. 936–942.

Six, J., Schultz, P.A., Jastrow, J.D., Merckx, R., 1999. Recycling of sodium polytungstate used in soil organic matter studies. Soil Biol. Biochem. 31, 1193–1196.

Six, J., Elliott, E.T., Paustian, K., 2000. Soil macroaggregate turnover and microaggregate formation: a mechanism for C sequestration under no-tillage agriculture. Soil Biol. Biochem. 32, 2099–2103.

Six, J., Bossuyt, H., De Gryze, S., Denef, K., 2004. A history of research on the link between microaggregates, soil biota, and soil organic matter dynamics. Soil Tillage Res. 79, 7–31.

Skjemstad, J.O., Clarke, P., Taylor, J.A., Oades, J.M., McClure, S.G., 1996. The chemistry and nature of protected carbon in soil. Aust. J. Soil Res. 34, 251–271.

Sleighter, R.L., Hatcher, P.G., 2007. Molecular characterization of dissolved organic matter (DOM) along a river to ocean transect of the lower Chesapeake Bay by ultrahigh resolution electrospray ionization Fourier transform ion cyclotron resonance mass spectrometry. Mar. Chem. 110, 140–152.

Sleutel, S., Cnudde, V., Masschaele, B., Vlassenbroek, J., Dierick, M., Van Hoorebeke, L., Jacobs, P., De Neve, S., 2008. Comparison of different nano- and micro-focus X-ray computed tomography set-ups for the visualization of the soil microstructure and soil organic matter. Comput. Geosci. 34, 931–938.

Solomon, D., Lehmann, J., Harden, J., Wang, J., Kinyangi, J., Heymann, K., Karunakaran, C., Lu, Y.S., Wirick, S., Jacobsen, C., 2012. Micro- and nano-environments of carbon sequestration: multi-element STXM-NEXAFS spectromicroscopy assessment of microbial carbon and mineral associations. Chem. Geol. 329, 53–73.

Stevenson, F.J., 1982. Humus Chemistry: Genesis, Composition and Reactions. J. Wiley, New York.

Stevenson, F.J., Cheng, C.N., 1970. Amino acids in sediments: recovery by acid hydrolysis and quantitative estimation by a colorimetric procedure. Geochim. Cosmochim. Acta 34, 77–88.

Strickland, T.C., Sollins, P., 1987. Improved method for separating light and heavy fraction organic material from soil. Soil Sci. Soc. Am. J. 51, 1390–1393.

Terhoeven-Urselmans, T., Michel, K., Helfrich, M., Flessa, H., Ludwig, B., 2006. Near-infrared spectroscopy can predict the composition of organic matter in soil and litter. J. Plant Nutr. Soil Sci. 169, 168–174.

Thévenot, M., Dignac, M.-F., Rumpel, C., 2010. Fate of lignins in soils: a review. Soil Biol. Biochem. 42, 1200–1211.

Thieme, J., Sedlmair, J., Gleber, S.C., Prietzel, J., Coates, J., Eusterhues, K., Abbt-Braun, G., Salome, M., 2010. X-ray spectromicroscopy in soil and environmental sciences. J. Synchrotron Radiat. 17, 149–157.

Torn, M.S., Trumbore, S.E., Chadwick, O.A., Vitousek, P.M., Hendricks, D.M., 1997. Mineral control of soil organic carbon storage and turnover. Nature 389, 170–173.

Totsche, K.U., Kögel-Knabner, I., 2004. Mobile organic sorbent affected by contaminant transport in soil: numerical case studies for enhanced and reduced mobility. Vadose Zone J. 3, 352–367.

Trumbore, S., 2009. Radiocarbon and soil carbon dynamics. Annu. Rev. Earth Planet. Sci. 37, 47–66.

Virto, I., BarrÈ, P., Chenu, C., 2008. Microaggregation and organic matter storage at the silt-size scale. Geoderma 146, 326.

von Lutzow, M., Kogel-Knabner, I., Ekschmittb, K., Flessa, H., Guggenberger, G., Matzner, E., Marschner, B., 2007. SOM fractionation methods: relevance to functional pools and to stabilization mechanisms. Soil Biol. Biochem. 39, 2183–2207.

Wan, J., Tyliszczak, T., Tokunaga, T.K., 2007. Organic carbon distribution, speciation, and elemental correlations within soil micro aggregates: applications of STXM and NEXAFS spectroscopy. Geochim. Cosmochim. Acta 71, 5439–5449.

Watteau, F., Villemin, G., Ghanbaja, J., Genet, P., Pargney, J.C., 2002. In situ ageing of fine beech roots (Fagus sylvatica) assessed by transmission electron microscopy and electron energy loss spectroscopy: description of microsites and evolution of polyphenolic substances. Biol. Cell. 94, 55–63.

Wershaw, R.L., Leenheer, J.A., Kennedy, K.R., Noyes, T.I., 1996. Use of C-13 NMR and FTIR for elucidation of degradation pathways during natural litter decomposition and composting. 1. Early stage leaf degradation. Soil Sci. 161, 667–679.

Wiesenberg, G.L.B., Schwark, L., Schmidt, M.W.I., 2004. Improved automated extraction and separation procedure for soil lipid analyses. Eur. J. Soil Sci. 55, 349–356.

Wilcke, W., Baumler, R., Deschauer, H., Kaupenjohann, M., Zech, W., 1996. Small scale distribution of Al, heavy metals and PAHs in an aggregated Alpine Podzol. Geoderma 71, 19–30.

Zelles, L., Bai, Q.Y., Beck, T., Beese, F., 1992. Signature fatty acids in phospholipids and lipopolysaccharides as indicators of microbial biomass and community structure in agricultural soils. Soil Biol. Biochem. 24, 317–323.

Zsolnay, A., 2003. Dissolved organic matter: artefacts, definitions, and functions. Geoderma 113, 187–209.

Chapter 14

Nitrogen Transformations

G.P. Robertson[1] and P.M. Groffman[2]

[1]*Department of Plant, Soil, and Microbial Sciences, Michigan State University, East Lansing, MI, USA*
[2]*Institute of Ecosystem Studies, Millbrook, NY, USA*

Chapter Contents

I INTRODUCTION

No other element essential for life takes as many forms in soil as nitrogen (N), and transformations among these forms are mostly mediated by microbes. Soil microbiology thus plays yet another crucial role in ecosystem function: in most terrestrial ecosystems N limits plant growth, and thus net primary production—the productive capacity of the ecosystem—can be regulated by the rates at which soil microbes transform N to plant-usable forms. Several forms of N are also pollutants, so soil microbial transformations of N also affect human and environmental health, sometimes far distant from the microbes that performed the transformation. Understanding N transformations and the soil microbes that perform them is thus essential for understanding and managing ecosystem health and productivity.

Nitrogen takes nine different chemical forms in soil corresponding to different oxidative states (Table 14.1). Dinitrogen gas (N_2) comprises 79% of our atmosphere and is by far the most abundant form of N in the biosphere, but it is unusable by most organisms, including plants. Biological N_2 fixation, whereby N_2 is transformed to organic N (described in Chapter 15), is the

Soil Microbiology, Ecology, and Biochemistry. http://dx.doi.org/10.1016/B978-0-12-415955-6.00014-1

TABLE 14.1 Main Forms of Nitrogen in Soil and Their Oxidation States

Name	Chemical Formula	Oxidation State
Nitrate	NO_3^-	+5
Nitrogen dioxide (g)	NO_2	+4
Nitrite	NO_2^-	+3
Nitric oxide (g)	NO	+2
Nitrous oxide (g)	N_2O	+1
Dinitrogen (g)	N_2	0
Ammonia (g)	NH_3	-3
Ammonium	NH_4^+	-3
Organic N	R_{NH_3}	-3

Gases (g) occur both free in the soil atmosphere as well as dissolved in soil water.

dominant natural process by which N enters soil biological pools. All subsequent soil N transformations are covered in this chapter: (1) *N mineralization*, which is the conversion of organic-N to inorganic forms; (2) *N immobilization*, which is the uptake or assimilation of inorganic N forms by microbes and other soil organisms; (3) *nitrification*, which is the conversion of ammonium (NH_4^+) to nitrite (NO_2^-) and then nitrate (NO_3^-); and (4) *denitrification*, which is the conversion of nitrate to nitrous oxide (N_2O) and to dinitrogen gas (N_2). Other forms of N (Table 14.1) are involved in these conversions primarily as intermediaries, and during conversion they can escape to the environment, where they can participate in chemical reactions or are transported elsewhere for further reactions.

Löhnis (1913) first formulated the concept of the N cycle, which formalizes the notion that N is converted from one form to another in an orderly and predictable fashion (Fig. 14.1), and that at global scale, the same amount of dinitrogen gas that is fixed each year by N_2 fixation must either be permanently stored in deep ocean sediments or converted back to N_2 gas via denitrification to maintain atmospheric equilibrium.

The fact that N_2 fixation—both biological and industrial—now far outpaces historical rates of denitrification is the principal reason N has become a major pollutant (Galloway et al., 2008). Making managed ecosystems more N conservative and removing N from wastewater streams, such as urban and industrial effluents, are major environmental challenges that require a fundamental knowledge of soil microbial N transformations (Robertson and Vitousek, 2009).

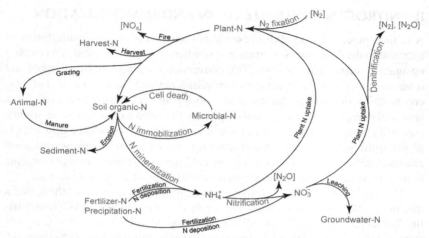

FIG. 14.1 Schematic representation of the major elements of the terrestrial nitrogen cycle. Those processes mediated by soil microbes appear in red. Gases appear in brackets.

Although the microbiology, physiology, and biochemistry of N cycle processes have been studied for over a century, much of our understanding of the N cycle has been derived from molecular and organismal scale studies in the laboratory. Laboratory observations and experiments have character-ized the nature and regulation of the processes discussed in this chapter, but their reductionist nature has caused us to sometimes overlook the surpris-ing possibilities for microbial activity in nature, thus impairing our ability to understand the ecological significance of these processes. The occurrence of denitrification (an anaerobic process) in dry and even desert soils is one exam-ple: theory and years of laboratory work suggest that denitrification ought to occur only in wetland and muck soils, but when new field-based methods became available in the 1970s, it became clear that almost all soils support active denitrifiers.

Key problems have also arisen from evaluating microbial N cycle pro-cesses in isolation from other biogeochemical processes (e.g., carbon (C) metabolism and plant nutrient uptake). This has resulted in an underestimation of the physiological flexibility of bacteria and archaea in nature (e.g., nitrify-ing denitrifiers, aerobic denitrifiers, *anaerobic ammonium oxidation* (ana-mmox)). The disconnect between laboratory-derived knowledge and what actually occurs in the field is a problem throughout soil microbial ecology, but is perhaps most acute in the area of N cycling, which has great practical importance at field, landscape, regional, and global scales. When we attempt to increase information from the microbial scale to address important ques-tions relating to plant growth, water pollution, and atmospheric chemistry at ecosystem, landscape, and regional scales, this problem becomes especially obvious and significant.

II NITROGEN MINERALIZATION AND IMMOBILIZATION

A critical process in any nutrient cycle is the conversion of organic forms of nutrients in dead biomass (detritus) into simpler, soluble forms that can be taken up again by plants and microbes. This conversion is carried out by microbes and other soil organisms that release, or mineralize, nutrients as a by-product of their consumption of detritus. Although microbes consume detritus primarily for a source of energy and C to support their growth, they also have a need for nutrients, especially N, to assemble proteins, nucleic acids, and other cellular components. If plant detritus is rich in N, microbial needs are easily met, and N release, or mineralization, proceeds. If plant detritus is low in N, microbes must scavenge inorganic N from their surroundings, leading to immobilization of N in their biomass.

The key to understanding mineralization-immobilization is to "think like a microbe," that is, attempt to make a living by obtaining energy and C from detritus. Sometimes the detritus has all the N that the microbe needs, so as C is consumed, any extra N is released (mineralized) to the soil solution. Sometimes the detritus does not have enough N to meet microbial needs, so as C is consumed, additional N must be immobilized from the soil solution. It has been shown that microbes invest more energy in the synthesis of enzymes (e.g., amidases to acquire N and phosphatases to acquire P) to obtain nutrients that they need when decomposing substrates of low quality. Microbial N uptake is also affected by organism growth efficiency. Fungi have wider C:N ratios in their tissues than bacteria and archaea and can grow more efficiently on low N substrates.

Mineralization results in an increase, whereas immobilization results in a decrease in plant available forms of N in the soil. Traditionally, ammonium has been viewed as the immediate product of mineralization, and in the older literature, mineralization is often referred to as ammonification. More recently, recognition of the fact that plants can take up simple, soluble organic forms of nutrients leads us to broaden our definition of mineralization products to include any simple, soluble forms of N that can be taken up by plants (see Schimel and Bennett, 2004). Plants from a variety of habitats have been shown to take up amino acids and other organic N forms; mycorrhizae can play a role in this uptake by absorbing amino acids, amino sugars, peptides, proteins, and chitin that are then used by their hosts as an N source.

Mineralization and immobilization occur at the same time within relatively small volumes of soil. Whereas one group of microbes might be consuming a protein-rich and therefore N-rich piece of organic matter (think seed or leguminous leaf tissue), another group, perhaps <100 µm away, might be consuming detritus rich in C, but low in N (think leaf stalk or wood). The first group is mineralizing N, while the second is immobilizing it, perhaps even immobilizing the same N that is being mineralized by the first. As a result of the simultaneous nature and small scale of these processes, it is important to make a distinction between gross and net mineralization and immobilization. Gross N mineralization is the total amount of soluble N produced by microorganisms, and gross N immobilization is the total amount of soluble N consumed. Net N mineralization is the balance between the two. When gross mineralization exceeds gross immobilization, inorganic N in the soil increases (i.e., there is net mineralization). When gross immobilization exceeds gross mineralization, inorganic N in the soil decreases (i.e., there is net immobilization).

Soil fauna also play an important role in mineralization and immobilization processes. They are responsible for much of the preliminary decomposition of detritus, they feed on and can regulate populations of bacteria and fungi, and they can create or modify habitats for a wide array of organisms. For example, earthworms create burrows, isopods shred leaf litter, and termites macerate wood. All heterotrophic soil organisms consume organic materials for energy and C and, at the same time, immobilize and mineralize N.

The widely distributed nature of mineralization and immobilization processes means that the environmental regulation of these processes is relatively straightforward. Rates of activity increase with temperature and are optimal at intermediate soil water contents, similar to respiration, as seen in Fig. 14.2, although it is important to recognize that significant activity often occurs at extremes of both temperature and moisture. In most soils, the quantity and quality of detrital inputs are the main factors that control the rates and patterns of mineralization and immobilization. When moisture and temperature are favorable, large inputs of organic matter lead to high rates of microbial activity and the potential for high rates of mineralization and immobilization. However, in soils that are waterlogged or very cold (think wetlands or Arctic tundra),

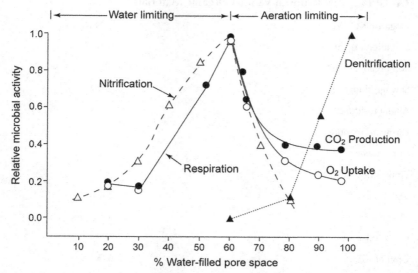

FIG. 14.2   The relationship between water-filled pore space (a measure of soil moisture availability) and relative amount of microbial activities. *Redrawn from Linn and Doran (1984).*

moisture and temperature can limit microbial activity, and soil organic matter and the organic N it contains will accumulate due to low rates of mineralization.

Water-filled pore space (WFPS) is a useful measure to examine moisture's influence on soil biological activity because it includes information about the impact of soil water on aeration in addition to information on water availability per se. The calculation of %WFPS is

$$\%\text{WFPS} = \frac{\text{soil water content} \times \text{bulk density} \times 100}{1 - (\text{bulk density}/2.65)}. \tag{14.1}$$

Soil water content is determined gravimetrically (g H_2O/g dry soil), bulk density (g cm^{-3}) is the oven dry weight of a given soil volume, and the value 2.65 is the density (g cm^{-3}) of sand grains and other soil mineral particles.

What controls the balance between N mineralization and immobilization? The answer is primarily organic matter quality—the availability of C in the material relative to its available N. Consider the effects of adding various organic materials with different C:N ratios to soil (Table 14.2). When one adds to soil manure with a relatively low C:N ratio (ca. 20:1), the microbes have no trouble obtaining N, and as a result, mineralization dominates over immobilization, and plant-available N increases in soil. This is why manure is frequently used as a fertilizer. On the other hand, were one to add sawdust to soil, a material with a high C:N ratio (625:1), the microbes would be keen to obtain the energy

TABLE 14.2 C:N Ratios in Various Organic Materials

Organic Material	C:N Ratio
Soil microorganisms	8:1
Soil organic matter	10:1
Sewage sludge	9:1
Alfalfa residues	16:1
Farmyard manure	20:1
Corn stover	60:1
Grain straw	80:1
Oak litter	200:1
Pine litter	300:1
Crude oil	400:1
Conifer wood	625:1

From Tisdale et al. (1993) and Hyvönen et al. (1996).

and C in the sawdust, but could not degrade this material without additional N because the sawdust does not have sufficient N to allow the microbes to build proteins. Thus, the microbes must immobilize N from their environment, resulting in a decrease in plant-available N in the soil. If there is no N to immobilize, microbial growth is slowed.

The balance between mineralization and immobilization is also affected by organism growth efficiency. For example, fungi have wider C:N ratios in their tissues than bacteria and therefore, have a lower need for N and will thus mineralize N more readily. As a general rule of thumb, materials with a C:N ratio > 25:1 stimulate immobilization, whereas those with a C:N ratio < 25:1 stimulate mineralization (Table 14.2). Highly decomposed substances such as soil organic matter, humus, and compost, in which labile C and N have been depleted, are the exception to this rule. Even though these substances may have a low C:N ratio, the undecomposed C is in complex forms inherently resistant to decomposition, so mineralization also proceeds slowly.

There are a wide variety of methods for measuring mineralization and immobilization (Hart et al., 1994; Robertson et al., 1999). Measurement of net mineralization and immobilization rates is much easier and more common than is the measurement of gross rates. Measurement of net rates usually involves measuring changes in inorganic N levels in some type of whole soil incubation. In most cases, these incubations are in containers, with no plant uptake or leaching losses, and changes in inorganic N levels are measured by periodic extractions of the soil. Incubation methods vary widely, from short (10-day) incubations of intact soil cores buried in the field to long (>52-week) incubations of sieved soils in the laboratory. Gross rates are measured using isotope dilution methods whereby small amounts of ^{15}N-labeled ammonium are added to the soil, and the subsequent dilution of the ^{15}N with natural ^{14}N from mineralized organic matter is used as a basis for calculating the gross production and consumption of ammonium.

III   NITRIFICATION

Nitrification is the microbial oxidation of ammonia to less reduced forms, principally NO_2^- and NO_3^-. Autotrophic bacteria, first isolated in the late 1800s, gain as much as 440 kJ of energy per mole of NH_3 oxidized when NO_3^- is the end product. We know now that archaea and heterotrophic microbes can also nitrify, although autotrophic nitrification appears to be the dominant process in most soils.

The importance of nitrifiers to ecosystem function is substantial: although some nitrate enters ecosystems in acid rain or as fertilizer, in most ecosystems, nitrate is formed *in situ* via nitrification. Because nitrate is an anion, it is more mobile than ammonium, the ionized source of NH_3 in soil water:

$$NH_4^+ (aq) \rightleftarrows NH_3 (aq) + H^+ (aq). \tag{14.2}$$

As a positively charged ion, ammonium can be held on cation-exchange sites associated with soil organic matter, clay surfaces, and variable-charge minerals. Nitrate, on the other hand, is mostly free in the soil solution and can be easily transported out of the rooting zone by water when precipitation exceeds evapotranspiration.

Nitrification in many soils is a major source of soil acidity, which can have multiple effects on ecosystem health, including the mobilization of toxic metals and the hydrologic loss of base cations as hydrogen ions displace other cations from exchange sites. In soils dominated by variable-charge minerals, which include most highly weathered tropical soils, soil acidity largely controls cation-exchange capacity (CEC), and nitrifier-generated acidity can drive CEC to very low levels. Further, some plants and microbes appear better able to take up ammonium than nitrate, and vice versa, implying a potential effect of nitrifiers on plant and microbial community composition. Finally, nitrifiers themselves can also be direct and important sources of the atmospheric gases NO_x and N_2O through nitrifier denitrification when O_2 is low (Zhu et al., 2013) or via by-product formation.

A The Biochemistry of Autotrophic Nitrification

Autotrophic nitrification is a two-step process, carried out by separate groups of bacteria and archaea—the ammonia and nitrite oxidizers, respectively. Autotrophic nitrifiers derive their C from CO_2 or carbonates, rather than from organic matter, and are obligate aerobes. The NH_3 oxidation is characterized as:

$$NH_3 + 1\tfrac{1}{2}O_2 \rightarrow NO_2^- + H^+ + H_2O. \tag{14.3}$$

The first step in this oxidation is mediated by the membrane-bound enzyme ammonia mono-oxygenase, which can also oxidize a wide variety of organic, nonpolar low-molecular-weight compounds, including phenol, methanol, methane, and halogenated aliphatic compounds, such as trichloroethylene:

$$NH_3^+ + 2H^+ + O_2 + 2e^- \xrightarrow{\text{ammonia mono-oxygenase}} NH_2OH + H_2O. \tag{14.4}$$

The reaction is irreversibly inhibited by small quantities of acetylene, which inhibits ammonia mono-oxygenase. This provides a means for experimentally differentiating autotrophic from heterotrophic nitrification in soil. Hydroxylamine is further oxidized to nitrite by the reaction

$$NH_2OH + H_2O \xrightarrow{\text{NH}_2\text{OH oxidoreductase}} NO_2^- + 4e^- + 5H^+. \tag{14.5}$$

Two of the four electrons released in this reaction are used in the prior NH_3 oxidation step; the remaining two are used in electron transport, generating energy for cell growth and metabolism:

$$2H^+ + \tfrac{1}{2}O_2 + 2e^- \xrightarrow{\text{terminal oxidase}} H_2O. \tag{14.6}$$

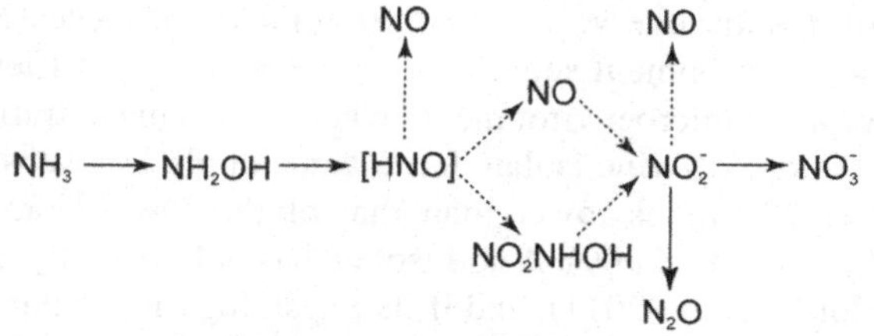

$$NO \qquad\qquad NO$$

$$NH_3 \longrightarrow NH_2OH \longrightarrow [HNO] \underset{NO_2NHOH}{\overset{NO}{\diamond}} NO_2^- \longrightarrow NO_3^-$$

$$N_2O$$

FIG. 14.3 Autotrophic nitrification pathways including pathways for gas loss. Broken lines indicate unconfirmed pathways. *From Firestone and Davidson (1989).*

Intermediary compounds formed during the oxidation of hydroxylamine to nitrite can result in the formation of NO (Fig. 14.3), which can escape to the atmosphere and influence the photochemical production of ozone (O_3) and the atmospheric abundance of hydroxyl (OH) radicals, primary oxidants for a number of tropospheric trace gases, including methane. Ammonia oxidizers are also able to produce NO via NO_2^- reduction, which results in the production of N_2O, an important greenhouse gas that can also escape to the atmosphere. Nitrite reduction occurs when ammonia oxidizers use NO_2^- as an electron acceptor when O_2 is limiting—effectively becoming denitrifying nitrifiers!

In most soils, the nitrite produced by ammonia oxidizers does not accumulate, but is quickly oxidized to nitrate by the nitrite-oxidizing bacteria when they perform nitrite oxidation:

$$NO_2^- + H_2O \xrightarrow{\text{nitrite oxidoreductase}} NO_3^- + 2H^+ + 2e^-. \qquad (14.7)$$

These reactions are membrane-associated, and because nitrite oxidoreductase is a reversible enzyme, the reaction can be reversed to result in nitrate reduction to nitrite. Up to 80% of the energy produced during nitrification is used to fix C; growth efficiencies of the nitrifiers are correspondingly low.

B The Diversity of Autotrophic Nitrifiers

Our taxonomic understanding of nitrifiers has been fundamentally transformed in the last few years by new molecular techniques that have revealed considerable taxonomic diversity, where before we thought there was little. The development and use of 16S rRNA gene primers and subsequent metagenomic techniques targeting genes for ammonia monooxygenase (*amoA*) have demonstrated both a greater diversity among bacterial nitrifiers as well as the presence of nitrifiers in a completely different kingdom, the Archaea. Remarkably, as first noted by Leininger et al. (2006), growing evidence suggests that archaeal nitrifiers may be as, or more, abundant than bacterial nitrifiers in many soils! Although, to date, we know little about the ecological significance of this discovery, inference from nitrifiers successfully isolated from the marine environment suggests that archaeal soil nitrifiers may dominate in oligotrophic microenvironments where NH_3 concentrations are very low. The best-studied marine isolate has a vanishingly low substrate affinity for NH_4^+—over 200 times lower than that of the lowest bacterial isolate (Martens-Habbena et al., 2009). A soil isolate has only recently been cultured (Lehtovirta-Morley et al., 2011), and if its physiology is similar to the above nitrifiers, may be far more competitive for NH_4^+ in soil than is currently assumed. Such a substrate affinity may also give them access to NH_3 even in acid soils where high H^+ concentrations favor NH_4^+ over NH_3 (aq; He et al., 2012).

Prior to 2000, the bacterial nitrifiers were viewed as the single family Nitrobacteraceae, defined by their characteristic ability to oxidize ammonia or nitrite. Early work beginning with Winogradsky (1892) classified the ammonia-oxidizing genera of Nitrobacteraceae on the basis of cell shape and the arrangement of intracytoplasmic membranes. This yielded five genera: *Nitrosomonas, Nitrosospira, Nitrosococcus, Nitrosolobus,* and *Nitrosovibrio.* Recent work with isolates, based principally on 16S rRNA oligonucleotide and gene sequence analysis, places terrestrial ammonia-oxidizing bacteria in the beta subclass of the Proteobacteria (Fig. 14.4); *Nitrosolobus* and *Nitrosovibrio* are no longer considered distinct from *Nitrosospira,* and *Nitrosococcus* is

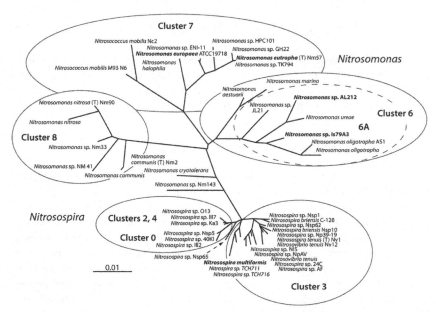

FIG. 14.4 A 16S ribosomal RNA guide tree for bacterial nitrifiers in the Betaproteobacteria based on isolates. The scale is substitutions per site. *Redrawn from Norton (2011).*

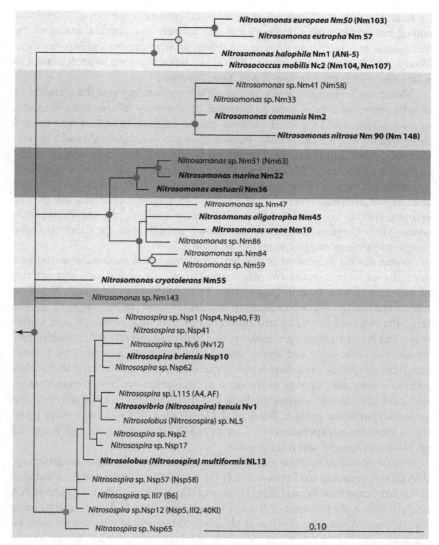

FIG. 14.5 16S rRNA-based phylogenetic tree of the betaproteobacterial ammonia oxidizers. The tree includes oxidizers of different genospecies (DNA-DNA similarity < 60%) with available 16S rRNA gene sequences longer than 1000 nucleotides. Strains with DNA-DNA similarity > 60% are in parentheses after the respective species name. Described species are depicted in bold. Scale bar represents 10% estimated sequence divergence. *From Koops* et al. *(2006).*

being reclassified to *Nitrosomonas* (Norton, 2011). Today, we have almost complete 16S rRNA gene sequences with >1000 nucleotides for the 14 described species of Betaproteobacteria ammonia oxidizers, which have a gene sequence similarity of 89% (Fig. 14.5; Koops et al., 2006).

In arable soils, the *Nitrosomonas communis* lineage is numerically dominant among culturable strains. Unfertilized soils usually also contain strains of the *Nitrosomonas oligotropha* lineage and strains of *Nitrosospira* and *Nitrosovibrio* (Koops and Pommerening-Röser, 2001). The latter two tend to be dominant in acid soils, which contain few if any *Nitrosomonas*.

Molecular techniques, such as 16S rRNA sequencing and the retrieval of *amoA* clones, have been used to examine the diversity of ammonia oxidizers *in vivo*, which avoid the need for pure-culture cultivation and its bias toward those species that are cultivatable outside their native habitat. Although molecular techniques can themselves be biased because of their dependence on the extraction of nucleic acid from soil, PCR amplification, primer bias, and cloning methods, they nevertheless suggest that most soils are dominated by *Nitrosospira* and archaeal species—and not by *Nitrosomonas* (Prosser, 2011). The archaeal species are currently found in the new archaeal phylum Thaumarchaeota. Their ubiquity, numerical dominance in soils thus far examined, and unique physiology suggest surprises in store.

Worth noting in general is that neither classical nor molecular techniques normally provide quantitative information about the abundance and activity of different species *in situ*. Quantitative PCR and newer techniques based on membrane or *in situ* hybridization in concert with rRNA-targeted probes (e.g., fluorescence *in situ* hybridization or FISH, as used in aquatic and wastewater treatment studies; Juretschko et al., 1998) can directly relate community structure with activity and spatial distribution of targeted organisms. Prosser and Embley (2002) have shown how these techniques can be used to discover nitrifier community change in response to changes in ecosystem management and land use. Stable isotope probing can also demonstrate the activity and growth of particular groups; Zhang et al. (2010) used $^{13}CO_2$ stable isotope probing to show the incorporation of ^{13}C-enriched CO_2 into the *amoA* and *hcd* genes of Thaumarchaea in soil microcosms.

Nitrite-oxidizing bacteria appear in a broader array of phylogenetic groupings than do the ammonia oxidizers, but only the genus *Nitrobacter* and the candidate genus *Nitrotoga* have been cultured from soil (Daims et al., 2011). The 16S rRNA analysis shows the presence of *Nitrospira* in most soils, which appear to be more diverse than *Nitrobacter* (Freitag et al., 2005). Members of *Nitrobacter* form an exclusive and highly related cluster in the Alphaproteobacteria. Though widely distributed in nature, pairwise evolutionary distance estimates are less than 1%, indicating little genetic diversity within the group, a finding supported by 16S rRNA sequence comparisons (Orso et al., 1994). The other nitrite-oxidizing genera are in the delta (*Nitrospina* and *Nitrospira*), gamma (*Nitrosococcus*), and beta (*Candidatus* Nitrotoga) subclasses of the Proteobacteria.

C   Heterotrophic Nitrification

A wide variety of heterotrophic bacteria and fungi have the capacity to oxidize NH_4^+. So-called heterotrophic nitrification is not linked to cellular growth, as it is for autotrophic nitrification. There is evidence for two pathways for heterotrophic ammonia oxidation. The first pathway is similar to that of autotrophic oxidation, in that the nitrifying bacteria have similar ammonia- and hydroxylamine-oxidizing enzymes. These enzymes can oxidize a number of different substrates, and it may be that ammonia oxidation is only secondary to these enzymes' main purpose of oxidizing propene, benzene, cyclohexane, phenol, methanol, or any of a number of other nonpolar organic compounds.

The second heterotrophic pathway is organic and appears limited to fungi. It involves the oxidation of amines or amides to a substituted hydroxylamine followed by oxidation to a nitroso and then a nitro compound with the following oxidation states:

$$RNH_2 \rightarrow RNHOH \rightarrow RNO \rightarrow RNO_3 \rightarrow NO_3^- \atop {-3 \qquad -1 \qquad\quad +1 \qquad +3 \qquad +5}$$ (14.8)

These reactions are not coupled to ATP synthesis and thus produce no energy. Alternately, N compounds may react with hydroxyl radicals produced in the presence of hydrogen peroxide and superoxide, which may happen when fungi release oxidases and peroxidases during cell lysis and lignin degradation.

Heterotrophic bacteria such as *Arthrobacter globiformis*, *Aerobacter aerogenes*, *Thiosphaera pantotropha*, *Streptomyces grisens*, and various *Pseudomonas* spp. have been found to nitrify. The fungi *Aspergillus flavus* was first isolated as a nitrifier in 1954 and is the most widely studied of the nitrifying heterotrophs. Interest in heterotrophic nitrification increased substantially in the late 1980s when it became clear that accelerated inputs of atmospheric ammonium to acid forest soils were being nitrified to nitrate with alarming effects on soil acidity, forest health, and downstream drinking water quality. Until recently, it was assumed that most of this nitrification was heterotrophic; we know now that most nitrification in acid soils is autotrophic (De Boer and Kowalchuk, 2001), and as noted earlier, may be chiefly performed by archaeal nitrifiers able to scavenge NH_3 in low pH soils (He et al., 2012). Heterotrophic nitrification thus appears important in some soils and microenvironments, perhaps where autotrophic nitrifiers are chemically inhibited (see following section), but are thought now to rarely dominate the soil nitrifier community.

D Environmental Controls of Nitrification

The single most important factor regulating nitrification in the majority of soils is ammonium supply (Fig. 14.6). Where decomposition and thus N mineralization is low, or where NH_4^+ uptake and thus N immobilization by heterotrophs or plants is high, nitrification rates will be low. Conversely, any ecosystem disturbance that increases soil NH_4^+ availability will usually accelerate nitrification unless some other factor is limiting. Examples are tillage, fire, clear cutting, waste disposal, fertilization, and atmospheric N deposition—all of which have well-documented effects on nitrate production in soils, mostly due to their effects on soil NH_4^+ pools.

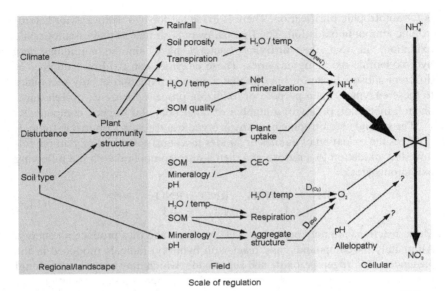

FIG. 14.6 Environmental controls on nitrification at different scales. *From Robertson (1989).*

Given that nitrification usually accelerates only when the NH_4^+ supply exceeds plant and heterotroph demand implies that nitrifiers are relatively poor competitors for NH_4^+ in the soil solution. In fact, this is the case: nitrification rates are typically low in midsuccessional communities and aggrading forests because of high plant demand for N. This also occurs following the addition of high C:N residues to agricultural soils because of high N demand by heterotrophic microbes (high immobilization; Fig. 14.1). In old-growth forests and mature grasslands, plant N demand has diminished and consequently, nitrification is usually higher than in midsuccessional communities where plant biomass is still accumulating, but not usually as high as in early successional and agricultural ecosystems, where N-supply often greatly exceeds demand (Robertson and Vitousek, 1981).

Oxygen is another important regulator of nitrification in soil. All known nitrifiers are obligate aerobes, and nitrification proceeds very slowly, if at all, in submerged soils. In flooded environments, such as wetlands and lowland rice, nitrifiers are active only in the oxidized zone around plant roots and at the water-sediment interface, which is usually only a few millimeters thick. In addition, even though some nitrifiers have the capacity to use nitrite rather than O_2 as an electron acceptor during respiration, O_2 is still required for ammonia oxidation.

Nitrifiers are little different from other aerobic microbes with respect to their response to temperature, moisture, and other environmental variables (see Fig. 14.2). Nitrification occurs slowly but readily under snow and in refrigerated soils, and soil transplant experiments (Mahendrappa et al., 1966) have demonstrated an apparent capacity for nitrifiers to adapt to different temperature and moisture regimes. For many decades, nitrifiers were thought to be inhibited in acid soils, probably because in many cases, especially in soils from cultivated fields, raising soil pH with calcium or magnesium carbonate stimulates nitrification, and culturable nitrifiers exhibit a pH optimum of 7.5-8 (Prosser, 2011). We now recognize that nitrification can be high even in very acid forest soils (pH < 4.5; Robertson, 1989), although the physiological basis for this is still not well understood (De Boer and Kowalchuk, 2001).

IV INHIBITION OF NITRIFICATION

Nitrification is unaccountably slow in some soils, and in some circumstances, it may be inhibited by natural or manufactured compounds. A wide variety of plant extracts can inhibit culturable nitrifiers *in vitro*, even though their importance *in situ* is questionable. Likewise, commercial products, such as nitrapyrin and diocyandimide, can be used to inhibit nitrification in soil with varying degrees of success. Most commercial compounds are pyridines, pyrimidines, amino triazoles, and sulfur compounds, such as ammonium thiosulfate. Another innovation is paraffin-coated calcium carbide (CaC_2; Freney et al., 2000). Calcium carbide reacts with water to form acetylene (C_2H_2), which inhibits nitrifiers at very low partial pressures, ca. 10 Pa. As the paraffin wears off, CaC_2 is exposed to soil moisture, and the C_2H_2 formed inhibits nitrification. Likewise, neem oil, extracted from the Indian neem tree (*Azadirachta indica*), has been used commercially to coat urea fertilizer pellets to slow its nitrification to NO_3^-.

The potential value of managing nitrifiers in ecosystems can be easily seen from the position of nitrification in the overall N cycle (Fig. 14.1). Nitrogen is lost from ecosystems mainly after its conversion to NO_3^- and prior to plant uptake, so keeping N in the NH_4^+ form keeps it from being lost via nitrate leaching and denitrification, the two principal pathways of unintentional N loss and subsequent atmospheric and water contamination in most ecosystems. Because many plants prefer to take up N as NO_3^-, it is not desirable to completely inhibit nitrification, even in intensively managed ecosystems such as fertilized row crops, but slowing nitrifiers or restricting their activity to periods of active plant growth is an attractive—if still elusive—management option.

V DENITRIFICATION

Denitrification is the reduction of soil nitrate to the N gases NO, N_2O, and N_2. A wide variety of mostly heterotrophic bacteria can denitrify, whereby they use NO_3^- rather than oxygen (O_2) as a terminal electron acceptor during respiration. Because nitrate is a less-efficient electron acceptor than O_2, most denitrifiers undertake denitrification only when O_2 is unavailable. In most soils, this mainly occurs following rainfall as soil pores become water saturated, and the diffusion of O_2 to microsites is drastically slowed. Typically denitrification starts to occur at water-filled pore space concentrations of 60% and higher (Fig. 14.2). In wetlands and lowland rice, soil diffusion may be restricted most of the time. Oxygen demand can also exceed supply inside soil aggregates and in rapidly decomposing litter.

Denitrification is the only point in the N cycle where fixed N reenters the atmosphere as N_2; it thus serves to close the global N cycle. In the absence of denitrification, N_2 fixers (see Chapter 15) would eventually draw atmospheric N_2 to nil, and the biosphere would be awash in nitrate. Denitrification is also significant as the major source of atmospheric N_2O, an important greenhouse gas that also consumes stratospheric ozone.

From a management perspective, denitrification is advantageous when it is desirable to remove excess NO_3^- from soil prior to its movement to ground or surface waters. Sewage treatment often aims to remove N from wastewater streams by managing nitrification and denitrification. Typically, wastewater is directed through sedimentation tanks, filters, and sand beds designed to remove particulates and encourage decomposition and the mineralization of organic N to NH_4^+, which is then nitrified under aerobic conditions to NO_3^-. The stream is then directed to anaerobic tanks, where denitrifiers convert the NO_3^- to N_2O and N_2, which is then released to the atmosphere. Part of the nitrification/denitrification management challenge is ensuring that the stream is exposed to aerobic conditions long enough to allow nitrifiers to convert most NH_4^+ to NO_3^-, but not so long as to remove all dissolved organic C (known as biological oxygen demand or BOD to wastewater engineers), which the denitrifiers need for substrate. Recently, anammox (described later in the chapter) has been utilized for wastewater N removal.

Denitrification can also remove nitrate from groundwater prior to its movement to streams and rivers. In most wetlands and riparian areas, nitrate-rich groundwater must move across a groundwater-sediment interface that is typically anaerobic and C-rich. As nitrate moves across this interface, it can be denitrified to N_2O and N_2, keeping it from polluting downstream surface waters.

It is usually desirable to minimize denitrification in managed ecosystems to conserve N for plant uptake. In regions with ample rainfall, N losses due to denitrification can rival or exceed losses by nitrate leaching. There are no technologies designed to inhibit denitrification *per se*; usually denitrifiers are best managed indirectly by manipulating water levels (e.g., in rice cultivation) or nitrate supply (e.g., nitrification inhibitors).

A Denitrifier Diversity

Denitrification is carried out by a broad array of soil bacteria, including organotrophs, chemo- and photolithotrophs, N_2 fixers, thermophiles, halophiles, and various pathogens. Over 50 genera with over 125 denitrifying species have been identified (Zumft, 1992). In soil, most culturable denitrifiers are facultative anaerobes from only 3-6 genera, principally *Pseudomonas* and *Alcaligenes*, and to a lesser extent, *Bacillus, Agribacterium*, and *Flavibacterium*. Typically, denitrifiers constitute 0.1-5% of the total culturable soil population and up to 20% of total microbial biomass (Tiedje, 1988).

Organisms denitrify to generate energy (ATP) by electron transport phosphorylation via the cytochrome system. The general pathway is

$$2NO_3^- \xrightarrow[nar]{} 2NO_2^- \xrightarrow[nir]{} 2\overset{\uparrow}{NO} \xrightarrow[nor]{} \overset{\uparrow}{N_2O} \xrightarrow[nos]{} \overset{\uparrow}{N_2}. \qquad (14.9)$$

Each step is enacted by individual enzymes: nitrate reductase (*nar*), nitrite reductase (*nir*), nitric oxide reductase (*nor*), and nitrous oxide reductase (*nos*). Each is inhibited by O_2, and the organization of these enzymes in the cell membrane for G^- bacteria is described in Fig. 14.7. At any step in this process, intermediate products can be exchanged with the soil environment, making denitrifiers a significant source of NO_2^- in soil solution and important sources of the atmospheric gases NO and N_2O.

Each denitrification enzyme is inducible, primarily in response to the partial pressure of O_2 and substrate (C) availability. Because enzyme induction is sequential and substrate dependent, there is usually a lag between the production of an intermediate substrate and its consumption by the next enzyme. In pure culture, these lags can be on the order of hours (Fig. 14.8); *in situ* lags in soil can be substantially longer, and differences in lags among different microbial taxa may significantly affect the contribution of denitrifiers to fluxes of NO and N_2O to the atmosphere. That induced enzymes degrade at different rates, and more slowly than they are induced, also leads to a complex response to the environmental conditions that induce denitrification; whether a soil has denitrified recently (whether denitrifying enzymes are present) may largely

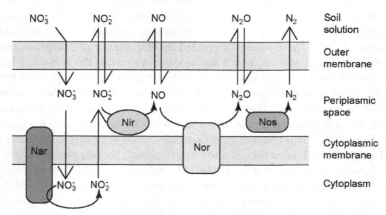

FIG. 14.7 The organization of denitrification enzymes in gram-negative bacteria. *Adapted from Ye et al. (1994).*

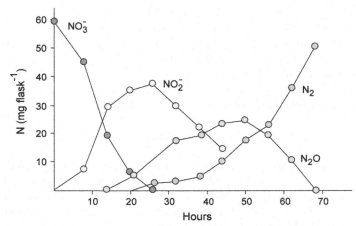

FIG. 14.8   The sequence of products formed during denitrification *in vitro* as different enzymes are sequentially induced. *Adapted from Cooper and Smith (1963)*.

determine its response to newly favorable conditions for denitrification. Rainfall onto soil that is moist, for example, will likely lead to a faster and perhaps stronger denitrification response than will rainfall onto the same soil when it is dry (Groffman and Tiedje, 1988) and will affect the proportion of N product that is N_2O vs. N_2 because of the presence of *nos* in recently wet soil (Bergsma et al., 2002).

B   Environmental Controls of Denitrification

For decades after its discovery as an important microbial process, denitrification was assumed to be important only in aquatic and wetland ecosystems. It was not until the advent of whole-ecosystem N budgets and the use of ^{15}N to trace the fate of fertilizer N in the 1950s that denitrification was found to be important in unsaturated soils. These studies suggested the importance of denitrification in fertilized agricultural soils, and with the development of the acetylene block technique in the 1970s, the importance of denitrification in well-drained forest and grassland soils was also confirmed. Acetylene selectively inhibits nitrous oxide reductase (*nos*; see Fig. 14.7), allowing the assessment of N_2 production by following N_2O accumulation in a soil core or monolith treated with acetylene. Unlike N_2, small changes in N_2O concentration are easily detected in air.

Today, denitrification is known to be an important N cycle process wherever O_2 is limiting. In unsaturated soils, this frequently occurs within soil aggregates, in decomposing plant litter, and in rhizospheres. Soil aggregates vary widely in size, but in general are composed of small mineral particles and pieces of organic matter <2 mm in diameter that are glued to one another with biologically derived polysaccharides. Like most particles in soil, aggregates are surrounded by a thin water film that impedes gas exchange. Modeling efforts in the 1970s (Smith, 1980) suggested that the centers of these aggregates ought to be anaerobic owing to a higher respiratory demand in the aggregate center than could be satisfied by O_2 diffusion from the bulk soil atmosphere. This was confirmed experimentally in 1985 (Sexstone et al., 1985), providing a logical explanation for active denitrification in soils that appeared otherwise to be aerobic, and an explanation for the almost universal presence of denitrifiers and denitrification enzymes in soils worldwide.

In addition to O_2, denitrification is also regulated by soil C and NO_3^-. Carbon is important because most denitrifiers are heterotrophs and require reduced C as the electron donor, although as noted earlier, denitrifiers can also be chemo- and photolithotrophs. Nitrate serves as the electron acceptor and must be provided via nitrification, rainfall, or fertilizer. However, O_2 is the preferred electron acceptor because of its high energy yield and thus must be largely depleted before denitrification occurs. In most soils, the majority of denitrifiers are facultative anaerobes that will simply avoid synthesizing denitrification enzymes until O_2 drops below some critical threshold.

In the field, O_2 is by far the dominant control on denitrification rates. Denitrification can be easily stimulated in an otherwise aerobic soil by removing O_2, and can be inhibited in saturated soil by drying or otherwise aerating it. The relative importance of C and NO_3^-, the other major controls, will vary by ecosystem. Under saturated conditions, such as those found in wetlands and lowland rice paddies, NO_3^- limits denitrification because the nitrifiers that provide NO_3^- are inhibited at low O_2 concentrations. Consequently, denitrification occurs only in the slightly oxygenated rhizosphere and at the sediment-water interface, places where there is sufficient O_2 for nitrifiers to oxidize NH_4^+ to NO_3^-, which can then diffuse to denitrifiers in the increasingly anaerobic zones away from the root surface or sediment-water interface. It is often difficult to find NO_3^- in persistently saturated soils, not only because of low nitrification, but also because of the tight coupling between nitrifiers and denitrifiers. In wetlands with fluctuating water tables or with significant inputs of NO_3^- from groundwater, NO_3^- may be more available.

On the other hand, the availability of soil C in unsaturated soils more often limits denitrification. In these soils, C supports denitrification both directly, by providing donor electrons to denitrifiers, and indirectly, by stimulating O_2 consumption by heterotrophs. It can be difficult to experimentally distinguish between these two effects; from a management perspective, there probably is no need to.

VI OTHER NITROGEN TRANSFORMATIONS IN SOIL

Several additional microbial processes transform N in soil, although none are thought to be as quantitatively important as mineralization, immobilization, nitrification, and denitrification. *Dissimilatory nitrate reduction to ammonium* (DNRA) refers to the anaerobic transformation of nitrate to nitrite and then to ammonium. Like denitrification, this process allows for respiration to go on in the absence of O_2 and is thought to be favored in environments where the ratio of C to nitrate is high because the process consumes more electrons than denitrification. A capacity for DNRA has been found in facultative and obligately fermentative bacteria and has long been thought to be restricted to high C, highly anaerobic environments, such as anaerobic sewage sludge bioreactors, anoxic sediments, and the bovine rumen. However, DNRA has been found to be common and important in some tropical forest soils (Silver et al., 2001) and in a variety of freshwater sediments (Burgin and Hamilton, 2007). In these soils, the flow of inorganic N through DNRA is as large as or larger than the flow through denitrification and nitrification and may help to conserve N in these ecosystems by shunting nitrate into ammonium rather than to N_2O or N_2.

Nonrespiratory denitrification, like respiratory denitrification, also results in the production of N gas (mainly N_2O), but the reduction does not enhance growth and can occur in aerobic environments. A variety of nitrate-assimilating bacteria, fungi, and yeast can carry out nonrespiratory denitrification, which may be responsible for some of the N_2O now attributed to nitrifiers in well-aerated soils (Robertson and Tiedje, 1987).

Anammox, in which ammonium and nitrite are converted to N_2 (Mulder et al., 1995; Jetten, 2001) is known to occur in sewage treatment plants and oceanic systems (Kuypers et al., 2005), where they can be the dominant source of N_2 flux. Anammox bacteria grow very slowly in enrichment culture and only under strict anaerobic conditions, and are thus likely to be part of a significant soil process only in periodically or permanently submerged soils (Strous, 2011).

Bacteria capable of performing anammox occur within the single order Brocadiales in the phylum Planctomycete. In these bacteria, anammox catabolism occurs in a specialized organelle called the anammoxosome, wherein

$$NH_4^+ + NO_2^- \rightarrow N_2H_2 \rightarrow N_2, \tag{14.10}$$

although much remains to be learned about the biochemistry and bioenergetics of the process, including intermediate compounds (Kartal et al., 2011).

Chemodenitrification occurs when NO_2^- in soil reacts to form N_2 or NO_x. This can occur by several aerobic pathways. In the Van Slyke reaction, amino groups in the α position to carboxyls yield N_2:

$$RNH_2 + HNO_2 \rightarrow ROH + H_2O + N_2. \tag{14.11}$$

In a similar reaction, NO_2^- reacts with NH_3, urea, methylamine, purines, and pyrimidines to yield N_2:

$$HNO_2 + NH_3 \rightarrow N_2 + 2H_2O. \tag{14.12}$$

Chemical decomposition of HNO_2 may also occur spontaneously:

$$3HNO_2 \rightarrow HNO_3 + H_2O + 2NO. \tag{14.13}$$

In general, chemodenitrification is thought to be a minor pathway for N loss in most ecosystems. It is not easily evaluated *in situ*, however, and in the lab requires a sterilization procedure that does not itself significantly disrupt soil N chemistry.

VII NITROGEN MOVEMENT IN THE LANDSCAPE

Microbial transformations of reactive N (Table 14.3) have great importance for soil fertility, water quality, and atmospheric chemistry at ecosystem, landscape, and regional scales. It is at these scales that differences between what we have learned in the laboratory and what we observe in the environment (see Introduction) become most obvious.

TABLE 14.3 Forms of N of Concern in the Environment

N Form	Sources	Dominant Transport Vectors	Environmental Effects
Nitrate (NO_3^-)	Nitrification	Groundwater	Pollution of drinking water
	Fertilizer		Coastal eutrophication
	Disturbance that stimulates nitrification		
	Combustion (acid rain)		
Ammonia (NH_3, NH_4^+)	Fertilizer	Surface runoff	Pollution of drinking water
	Animal waste	Atmosphere	Eutrophication
Nitrous oxide (N_2O)	By-product of nitrification, denitrification, anammox	Atmosphere	Greenhouse gas
		Groundwater	Ozone destruction in stratosphere
Nitric oxide (NO)	By-product of nitrification, denitrification, anammox	Atmosphere	Ozone precursor in troposphere
Dissolved organic N	By-product of mineralization	Surface runoff	Eutrophication (?)
		Groundwater	

TABLE 14.4 Criteria for Determining if a Site is a Source or Sink of N in the Landscape

Criteria	Determinants
Is the site N rich?	Fertilized
	Fine texture (clay)
	Legumes
	Wet tropics
Is the site highly disturbed?	Disturbance of plant uptake (e.g., harvest)
	Stimulation of mineralization (e.g., tillage)
	Disturbance of links between plant and microbial processes (e.g., tillage)
Does the site have a high potential for denitrification?	Wet soil
	Well-aggregated
	High available organic matter
Does the site have a high potential for NH_3 volatilization?	High pH (>8.0)

From Groffman (2000).

One approach to thinking about microbial N cycle processes at large scales is to ask a series of questions that attempt to determine if a particular ecosystem is a source or sink of particular N species of environmental concern (Table 14.4). Sites that are N-rich either naturally or following disturbance have a high potential to function as sources of most of the reactive N forms identified in Table 14.1 because mineralization and nitrification, the processes that produce most of these reactive forms, occur at high rates.

Nitrogen sinks are defined as habitats that have a high potential to remove reactive N from the environment, preventing its movement into adjacent ecosystems. Ecosystems such as wetlands that are wet and rich in organic materials, for example, have a great potential to function as sinks because of their ability to support denitrification. In many cases these sink areas retain reactive N produced in source areas of the landscape. Riparian buffer zones next to streams, for example, can be managed to retain nitrate moving out of crop fields in groundwater (Lowrance et al., 1984). This nitrate can be stored in plant tissue or in soil organic matter as organic N or can be denitrified to N gas and thereby released to the atmosphere, preferably as N_2, a nonreactive form.

Humans have doubled the circulation of reactive N on Earth, creating a nitrogen cascade in which added N flows through the environment, leading to degradation of air and water quality and coastal ecosystems in many areas (Galloway et al., 2008). Solutions to landscape, regional, and global N enrichment problems often rely heavily on managing microbial N transformations. For example, coastal areas of the Gulf of Mexico suffer from eutrophication and hypoxia that have been linked to excess N from the Mississippi River Basin (Rabalais et al., 2002). Proposed solutions to this problem include better management of microbial N-transformations in crop fields as well as the creation of denitrifying wetland sinks for excess N moving out of agricultural areas (Mitsch et al., 2001).

Source-sink dynamics of N ultimately depend on the juxtaposition of different ecosystems in the landscape and the hydrologic and atmospheric transport vectors that link them—a complex topic that requires knowledge of hydrology and atmospheric chemistry in addition to soil ecology and microbiology. Because soil microbes play a crucial role in forming and consuming reactive N in the environment, their management can be an important and even crucial means for regulating N fluxes at local, regional, and global scales.

REFERENCES

Bergsma, T.T., Robertson, G.P., Ostrom, N.E., 2002. Influence of soil moisture and land use history on denitrification end products. J. Environ. Qual. 31, 711–717.

Burgin, A.J., Hamilton, S.K., 2007. Have we overemphasized the role of denitrification in aquatic ecosystems? A review of nitrate removal pathways. Front. Ecol. Environ. 5, 89–96.

Cooper, G.S., Smith, R., 1963. Sequence of products formed during denitrification in some diverse Western soils. Soil Sci. Soc. Am. Proc. 27, 659–662.

Daims, H., Lücker, S., Le Paslier, D., Wagner, M., 2011. Diversity, environmental genomics, and ecophysiology of nitrite-oxidizing bacteria. In: Ward, B.B., Arp, D.J., Klotz, M.G. (Eds.), Nitrification. American Society for Microbiology Press, Washington, DC, pp. 295–322.

De Boer, W., Kowalchuk, G.A., 2001. Nitrification in acid soils: micro-organisms and mechanisms. Soil Biol. Biochem. 33, 853–866.

Firestone, M.K., Davidson, E.A., 1989. Microbiological basis of NO and N_2O production and consumption in soil. In: Andreae, M.D., Schimel, D.S. (Eds.), Trace Gas Exchange between Terrestrial Ecosystems and the Atmosphere. John Wiley, Berlin, pp. 7–22.

Freitag, T.E., Chang, L., Clegg, C.D., Prosser, J.I., 2005. Influence of inorganic nitrogen-management regime on the diversity of nitrite oxidizing bacteria in agricultural grassland soils. Appl. Environ. Microbiol. 71, 8323–8334.

Freney, J.R., Randall, P.J., Smith, J.W.B., Hodgkin, J., Harrington, K.J., Morton, T.C., 2000. Slow release sources of acetylene to inhibit nitrification in soil. Nutr. Cycl. Agroecosyst. 56, 241–251.

Galloway, J.N., Townsend, A.R., Erisman, J.W., Bekunda, M., Cai, Z.C., Freney, J.R., Martinelli, L.A., Seitzinger, S.P., Sutton, M.A., 2008. Transformation of the nitrogen cycle: recent trends, questions, and potential solutions. Science 320, 889–892.

Groffman, P.M., 2000. Nitrogen in the environment. In: Sumner, M.E. (Ed.), Handbook of Soil Science. CRC Press, Boca Raton, pp. C190–C200.

Groffman, P.M., Tiedje, J.M., 1988. Denitrification hysteresis during wetting and drying cycles in soil. Soil Sci. Soc. Am. J. 52, 1626–1629.

Hart, S.C., Stark, J.M., Davidson, E.A., Firestone, M.K., 1994. Nitrogen mineralization, immobilization, and nitrification. In: Weaver, R.W., Angle, J.S., Bottomley, P.J., Bezdicek, D.F., Smith, M.S., Tabatabai, M.A., Wollum, A.G. (Eds.), Methods of Soil Analysis, Part 2—Microbiological and Biochemical Properties. Soil Science Society of America, Madison, pp. 985–1018.

He, J.-Z., Hu, H.-W., Zhang, L.-M., 2012. Current insight into the autotrophic thaumarchaeal ammonia oxidation in acidic soils. Soil Biol. Biochem. 55, 146–154.

Hyvönen, R., Agren, G.I., Andren, O., 1996. Modeling long-term carbon and nitrogen dynamics in an arable soil receiving organic matter. Ecol. Appl. 6, 1345–1354.

Jetten, M.S.M., 2001. New pathways for ammonia conversion in soil and aquatic systems. Plant Soil 230, 9–19.

Juretschko, S., Timmermann, G., Schmid, M., Schleifer, K.H., Pommerening-Roser, A., Koops, H.-P., Wagner, M., 1998. Combined molecular and conventional analyses of nitrifying bacterium diversity in activated sludge: *Nitrosococcus mobilis* and *Nitrospira*- like bacteria as dominant populations. Appl. Environ. Microbiol. 64, 3042–3051.

Kartal, B.K., Keltjens, J.T., Jetten, M.S.M., 2011. Metabolism and genomics of anammox bacteria. In: Ward, B.B., Arp, D.J., Klotz, M.G. (Eds.), Nitrification. American Society for Microbiology Press, Washington, DC, pp. 181–200.

Koops, H.-P., Pommerening-Röser, A., 2001. Distribution an ecophysiology of the nitrifying bacteria emphasizing cultured species. FEMS Microbiol. Ecol. 37, 1–9.

Koops, H.-P., Purkhold, U., Pommerening-Röser, A., Timmermann, G., Wagner, M., 2006. The lithoautotrophic ammonia-oxidizing bacteria. In: Dworkin, M., Falkow, S., Rosenberg, E., Schleifer, K.-H., Stackebrandt, E. (Eds.), The Prokaryotes: A Handbook on the Biology of Bacteria. Springer, Berlin, pp. 778–788.

Kuypers, M.M.M., Lavik, G., Woebken, D., Schmid, M., Fuchs, B.M., Amann, R., Jorgensen, B.B., Jetten, S.M., 2005. Massive nitrogen loss from the Benguela upwelling system through anaerobic ammonium oxidation. Proc. Natl. Acad. Sci. U. S. A. 102, 6478–6483.

Lehtovirta-Morley, L.E., Stoecker, K., Vilcinskas, A., Prosser, J.I., Nicol, G.W., 2011. Cultivation of an obligate acidophilic ammonia oxidizer from a nitrifying acid soil. Proc. Natl. Acad. Sci. U. S. A. 108, 15892–15987.

Leininger, S., Urich, T., Schloter, M., Schwark, L., Qi, J., Nicol, G.W., Prosser, J.I., Schuster, S.C., Schleper, C., 2006. Archaea predominate among ammonia-oxidizing prokaryotes in soils. Nature 442, 806–809.

Linn, D.M., Doran, J.W., 1984. Effect of water-filled pore space on CO_2 and N_2O production in tilled and non-tilled soils. Soil Sci. Soc. Am. J. 48, 1267–1272.

Löhnis, F., 1913. Vorlesungen uber Landwortschaftliche Bacterologia. Borntraeger, Berlin.

Lowrance, R.R., Todd, R.L., Fail, J., Hendrickson, O., Leonard, R., Asmussen, L., 1984. Riparian forests as nutrient filters in agricultural watersheds. Bioscience 34, 374–377.

Mahendrappa, M.K., Smith, R.L., Christiansen, A.T., 1966. Nitrifying organisms affected by climatic region in western U.S. Proc. Soil Sci. Soc. Am. 30, 60–62.

Martens-Habbena, W., Berube, P.M., Urakawa, H., de la Torre, J.R., Stahl, D.A., 2009. Ammonia oxidation kinetics determine niche separation of nitrifying Archaea and Bacteria. Nature 461, 976–979.

Mitsch, W.J., Day, J.W., Gilliam, J.W., Groffman, P.M., Hey, D.L., Randall, G.W., Wang, N., 2001. Reducing nitrogen loading to the Gulf of Mexico from the Mississippi River basin: strategies to counter a persistent ecological problem. Bioscience 51, 373–388.

Mulder, A., van de Graaf, A.A., Robertson, L.A., Kuenen, J.G., 1995. Anaerobic ammonium oxidation discovered in a denitrifying fluidized bed reactor. FEMS Microbiol. Ecol. 16, 177–184.

Norton, J.M., 2011. Diversity and environmental distribution of ammonia-oxidizing bacteria. In: Ward, B.B., Arp, D.J., Klotz, M.G. (Eds.), Nitrification. American Society for Microbiology Press, Washington, DC, pp. 39–55.

Orso, S., Guoy, M., Navarro, E., Normand, P., 1994. Molecular phylogenetic analysis of *Nitrobacter* spp. Int. J. Syst. Bacteriol. 44, 83–86.

Prosser, J.I., 2011. Soil nitrifiers and nitrification. In: Ward, B.B., Arp, D.J., Klotz, M.G. (Eds.), Nitrification. American Society for Microbiology Press, Washington, DC, pp. 347–383.

Prosser, J.I., Embley, T.M., 2002. Cultivation-based and molecular approaches to characterisation of terrestrial and aquatic nitrifiers. Anton van Leeuwenhoek 81, 165–179.

Rabalais, N.N., Turner, E.R., Scavia, D., 2002. Beyond science into policy: Gulf of Mexico hypoxia and the Mississippi River. Bioscience 52, 129–152.

Robertson, G.P., 1989. Nitrification and denitrification in humid tropical ecosystems. In: Proctor, J. (Ed.), Mineral Nutrients in Tropical Forest and Savanna Ecosystems. Blackwell Scientific, Cambridge, pp. 55–70.

Robertson, G.P., Tiedje, J.M., 1987. Nitrous oxide sources in aerobic soils: nitrification, denitrification, and other biological processes. Soil Biol. Biochem. 19, 187–193.

Robertson, G.P., Vitousek, P.M., 1981. Nitrification potentials in primary and secondary succession. Ecology 62, 376–386.

Robertson, G.P., Vitousek, P.M., 2009. Nitrogen in agriculture: balancing the cost of an essential resource. Ann. Rev. Environ. Res. 34, 97–125.

Robertson, G.P., Wedin, D.A., Groffman, P.M., Blair, J.M., Holland, E., Harris, D., Nadelhoffer, K., 1999. Soil carbon and nitrogen availability: nitrogen mineralization, nitrification, and soil respiration potentials. In: Robertson, G.P., Bledsoe, C.S., Coleman, D.C., Sollins, P. (Eds.), Standard Soil Methods for Long-Term Ecological Research. Oxford University Press, New York, pp. 258–271.

Schimel, J.P., Bennett, J., 2004. Nitrogen mineralization: challenges of a changing paradigm. Ecology 85, 591–602.

Sexstone, A.J., Revsbech, N.P., Parkin, T.P., Tiedje, J.M., 1985. Direct measurement of oxygen profiles and denitrification rates in soil aggregates. Soil Sci. Soc. Am. J. 49, 645–651.

Silver, W.L., Herman, D.J., Firestone, M.K., 2001. Dissimilatory nitrate reduction to ammonium in upland tropical forest soils. Ecology 82, 2410–2416.

Smith, K.A., 1980. A model of the extent of anaerobic zones in aggregated soils, and its potential application to estimates of denitrification. J. Soil Sci. 31, 263–277.

Strous, M., 2011. Beyond denitrification: alternative routes to dinitrogen. Nitrogen Cycling in Bacteria: Molecular Analysis. Caister Academic Press, Norfolk, U.K.

Tiedje, J.M., 1988. Ecology of denitrification and dissimilatory nitrate reduction to ammonium. In: Zehnder, A.J.B. (Ed.), Biology of Anaerobic Microorganisms. John Wiley and Sons, New York, pp. 179–244.

Tisdale, S.L., Nelson, W.L., Beaton, J.D., Havlin, J.L., 1993. Soil Fertility and Fertilizers, fifth ed. MacMillan, New York.

Winogradsky, S., 1892. Contributions a la morphologie des organismes de la nitrification. Arch. Biol. Sci. 1, 86–137.

Ye, R.W., Averill, B.A., Tiedje, J.M., 1994. Denitrification of nitrite and nitric oxide. Appl. Environ. Microbiol. 60, 1053–1058.

Zhang, L.-M., Offre, P.R., He, J.-Z., Vernamme, D.T., Nicol, G.W., Prosser, J.I., 2010. Autotrophic ammonia oxidation by soil thaumarchaea. Proc. Natl. Acad. Sci. U. S. A. 107, 17240–17245.

Zhu, X., Burger, M., Doane, T.A., Horwath, R.W., 2013. Ammonia oxidation pathways and nitrifier denitrification are significant sources of N_2O and NO under low oxygen availability. Proc. Natl. Acad. Sci. U. S. A. 110, 6328–6333.

Zumft, W.G., 1992. The denitrifying prokaryotes. In: Balows, A. (Ed.), The Prokaryotes. Springer-Verlag, New York.

FURTHER READING

Davidson, E.A., David, M.B., Galloway, J.N., Goodale, C.L., Haeuber, R., Harrison, J.A., Howarth, R.W., Jaynes, D.B., Lowrance, R.R., Nolan, B.T., Peel, J.L., Pinder, R.W., Porter, E., Snyder, C.S., Townsend, A.R., Ward, M.H., 2012. Excess nitrogen in the U.S. environment: trends, risks, and solutions. Iss. Ecol. 15, 1–16.

Groffman, P.M., Tiedje, J.M., Robertson, G.P., Christensen, S., 1988. Denitrification at different temporal and geographical scales: proximal and distal controls. In: Wilson, J.R. (Ed.), Advances in Nitrogen Cycling in Agricultural Ecosystems. CAB International, Wallingford, U.K., pp. 174–192.

Robertson, G.P., 1997. Nitrogen use efficiency in row-crop agriculture: crop nitrogen use and soil nitrogen loss. In: Jackson, L. (Ed.), Ecology in Agriculture. Academic Press, New York, pp. 347–365.

Robertson, G.P., 2000. Denitrification. In: Sumner, M.E. (Ed.), Handbook of Soil Science. CRC Press, Boca Raton, pp. C181–C190.

Teske, A., Alm, E., Regan, J.M., Toze, S., Rittman, B.E., Stahl, D.A., 1994. Evolutionary relationships among ammonia- and nitrite-oxidizing bacteria. J. Bacteriol. 176, 6623–6630.

Chapter 15

Biological N Inputs

Peter J. Bottomley and David D. Myrold
Department of Crop and Soil Science, Oregon State University, Corvallis, OR, USA

Chapter Contents

I GLOBAL N INPUTS

The nonmetallic element N is essential for life. Although a huge amount of N $(4 \times 10^{21}$ g) exists in the atmosphere, soils, and waters of Earth, more than 99% of this is in the form of N_2 and is unavailable to almost all living organisms (Table 15.1). To transform N_2 into "reactive N," the triple bond must be broken so that N can bond with C, H, and O and form the building blocks of life. In the pre-human world, N_2 was transformed into reactive N via lightning discharge (5 Tg N y^{-1}; 1 teragram equals 10^{12} g) and by biological nitrogen fixation (BNF; 100-140 Tg N y^{-1}). In the past, the microbial process of denitrification (whereby NO_3^- is converted to N_2) occurred at approximately the same rate as BNF. During the past 100 years, however, the annual input of anthropogenically created, reactive N has increased dramatically due to: (a) modest increases in legume use in agriculture (15-30 Tg N y^{-1}), (b) increases in fossil fuel combustion (1 Tg N y^{-1} in 1860 vs. 25 Tg N y^{-1} in 2000), and (c) enormous increases in the use of Haber-Bosch process derived fertilizer N for food production (zero, pre-twentieth century; currently ~110 Tg N year^{-1}; Fig. 15.1). Enhanced use of reactive N for food production has had secondary effects on the N cycle. The intensification of

Soil Microbiology, Ecology, and Biochemistry. http://dx.doi.org/10.1016/B978-0-12-415955-6.00015-3

TABLE 15.1 Global Nitrogen Pool Sizes

Nitrogen Pool	Pool Size (g of N)
Lithosphere	1.0×10^{23}
Atmosphere	3.9×10^{21}
Coal	1.0×10^{17}
Hydrosphere	2.3×10^{19}
Soil organic N	1.0×10^{17}
Soil fixed NH_4^+	2.0×10^{16}
Biota N	3.5×10^{15}
Microbial N	1.5×10^{15}

Reprinted from Soil Microbiology, Ecology, and Biochemistry, third ed., Paul, E.A., Biological N inputs. (2007). pp. 365–387.

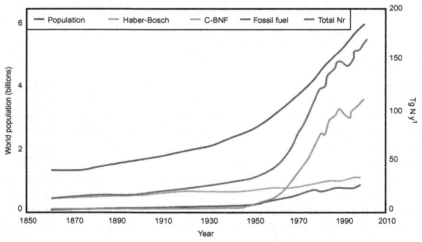

FIG. 15.1 Global population trends from 1860 to 2000 (billions, left axis) and reactive nitrogen (Nr) creation (teragrams nitrogen [Tg N] per year, right axis). "Haber-Bosch" represents Nr creation through the Haber-Bosch process, including production of ammonia for nonfertilizer purposes. "C-BNF" (cultivation-induced biological nitrogen fixation) represents Nr creation from cultivation of legumes, rice, and sugarcane. "Fossil fuel" represents Nr created from fossil fuel combustion. "Total Nr" represents the sum created by these three processes. *From Galloway et al. 'The Nitrogen Cascade' (2003). Copyright, American Institute of Biological Sciences.*

agriculture and its expansion into forest and grasslands has released reactive N from long-term storage in soil organic matter. It has been estimated that the burning of forests and grasslands, draining of wetlands, and soil tillage liberates approximately 40 Tg N y^{-1} of reactive N (Vitousek et al., 1997).

The distribution of BNF has changed as a result of urbanization and intensive agriculture. Large areas of diverse natural vegetation, that once included N$_2$-fixing species as part of their floral composition, have been replaced by monocultures of non-N$_2$-fixing, high-yielding crop species that require addition of reactive N to achieve their yield potentials. In contrast, in the vast acreage planted to soybeans in Brazil and Argentina, 70% to >90% of the N is derived from BNF (Herridge et al., 2008). Increased use of N$_2$-fixing grain legumes can result in loss of N from agricultural soils if BNF performs suboptimally. Furthermore, the use of fertilizer N on nonleguminous crops used in a rotation cycle can reduce the %N derived from BNF during the legume phase. Another major consequence of the human-driven alteration of the N cycle involves the movement of reactive N from sites of application to remote sites. BNF occurs inside living organisms where fixed N is quickly assimilated into cell constituents. Reactive N applied as fertilizer or animal waste can lead to losses to the environment. There are both positive and negative effects associated with the deposition of reactive N at sites remote from its source. Plant and microbe growth may be stimulated and N immobilized. Scientists have estimated that an extra 100-1000 Tg C y^{-1} may be fixed globally because of atmospheric transfer and remote deposition of reactive N. Too much N deposition, however, can result in soil and aquatic acidification, greater NO$_3$ leaching, and loss of diversity. The justification for studying BNF so intensively over the past 100 years was based on the need to increase inputs of reactive N via BNF into N-limited agroecosystems. Scientists work to thoroughly understand the mechanism of BNF to increase the efficiency of manufacturing N fertilizer. There is an urgent need to learn how certain N$_2$-fixing plants and specific microorganisms establish symbiotic and endophytic associations whereas others do not. There is hope that one day scientists will create N$_2$-fixing corn, wheat, and rice and thereby reduce the use of fertilizer N.

II BIOLOGICAL NITROGEN FIXATION

BNF is a process restricted to the prokaryotes of the domains Archaea and Bacteria. However, there are many examples of symbiotic associations that have developed between eukaryotes and prokaryotes that circumvent the eukaryotes' needs for an exogenous supply of fixed N. Bacteria that use N$_2$ as sole source of N are called diazotrophs. Although a relatively limited number of bacterial species fix N$_2$, they represent a wide variety of phylogenetically and physiologically distinct types that occupy different ecological niches (see Supplemental Table 15.1 at http://booksite.elsevier.com/9780124159556). These bacteria use diverse energy sources including sunlight (phototrophs), reduced inorganic

elements and compounds (lithotrophs), and a plethora of different organic substrates (heterotrophs). They are represented by obligate aerobes and facultative
and obligate anaerobes. This metabolic diversity indicates that diazotrophs can
contribute fixed N to other life forms in a wide variety of environments.

BNF is mediated by an enzyme complex called nitrogenase, which is composed of two proteins (dinitrogenase and nitrogenase reductase). Phylogenetic
analysis of the sequences of the gene that encodes for nitrogenase reductase
(*nif*H) shows that there are several distinct forms of this gene distributed among
Archaea and different bacterial phyla (proteobacteria, cyanobacteria, firmicutes, and actinobacteria) and also among different classes within the proteobacteria (Fig. 15.2). There are three biochemically distinct forms of
nitrogenase that differ in their requirements for either molybdenum (Mo), vanadium (V), or iron (Fe) as a critical metallic component of the cofactor associated
with the catalytic site. Most of the nitrogenases that have been studied extensively possess a Mo-containing cofactor. N_2 fixation is an energetically expensive process. In BNF, two ATP molecules are required for each electron
transferred from nitrogenase reductase to dinitrogenase, which contains the catalytic site. Nitrogenase reductase is recharged with electrons provided by a protein called ferredoxin, or other strongly reducing variants such as flavodoxin.
A total of 16 ATP molecules are required to provide the six electrons necessary
to reduce one N_2 molecule into two NH_3 because 25% of the energy used to
reduce N_2 is "lost" in the reduction of $2H^+$ to H_2. One mole of H_2 is formed

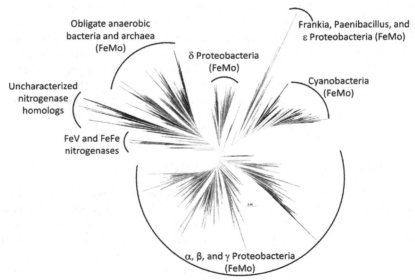

FIG. 15.2   Phylogenetic distribution of *nifH* sequences with red branches designating sequences
associated with N_2-fixing bacteria and archaea of soil. *Courtesy of Gaby and Buckley (2011). Used
with permission.*

per mole of N_2 transformed to $2NH_3$. Considerable interest has been shown in those diazotrophs that contain an enzyme called uptake hydrogenase, which reoxidizes H_2 to protons and electrons and salvages some of the reductant and energy lost in nitrogenase-dependent H_2 formation.

The high energy costs of BNF can also be illustrated by comparing the amount of reduced C oxidized g^{-1} of N assimilated from either N_2 or NH_4^+-N. (Table 15.2). It can be seen that the cost of assimilating N_2-N is about twice that of assimilating NH_4^+-N. Several factors contribute to this high energy cost. In addition to the high ATP cost of reducing N_2 to NH_3, at least 82 genes and their products are needed to synthesize and support a completely functional N_2-fixing enzyme system in the soil bacterium *Azotobacter vinelandii* (Supplemental Table 15.2, see online supplemental material at http://booksite.elsevier.com/9780124159556). The genes involved collectively in the biosynthesis of nitrogenase and development of the catalytic process of N_2 fixation are called *nif* genes. Accessory genes are identified as *rnf* or *fix* genes, and they are also necessary for the function and regulation of nitrogenase in many diazotrophic bacteria. The catalytic process of N_2 fixation is remarkably slow and inefficient. The two proteins must come together and dissociate eight times to reduce one molecule of N_2 to two NH_3. To compensate for this sluggish mechanism, nitrogenase can account for ~ 10% of total cell protein in a

TABLE 15.2 Energy Costs for Biological N_2 Fixation Relative to Assimilation of NH_4^+-N

Bacterial Species	Energy Source	mg N_2-N Fixed/g C Source Used	mg NH_4^+-N Assimilated/g C Source Used
Anaerobic growth			
Clostridium pasteurianum	Sucrose	11	22
Microaerophilic growth			
Azospirillum brasilense	Malate	26	48
Aerobic growth			
Azotobacter vinelandii	Sucrose	7	38

Adapted from Hill (1992). With kind permission of Springer Science + Business Media B.V. From Soil Microbiology, Ecology, and Biochemistry, third ed., Paul, E.A., Biological N inputs. (2007). pp. 365–387.

diazotrophic microorganism. Finally, many of the *nif* gene proteins are denatured by oxygen and turn over quickly in an aerobic environment. Replacement of these denatured proteins constitutes an additional energy expense.

Because of the extreme sensitivity of nitrogenase to irreversible denaturation by O_2, aerobic N_2-fixing bacteria have developed various O_2 protection mechanisms. Some diazotrophic O_2-evolving cyanobacteria produce specialized cells called heterocysts under N_2-fixing conditions (Fig. 15.3). Heterocysts do not divide like their neighboring vegetative cells, do not evolve O_2 or fix CO_2, and use light energy to provide the energy needed for BNF. A combination of respiration and thick cell walls exclude O_2 sufficiently well to protect the dinitrogenase enzyme complex. In some N_2-fixing nonheterocystous cyanobacteria, temporal separation of N_2 fixation and O_2 evolution results in BNF occurring primarily during darkness in the absence of O_2 production. At least under C-rich laboratory conditions, aerobic, N_2-fixing bacteria such as *Azotobacter* express a high respiratory rate that maintains a low O_2 concentration inside the cell. If the O_2 level increases quickly, however, nitrogenase interacts with other redox-sensitive protein(s) in the cell that invoke a conformational change that inactivates nitrogenase and prevents its irreversible damage.

Because NH_3, the primary product of N_2 fixation, is toxic to cells in high concentrations, N_2-fixing bacteria possess highly efficient mechanisms of NH_3 assimilation that prevent its accumulation. In most diazotrophs, the enzyme combination of glutamine synthetase (GS) and glutamate synthase (GOGAT) carry out this task (Equations 15.1 and 15.2).

$$\text{Glutamate} + \text{NH}_3 + \text{ATP} \xrightarrow{\text{Glutamine synthetase}} \text{Glutamine} + \text{ADP} + \text{Pi} \quad (15.1)$$

$$\text{Ketoglutarate} + \text{Glutamine} + \text{NADH} \xrightarrow{\text{GOGAT}} 2\ \text{Glutamate} + \text{NADox} \quad (15.2)$$

Both ATP and reductant are required at this stage of NH_3 assimilation adding an additional energy burden to the N_2-fixing organism. Because BNF is energetically expensive, bacteria do not usually fix N_2 in the presence of reactive N sources. A number of mechanisms have been identified whereby

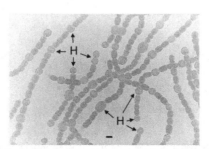

FIG. 15.3 N_2-fixing heterocystous cyanobacteria recovered from the water fern, *Azolla*. Heterocysts are designated with the letter *H*. Note the high percentage of heterocysts relative to vegetative cells. This reflects a high rate of N_2 fixation by the cyanobacterium in the symbiotic state with the Azolla. *Courtesy of J.C. Meeks. Used with permission.*

nitrogenase activity is inhibited, and *nif* gene expression is down-regulated, in response to increasing levels of NO_3^-, NH_4^+, and/or amino acid-N in the environment (Merrick, 1992).

A Measuring BNF

Many attempts have been made to quantify BNF by free-living diazotrophs and their symbiotic associations in both natural and agricultural ecosystems throughout the earth, and the values vary widely (Fig. 15.4; Table 15.3). Two methods are generally favored: (1) the acetylene reduction method, and (2) incorporation of the stable isotope of N, ^{15}N. Like many other metalloenzymes that utilize gaseous substrates, nitrogenase has a broad substrate range and will reduce many small triple-bonded molecules including acetylene ($HC\equiv CH$). About 50 years ago, it was discovered that the enzyme nitrogenase reduced acetylene to ethylene ($CH_2=CH_2$), which can be easily measured by gas chromatography. This method was much easier to use and less expensive than the traditional method of quantifying the incorporation of $^{15}N_2$ into bacterial or plant biomass. Although the advent of acetylene reduction methodology accelerated progress in our understanding of BNF, acetylene reduction is limited in its ability to accurately quantify N_2 fixed in the field over a growing season, and the stoichiometry of $C_2H_2:N_2$ reduced varies with environmental conditions. Indeed, most soil microbiologists and field scientists generally favor the use of ^{15}N for quantifying N_2 fixation under field conditions using the isotope dilution approach (Weaver and Danso, 1994).

The N in biological systems is composed predominantly of two stable isotopes, ^{15}N and ^{14}N, which constitute 0.3663% and 99.6337%, respectively, of the N in the atmosphere. An N_2-fixing organism will incorporate ^{14}N and ^{15}N from an N source in proportion to the concentrations of the two isotopes in that source. If one of these N sources is atmospheric N_2, and because atmospheric N_2 generally has a different percentage of ^{15}N than other sources (soil or fertilizer N, for example), the proportion of soil or fertilizer ^{15}N assimilated by the N_2 fixer will be "diluted" by the relative contribution of N_2 to the cell N budget. The "isotope dilution" approach to measuring BNF involves the addition of a small amount of ^{15}N-enriched inorganic

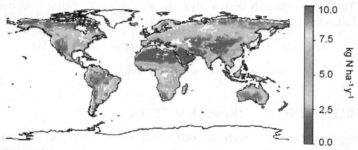

FIG. 15.4 Global distribution of natural BNF based on conservative modeled estimates (Cleveland et al., 1999). *Courtesy of C.C. Cleveland. Used with permission.*

TABLE 15.3 Range of Rates of BNF Measured under Field Conditions by Different Diazotrophic Organisms and Their Associations

Diazotrophic Bacteria and Their Associations	N_2 Fixed (kg N ha^{-1} y^{-1})
Free-living, N_2-fixing bacteria	
Soil, decaying wood and litter, biological soil crusts	1-10
Associative N_2-fixing bacteria	
Azolla-cyanobacteria	33
Endophytic, root associations	0-15
Endophytic, sugarcane association	25
Legume-rhizobial symbioses	
Grain and oilseed crops (peas, chickpeas, lentils, soybeans, etc.)	20-180
Forage crops (alfalfa, clover, vetch, etc.)	110-230
Trees (*Acacia, Gliricidia, Leucaena, Sesbania,* etc.)	50-300
Actinorhizal-*Frankia* symbioses	
Alder	50-300
Casuarina	75
Ceanothus	10-100

Data from Hibbs and Cromack (1990), Herridge et al. (2008), Reed et al. (2011), and Nygren et al. (2012).

N to the soil to increase the ^{15}N/^{14}N ratio of available soil N. In general, the ^{15}N/^{14}N ratio of organisms that assimilate soil N will be the same as the ^{15}N/^{14}N ratio of the available soil N. In the case of an organism that is fixing atmospheric N_2, and assimilating soil N, its ^{15}N/^{14}N ratio will be lowered proportional to the amount of N_2 being fixed (i.e., the ^{15}N content of the biomass will be "diluted" as a result of BNF). The percentage of N derived from the atmosphere (%Ndfa) is represented by the following equation.

$$\%\text{Ndfa} = \left(1\text{-atom}\%^{15}\text{N excess in } N_2\text{-fixing plant}\right) \times 100$$

$$\text{atom}\%^{15}\text{N excess in nonfixing plant} \quad (15.3)$$

III FREE LIVING N_2-FIXING BACTERIA

Diazotrophic prokaryotes can be divided into those that carry out BNF in a symbiotic or commensalistic relationship with a eukaryote and those that fix N_2 in a free-living state. Because the major factor usually limiting BNF is the C energy supply, it is reasonable to predict that free-living photosynthetic diazotrophs would fix greater amounts of N_2 under some soil conditions than free-living heterotrophs, and that the latter would only fix measurable amounts of N_2 in the presence of readily available plant-derived C (e.g., labile C in the rhizosphere zone of an actively growing plant or during the decomposition of herbaceous and woody plant residues of high C:N ratio). Facultative and obligately anaerobic diazotrophic bacteria are often found in decaying wood, where it is assumed that cellulolytic and ligninolytic fungi depolymerize the sugars and phenolics necessary to support diazotrophs (Silvester and Musgrave, 1991). A similar situation exists in agricultural straw residues, where it has been shown that additions of both cellulolytic and diazotrophic bacteria enhance BNF and accelerate the decomposition of the N-deficient straw. It remains extremely difficult to quantify the contribution of BNF by free-living diazotrophs to an ecosystem, even in environments where BNF can be detected by the acetylene reduction method. Estimates of free-living BNF in soil and plant residue-enriched environments usually range between <1 and 10 kg N ha^{-1} y^{-1}, with many natural systems occurring in the lower range of these estimates.

IV ASSOCIATIVE N_2-FIXING BACTERIA

Root secretions and other rhizodepositions are a major source of plant C input to soil, and diazotrophic bacteria are associated with the roots of plants (Reinhold-Hurek and Hurek, 1998). During the past 30 years, many studies have shown that a variety of diazotrophic bacteria associate with the roots of tropical grasses, notably *Paspalum* and *Digitaria* species, where they fix measurable amounts of N_2. These bacteria belong to the genera *Azospirillum*, *Gluconacetobacter*, *Herbaspirillum*, and *Burkholderia*. They are found both in the rhizosphere and in the intercellular spaces of the root cortex. It has been claimed these bacteria can provide between 5% and 30% of the total N accumulated by the plants. Urquiaga et al. (1992) showed that some pot-grown Brazilian sugar cane cultivars could derive >60% of plant N from BNF. A diazotrophic bacterium, *Gluconacetobacter diazotrophicus*, was isolated from sugar cane, where it was found to occupy internal tissues at high cell densities (10^6 to 10^7 cells g^{-1}; Fig. 15.5) and could grow on high sucrose concentrations (10%) and fix N_2. Data obtained from field studies have provided more conservative estimates of the percentage of sugar cane N derived from BNF, ranging between 13% and 30% (Boddy et al., 2001). Herridge et al. (2008) concluded that given the variability in the field data, it is difficult to extrapolate to an accurate global estimate of BNF occurring in the global sugar cane crop.

V PHOTOTROPHIC BACTERIA

Because BNF is an energy-expensive process, it is not surprising that photosynthetic microorganisms are major suppliers of newly fixed N in certain soil ecosystems. Cyanobacteria and other photosynthetic bacteria provide substantial

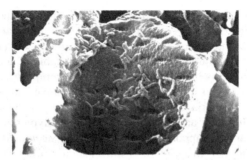

FIG. 15.5 Scanning electron micrograph (SEM) of *Herbaspirillum seropedicae* within the meta-xylem of a sugar cane stem. Note the bacteria are associated closely with the walls of the vessel. *Courtesy of F. Olivares. Used with permission.*

inputs of N in rice paddies early in the growing season because of the presence of abundant light until canopy closure, adequate phosphate fertilization, and the low O_2 conditions found at the sediment–water interface. There is a long history of BNF in rice production in Asia through use of the water fern *Azolla* and its cyanobacterial microsymbiont, *Anabaena azollae*, either used as a green manure or cocultured with the rice crop. Conservative N inputs from BNF of ~ 30 kg N ha^{-1} y^{-1} have been estimated (Giller, 2001). Biological soil crusts are highly specialized photosynthetic communities of cyanobacteria, algae, lichens, and mosses, which are commonly found in arid and semiarid environments throughout the world (Belnap and Lange, 2003). Because they are concentrated in the top few millimeters of soil, biological crusts contribute to soil stability, water infiltration, and the soil N status. Estimates of annual BNF rates in biological soil crusts range widely (Table 15.3), but generally fall in the range of $1–10$ kg N ha^{-1} y^{-1}. Because of rapid wetting and drying, the N_2 fixed tends to be released quickly and provides immediate benefit to the vascular plants in the surrounding ecosystem.

VI SYMBIOTIC N$_2$-FIXING ASSOCIATIONS BETWEEN LEGUMES AND RHIZOBIA

Most terrestrial N_2-fixing symbioses involve a N_2-fixing bacterium and a photosynthetic host. Because the prokaryote gains its energy from the photosynthetic host, the energy cost of BNF is compensated, and considerable amounts of N can be fixed if other factors are not limiting (Table 15.3). In agroecosystems, an intensively managed perennial legume, such as alfalfa (*Medicago sativa*), can fix several hundred kg N ha^{-1} y^{-1}. In 1886, Hellreigel and Willfarth demonstrated the ability of legumes to convert N_2 into organic N. Beijerinck in 1888 isolated bacteria (rhizobia) from legume root nodules and showed they were able to reinfect the legume, form nodules, and fix N_2 in the symbiosis. Studies carried out in the early part of the twentieth century illustrated that rhizobia recovered from root nodules of different legume species possessed different phenotypic characteristics and were characterized as having either promiscuous or very narrow host range specificity. Today we recognize that rhizobia fall into many genera and species within both the alpha- and beta-proteobacteria. They are currently subdivided into approximately 13 genera and more than 90 species, and these numbers are increasing (Supplemental Table 15.3, see online supplemental material at http://booksite.elsevier.com/9780124159556). Most species are found in the *Rhizobiaceaea* family of the alpha-proteobacteria class, and most belong to five genera: *Azorhizobium*, *Bradyrhizobium*, *Mesorhizobium*, *Rhizobium*, and *Ensifer* (previously named *Sinorhizobium*). However, rhizobia have also been identified among other genera of the alpha-proteobacteria and even among two genera of beta-proteobacteria (*Burkholderia* and *Cupriavidus*).

A Formation of the Legume Symbiosis

The establishment of the mutualistic relationship between rhizobia and legumes is accomplished via a series of developmental stages, all of which are mediated by a chronologically orchestrated cascade of molecular signals from both participants. The process of establishing the symbiotic relationship is highly specific (i.e., a specific bacterial species with one, or a limited number, of legume species). At each stage of the formation of this relationship, the chemical signals released by the plant and the bacterium reciprocally induce unique genetic programs that lead to the formation of a nodule and result in BNF.

Stage 1. Rhizobia exist primarily as soil saprophytes that are widely distributed and are found in the rhizospheres of plant roots. They initiate nodule formation by penetrating a legume root either by infection of root hairs (e.g., clover and pea), more rarely by entry at the sites of lateral root emergence (peanut), or through penetration of root primordia found on the stems of some legumes such as *Sesbania* and *Aeschynomene*. In the case of root hair infection, the rhizobia attach to the cell wall of the root hair and trigger a series of morphological and physiological changes in the latter. Inhibition of cell expansion on one side of the root hair causes it to curl back on itself and trap the rhizobia in an infection focus on the inside surface of the curled root hair wall.

Stage 2. Rhizobia pass through the root hair inside structures known as infection threads or tubes (ITs), which are invaginations of the root hair wall and are clearly visible within two days of seedling exposure to rhizobia (Fig. 15.6). ITs are unique plant-made invaginations capable of crossing cell boundaries. The IT penetrates into the root cortex, where cortical cells have been signaled to divide and enlarge to form a prenodule in response to the rhizobial invasion.

Stage 3. As the IT reaches the base of the root hair epidermal cell, a "cytoplasmic bridge" forms in the adjacent plant cell, which aligns with the IT of the already infected root hair cell. This is accompanied by localized degradation of the plant cell wall and permits the IT to grow into the next cell. The process is

FIG. 15.6   *Medicago* root hair curling and infection thread invasion by rhizobia on inoculation with *E. meliloti. Courtesy of R. Geurts. Used with permission.*

repeated at each cell junction, allowing the IT to propagate into the cortex of the root. As the IT gets near the dividing cortical cells, it forms various side branches, each of which eventually comes to a halt at the cell wall of a cortical cell. At the point of contact, the cell wall dissolves and rhizobia (10-100 bacterial cells) are delivered into the nodule cell by endocytosis.

Stage 4. Rhizobia are enclosed within a plant-derived membrane called the peribacteroid membrane. They remain physically isolated from the host cell cytoplasm. Each membrane-enclosed bacterium is referred to as a "symbiosome." As the plant cortical cells divide and become infected to form the nodule structure, the vascular strands of the plant extend into the nodule to permit exchange of nutrients.

Stage 5. The rhizobia undergo further pleiomorphic, physiological, and biochemical changes in the symbiosome and are now referred to as "bacteroids" and commence BNF.

Root nodules differ among legume species. Some plants develop determinate nodules that are generally spherical in shape, reach a finite size, and do not possess active meristems (e.g., *Phaseolus* [beans], *Vigna* [cowpea], and *Glycine* [soybean]). The bacteroids of determinate nodules remain culturable and can recover from the bacteroid state into a replicative vegetative form. Other legumes produce cylindrically shaped, indeterminate nodules with persistent, active meristems that enable the nodule to grow, enlarge, and produce new cortical cells in response to plant growth (e.g., *Medicago* [alfalfa], *Trifolium* [clover], and *Vicia* [vetches]). Bacteroids of indeterminate nodules cannot be cultured, but there are usually sufficient nonbacteroid-like, viable rhizobia remaining in the ITs that can infect new cortical cells as the meristem of the nodule continues to divide. Each bacteroid in an indeterminate nodule undergoes chromosome replication (endoreduplication) to yield a chromosome count of 24 per bacteroid. Bacteroids of determinate nodules retain a chromosome count of 1-2 copies per cell. Legumes that form indeterminate nodules induce production of ~400 nodule-specific, cysteine-rich proteins that are absent from determinate nodules. These polypeptides are responsible for the changes that occur in the transformation of the rhizobial cell into the nonculturable N_2-fixing bacteroid of the indeterminate nodule.

B   Rhizobial Nodulation Genes

The rhizobial genes needed to establish a symbiosis have been identified, and some are listed in Table 15.4. Genes referred to as common nodulation genes (*nod* A, B, C, and D) are found in all rhizobia and are essential for initiation of

TABLE 15.4 Examples of Nodulation (*nod*) Genes and Their Proposed Functions

Gene	Proposed Function
Regulatory genes	
*nod*D$_1$	Transcriptional activator
*nod*V	Two-component regulator
*nod*W	Two-component regulator
*nol*A	Transcriptional regulator
*nol*R	Repressor, DNA binding protein
*syr*M	Transcriptional regulator
Nod factor core synthesis	
*nod*A	Acetyltransferase
*nod*B	Deacetylase
*nod*C	Chitin synthase
*nod*M	D-glucosamine synthase
Nod factor core modifications	
*nod*E	Ketoacyl synthase
*nod*F	Acyl carrier protein
*nod*H	Sulfotransferase
*nod*L	Acetyl transferase

Continued

TABLE 15.4 Examples of Nodulation (*nod*) Genes and Their Proposed Functions—Cont'd

Gene	Proposed Function
nodS	Methyl transferase
nodL	ATP-sulfurylase
nodP	ATP-sulfurylase
nodZ	Fucosyl transferase
nolL	*O*-acetyl transferase
nolO	Carbamoyl transferase
noeC	Arabinosylation
noeI	2-*O*-methylation
Nod factor transport	
nodI	Integral membrane protein
nodJ	Outer membrane transport
nodO	Pore-forming protein

From Soil Microbiology, Ecology, and Biochemistry, third ed., Paul, E.A., Biological N inputs. (2007). pp. 365–387.
This list of genes is not inclusive; some genes coding for the same function in different organisms are named differentially; see online supplemental material at http://booksite.elsevier.com/9780124159556.

nodulation. Many other genes have been identified that define the legume host range of specific rhizobia and are referred to as either *nod*, *nol*, or *noe* genes. These genes are involved in the synthesis and transport of an unusual class of compounds referred to as "*nod* factors" that induce nodule formation. *Nod* factors are composed of an oligosaccharide backbone of β_{1-4} linked *N*-acetyl glucosamine residues with a fatty acid acyl group attached to the N atom of the nonreducing acetyl glucosamine residue (Supplemental Fig. 15.1; see online supplemental material at http://booksite.elsevier.com/9780124159556). These compounds are referred to as lipo-chitin oligosaccharides or LCOs. LCOs usually contain 3 to 6 *N*-acetylglucosamine residues. The common *nod* genes A, B, and C encode enzymes that play key roles in synthesis of the LCO backbone structure and include chitin oligosaccharide synthase (*nod* C), chitin oligosaccharide deacetylase (*nod* B), and acyl transferase (*nod* A). A large number of structural variants exist among the LCOs (Table 15.5). Host specificity is based on the structural variations among the *nod* factors. A number of specific substituents ("decorations") are found on positions R1 through R9, and different

TABLE 15.5 Examples of Modifications of the Rhizobial Core *nod* Factors and the Genes and Their Products Involved

Bacterium	Core Substituents	Genes
E. meliloti	R4:Ac, R5:S, C16:2	R4:*nod*L, R5:*nod*H, fatty acid: *nod*AFEG
M. loti	R1:Me, R3:Cb, R5:AcFuc	R1:*nod*S, R3:*nol*O,R5: *nod*Z & *nol*L
B. japonicum	R5:MeFuc	R5:*nod*Z & *noe*l
A. caulinodans	R1:Me, R4:Cb, R5:Fuc,R8: Ara	R1:*nod*S, R4:*nod*U, R5:*nod*Z, R8: *noe*C

From Soil Microbiology, Ecology, and Biochemistry, third ed., Paul, E.A., Biological N inputs. (2007). pp. 365–387.
These examples do not represent *nol* factors produced by all species of a named genus, nor all strains of the same species.

rhizobial genes (referred to as host range genes) are involved in the attachment of these moieties. For example, the placement of a SO_4 group at position R5 of the LCO from *Ensifer meliloti* is essential for nodulation of alfalfa (*Medicago sativa*), whereas O-methyl fucosylation of R5 is essential for nodulation of soybean (*Glycine max*) by *Bradyrhizobium japonicum*. Nod factors are produced by rhizobia in response to inducers secreted by germinating seedlings and by plant roots. The most potent inducers are flavones and isoflavones, which are phenolic compounds collectively referred to as flavanoids. The flavanoids bind to the *nod* D gene product, which acts as a transcriptional activator of the other *nod* genes. Different types of flavanoid molecules are involved in *nod* gene induction in different legume species. For example, luteolin in *Medicago sativa* (alfalfa) and genistein in *Glycine max* (soybeans).

C Plant Nodulation Genes

Several plant genes (referred to as *nodulin* genes) have been identified that are involved in the early stages of root hair infection and nodule development (Limpens et al., 2003). Other nodulins are expressed at later stages of nodule functional development and include hemoglobin and enzymes involved in N assimilation. In the "molecular model" legume, *Lotus japonicus*, two genes, *NFR*1 and *NFR*5, are absolutely critical to initiating nodulation because plants with mutations in these genes lack all known responses to *nod* factor. *NFR*1 and *NFR*5 code for transmembrane receptor-like kinases (an enzyme that acts as a "molecular switch" by covalently attaching phosphate groups to other proteins), which turn enzymatic pathways either on or off. It is thought that a specific part of the kinase protein binds to the *N*-acetyl glucosamine backbone of *nod* factor

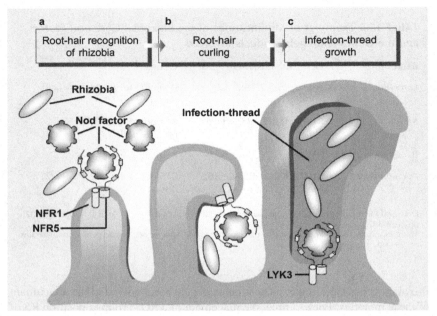

FIG. 15.7 Proposed model of how rhizobial nod factor interacts with the early nodulin proteins NFR1 and NFR5 to promote root hair curling and infection thread growth. *From Parniske and Downie (2003). Used with permission of the Nature Publishing Group.*

and initiates the nodulation process (Fig. 15.7). This interaction activates oscillations of the Ca concentration in the nuclear region of the epidermal root hair cell. Perception of the Ca oscillations by a calcium/calmodulin dependent protein kinase (CCaMK) activates the initiation of bacterial infection into the root hair epidermal cell and also promotes cortical cell division deeper in the root cortex. A role for the plant hormone, cytokinin, in nodule development has been inferred by "knocking out" a gene in *L. japonicus* that codes for a specific cytokinin receptor protein, which then blocks nodule formation without affecting rhizobial infection.

In the legume *Medicago truncatula*, mutants have been obtained in genes analogous to NFR1 and NFR5 (*DMI*1 and *DMI*2) that also prevent most *nod* factor responses (e.g., no root hair branching, nodulin gene expression, Ca oscillations, or cortical cell division). Interestingly, these mutants cannot form mycorrhizal associations with the arbuscular mycorrhizal (AM) fungus, *Glomus*, implying there are common molecular signals for root infection of plants by AM fungi and rhizobia. Recently, AM fungi have been shown to produce nod factor-like molecules (Maillet et al., 2011). The temporal spectrum of physiological and genetic changes occurring during nodule organogenesis is beginning to be understood. The emerging picture seems to indicate that *nod*-signal activities result in changes of endogenous phytohormone levels and changes in the sensitivity of specific plant tissues to the actions of these phytohormones. For an in-depth review of the molecular biology surrounding this sophisticated process the reader is referred to Oldroyd et al. (2011).

D Development of BNF and Nitrogen Assimilatory Processes in Nodules

Most of the bacterial genes that are specifically induced in bacteroids appear to be regulated, at least in part, by O_2. This is not surprising because the nodule structure and function are designed to provide a microaerobic environment for BNF. The outer cortex of the nodule is aerobic, but beyond this region toward the interior of the nodule, there is a significant drop in O_2 concentration. A low O_2 level within the nodule is essential because the nitrogenase enzyme is very O_2 labile. However, the low O_2 level of the nodule interior creates a paradoxical situation for the rhizobial symbiont, which needs O_2 to generate ATP. The infected cortical cells of the plant produce an O_2-carrying protein called leghemoglobin, which facilitates transfer of O_2 to the bacteroid. In addition, the bacteroid produces a specialized terminal cytochrome oxidase that enables it to respire under low O_2 conditions.

As the rhizobia transform from free-living bacteria into bacteroids, their C metabolism is modified so that C is directed into the bacteroid to provide reductant and energy for BNF, and N assimilation is modified so that the plant, not the rhizobia, benefits primarily from the BNF (Fig. 15.8). Malate and/or succinate are the C sources transported into the bacteroid and used in energy production and as C skeletons for bacteroid biosynthetic processes. Expression of plant genes involved with sucrose metabolism is enhanced in the rhizobia-infected

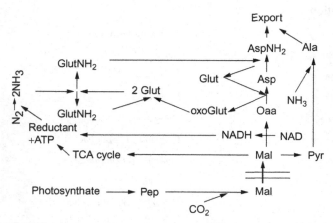

FIG. 15.8 Simplified model of carbon and nitrogen metabolism in the bacteroid and the infected plant cell to illustrate the supply of C and energy from the plant and the assimilation of NH_4^+ produced by BNF.

plant cells. There is a several-fold increase in expression of sucrose synthase, which converts the photosynthate (sucrose) into precursors for the glycolytic pathway. Glycolysis breaks down the photosynthate to phosphoenolpyruvate (PEP), which can be processed to malate (Mal) by high levels of phosphoenol-pyruvate carboxylase (PEPC) and malic dehydrogenase in the nodular plant cells. The bacteroids catabolize malate and succinate and use the energy and reductant to produce ammonia via BNF. In legumes that form indeterminate nodules, N is transferred to growth sinks of the plant in the form of the amino acids, alanine (Ala), asparagine Asp(NH_2) and glutamine (glutNH_2). In contrast, legumes that produce determinate nodules further convert fixed N from amino acids into compounds called ureides (allantoic acid, allantoin) that are transported to the growth sinks. Both types of molecules (amino acids and ureides) have low C:N ratios and represent efficient N-transporting molecules.

E Symbiotic Associations between Actinorhizal Plants and *Frankia*

Symbiotic associations are formed between the bacterial genus *Frankia* (Gr^+, high $G+C$, filamentous *Actinobacteria*) and many nonleguminous plant species widely distributed among eight different families and commonly referred to as actinorhizal species (Table 15.6). Plant genera include *Alnus*, *Casuarina*, *Myrica*, *Hippophae*, *Elaeagnus*, *Ceanothus*, and *Purshia*. These plants are distributed worldwide and tend to be woody shrubs or trees that colonize N-limited landscapes that are usually in the initial stages of recovery

TABLE 15.6 Families and Genera of N_2-Fixing Plant-Frankia Actinorhizal Associations

Family	N_2-Fixing Genera
Betulaceae	*Alnus*
Casuarinaceae	*Casuarina*
Coriariaceae	*Coriaria*
Datiscaceae	*Datisca*
Elaeagnaceae	*Elaeagnus, Hippophae*
Myricaceae	*Myrica* and *Comptonia*
Rhamnaceae	*Ceanothus*
Rosaceae	*Cercocarpus* and *Purshia*

From Soil Microbiology, Ecology, and Biochemistry, third ed., Paul, E.A., Biological N inputs. (2007). pp. 365–387.

from some disturbance (wildfire, landslide, flood, or logging). These plants colonize sites that range widely in latitude, elevation, water, and nutrient availability, and the amounts of N_2 fixed vary considerably (<10 to 300 kg N ha^{-1} y^{-1}; Table 15.3).

Frankia isolates were first obtained in culture in 1978 (Callaham et al., 1978). *Frankia* has continued to be difficult to isolate from nodules and to study in the laboratory, yet, considerable information has been obtained about it by using molecular biological methods that circumvent the need for obtaining pure cultures. Phylogenetic analysis of 16S rRNA and glutamine synthetase gene sequences of *Frankia* strains indicate that the genus belong to three phylogenetically distinct clusters (Fig. 15.9). Cluster 1 isolates are associated with a few genera of actinorhizal plants (e.g., *Casuarina*, *Alnus*, *Myrica*) that establish their symbioses through a root hair infection process similar to that of legumes. Cluster 2 is represented exclusively by microsymbionts that have been detected in nodules of a wide variety of hosts, but never obtained into culture. Cluster 3 strains tend to range widely across hosts that primarily practice the intercellular infection process.

There are many similarities between legumes and actinorhizals in the mode of plant infection, symbiosis establishment, and N assimilation, but there are also many differences. Whereas legume nodules represent stem-like organs, with a peripheral vascular system and nodule primordia that originate in the root cortex, actinorhizal nodules consist of modified lateral roots, possess a central vascular system, and arise from nodule primordia originating in the root pericycle. In *Frankia*, N_2 fixation occurs in swellings located at the tips of filaments referred to as vesicles. In the symbiotic state, vesicles are produced in large numbers, with vesicle shape and the degree of septation differing among

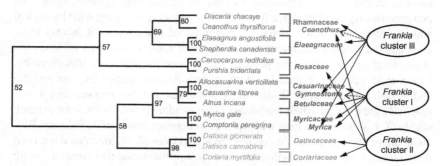

FIG. 15.9 Relationships between actinorhizal host genera and three phylogenetically distinct clusters of *Frankia*. Actinorhizal hosts are color coded: *Rosales* in blue, *Fagales* in red, and *Cucurbitales* in green. Solid arrows connecting a *Frankia* cluster with a group of plants indicate that members of this cluster are commonly associated with the plants; dashed arrows indicate that members of this cluster have been isolated from or detected in an effective or ineffective nodule of a member of the plant group at least once. *Modified from Pawlowski and Demchenko (2012) by permission from the publisher; copyright (2012) Springer.*

different host plant species. For example, spherical, septated vesicles are found in *Alnus*, and elongated, nonseptated vesicles in *Coriaria*. Interestingly, vesicles are not found in infected cells of *Casaurina* nodules.

Similar to the legume-rhizobia associations, some actinorhizal plants, such as *Casuarina* and *Myrica*, produce an O_2-transporting hemoglobin that is associated with the presence of an O_2 diffusion barrier surrounding infected cells. Most actinorhizal plants do not produce hemoglobin, and it is generally felt that the vesicle wall of the *Frankia* itself is the major diffusion barrier to O_2. In *Coraria*, infected cells are surrounded by a suberized periderm layer whose thickness varies with O_2 concentration.

VII   MICROBIAL ECOLOGY OF BNF

BNF is controlled by a range of biological and environmental factors and their interactions. The ability to fix N_2 is spread across a wide range of microbial phyla and physiological types (Fig. 15.2; Supplemental Table 15.1; see online supplemental material at http://booksite.elsevier.com/9780124159556), but there is also a high genetic diversity among different strains within a single species of a N_2-fixing organism, which is particularly noticeable in symbiosis-forming bacteria. In the case of rhizobia and *Frankia*, it is well established that there is a wide range of symbiotic effectiveness among different species, and within strains of an individual species that can establish a partnership with a particular plant species. This phenomenon can range from the extreme case of some strains being completely ineffective at fixing N_2, to the other extreme of strains having an N_2-fixing potential that supports plant growth as well as, or even better than, a supply of fertilizer N. There has been considerable interest shown in this phenomenon by agronomists whose goal it is to optimize the match between host plant and microsymbiont to maximize the contribution of BNF by agricultural legumes. Ecologists and population geneticists have focused on the question by asking what biological and environmental factors drive the evolution of mutual benefit among these associations. Nodule occupancy has often been used as a means of assessing strain diversity as well as the plant or site factors that influence this diversity. Bever et al. (2013) conducted a study of two *Acacia* species (*A. stemophylla* and *A. salicia*) growing at 58 sites in SE Australia, and they showed that *A. stemophylla* grew larger when inoculated (and nodulated) with soils from beneath stands of *A. stemophylla* than with soils from beneath stands of *A. salicina*. Rhizobial diversity was greater in nodules recovered from plants inoculated with soil from under *A. salicia* stands than under *A. stemophylla,* and this correlated with the greater negative effect on the growth of *A. stemophylla* than it did on *A. salicina*. The authors hypothesized that *A. salicia* was a less selective and less symbiotically dependent host than *A. stemophylla* and developed a more diverse rhizobial soil population containing less effective symbionts.

Anderson et al. (2009) studied nodules of two sympatric *Alnus* species in Alaska and found a strong influence of host species on nodule occupancy, suggesting that the host selected for certain genotypes (based on sequence differences in *nif* genes) of *Frankia*. Studies in Oregon and Alaska of sites dominated by a single *Alnus* species have shown that genotypic diversity of *Frankia* strains range from 2 to 5 within a site, with one strain normally dominating nodule occupancy and the dominant strain varying among sites (Anderson et al., 2009; Kennedy et al., 2010). Thus, soil and environmental factors also affect the success of a given *Frankia* strain.

Molecular methods based on *nif* genes have also been applied to study the community composition of free-living N_2-fixing bacteria in soils. Studies have generally shown that the relative abundances of free-living N_2-fixing bacteria vary by soil type and respond to agricultural management practices (Hsu and Buckley, 2009; Wakelin et al., 2010). Although not universal, *in situ* rates of N_2 fixation have been found to correlate with either the copy number of *nif* genes or the diversity of the free-living N_2-fixing bacteria (Hsu and Buckley, 2009; Reed et al., 2011).

N_2-fixing bacteria and N_2 fixation rates can be influenced by many other factors, such as the availability of C, N, and other nutrients. Silvester (1989) showed that Mo (the FeMo cofactor in the most common form of nitrogenase) was often limiting to N_2 fixation rates in litter of Pacific Northwest forests, an observation that has also been made in highly weathered forest soils of the tropics (Supplemental Fig. 15.2; see online supplemental material at http://booksite.elsevier.com/9780124159556).

VIII BIOTECHNOLOGY OF BNF

With the world's population projected to increase to ~10 billion by 2050, and the accompanying demand for plant nutrient N, there will continue to be a need to increase the contribution of BNF to food and fiber production. As mentioned earlier, rates of BNF under field conditions vary widely, and there are many environmental/agronomic factors associated with plant nutrition, water availability, and plant disease that hinder N_2-fixing species from reaching their yield potential. Some undesirable attributes of N_2-fixing plants, such as toxin production, along with ethnical/historical biases, disfavor their use as major food and fiber sources in some parts of the world. The Leguminosae is an enormous plant family distributed worldwide with 16,000 to 19,000 species in about 750 genera (Allen and Allen, 1981). Despite this diversity, only a few legumes have been successfully domesticated and used by humans. Attempts to expand the agricultural use of leguminous species beyond their regions of origin have had a checkered history because the appropriate rhizobia were not always present in the soil. As a consequence, it has been standard practice to inoculate legume seed with appropriate rhizobia before planting a legume species in regions where: (1) it has not been grown previously, (2) the native flora of the region did not contain legumes that were close relatives of the crop, and (3) an interval of several years exists between the use of a legume in the rotation. Another example of a situation where inoculation is recommended includes replacement of older established legume species with newer ones more suitable to the current agricultural practices of the region. The new legume species/varieties might not form highly effective symbiotic association with native rhizobial strains that were effective on the older ones. The reader with an interest in the practical use of legumes and BNF is referred to the article by Nichols et al. (2007), which describes the needs and challenges of introducing new legume species into Australian agriculture. During the past 100 years, agronomists and soil microbiologists have recognized that many soils where crop production would benefit tremendously from BNF are exposed to stresses hostile for survival of rhizobia. As a result, rhizobial strains used as inoculants are often prescreened for their ability to tolerate, persist, and nodulate under environmental stresses. Howieson and Ballard (2004) present a synopsis of thoughts and ideas on these issues. Despite our best efforts, we have not yet been able to overcome the fact that many soils contain indigenous or naturalized populations of rhizobia that are suboptimally effective at BNF on legumes of agronomic importance and that are extremely difficult to displace with superior N_2-fixing strains. Further complexity to this challenging problem has emerged recently from studies of the pasture legume *Biserula pelecinus* introduced to Australian agriculture about 20 years ago. This legume requires a specific type of rhizobia for its mutualism (*Mesorhizobium ciceri* bv. Biserrulae). Six years after establishment of a stand of *B. pelecinus*, Nandasena et al. (2007) recovered rhizobial isolates from root nodules and found that the "symbiotic island" of *nod* genes had been transferred from the original inoculant strain into some native soil rhizobia that were now capable of nodulating field-grown *Biserula*. Unfortunately, these natives had reduced N_2-fixing effectiveness relative to the original inoculant strain. As scientists make more progress in elucidating the molecular signals needed for nodulation by rhizobia and infection by mycorrhyza, and with the genomic sequences of more bacteria and plants becoming available, it is not outside the realm of possibility that we will learn how to develop other mutualistic associations between plants and N_2-fixing bacteria. The example of sugar cane and *G. diazotrophicus* gives hope for a future that includes N_2-fixing cereals and other food and fiber crops.

ACKNOWLEDGMENTS

The authors wish to acknowledge the contributions of photographs and figures from many colleagues who are research specialists in various aspects of BNF. It would have been very difficult to produce this chapter without their assistance.

REFERENCES

Allen, O.N., Allen, E.K., 1981. The Leguminosae: A Source Book of Characteristics, Uses, and Nodulation. The University of Wisconsin Press, Madison, 812 pp.

Anderson, M.D., Ruess, R.W., Myrold, D.D., Taylor, D.L., 2009. Host species and habitat affect nodulation by specific *Frankia* genotypes in two species of *Alnus* in interior Alaska. Oecologia 160, 619–630.

Belnap, J., Lange, O.L., 2003. Biological Soil Crusts: Structure, Function and Management. In: Ecological Studies Series #150. Springer-Verlag, New York.

Bever, J.D., Broadhurst, L.M., Thrall, P.H., 2013. Microbial phylotype composition and diversity predicts plant productivity and plant-soil feedbacks. Ecol. Lett. 16, 167–174.

Boddy, R.M., Polidoro, J.C., Resende, A.S., Alves, B.J.R., Urquiaga, S., 2001. Use of the ^{15}N natural abundance technique for the quantification of the contribution of N_2 fixation to sugar cane and other grasses. Aust. J. Plant Phys. 28, 889–895.

Callaham, D., Del Tredici, P., Torrey, J.G., 1978. Isolation and cultivation in vitro of the actinomycete causing root nodulation of *Comptonia*. Science 199, 899–902.

Cleveland, C.C., Townsend, A.R., Schimel, D.S., Fisher, H., Howarth, R.W., Hedin, L.O., Perakis, S.S., Latty, E.F., Von Fischer, J.C., Elseroad, A., Wasson, M.F., 1999. Global patterns of terrestrial biological nitrogen (N_2) fixation in natural ecosystems. Global Biogeochem. Cycles 13, 623–645.

Gaby, J.C., Buckley, D.H., 2011. A global census of nitrogenase diversity. Environ. Microbiol. 13, 1790–1799.

Galloway, J.N., Aber, J.D., Erisman, J.W., Seitzinger, S.P., Horwath, R.W., Cowling, E.B., Cosby, B.J., 2003. The nitrogen cascade. Bioscience 53, 341–356.

Giller, K.E., 2001. Nitrogen Fixation in Tropical Cropping Systems, second ed. CABI, Wallingford.

Herridge, D.F., Peoples, M.B., Boddey, R.M., 2008. Global impacts of biological nitrogen fixation in agricultural systems. Plant Soil 311, 1–18.

Hibbs, D.E., Cromack Jr., K., 1990. Actinorhizal plants in Pacific Northwest Forests. In: Schwintzer, C.R., Tjepkema, J.D. (Eds.), The Biology of *Frankia* and Actinorhizal Plants. Academic Press, Inc., San Diego, pp. 343–363.

Hill, S., 1992. Physiology of nitrogen fixation in free-living heterotrophs. In: Stacey, G., Burris, R.H., Evans, H.J. (Eds.), Biological Nitrogen Fixation. Chapman and Hill, New York, pp. 87–134.

Howieson, J.G., Ballard, R., 2004. Optimising the legume symbiosis in stressful and competitive environments within southern Australia—some contemporary thoughts. Soil Biol. Biochem. 36, 1261–1273.

Hsu, S.-F., Buckley, D.H., 2009. Evidence for the functional significance of diazotroph community structure in soil. ISME J. 3, 124–136.

Kennedy, P.G., Weber, M.G., Bluhm, A.A., 2010. *Frankia* bacteria in *Alnus rubra* forests: genetic diversity and determinants of assemblage structure. Plant Soil 335, 479–492.

Limpens, E., Franken, C., Smit, P., Willemse, J., Bisseling, T., Geurts, R., 2003. Lys M domain receptor kinases regulating rhizobial nod factor-induced infection. Science 302, 631–633.

Maillet, F., Poinsot, V., Andre, O., Puech-Pages, V., Haouy, A., Gueunier, M., Cromer, L., Giraudet, D., Formey, D., Niebel, A., Martinez, E.A., Driguez, H., Becard, G., Denarie, J., 2011. Fungal lipochitooligosaccharide symbiotic signals in arbuscular mycorrhiza. Nature 469, 58–63.

Merrick, M.J., 1992. Regulation of nitrogen fixation genes in free-living and symbiotic bacteria. In: Stacey, G., Burris, R.H., Evans, H.J. (Eds.), Biological Nitrogen Fixation. Chapman and Hall, New York, pp. 835–876.

Nandasena, K.G., O'Hara, G.W., Tiwari, R.P., Sezmis, E., Howieson, J.G., 2007. In situ lateral transfer of symbiosis islands results in rapid evolution of diverse competitive strains of mesorhizobia suboptimal in symbiotic nitrogen fixation on the pasture legume *Biserrula pelecinus* L. Environ. Microbiol. 9, 2496–2511.

Nichols, P.G.H., Loi, A., Nutt, B.J., Evans, P.M., Craig, A.D., Pengelly, B.C., Dear, B.S., Lloyd, D.-L., Revell, C.K., Nair, R.M., Ewing, M.A., Howieson, J.G., Auricht, G.A., Howie, J.H., Sandral, G.A., Carr, S.J., de Koning, C.T., Hackney, B.F., Crocker, G.J., Snowball, R.,

Hughes, S.J., Hall, E.J., Foster, K.J., Skinner, P.W., Barbetti, M.J., You, M.P., 2007. New annual and short-lived perennial pasture legumes for Australian agriculture—15 years of revolution. Field Crop. Res. 104, 10–23.

Nygren, P., Fernádez, M.P., Harmand, J.-M., Leblanc, H.A., 2012. Symbiotic dinitrogen fixation by trees: an underestimated resource in agroforestry systems? Nutr. Cycl. Agroecosyst. 94, 123–160.

Oldroyd, G.E.D., Murray, J.D., Poole, P.S., Downie, J.A., 2011. The rules of engagement in the legume rhizobial symbiosis. Annu. Rev. Genet. 45, 119–144.

Parniske, M., Downie, J.A., 2003. Locks, keys and symbioses. Nature 425, 569–570.

Pawlowski, K., Demchenko, K.N., 2012. The diversity of actinorhizal symbioses. Protoplasma 249, 967–979.

Reed, S.C., Cleveland, C.C., Townsend, A.R., 2011. Functional ecology of free-living nitrogen fixation: a contemporary perspective. Annu. Rev. Ecol. Evol. Syst. 42, 489–512.

Reinhold-Hurek, B., Hurek, T., 1998. Life in grasses: diazotrophic endophytes. Trends Microbiol. 6, 139–144.

Silvester, W.B., 1989. Molybdenum limitation of asymbiotic nitrogen fixation in forests of Pacific Northwest America. Soil Biol. Biochem. 21, 283–289.

Silvester, W.B., Musgrave, D.R., 1991. Free-living diazotrophs. In: Dilworth, M.J., Glenn, A.R. (Eds.), Biology and Biochemistry of Nitrogen Fixation. Elsevier, New York, pp. 162–186.

Urquiaga, S., Cruz, K.H.S., Boddy, R.M., 1992. Contribution of nitrogen fixation to sugar cane: nitrogen-15 and nitrogen-balance estimates. Soil Sci. Soc. Am. J. 56, 105–114.

Vitousek, P.M., Aber, J., Horwath, R.W., Likens, G.E., Matson, P.A., Schindler, D.W., Schlesinger, W.H., Tilman, D., 1997. Human alterations of the global nitrogen cycle: causes and consequences. Iss. Ecol. 1, 1–16.

Wakelin, S.A., Gupta, V.V.S.R., Forrester, S.T., 2010. Regional and local factors affecting diversity, abundance and activity of free-living N_2-fixing bacteria in Australian agricultural soils. Pedobiologia 53, 391–399.

Weaver, R.W., Danso, S.K.A., 1994. Dinitrogen fixation. In: Weaver, R.W., Angle, J.S., Bottomley, P.J. (Eds.), Methods of Soil Analysis, Part 2, Microbiological and Biochemical Properties. Soil Science Society of America, Inc., Madison, pp. 1019–1045.

SUGGESTED READING

Barron, A.R., Wurzburger, N., Bellenger, J.P., Wright, S.J., Kraepiel, A.M.L., Hedin, L.O., 2009. Molybdenum limitation of asymbiotic nitrogen fixation in tropical forest soils. Nat. Geosci. 2, 42–45.

Dean, D.R., Jacobson, M.R., 1992. Biochemical genetics of nitrogenase. In: Stacey, G., Burris, R.H., Evans, H.J. (Eds.), Biological Nitrogen Fixation. Chapman and Hall, New York, pp. 763–834.

Dos Santos, P.C., Dean, D.R., 2011. Coordination and fine-tuning of nitrogen fixation in *Azotobacter vinelandii*. Mol. Microbiol. 79, 1132–1135.

Setubal, J.C., Dos Santos, P., Goldman, B.S., Ertesvag, H., Espin, G., Rubio, L.M., Valla, S., Almeida, N.F., et al., 2009. Genome sequence of *Azotobacter vinelandii*, an obligate aerobe specialized to support diverse anaerobic metabolic processes. J. Bacteriol. 191, 4534–4545.

Spaink, H.P., 2000. Root nodulation and infection factors produced by rhizobial bacteria. Annu. Rev. Microbiol. 54, 257–288.

Young, J.P.W., 1992. Phylogenetic classification of nitrogen fixing organisms. In: Stacey, G., Burris, R.H., Evans, H.J. (Eds.), Biological Nitrogen Fixation. Chapman and Hall, New York, pp. 43–86.

Chapter 16

Biological Cycling of Inorganic Nutrients and Metals in Soils and Their Role in Soil Biogeochemistry

Michael A. Kertesz[1] and Emmanuel Frossard[2]
[1]*Department of Environmental Sciences, The University of Sydney, Sydney, Australia*
[2]*Institute of Agricultural Sciences, ETH Zurich, Zurich, Switzerland*

Chapter Contents

I INTRODUCTION

Microorganisms play a major role in element cycling in terrestrial systems (Ehrlich and Newman, 2009; Gadd et al., 2012). Microbes inhabit a wide range of environmental niches in soils, populating oxygen gradients from fully aerobic in aerated surface soils to entirely anaerobic conditions within soil aggregates or in sediments and deep subsurfaces. They tolerate a broad range of moisture content, salinities, temperatures, and chemical compositions, providing 80-90% of the life support systems for many different types of soil (Silva et al., 2013). Most microbes inhabit soil surfaces, are able to access nutrients in the soil solution and in mineral and organic substrates, and are then able to interconvert them. They can also form multispecies biofilms on these surfaces in which organisms with different capabilities can interact. The resulting redundancy of this microbial function helps ensure that nutrient transformations and soil fertility are resilient to environmental stresses (Griffiths and Philippot, 2013). The importance of soil microorganisms in C and N cycles has been explored in other chapters of this volume. This chapter will focus on the role of prokaryotes and fungi on cycles of other elements in soil systems.

II NUTRIENT NEEDS OF SOIL MICROORGANISMS

Because microbes provide the lowest trophic level in soil foodwebs, their elemental composition provides a clear indication of their production dynamics, and the biogeochemical cycles in which they are involved. Microbial cells are predominantly made up of C and N. They also show structural requirements for phosphorus (P) and sulfur (S), which are used for the production of nucleic acids/phospholipids and proteins. These elements are enriched in the microbial biomass above their content in the soil (Table 16.1). A recent work claimed that arsenic could replace P in a bacterium (Wolfe-Simon et al., 2011). However, this hypothesis was soon rejected as it was shown that the studied bacteria were arsenate resistant and could grow in the presence of very low available P concentrations (Erb et al., 2012).

The preferred forms of P and S for microbial growth are free inorganic phosphate and sulfate, respectively, although almost all of the P and S are sequestered either as organic compounds or are strongly sorbed on soil surfaces. The microbial preference for the simple anionic forms of these nutrients is clearly seen in the networks of "P-starvation-induced" and "sulfate-starvation induced" genes that control the uptake and assimilation of alternative P- and S-containing compounds (Hsieh and Wanner, 2010; Kertesz, 1999).

A compilation of studies that evaluate the organic P and S contents of nearly 1000 different soils from around the world revealed that: (1) C:N ratios varied between 9.8 and 17.5 (598 soils), (2) C:P ratios lay in the range 44-287 (408 soils), and (3) C:S ratios were between 54 and 132 (527 soils; Kirkby et al., 2011). In this chapter, the transformations of these elements have been subdivided into those carried out by prokaryotes (bacteria and archaea) and those carried out by fungi, in part because the ratios of these elements are fundamentally different for the various groups of organisms. Soil bacteria have a C:P ratio of about 59.5, whereas fungi contain much less P, with C:P in the range of 300-1190 (Kirchman, 2012). These differences reflect different biological roles for these organisms in the soil environment and are so pronounced that the C:N ratios have been successfully used to help predict the bacterial:fungal ratios in microbial communities of soils over a range of habitats and environments (Fierer et al., 2009).

Other elements are also essential for microbial growth, yet are required in smaller amounts than for major biogenic elements. The primary role they play is in biochemical processes, and they may be needed for functioning of specific enzymes or for redox processes. The most important of these elements is iron (Fe), which is abundant in the earth's crust and is required in small amounts in microbial cells, mainly in electron transfer reactions. Additional roles for other

TABLE 16.1 Elemental Composition of a Representative Soil Bacterium, *Pseudomonas putida* (Passman and Jones, 1985), of a Representative Soil Archaeon, *Methanosarcina barkeri* (Scherer et al., 1983), and of a Representative Soil Fungus, *Glomus intradices* (Olsson et al., 2008)

Element	*Pseudomonas putida* (% w/w)	*Methanosarcina barkeri* (% w/w)	*Glomus intradices* (Young Hyphae) (% w/w)	*Glomus intradices* (Spores) (% w/w)
C	51-53	37-44	nm	nm
N	41-42	9.5-12.8	nm	nm
P	1.3-2.2	0.5-2.8	0.24-0.96	0.13-0.8
S	0.5-0.54	0.56-1.2	0.26-0.92	0.008-0.02
Mg	0.26-0.53	0.09-0.53	nm	nm
Na	0.16-0.39	0.3-4	nm	nm
K	0.18-0.3	0.13-5	1.2-3.6	0.09-0.28
Fe	0.012-0.023	0.07-0.28	0.06-0.13	0.03
Ca	0.19-0.33	0.0085-0.055	0.57-2.2	0.23-0.6
Zn	0.006-0.013	0.005-0.063	0.07	0.008-0.04
Cu	0.002-0.003	0.001-0.016	nm	0.004
Mo	nm	0.001-0.007	nm	nm
Mn	nm	0.0005-0.0025	0.02	0.003-0.04

nm, not measured.

TABLE 16.2 Major and Trace Elements Used by Bacteria, Archaea, and Fungi with Examples of Their Roles

Element	Inorganic Form Taken Up	Examples of Element's Role in Cell Metabolism
Major elements		
C	HCO_3^-	All organic compounds (bacteria/archaea/fungi)
N	N_2, NO_3^-, NH_4^+	Proteins, nucleic acids (bacteria/archaea/fungi)
P	PO_4^{3-}	Nucleic acids, phospholipids (bacteria/archaea/fungi)
S	SO_4^{2-}	Proteins, coenzymes (bacteria/archaea/fungi)
K	K^+	Cofactor for enzymes (bacteria/archaea/fungi)
Ca	Ca^{2+}	Intracellular signaling (bacteria/archaea/fungi)
Si	$Si(OH)_4$	Diatom frustules
Trace elements		
Fe	Fe^{3+} and Fe^{2+} complexes	Electron transfer systems (bacteria/archaea/fungi)
Mn	Mn^{2+}, MnO_2	Superoxide dismutase (bacteria/archaea/fungi), oxygenic photosynthesis in cyanobacteria
Mg	Mg^{2+}	Chlorophyll (bacteria/archaea)
Ni	Ni^{2+}	Urease, hydrogenase (bacteria/archaea/fungi)
Zn	Zn^{2+}	Carbonic anhydrase, alkaline phosphatase, RNA/DNA polymerase (bacteria/archaea/fungi)
Cu	Cu^{2+}	Electron transfer system, superoxide dismutase (bacteria/archaea/fungi)
Co	Co^{2+}	Vitamin B_{12} (bacteria/archaea)/antimicrobial compounds (fungi)
Se	SeO_4^{2-}	Formate dehydrogenase, betaine reductase, sarcosine reductase (bacteria/archaea)
Mo	MoO_4^{2-}	Nitrogenases, sulfite oxidases, nitrite reductase (bacteria/archaea/fungi)
Cd	Cd^{2+}	Carbonic anhydrase in diatoms
W	WO_4^{2-}	Hyperthermophilic enzymes (bacteria/archaea)
V	VO_4^{3-}	Nitrogenases (bacteria/archaea/fungi)

essential elements include (1) copper in some redox enzymes, (2) zinc in alkaline phosphatase and carbonic anhydrase, (3) molybdenum in the enzyme that catalyzes N fixation in bacteria, and (4) nickel present in urease and in several enzymes that are important in anaerobic environments, including hydrogenase, methyl coenzyme M reductase, and C monoxide dehydrogenase. Other microbes require a range of even less common metals, including tungsten and vanadium for specific enzymes (Table 16.2).

Attempts have been limited for both defining "typical" elemental compositions of bacteria and archaea *in vitro* and then linking them to the compositions observed in natural environments. This is largely because the actual composition of cells *in situ* reflects both the biochemical requirements of the cell under the given conditions and the availability of particular nutrients in the soil environment. A corollary of this is that by measuring the ratios of C to N and P in soil microbial biomass using fumigation techniques, it should be possible to extract direct information about the nutrient limitations that microbes experience in soils (Cleveland and Liptzin, 2007). However, showing nutrient limitation for microorganisms would require, in addition to information on nutrient content and ratio, an indication of microbial activity (Ehlers et al., 2010). In principle, fumigation techniques could be extended to minor and trace elements, which to date has not yet been attempted. However, there are known sources of error for this technique, which limit its usefulness (Ross, 1990).

Several inorganic nutrients are used by a range of bacteria to provide energy for metabolism and growth. Different compounds of S (sulfate, sulfide, sulfite, and elemental sulfur) can be used by a range of organisms, either as electron donors or as terminal electron acceptors, in redox pathways that supply reducing equivalents to the cell. Metals such as Fe and Mn are used as energy sources by chemoautotrophic bacteria and also as terminal electron acceptors by a range of heterotrophic bacteria. The transformations of these metals between different redox states often have a significant effect on their solubility and hence on both their mobility within the soil environment and their bioavailability to microbes in the soil and to plants. Ferric (Fe^{3+}) compounds, for example, are much less soluble and bioavailable than ferrous (Fe^{2+}) compounds. Ferrous iron may also be said to be the preferred form of Fe for soil microbes (Cartron et al., 2006), despite the fact that Fe^{3+} predominates in all aerobic environments, and levels of Fe required for optimal microbial growth (10^{-7} to 10^{-5} molar, Loper and Buyer, 1991) are rarely present. Many microbes use extracellular reductases to convert Fe^{3+} to Fe^{2+} for uptake. In addition, Fe^{2+} is the active form controlling iron metabolism within the cell.

III   EFFECT OF MICROORGANISMS ON ELEMENT CYCLES

A   Phosphorus Cycle

Phosphorus ultimately derives from phosphate-containing minerals in the bedrock, such as apatite, which are progressively released into the soil by chemical

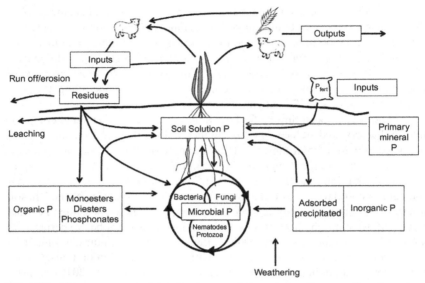

FIG. 16.1 Major components of the phosphorus cycle in agricultural systems. Transformations between major P components are indicated. *Adapted from Richardson et al. (2009).*

and biological weathering. The levels of P found in soils vary considerably. Young soils on bedrocks containing high levels of apatite are often quite rich in total P, whereas many highly weathered tropical soils are low in total P and are particularly deficient in soluble P. The total P content of agricultural soils ranges from 150 to 2000 $\mu g\ P\ g^{-1}$. Chemical weathering releases inorganic phosphate (Pi) into the soil solution (Fig. 16.1), but the levels of soluble ortho-phosphate present are very low in most soils (often below 0.1 mg P L^{-1} and in highly weathered soils below 0.01 mg P L^{-1}, Randriamanantsoa et al., 2013) because it sorbs to soil surfaces and forms precipitates with Ca salts in alkaline soils. These processes also affect phosphate applied to the soil as chemical fer-tilizer, with generally only 10-30% of applied fertilizer P taken up by crop plants in the year following application (Doolette and Smernik, 2011), and the remainder transferred to less soluble pools.

Soils also contain a large pool of organic forms of P. These are partly derived from biological compounds released into the soil as animal waste, plant litter, and microbes. The exact nature of organic P compounds is difficult to assess, but they have been classified (in decreasing order of abundance) as: (1) mono-ester phosphates, such as phytate (myo-inositol-hexakisphosphate); (2) diester phosphates, such as nucleic acids; and (3) phosphonates containing a direct C-P bond, probably derived from phosphonolipids, which replace phospholipids in some microbes (White and Metcalf, 2007). In marine systems, methylphospho-nate is synthesized by the Thaumarchaeota (Metcalf et al., 2012), a major group in soils for which the functions are not yet well defined (Pester et al., 2011).

Phosphorus speciation has largely been conducted in recent years by two spectroscopic methods, nuclear magnetic resonance (NMR) and X-ray absorption near edge spectroscopy (XANES). Each of these characterizes the chemical environment surrounding the P nucleus to provide information about the nature of the P atom. NMR may be performed either on solid samples or soil extracts and is particularly useful for differentiating organic P forms, whereas XANES is usually done with solid samples, providing useful information on inorganic P forms (Kizewski et al., 2011).

Phosphorus availability is mediated by mineralization and immobilization from organic fractions, whereas sorption/desorption and precipitation/solubilization processes are mediated from the inorganic fractions (Frossard et al., 2000). According to McGill and Cole (1981), organic P can be mineralized either biologically, when microbes mineralize organic compounds in search of C and thereby release P associated with C, or biochemically, when microbes specifically scavenge P through the release of phosphatase enzymes. Work by Bünemann et al. (2012) suggests that the release of P through enzymatic hydrolysis constitutes a major part of net organic P mineralization in a pasture soil.

Phosphatase enzymes catalyze the hydrolysis of the phosphate ester bonds to release inorganic phosphate. These phosphatases are specific to particular forms of phosphate esters, many of which may be specific to certain compounds. Thus, phosphomonoesterases cleave phosphate from monoester forms, such as phospholipids or nucleotides. Phosphodiesterases release phosphate from diester forms, such as nucleic acids, and phytases release phosphate from inositol phosphates (Keller et al., 2012). To mobilize the amounts of phosphate needed for cell growth, most of these enzymes are synthesized by the microbial cell with a signal sequence that targets them for extracellular localization. Inorganic phosphate is therefore released outside the cell and taken up by specific high-affinity phosphate transporters in the cell membrane. This allows the cell to scavenge phosphate from a broad range of phosphate-containing substrates without the need to synthesize specific transporters for multiple different compounds, many of which are of high molecular weight. By contrast, phosphonates are almost certainly degraded within the bacterial cell, as the enzymes responsible, the C-P lyases, are unstable and require several cofactors (Zhang and van der Donk, 2012). Because the cells also require other nutrients for growth, the rate of incorporation of the released phosphate into microbial biomass depends on the availability of C and N in the soil. Generally, if the C:P ratio is $>\sim300{:}1$, net immobilization of P into microbial biomass occurs, whereas for C:P ratios $<\sim200$, microbial growth yields a surplus of P, and net mobilization of orthophosphate into the soil solution is observed.

Although the process described immobilizes P into the soil microbial biomass, this P is also readily remineralized. Bacterial cell turnover in the soil can be rapid, especially in the regions surrounding plant roots (the rhizosphere), where there is organic C for growth. Microbial cells die both through predation by protists and by viral attack, releasing their cell contents into the soil solution. In addition, the processes of soil drying and rewetting cause changes in cell turgor, which can both release cell contents during the drying process and cause cell lysis through osmotic shock on rewetting (Bünemann et al., 2013). This leads to the release of a significant proportion of the P from microbial biomass during events such as rainfall after a dry period (Blackwell et al., 2010). Similar effects are also caused by cycles of freezing and thawing, although the quantity and form of biomass P that is released will change over time as these environmental stresses lead to selective changes in the microbial community (Blackwell et al., 2010).

Bacteria play an important role in both the solubilization of phosphate from the precipitated inorganic fraction of soil P and in weathering of minerals to release P (Uroz et al., 2009). This process is mediated by the release of organic acids from the cell, which causes a localized decrease in the soil pH. Acids that are released in this way include gluconic, oxalic, malonic, succinic, lactic, isovaleric, isobutyric, and acetic acids (Rodriguez and Fraga, 1999). The "phosphate-solubilizing bacteria" (PSB) that mediate this process include many strains of *Pseudomonas*, *Bacillus*, and *Rhizobium* (Rodriguez and Fraga, 1999), the best-characterized of which is *Bacillus megaterium*, which has been commercially applied in biofertilizers. The PSBs are readily isolated from the soil by screening for the ability to solubilize Ca or Fe phosphates in an agar-plate test. Specific genes have also been linked with their ability to solubilize bound phosphate (Gyaneshwar et al., 2002). These genes are also found in many bacteria that do not catalyze the phosphate solubilization process *in vitro*, and their presence is therefore not a reliable diagnostic test for phosphate solubilization ability. In addition, supplementation of soils with PSBs does not reliably increase soluble soil P or increase plant P-uptake (Gyaneshwar et al., 2002), so it is not yet clear which bacteria are most important for this process *in situ*.

All fungi take up P as orthophosphate ions using both low- and high-affinity transporters (Plassard and Dell, 2010; Plett and Martin, 2011). Glomeromycota take up mostly orthophosphate ions from the soil solution and have limited direct effect on soil P solubility (Smith and Smith, 2011). The presence of P-solubilizing bacteria around the hyphae can allow the uptake of P that was not originally available, although the importance of this remains to be clarified. Ectomycorrhizal fungi and other saprotrophs can release phosphatase enzymes able to cleave Pi from organic P sources. For instance, some fungi (e.g., *Aspergillus fumigatus*) release high quantities of phytase that cleave Pi from phytate, which can be present in high concentration in soils (Jennings, 1995; Plassard et al., 2011). Fungi can also release low-molecular-weight organic acids, such as oxalate, which help the release of P from inorganic forms as the result of acidification and chelation of the cations bonding P by the organic acids (Plassard et al., 2011). Smits et al. (2012) demonstrated that *Paxillus involutus* in symbiosis with *Pinus sylvestris*, growing under low available P conditions, was able to deliver plant C to apatite grains and to accelerate the rate of P release from these grains. Similarly, *Aspergillus niger*, which is known to produce large amounts of citric acid, can dissolve large amounts of apatite (Bojinova et al., 2008). Finally, siderophores produced by fungi for Fe acquisition can also significantly increase phosphate solubility (Reid et al., 1985), probably by displacing phosphate associated with Fe. Phosphate taken up in excess to growth is stored in the fungal vacuole as polyphosphates. Compared to plants and bacteria, fungi contain more polyphosphate and less diester P (Bünemann et al., 2011). In a recent review, Richardson et al. (2009) suggest that fungi (Glomeromycota, but also *Penicillium* spp. or *Aspergillus* spp.) could be used to improve the use efficiency of soil P for agricultural plants, yet their successful application in the field on a large scale remains to be demonstrated.

B Sulfur Cycle

Most global reserves of S are tied up in the lithosphere, from which they are slowly released by weathering processes. Weathering provides sulfate input to the oceans (the inorganic sulfate concentration in seawater is *ca* 27 mM) and to soils. Agricultural soils contain 2-2000 μg S g^{-1}, but in contrast to marine environments, inorganic sulfate usually makes up only a small proportion of the total S, with organic forms making up >90% of total soil S. Uncultivated soils, such as forest and grassland, contain different proportions of organic S compounds, but inorganic sulfate content is similar to that of farmed soils (Autry and Fitzgerald, 1990; Chen et al., 2001). Unlike phosphate, inorganic sulfate is relatively soluble at soil pH values, and ground water can contain considerable sulfate concentrations (Miao et al., 2012). Available levels of inorganic sulfate in soil therefore vary considerably throughout the year with changes in the water table.

The bulk of the S bound to soil organic matter (SOM) is found in the high-molecular-weight fraction (MW > 100,000 kDa; Eriksen, 2009). Much of this is physically protected within soil microaggregates and therefore not readily mobilizable in the short term (Eriksen et al., 1995). The precise chemical structure of the larger S-containing molecules in SOM is not known, but the chemical environment of the S atoms has been defined in a number of ways. Historically, organic S has been classified according to its chemical reactivity with reducing agents (Freney et al., 1975). Two major groups of organic S compounds can be distinguished: those that yield H_2S on reduction with hydriodic acid predominantly contain sulfate ester-S bonds (C—O—S), whereas those that are unreactive to this treatment contain a direct C—S linkage. Further treatment of the nonextracted soil S with Raney-Ni releases H_2S from amino acid S (cysteine/methionine/peptides), and the residual, unreduced S is considered to be predominantly sulfonate-S, containing a —SO$_3$H moiety (Autry and Fitzgerald, 1990). More recently, S speciation within SOM has been defined in terms of the S oxidation state, using X-ray spectroscopy (K-edge XANES). This can be done by using either bulk soils directly (Prietzel et al., 2011) with humic extracts of the soils (Zhao et al., 2006) or with organic S compounds extracted from soil with acetylacetone (Boye et al., 2011). This provides specific information on S functional groups within the sample, distinguishing between reduced S, oxidized S (sulfonates and sulfate esters), and S with inter-mediate redox state, such as sulfones and sulfoxides (Prietzel et al., 2011).

Inorganic sulfate makes up <5% of the S in most aerobic soils, with the organic fraction comprised of sulfate esters (30-75% of total S) and sulfonates (20-50% of total S). There is considerable variation in the organic S composi-tion between different types of soil, but on the whole, pastures and grasslands tend to be higher in sulfate esters, and forest soils contain more of the sulfonated fraction. Organic S in the soil is derived from plant litter and animal waste inputs, such as: (1) sheep urine containing 30% of its S as sulfate esters, (2) sheep dung containing about 80% of its S as carbon-bound-S (Williams and Haynes, 1993), and (3) 60-90% of the S in decaying plant material is C-bound S (Zhao et al., 1996), much of it derived from sulfolipid, a common lipid in the thylakoid membranes of the plant chloroplast (Benning, 1998). Carbon-bound S also enters soil as methanesulfonate during rainfall, derived in part from atmo-spheric dimethylsulfide (Kelly and Murrell, 1999).

Sulfur-containing compounds in aerobic soils undergo a range of transforma-tions as part of the soil S cycle, including immobilization of inorganic sulfate and organic S compounds into microbial biomass and SOM and mineralization of soil organic S by means of sulfatase and sulfonatase enzymes in the soil (Fig. 16.2).

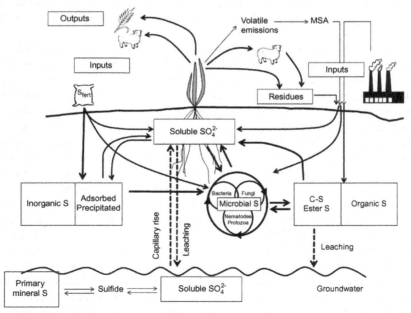

FIG. 16.2 Schematic diagram of the sulfur cycle in agricultural systems. Note that the S cycle is remarkably similar to the P cycle, but with the key differences of enhanced sulfate exchange with groundwater and the redox transformations of S compounds.

Immobilization of sulfate into organic matter has been studied primarily by experiments with ^{35}S-sulfate, which is initially incorporated into the sulfate ester pool and then slowly transformed into C-bonded S (Ghani et al., 1993b). This immobilization of sulfate is microbially mediated because it is stimulated by preincubation under moist conditions to stimulate bacterial growth (Ghani et al., 1993b) and also by addition of C or N in the form of glucose, organic acids, or model root exudates (sugars, organic, and amino acids; Dedourge et al., 2004; Vong et al., 2003). Incorporation of sulfate into the organic S pool is also dramatically increased by the addition of cellulose as a C source (Eriksen, 1997).

Early studies on mineralization of sulfate from the organic S fraction (Freney et al., 1975) suggested that sulfate ester S and C-bonded S contributed about equally to S mineralization. However, other chemical-speciation studies (Ghani et al., 1992, 1993a) showed that almost all the S released in short-term incubation studies was derived from C-bonded S. This agrees with spectroscopic (XANES) data, which show a good correlation between S mineralization and the amount of peptide-S and sulfonate-S in the soil humic fraction and a less clear relationship with sulfate ester-S (Zhao et al., 2006). Nonetheless, soil enzymes, such as sulfatases, are thought to be important in soil S transformations. These enzymes catalyze the hydrolysis of sulfate esters and are common and easily measured in soils. Arylsulfatase activity has been widely used as a measure of soil health and soil microbial activity (Taylor et al., 2002). Until recently, arylsulfatases have been thought to be largely extracellular enzymes (Gianfreda and Ruggiero, 2006). However, the arylsulfatase of one common soil genus, *Streptomyces*, has now been shown to be located in the cell membrane (Cregut et al., 2012), and the arylsulfatase genes in several *Pseudomonas* species do not appear to encode a signal peptide for extracellular secretion (Kertesz et al., 2007). Arylsulfatase activity is correlated with soil microbial biomass and the rate of S immobilization (Vong et al., 2003), along with other factors that affect microbial activity, such as pH and organic C levels (Goux et al., 2012). The most common cultivable, sulfatase-producing bacteria in agricultural soils belong to the *Actinobacteria* and *Pseudomonas* clades (Cregut et al., 2009).

Sulfonate-S comprises up to 80% of total organic S in some soils, with a large proportion of soil and water bacteria able to mineralize low-molecular-weight sulfonates (King and Quinn, 1997). The desulfurization reaction is catalyzed by a family of flavin-dependent monooxygenase enzymes, which cleave the sulfonate moiety to yield sulfite, which can then be assimilated into bacterial biomass. The key groups of bacteria responsible for this process in agricultural and grassland soils are members of the *Variovorax* and *Polaromonas* genera and *Rhodococcus* species, which have been studied using molecular methods to assess *in situ* diversity of the sulfonatase genes (Schmalenberger et al., 2008, 2009, 2010). The sulfonatase enzymes are thought to be entirely intracellular in location due to the mechanistic requirement for flavin and nucleotide cofactors. It is not clear how high-molecular-weight humic substances carrying sulfonate moieties can enter the cell for mineralization to occur.

Unlike P, S is subject to a range of different oxidation and reduction transformations in soil environments, almost all of which are mediated by bacteria. The main microbially catalyzed oxidation sequence for inorganic S compounds involves the sequential conversion of sulfide to sulfate as follows:

$$\underset{\text{Sulfide}}{S^{2-}} \rightarrow \underset{\text{Sulfur}}{S^0} \rightarrow \underset{\text{Thiosulfate}}{S_2O_3^{2-}} \rightarrow \underset{\text{Tetrathionate}}{S_4O_6^{2-}} \rightarrow \underset{\text{Sulfate}}{SO_4^{2-}}$$

Oxidation of these S-containing compounds is catalyzed by two main groups of microbes. Chemoautrophic bacteria use the S-substrates as a source of energy, while using inorganic C sources (lithotrophs). These are primarily bacteria of the genera *Thiobacillus* and *Acidithiobacillus*, most of which are obligate aerobes and use oxygen as a terminal electron acceptor for growth. However, *Thiobacillus denitrificans* is also able to grow anaerobically using nitrate as an electron acceptor. In addition, some of the thiobacilli are able to oxidize inorganic S compounds while growing either heterotrophically or autotrophically (facultative heterotrophy), or with a mixture of organic and inorganic C sources (mixotrophy). Chemoautrophic growth with inorganic S compounds is best characterized for organisms isolated from environments with low organic C (organic C $<<<$ inorganic S), such as hot spring sediments; many of which are thermophiles. However, the levels of organic C in soil are usually much higher (organic C $>>>$ inorganic S), with bacterial populations being dominated by heterotrophs. A wide range of heterotrophic bacteria are also able to oxidize inorganic S compounds, including *Arthrobacter, Bacillus, Micrococcus*, and *Pseudomonas* species, many of which only carry out a partial oxidation (e.g., oxidize S to thiosulfate, or thiosulfate to sulfate, rather than the complete oxidation of S to sulfate). The role of these organisms is unclear because they do not appear to gain any energy from the oxidation process, with the possibility that the transformation is entirely cometabolic. On addition of elemental S to agricultural soils, populations of both thiobacilli and heterotrophic S-oxidizers are strongly stimulated, but the increase in thiobacilli population is only transient, suggesting that heterotrophic S-oxidation is more important in these environments (Yang et al., 2010).

Whereas microbial oxidation of inorganic S compounds almost always requires oxygen, the dissimilatory reduction of these compounds only takes place under anaerobic conditions. It is therefore not usually an important process in well-aerated soils (except in the interior of soil aggregates), but becomes very significant in sediments and when soils are flooded, especially when large amounts of organic C are present in the form of, for example, plant residues. This converts sulfate to sulfide and is catalyzed by the anaerobic sulfate-reducing bacteria, which are organotrophic organisms that use other low-molecular-weight organic compounds (e.g., propionate, butyrate, or lactate) or H_2 as an electron donor and sulfate as terminal acceptor. Marine sediments are particularly high in sulfate-reducing bacteria due to the high levels of inorganic sulfate present in seawater. Dissimilatory sulfate reduction is known for five major groups of bacteria and two archaeal groups (Muyzer and Stams, 2008), many of which are also able to reduce sulfite and thiosulfate. Among the bacteria, common sulfate-reducing genera include *Desulfovibrio*, *Desulfobacter*, *Desulfococcus*, and *Desulfotomaculum*, whereas the best-known sulfate-reducing archaeal genus is *Archaeoglobus*. The key genes involved in the dissimilatory sulfate reduction process are *dsrAB*, which encodes the dissimilatory sulfite reductase, and *aprA*, which encodes adenosine phosphosulfate reductase. Sulfur-reducing populations have been identified in hydrothermal vents, mud volcanoes, acid mine drainage at high pH values, soda lakes, oilfields, agricultural soils, plant rhizospheres, and waste water treatment plants (Muyzer and Stams, 2008). Because the *aprA* gene is an essential part of the S oxidation mechanism, molecular analysis of *aprA* diversity yields detailed information on the activity of both S oxidizing and reducing pathways in any given environment, without the need to cultivate the bacteria in the laboratory (Meyer and Kuever, 2007). The ability of sulfate-reducing bacteria (SRB) to adapt to a wide range of anaerobic environmental conditions has been investigated further using systems biology approaches and appears to be largely due to flexibility in their energy metabolism and their readiness to form syntrophic associations with other microbes, especially H_2-producers (Zhou et al., 2011).

There is little information on S in soil fungi. Studies carried out on *Neurospora crassa*, *Aspergillus*, and *Penicillium* spp. show the existence of two types of sulfate transporters for intracellular uptake (Jennings, 1995; Marzluf, 1997). When sulfate availability decreases, fungi can synthesize arylsulfatase, methionine permease, extracellular protease, and more enzymes of the two transport systems (Jennings, 1995). After uptake, sulfate is reduced stepwise to 3′-phosphoadenosine 5′-phosphosulfate (PAPS), thiosulfonate, and sulfide, from which it is incorporated into cysteine and homocysteine. Some fungi (*Puccinia* spp.) are not able to use sulfate and require the addition of methionine for growth. Some of the aforementioned reduction steps are probably blocked in these fungi. When sulfate is taken up in excess, it can be stored as sulfate, choline sulfate, or glutathione in the vacuole. Fungi can also oxidize inorganic S, but it is not known whether they gain sulfate or energy from oxidizing inorganic S (Jennings, 1995).

C Iron Cycle

Iron is the fourth most abundant element in the earth's crust. However, it is poorly available to microorganisms because it is either bound up in primary minerals (either as Fe(II) or Fe(III)) or is in a sparingly soluble form in Fe oxides and oxyhydroxides, which may be strongly sorbed to clays and organic compounds. In aerobic soils, these oxides form Fe(III), and their solubility is governed by equilibria related to $Fe(OH)_3 \rightleftarrows Fe^{3+} + 3OH^-$ and is therefore strongly dependent on the soil pH. At acidic soil pH values (~3.5), the concentration of Fe^{3+} generated by this equilibrium is about 10^{-9} M, whereas at higher pH values (~8.5) this decreases to ~10^{-24} M, well below the levels required for

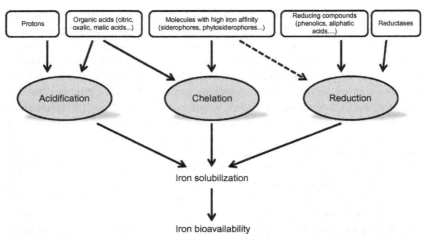

FIG. 16.3 Mechanisms affecting iron availability in soil and rhizosphere. Robin et al. (2008).

bacterial growth (Robin et al., 2008). Soil bacteria and fungi have developed several different mechanisms to increase Fe solubilization in the soil and to enhance the levels of bioavailable Fe, based on acidification, chelation, and reduction (Fig. 16.3, from Robin et al., 2008).

In the rhizosphere, microbial strategies are often assisted by corresponding mechanisms in plants, and there is some evidence for synergism in Fe solubilization, rather than competition for a limiting resource (Robin et al., 2008). Indeed, for aerobic soils, most studies of Fe and metal metabolism by bacteria have focused on the influence of soil bacteria in promoting uptake of Fe by plants for plant health and nutrition (Robin et al., 2008) or in facilitating the sequestration of heavy metals by plants in phytoremediation applications (Glick, 2010).

The most common mechanism used by prokaryotes to enhance Fe solubility is the production of low-molecular-weight Fe-chelating compounds known as siderophores, which are released into the soil solution. These are diverse molecules with affinities for ferric iron of up to 10^{52} (Albrecht-Gary and Crumbliss, 1998) that are able to sequester ferric iron from the soil environment, even when the metal is tightly sorbed onto solid surfaces or metal oxides (Kraemer, 2004). The nature of the chelating group varies between different siderophores and is classified according to the nature of functional ligands as catecholates, hydroxamates, hydroxypyridonates, and hydroxyl- or amino-carboxylates (Robin et al., 2008). A great deal is known about the siderophores of *Pseudomonas* species, called pyoverdins, much of which can be generalized to other aerobic Gr⁻ bacteria (Cornelis, 2010). These are polyketide and peptide-based molecules, whose synthesis is strongly upregulated by low iron availability and repressed under conditions in which soluble iron is plentiful. Iron is chelated to the siderophore via hydroxy and keto groups, and the Fe-siderophore complex then binds to a siderophore-specific outer membrane receptor on the bacterium. In pseudomonads and other Gr⁻ bacteria, this leads to transport of the complex across the outer cell membrane. The metal is released from the siderophore by reduction to Fe^{2+} and then enters the cell via a specific inner-membrane transporter. The Fe-free siderophore molecule (or *apo-siderophore*) is released back into the extracellular environment using a specific exporter (Cornelis, 2010). In Gr⁺ bacteria soil bacteria such as *Streptomyces*, the mechanism of Fe uptake is similar, except that the Fe-siderophore complex is reduced directly on binding to the cell. Plant roots also release Fe-binding molecules, called phytosiderophores, which are structurally distinct from bacterial siderophores (Robin et al., 2008) and have a lower affinity for Fe. Many bacteria are also able to obtain iron from phytosiderophores or by "siderophore piracy" from siderophore complexes released by other bacteria (Barona-Gomez et al., 2006; Cornelis, 2010; Traxler et al., 2012). This enhances competition for Fe in the soil environment, and because bacterial siderophores have considerably higher affinity for Fe than do the siderophores released by many fungi, siderophore-synthesizing bacteria are often selected as biocontrol agents against fungal pathogens in agricultural applications (Laslo et al., 2012).

Acidification of the soil increases the solubility of ferric salts and their bioavailability. Plant roots have developed an active strategy of releasing protons to increase the mobilization of Fe. The resulting pH changes have a considerable effect on Fe solubilization in the rhizosphere (Hinsinger et al., 2003). Both microbes and roots release organic acids in considerable quantities, which contribute to decreases in soil pH and also directly help in the mobilization of Fe because carboxylic acids (e.g., citrate and malate) act as lower-affinity ligands for Fe. A large proportion of organic acids in soil solution are complexed to metals, including Fe.

Oxidation and reduction reactions of iron also play a key part in cycling of Fe in soils and sediments. Fe^{2+} is rapidly chemically oxidized to Fe^{3+} at neutral or alkaline pH. For many years, abiotic reactions were thought to dominate Fe redox reactions. More recently, it has become clear that in many environments, microbial metabolism is extremely important in Fe redox transformations (Fig. 16.4; Weber et al., 2006). At acidic pH values, Fe^{2+} is oxidized to Fe^{3+} by chemoautrophic bacteria such as *Acidithiobacillus ferrooxidans*. At neutral pH values, Fe(II) functions as an electron donor for a wide range of lithotrophic Fe-oxidizing organisms, which belong to a diverse range of bacterial and archaeal phyla, the best-characterized of which are *Gallionella* and *Leptothrix* species (Emerson et al., 2010; Hedrich et al., 2011). Fe-oxidizing bacteria typically inhabit environments at redox boundaries, including wetlands, stream sediments, and waterlogged roots, where their activity is revealed by the presence of red Fe oxide precipitates (Emerson et al., 2010).

Insoluble Fe compounds may also be mobilized by dissimilatory Fe(III) reduction, in which Fe(III) is used as a terminal electron acceptor for microbial growth. Given the high levels of Fe(III) in the environment, this reaction is of great importance both for cell bioenergetics in the soil and for cycling of Fe

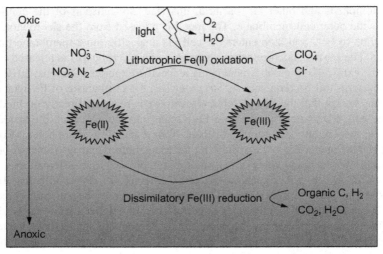

FIG. 16.4 The microbially mediated iron redox cycle. Reduction of Fe(III) to Fe(II) occurs only under anoxic conditions. Lithotrophic Fe oxidation is carried out by a range of organisms under different environmental conditions. Weber et al. (2006).

between different redox states. The best studied organisms that carry out this process belong to the *Geobacter* and *Shewanella* genera, although Fe reducers are also known from many other bacterial and archaeal phyla (Weber et al., 2006). These microbes face a difficult problem in transferring reducing equivalents to an insoluble ferric mineral or salt. *Shewanella* species have solved this problem by using soluble quinones or flavins as electron shuttles between the cell and the substrate, which may be some distance from the cell itself (Roden, 2012). By contrast, *Geobacter* species require direct contact between the cell and the substratum, allowing electrons to be passed from a bacterial biofilm directly to the metal. However, this contact may also be mediated by pili or microbial nanowires, which give metallic-like conductivity and facilitate electron transfer to the solid mineral phase (Lovley, 2012).

Iron is also needed in small amounts by fungi. It is essential as an acceptor and donor of electrons (e.g., in the cytochrome system). Fungi can also acquire Fe by releasing siderophores or by acidifying their environment (Philpott, 2006). The siderophores can be of either the hydroxamate type or the polycarboxylate type. Three families of hydroxamate siderophores have been identified thus far as ferrichromes, fusarinines, and coprogens (Gadd et al., 2012). The fungus can take up the Fe^{3+}-siderophore as a complex, or it can reduce Fe^{3+} to Fe^{2+}, which is then taken up by low- or high-affinity Fe^{2+} transporters (Philpott, 2006). Some of these transporters need Cu to be functional. Some Fe^{2+} transporters can also transport Mn^{2+} and Cu^{2+} in the cell (Bolchi et al., 2011; Philpott, 2006).

D Cycles of Other Elements

Many other elements are subject to microbial transformations in soils and sediments. Microbes have high potassium requirements, whereas assimilation of elements like Ca, Zn, Cr, Se, or Mn is required by microbes in small amounts for specific biochemical functions (Table 16.2). In some cases the uptake of higher concentrations of these elements may be toxic to the cell. Microbes have developed highly regulated mechanisms to ensure that only appropriate amounts of the element are taken up, and that only the correct metal is incorporated into the relevant cofactor or enzyme.

Bacteria and archaea use a range of mechanisms for this purpose, including metal-binding proteins (metallochaperones) that selectively bind the metal ions on entry into the cell. Systems that interact effectively with the target enzyme assemble proteins to insert the metal into its required site (Waldron and Robinson, 2009). The desired metal selectivity is achieved by variation in the number and type of metal ligands on the binding protein, by binding the metal in a cellular compartment that selects for specific redox qualities (the cytoplasm is a much more reducing environment than the periplasm) and by selective protein folding that depends on the type of metal ion bound (Waldron and Robinson, 2009). Comparative genomic studies of soil bacteria and archaea have revealed that these mechanisms are largely conserved (Zhang and Gladyshev, 2009).

In many soils, trace elements (e.g., Cu, Ni, Zn, or Mn) are present at much higher levels than soil microbes can assimilate, either because the soils are derived from weathering of rocks containing these elements or because of industrial contamination with metal-containing effluents or wastes. Most of these elements exist in a number of different oxidation states, and bacteria are involved in conversion between these redox forms. Because the solubility of metal salts often varies considerably between different oxidation states, microbial conversion of a contaminant metal compound to an insoluble form by oxidation or reduction may provide an effective method to immobilize it for bioremediation purposes (Kidd et al., 2009). Conversely, solubilization of insoluble metals by redox transformation may also be used for "biomining" to mobilize valuable metals from low-grade ores (Sorokin, 2003). Examples of organisms involved in the oxidation or reduction of a range of soil elements are given in Table 16.3.

Manganese redox cycling is an extremely dynamic process in soil environments. Manganese-oxidizing bacteria can be isolated from almost any soil or sediment sample, and the Mn oxides they generate are among the most reactive natural oxidizing agents, reacting rapidly with other reduced substrates, including Fe, S, and C compounds. Microbial Mn^{2+} oxidation occurs above pH 5, increases up to pH 8, and is catalyzed by a multicopper oxidase system. It does not appear to provide energy for the bacteria (no Mn^{2+}-dependent chemoautotrophs are known), but it may provide protection against damage by reactive oxygen intermediates, such as hydrogen peroxide or superoxide (Tebo et al., 2005). In many soils that

TABLE 16.3 Examples of Bacteria Involved in the Oxidation or Reduction of Manganese, Chromium, Arsenic, Mercury, and Selenium

Element	Redox States	Reaction Type	Examples of Bacterial Genera Involved
Manganese (Mn)	Mn^{4+}, Mn^{2+}	Oxidation	Arthrobacter, Bacillus, Pseudomonas, Leptothrix
		Reduction	Bacillus, Geobacter, Pseudomonas
Chromium (Cr)	Cr^{6+}, Cr^{3+}	Reduction	Aeromonas, Shewanella, Pseudomonas
Arsenic (As)	As^{5+}, As^{3+}	Oxidation	Alcaligenes, Pseudomonas, Thiobacillus
		Reduction	Alcaligenes, Pseudomonas, Micrococcus
Mercury (Hg)	Hg^{2+}, Hg^{0}	Oxidation	Bacillus, Pseudomonas
		Reduction	Pseudomonas, Streptomyces
Selenium (Se)	Se^{6+}, Se^{4+}, Se^{0}, Se^{2-}	Oxidation	Bacillus, Thiobacillus
		Reduction	Clostridium, Desulfovibrio, Micrococcus

Data from Sylvia et al. (1999) and Tebo et al. (2005).

are subject to cycles of oxidizing and reducing conditions, such as flooding or a fluctuating water table, the products of Mn^{2+} oxidation can be seen in the form of manganese dioxide (MnO_2) nodules. Microbial dissimilatory reduction of Mn^{4+} is also a common reaction. Most microbes that can reduce Fe^{3+} are also able to reduce Mn^{4+} (Lloyd et al., 2003).

Chromium is required by bacteria as a trace element, but its common use in the metal and tanning industries has made it a priority pollutant in many countries because Cr^{6+} salts are readily mobile, toxic, and carcinogenic. A wide range of microbes are able to reduce the toxic Cr^{6+} form to the Cr^{3+} form. Cr^{3+} is 1000 times less toxic than Cr^{6+}, and its salts are insoluble at neutral pH. The reduction process normally requires anaerobic conditions and has been studied in both facultative anaerobes (e.g., *Shewanella* and *Aeromonas*) and in sulfate reducing bacteria, yet has also been observed in aerobic species of *Pseudomonas*. The bacteria use the Cr as a terminal electron acceptor. The rate of reduction depends on a variety of environmental conditions, including the availability of appropriate electron donors, pH, temperature, and the presence of other metals (Lloyd et al., 2003).

Arsenic is widely distributed in soils and groundwater and is commonly associated with pyrite and other minerals containing sulfides. It is released in high concentrations into geothermal waters and groundwater, where it is a major health risk in several parts of the world, with over 40 million people considered to be in danger from drinking arsenic-containing water. For many years, arsenic-containing compounds were used as pesticides, and although their use has now been banned, they have left a legacy of contamination in agricultural soils. Arsenic exists as two major forms in soils: arsenate (As^{5+}) and arsenite (As^{3+}), with As^{3+} being more toxic and prevalent under reduced conditions. Arsenic is not associated with any essential intracellular microbial process, although its biochemistry resembles that of phosphate and can interfere with many aspects of phosphate metabolism. Bacterial oxidation and reduction of As occurs as a detoxification mechanism and also provides electron donors and acceptors. Oxidation of As^{3+} as an energy source has been seen for bacteria in the *Agrobacterium/Rhizobium* family, in which the *aox* genes encoding the arsenite oxidase enzyme are quite widespread. By contrast, As^{5+} reduction, as a terminal electron acceptor, is more common and is carried out by a range of different bacteria. The isolation of arsenate-dissimilating bacteria belonging to widely different phylogenetic groups suggests that they are spread throughout the whole bacterial domain. The electron donors coupled with arsenate reduction vary between strains and environments. Arsenic tolerance in bacteria is mediated primarily by active export of arsenic. The *ars* genes encoding this are widespread in soil environments. However, bacteria also carry out oxidative methylation of arsenic to produce methylarsenite, dimethylarsenate, dimethylarsenite, and trimethylarsine oxide. The ability to carry out this process has been described in a limited number of genera (*Clostridium, Desulfovibrio, Methanobacterium*), although the *arsM* gene associated with arsenic methylation has been found in over 120 different bacteria and archaea (Slyemi and Bonnefoy, 2012; Stolz et al., 2006).

Fungi can also strongly affect other elemental cycles in soil. Fungi have high requirements in potassium and take it up by two types of transport systems: high affinity and low affinity (Corratge et al., 2007). Paris et al. (1995) showed that two ectomycorrhizal fungi (*Pisolithus tinctorius* and *Paxillus involutus*) were able to solubilize phlogopite (a mica) and to access potassium trapped in the 2:1 layers. Although *Paxillus involutus* irreversibly transformed a fraction of the mica in hydroxy-aluminous vermiculite, the transformations caused by *Pisolithus tinctorius* remained reversible. More recently, Bonneville et al. (2009) studied the weathering of biotite at the nanoscale level by *Paxillus involutus* grown in symbiosis with *Pinus sylvestris*. They showed that the adherence of the hyphae on the mineral surface causes a mechanical distortion of the lattice structure of the mineral, and that chemical weathering leading to the development of a potassium depletion zone only starts later, leading to the oxidation of Fe^{2+} to Fe^{3+} and to the formation of vermiculite and Fe oxides.

The role of Ca for fungal growth has been debated for a long time, but it is now recognized as a very important secondary messenger within the fungal cell that can transmit a primary stimulus at the outer membrane into intracellular events. The concentration of Ca must be kept at a very low level within the cytoplasm (100 nM), whereas Ca is stored at high concentrations in the vacuole (1 mM). Oxalate-Ca precipitates are often seen on fungal hyphae. Beside the probable role of oxalate in mineral weathering and in softening cell walls, oxalate also regulates the concentration of free Ca and the pH in the cytoplasm (Jennings, 1995). The degradation of oxalate-Ca produced by fungi and plants in the presence of bacteria by the so-called oxalate-carbonate pathway can result in the precipitation of calcite (Martin et al., 2012). The oxalate-carbonate pathway has been shown to produce significant amounts of calcite under trees such as iroko (*Milicia excelsa*) in tropical soils (Cailleau et al., 2011).

Essential heavy metals (Zn^{2+}, Cu^{2+}, Mn^{2+}) are taken up in fungi by transporters. These metals are used in enzymatic reactions, but their concentrations are tightly regulated to avoid toxicity. Bolchi et al. (2011) recently identified the genes involved in metal homeostasis in *Tuber melanosporum*. Mycorrhizal fungi have been shown to alleviate stress to plants growing on soils containing large amounts of heavy metals (Colpaert et al., 2011; Hildebrandt et al., 2007). The mechanisms involved in metal homeostasis are extracellular precipitation (e.g., in the presence of oxalate), sorption on the cell wall and within the cell, binding to low-molecular-weight compounds (glutathione, phytochelatins, metallothioneins, nicotianamine), and transport into the vacuole or extrusion back to the soil solution (Bolchi et al., 2011; Colpaert et al., 2011; Jennings, 1995). Systems allowing the detoxification of reactive-oxygen species resulting from the presence of excessive metal concentrations also contribute to the tolerance of fungi exposed to high metal concentrations (Hildebrandt et al., 2007).

IV EXAMPLES OF INTERCONNECTIONS BETWEEN MICROBIAL COMMUNITY/ACTIVITY AND ELEMENT CYCLES

A Element Cycles During Early Soil Development

The role of living organisms on weathering and soil development has been extensively discussed (e.g., see Finlay et al., 2009; Peltzer et al., 2010). However, there is less information on how nutrient cycles develop during the early stages of soil formation (10s to 100s of years). In this section, we show how the interactions of soil biota and plants affect cycling of nutrients along a soil chronosequence developed in the forefield of the Damma Glacier in the Swiss Alps (Bernasconi et al., 2011). It includes unvegetated soils that are as young as 10 years to vegetated soils deglaciated 150 years ago. The forefield is located at 2000 m altitude on a granitic parent material and subjected to an alpine climate with very strong microclimatic heterogeneity caused by the different exposure of the slopes (south slope vs. north slope) and topographic heterogeneity (presence of two frontal moraines, 1928 and 1992, and presence of dry and wet sites).

The vegetation of the youngest sites located behind the 1992 moraine (soils between 6 and 14 years) is scarce and patchy. It is dominated by *Agrostis gigantea* Roth, *Rumex scutatus* L., *Cerastium uniflorum* Thom. ex Reichb., and *Oxyria digyna* (L.) Hill. The vegetation of the intermediate sites located between the 1992 and 1928 moraines (soils between 57 and 79 years) cover the ground from partially to fully. The dominant species are *Agrostis gigantea*, *Salix* spp., *Deschampsia cespitosa* (L.) Roem. and Schult., and *Athyrium alpestre* (Hoppe) Milde. The vegetation of the sites located in front of the 1928 moraine (soils between 108 and 140 years) fully covers the soil and is characterized by the presence of woody plants (e.g., *Rhododendron ferrugineum* L. and *Salix* spp.) and grasses (e.g., *Agrostis gigantea* and *Festuca rubra* L.). Nitrogen-fixing plants (e.g., *Alnus viridis* [Chaix] and *Lotus alpinus*) are also present on the chronosequence (Bernasconi et al., 2011; Brankatschk et al., 2011).

Plant biomass production, together with soil total organic C, N, and total P content increase with soil age. Total clay increases with soil age as soil pH decreases. Soil microbial biomass, as estimated by total DNA, phospholipid fatty acids, and microbial C content, also increases with soil age (Bernasconi et al., 2011). Similarly, the abundance and richness of testate amoebae also increase with soil age.

Zumsteg et al. (2011) conducted genetic profiling and clone library sequencing to characterize the microbial communities of soils along the Damma Glacier forefield. The major bacterial lineages were Proteobacteria, Actinobacteria, Acidobacteria, Firmicutes, and Cyanobacteria. The bacterial diversity was high, but no trend in the Shannon diversity index could be detected across the chronosequence. Euryarchaeota were found to predominantly colonize younger soils, whereas Crenarchaeota colonized mainly older soils. Ascomycota dominated the fungal community in younger soils, whereas Basidiomycota were more general in older soils. Welc et al. (2012) characterized soil bacterial and fungal communities along the forefield using fatty acid profiling. These authors found that the ratio of arbuscular mycorrhizal fungi to bacteria, and of AMF to other fungi, decreased with soil age, whereas the ratio of other fungi to bacteria remained constant with soil age. This suggests that AMF are more important in younger soils and become less important in older soils as they become more acidic and enriched in organic matter.

Goeransson et al. (2011) studied bacterial growth limitation in soils from the Damma using leucine incorporation under laboratory conditions. Bacterial growth increases with soil age (with total soil C), but as soil gets older, soil C availability to bacteria decreases. In the younger soils, bacterial growth is limited by low C and N availability, whereas in the older soils, bacterial growth is only limited by C. Bacterial growth is never limited by P. Esperschuetz et al. (2011) studied the degradation of ^{13}C enriched litter of *Leucanthemopsis alpina*

L, which is to be found along the entire chronosequence. Their results suggest that the contribution of bacteria (actinomycetes) to litter turnover becomes higher with soil age, whereas archaea, fungi, and protozoa play a more important role in recently deglaciated soil.

Duc et al. (2009) and Brankatschk et al. (2011) studied the role of the soil microbial biomass on the N cycle in the soils from the Damma forefield. Duc et al. (2009) observed a higher rate of biological N fixation by free-living microorganisms in the rhizosphere soil of pioneer plants (*Leucanthemopsis alpina, Poa alpina, Agrostis* sp.) compared to the bulk soil in a younger soil (8 years) and an older soil (70 years). These authors observed a very high diversity of NifH (the iron protein of the nitrogenase complex) in these soils. Altogether, these results suggest that diazotrophs play a significant role for N input in the Damma forefield. The high diversity of NifH protein was related to the very low soil inorganic N content (Duc et al., 2009). Brankatschk et al. (2011) analyzed the abundance of different genes related to the N cycle and the potential enzyme activities related to these genes in the soils of the Damma forefield. Their results show that in the youngest soils (10 years), N mineralization (chitinase and protease) was the main driver of the soil N turnover. In the intermediate soils (50-70 years), genes related to biological N fixation by free-living organisms were highest. Finally, in the older soils (120-2000 years), genes coding for nitrification (AOA) and denitrification (*nosZ*) enzymes and their potential activities were highest. From their results, Brankatschk et al. (2011) suggest that N and C are provided in sufficient amounts for initial microbial development at the youngest sites, whether through the deposition of allochthonous organic material on the forefield (e.g., animal dung), atmospheric deposition on the surface of the glacier, or the presence of microbes able to feed on ancient C.

Schmalenberger and Noll (2010) analyzed the diversity of the sulfonate-desulfurizing bacteria in soils of the Damma forefield based on the sequences of the oxidoreductase *asfA* gene. They studied bulk soils and rhizosphere soils from *Agrostis rupestris* and *Leucanthemopsis alpina*. Their study shows a wide diversity of this gene, which is affected by both soil age and the plant species. They related this high diversity to both the low level of soil available sulfate and to the inoculation of bacteria by atmospheric deposition. *asfA* associated with *Polaromonas*, a genus associated with very low sulfate availability environments, was found in the younger soils, whereas unidentified species were found in the older soils.

Frey et al. (2010) grew bacteria isolated from unvegetated granitic sand from the Damma forefield in the presence of glucose, ammonium, and granite powder from the forefield. They found that isolates of *Arthrobacter* sp., *Janthinobacterium* sp., *Leifsonia* sp., and *Polaromonas* sp. were able to cause a significant increase of granite dissolution as measured by Fe, Ca, K, Mg, and Mn release. The effects of these bacteria on weathering were related to the fact that: (1) they could attach to the granite surfaces, (2) they secreted high amounts of oxalic acid, (3) they lowered the pH of the solution, and (4) they produced HCN. However, these bacteria had a weak impact on P solubilization. Brunner et al. (2011) iso-lated fungi from unvegetated granitic sediments in the Damma forefield and con-ducted granite dissolution studies. They showed that *Mucor hiemalis, Umbelopsis isabellina*, and *Mortierella alpina* were able to exude large amounts of citrate, malate, and oxalate and could thereby release significant amounts of Ca, Cu, Fe, Mg, Mn, and P from granite powder. This work shows that fungi are also able to weather minerals in the absence of plants. By analyzing the isotopic compo-sition of oxygen associated with phosphate in different soil and plant pools along the Damma chronosequence, Tamburini et al. (2012) found that at each sampled site the soil available P had been fully processed by soil microbial biomass. These results suggest that at the youngest site, P was taken up by soil microorganisms as soon as it was released from the apatite. The turnover time of microbial P at this site might be as short as a few weeks.

The insights from the publications above are summarized in Fig. 16.5.

The parent material is colonized by microorganisms using C from exoge-nous inputs to fix N and which release P, Ca, Mg, Fe, and so on from the parent material. These elements arrive in the soil solution in forms that can be taken up by the plant. This allows the development of different plant species and an increase in plant biomass. In the younger stages, AMF mediate an efficient cap-ture of nutrients by plants. The development of N-fixing plants (*Alnus, Lotus*) allows a higher rate of N inputs into the ecosystem. At a later stage, the litter produced by the plants and the SOM can be degraded by soil microorganisms and microfauna, and the elements are then recycled in the microbial-plant loop. This organic matter or direct C inputs from the plant through rhizodeposition can be used by microorganisms to accelerate the weathering of minerals. This leads to deeper, more developed soils. Finally, as the ecosystem becomes richer in available nutrients, the tendency to lose a fraction of these to water and the atmosphere will increase.

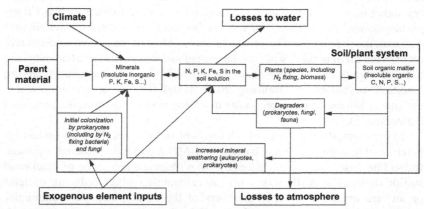

FIG. 16.5 Element cycling during the early stages of soil development as affected by the inter-actions between soil prokaryotes, fungi, fauna, and plants.

B   Coupling of Iron and Sulfur Transformations in Acid Mine Drainage

The leachate that comes from the waste rocks and mineral tailings left behind after mining operations is known as *acid mine drainage* (AMD), and its release has economic and environmental impacts worldwide. As the name implies, AMD is extremely acidic, with pH values as low as 2-3, and it also contains high concentrations of metal ions, both Fe and more toxic heavy metals. In ores, metals such as Cu, Pb, and Zn are commonly found as sulfide minerals together with pyrite (FeS_2), and coal deposits can contain up to 20% of S by weight, largely as sulfides. The combination of microbial Fe and S transformations leads to the production of sulfuric acid that characterizes AMD and is environmentally damaging.

The overall reaction of pyrite minerals involves reaction of FeS_2 with oxygen to generate sulfuric acid (Reaction 16.1). However, this process requires higher concentrations of oxygen than are commonly present in the subsurface environment. This reaction is much more efficient when the oxidant is not molecular oxygen, but Fe^{3+} (Reaction 16.2), which is thereby converted to the ferrous form (Fe^{2+}). Under the acidic conditions of AMD, the Fe^{2+} is then reoxidized to Fe^{3+} by Fe-oxidizing bacteria, the best-characterized of which is *Acidithiobacillus ferrooxidans* (Reaction 16.3).

$$FeS_2 + 3.5O_2 + H_2O \rightarrow Fe^{2+} + 2H^+ + 2SO_4^{2-} \tag{16.1}$$

$$FeS_2 + 14Fe^{3+} + 8H_2O \rightarrow 15Fe^{2+} + 16H^+ + 2SO_4^{2-} \tag{16.2}$$

$$14Fe^{2+} + 3.5O_2 + 14H^+ \rightarrow 14Fe^{3+} + 7H_2O \tag{16.3}$$

Due to the nature of AMD, the key organisms catalyzing this process are necessarily acidophiles. These include not just autotrophs like *A. ferrooxidans* and *Leptospirillum ferrooxidans*, but also a range of acidophilic heterotrophs belonging to the *Acidophilum* genus, and heterotrophic archaea such as *Ferroplasma acidiphilum*. Because the inputs of organic C and N to these systems are minimal, the population of heterotrophs is low. Most of the known acid-tolerant archaea are thermophilic, whereas the temperatures in AMD environments tend to be constantly low. The dominant organisms in these environments therefore tend to be psychrophiles (Hallberg, 2010; though there are a few mines where the energy released from rapid pyrite oxidation is sufficient to cause increased temperatures).

In principle, the presence of high levels of sulfate in AMD might select for heterotrophic sulfate-reducing bacteria in AMD sediments. Although evidence for sulfate reduction has occasionally been observed (presence of blackened sulfide deposits in AMD sediment), no acidophilic sulfate-reducing bacteria or archaea are known. Many members of the AMD microbial community can carry out S oxidation, using either sulfide, S, or reduced S compounds, such as trithionate or tetrathionate, as an electron donor. The common acidophile

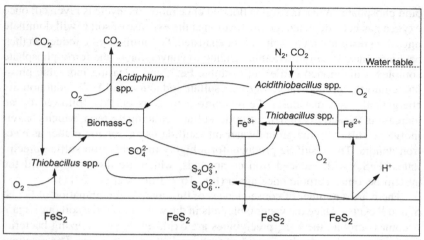

FIG. 16.6 Simplified overview of the processes in acid mine drainage and the key groups of organisms that control these transformations at temperatures <30°C. *Redrawn from Baker and Banfield (2003)*.

A. ferrooxidans can grow autrophically with elemental S as the electron donor and Fe^{3+} as the electron acceptor, and thiobacilli, like *Thiobacillus acidophilus*, use tetrathionate or trithionate for autotrophic growth. Overall, AMD ecosystems contain a fairly limited range of bacterial and archaeal taxa, but there is significant variation within these taxa due to the availability of a range of specialist niches that provide slightly different conditions of pH, temperature, and metal content (Fig. 16.6).

C   Integration of Element Cycles in Wetland Systems

Wetland ecosystems provide excellent examples of the interplay between plants and different groups of microbes under a range of environmental conditions. The deeper sediments tend to be anoxic, but this varies through the year, both in natural wetland systems (changes in groundwater flow at different seasons) and in agricultural environments (temporary flooding of rice fields or other irrigation). The upper layers of the soils tend to be high in organic C from plant inputs and are dominated by interactions with the plant roots themselves, providing C inputs in the form of root-derived materials and a partially aerobic environment in the immediate vicinity of the roots due to oxygen release from the root tissues.

In the anoxic environment of flooded wetland soils, ferric compounds are utilized by the resident microbial communities as electron acceptors; reduced Fe compounds therefore accumulate over time. However, when such soils are drained and become aerobic, Fe-oxidizing bacteria become more active, leading to the production of ferric oxyhydroxides (FeOOH), which strongly bind phosphate. When the soil is flooded once more, the cycle is reversed; once oxygen has been depleted, denitrification at the expense of soil C will dominate microbial respiration until nitrate is exhausted. Dissimilatory Fe reduction then becomes energetically favorable, leading to conversion of the ferric-phosphate complexes mentioned earlier into soluble Fe^{2+}, and releasing inorganic phosphate into the groundwater. Because sulfate reduction and Fe^{3+} reduction are energetically not dissimilar, Fe-reducing activity is quickly followed by an increase in the activity of heterotrophic sulfate reducing bacteria, which convert inorganic sulfate in the groundwater into sulfide using organic matter as electron donors. These sulfide products form highly insoluble iron sulfide precipitates (FeS_x) with reduced iron compounds, which are then stable until the system becomes aerobic once again (Fig. 16.7; Burgin et al., 2011).

There are a number of consequences of this complex combination of Fe, N, S, and P cycles. Once the water table falls in the wetland and the soil once again becomes aerobic, the FeS_x precipitates are oxidized by Fe-oxidizing bacteria, producing a pulse of sulfuric acid as a by-product of the process. The resultant acidification of the wetland soil can cause changes in both plant and microbial communities and can be sufficient to mobilize toxic metals, such as aluminum. Leachate from the wetland can then have damaging environmental effects on ecosystems downstream (Burgin et al., 2011).

In the past century, the application of synthetic fertilizers has had a dramatic effect on both natural and agricultural wetlands. Nitrate is one of the most common groundwater contaminants, and nitrate runoff contributes to surface water eutrophication. However, the impact of nitrate in wetlands goes beyond the N cycle. Although nitrate concentrations are high, ferrous concentrations in the groundwater tend to remain low because nitrate is energetically more favorable as an electron acceptor and therefore provides a redox buffer before ferric compounds are reduced. However, the presence of increased nitrate levels also

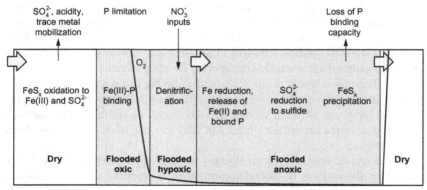

FIG. 16.7   Hypothetical cycle of Fe and S transformations in wetlands and their implications for P and N metabolism. Note that the length of the Fe and sulfate reduction phases under anoxic conditions will vary depending on nitrate inputs to the system. *Adapted from Burgin et al. (2011).*

stimulates what has been described as the "ferrous wheel," where chemolithoautotrophic nitrate reduction is coupled to the oxidation of Fe sulfide deposits, causing an increase in sulfate concentrations in the groundwater. This effect can be considerable; for example, between 1960 and 2000, sulfate concentrations in Dutch groundwater increased nearly threefold, where up to 70% of this sulfate is thought to be derived from pyrite (FeS_x). Sulfate-reducing bacteria convert this sulfate to sulfide, which displaces even more phosphate from Fe-P complexes, causing a large pulse of phosphate to be released. Hence the nitrate runoff and leaching from agricultural soils in this case can also cause phosphate-induced eutrophication within the wetland environment and reduced phosphate binding capacity of the soil in subsequent oxic/anoxic cycles (Smolders et al., 2010).

Nitrate pollution of wetland groundwater also affects the vegetation composition as a result of these coupled element cycles. Wetland plants are largely adapted to prefer ammonium (NH_4^+) as an N source, although this depends on the pH of the surface water. Changes in pH and in the availability of N sources can have a selective effect on the plant community. More importantly, the sulfide produced by sulfate-reducing bacteria and archaea is toxic to plants and also inhibits the growth of many plant species. Precipitation of iron-sulfide as a plaque around the roots interferes with root physiology and P uptake and generates significant problems, yet these problems are to some extent alleviated by diffusion of oxygen from the plant root, which stimulates the activity of S-oxidizing bacteria in the immediate vicinity of the root, converting toxic sulfide to elemental S (Lamers et al., 2012).

REFERENCES

Albrecht-Gary, A.M., Crumbliss, A.L., 1998. Coordination chemistry of siderophores: thermodynamics and kinetics of iron chelation and release. Met. Ions Biol. Syst. 35, 239–327.

Autry, A.R., Fitzgerald, J.W., 1990. Sulfonate S—a major form of forest soil organic sulfur. Biol. Fertil. Soils 10, 50–56.

Baker, B.J., Banfield, J.F., 2003. Microbial communities in acid mine drainage. FEMS Microbiol. Ecol. 44, 139–152.

Barona-Gomez, F., Lautru, S., Francou, F.X., Leblond, P., Pernodet, J.L., Challis, G.L., 2006. Multiple biosynthetic and uptake systems mediate siderophore-dependent iron acquisition in *Streptomyces coelicolor* A3(2) and *Streptomyces ambofaciens* ATCC 23877. Microbiology 152, 3355–3366.

Benning, C., 1998. Biosynthesis and function of the sulfolipid sulfoquinovosyl diacylglycerol. Annu. Rev. Plant Physiol. Plant Mol. Biol. 49, 53–75.

Bernasconi, S.M., Bauder, A., Bourdon, B., et al., 2011. Chemical and biological gradients along the Damma glacier soil chronosequence, Switzerland. Vadose Zone J. 10, 867–883.

Blackwell, M.S.A., Brookes, R.C., de la Fuente-Martinez, N., Gordon, H., Murray, P.J., Snars, K.E., Williams, J.K., Bol, R., Haygarth, P.M., 2010. Phosphorus solubilization and potential transfer to surface waters from the soil microbial biomass following drying-wetting and freezing-thawing. Adv. Agron. 106, 1–35.

Bojinova, D., Velkova, R., Ivanova, R., 2008. Solubilization of Morocco phosphorite by *Aspergillus niger*. Bioresour. Technol. 99, 7348–7353.

Bolchi, A., Ruotolo, R., Marchini, G., Vurro, E., di Toppi, L.S., Kohler, A., Tisserant, E., Martin, F., Ottonello, S., 2011. Genome-wide inventory of metal homeostasis-related gene products including a functional phytochelatin synthase in the hypogeous mycorrhizal fungus *Tuber melanosporum*. Fungal Genet. Biol. 48, 573–584.

Bonneville, S., Smits, M.M., Brown, A., Harrington, J., Leake, J.R., Brydson, R., Benning, L.G., 2009. Plant-driven fungal weathering: early stages of mineral alteration at the nanometer scale. Geology 37, 615–618.

Boye, K., Almkvist, G., Nilsson, S.I., Eriksen, J., Persson, I., 2011. Quantification of chemical sulphur species in bulk soil and organic sulphur fractions by S K-edge XANES spectroscopy. Eur. J. Soil Sci. 62, 874–881.

Brankatschk, R., Toewe, S., Kleineidam, K., Schloter, M., Zeyer, J., 2011. Abundances and potential activities of nitrogen cycling microbial communities along a chronosequence of a glacier forefield. ISME J. 5, 1025–1037.

Brunner, I., Ploetze, M., Rieder, S., Zumsteg, A., Furrer, G., Frey, B., 2011. Pioneering fungi from the Damma glacier forefield in the Swiss Alps can promote granite weathering. Geobiology 9, 266–279.

Bünemann, E.K., Prusziz, B., Ehlers, K., 2011. Characterization of phosphorus forms in soil micro-organisms. In: Bünemann, E.K., Oberson, A., Frossard, E. (Eds.), Phosphorus in Action: Biological Processes in Soil Phosphorus Cycling. Springer, Berlin, pp. 37–57.

Bünemann, E.K., Oberson, A., Liebisch, F., Keller, F., Annaheim, K.E., Huguenin-Elie, O., Frossard, E., 2012. Rapid microbial phosphorus immobilization dominates gross phosphorus fluxes in a grassland soil with low inorganic phosphorus availability. Soil Biol. Biochem. 51, 84–95.

Bünemann, E.K., Keller, B., Hoop, D., Jud, K., Boivin, P., Frossard, E., 2013. Increased availability of phosphorus after drying and rewetting of a grassland soil: processes and plant use. Plant Soil 370, 511–526.

Burgin, A.J., Yang, W.H., Hamilton, S.K., Silver, W.L., 2011. Beyond carbon and nitrogen: how the microbial energy economy couples elemental cycles in diverse ecosystems. Front. Ecol. Environ. 9, 44–52.

Cailleau, G., Braissant, O., Verrecchia, E.P., 2011. Turning sunlight into stone: the oxalate-carbonate pathway in a tropical tree ecosystem. Biogeosciences 8, 1755–1767.

Cartron, M.L., Maddocks, S., Gillingham, P., Craven, C.J., Andrews, S.C., 2006. Feo - Transport of ferrous iron into bacteria. Biometals 19, 143–157.

Chen, C.R., Condron, L.M., Davis, M.R., Sherlock, R.R., 2001. Effects of land-use change from grassland to forest on soil sulfur and arylsulfatase activity in New Zealand. Aust. J. Soil Res. 39, 749–757.

Cleveland, C.C., Liptzin, D., 2007. C:N:P stoichiometry in soil: is there a "Redfield ratio" for the microbial biomass? Biogeochemistry 85, 235–252.

Colpaert, J.V., Wevers, J.H.L., Krznaric, E., Adriaensen, K., 2011. How metal-tolerant ecotypes of ectomycorrhizal fungi protect plants from heavy metal pollution. Ann. Forest Sci. 68, 17–24.

Cornelis, P., 2010. Iron uptake and metabolism in pseudomonads. Appl. Microbiol. Biotechnol. 86, 1637–1645.

Corratge, C., Zimmermann, S., Lambilliotte, R., Plassard, C., Marmeisse, R., Thibaud, J.-B., Lacombe, B., Sentenac, H., 2007. Molecular and functional characterization of a Na^+-K^+ transporter from the Trk family in the ectomycorrhizal fungus *Hebeloma cylindrosporum*. J. Biol. Chem. 282, 26057–26066.

Cregut, M., Piutti, S., Vong, P.-C., Slezack-Deschaumes, S., Crovisier, I., Benizri, E., 2009. Density, structure, and diversity of the cultivable arylsulfatase-producing bacterial community in the rhizosphere of field-grown rape and barley. Soil Biol. Biochem. 41, 704–710.

Cregut, M., Piutti, S., Slezack-Deschaumes, S., Benizri, E., 2012. Compartmentalization and regulation of arylsulfatase activities in *Streptomyces* sp., *Microbacterium* sp. and *Rhodococcus* sp. soil isolates in response to inorganic sulfate limitation. Microbiol. Res. 168, 12–21.

Dedourge, O., Vong, P.C., Lasserre-Joulin, F., Benizri, E., Guckert, A., 2004. Effects of glucose and rhizodeposits (with or without cysteine-S) on immobilized-S-35, microbial biomass-S-35 and arylsulphatase activity in a calcareous and an acid brown soil. Eur. J. Soil Sci. 55, 649–656.

Doolette, A.L., Smernik, R.J., 2011. Soil organic phosphorus speciation using spectroscopic techniques. In: Bünemann, E.K., Oberson, A., Frossard, E. (Eds.), Phosphorus in Action Biological Processes in Soil Phosphorus Cycling. Springer, Berlin, pp. 3–36.

Duc, L., Noll, M., Meier, B.E., Buergmann, H., Zeyer, J., 2009. High diversity of diazotrophs in the forefield of a receding alpine glacier. Microb. Ecol. 57, 179–190.

Ehlers, K., Bakken, L.R., Frostegard, A., Frossard, E., Buenemann, E.K., 2010. Phosphorus limitation in a Ferralsol: impact on microbial activity and cell internal P pools. Soil Biol. Biochem. 42, 558–566.

Ehrlich, H.L., Newman, D.K., 2009. Geomicrobiology, fifth ed. CRC Press/Taylor and Francis, Boca Raton.

Emerson, D., Fleming, E.J., McBeth, J.M., 2010. Iron-oxidizing bacteria: an environmental and genomic perspective. Annu. Rev. Microbiol. 64, 561–583.

Erb, T.J., Kiefer, P., Hattendorf, B., Guenther, D., Vorholt, J.A., 2012. GFAJ-1 is an arsenate-resistant, phosphate-dependent organism. Science 337, 467–470.

Eriksen, J., 1997. Sulphur cycling in Danish agricultural soils: turnover in organic S fractions. Soil Biol. Biochem. 29, 1371–1377.

Eriksen, J., 2009. Soil sulfur cycling in temperate agricultural systems. Adv. Agron. 102, 55–89.

Eriksen, J., Lefroy, R.D.B., Blair, G.J., 1995. Physical protection of soil organic S studied by extraction and fractionation of soil organic matter. Soil Biol. Biochem. 27, 1011–1016.

Esperschuetz, J., Perez-de-Mora, A., Schreiner, K., Welzl, G., Buegger, F., Zeyer, J., Hagedorn, F., Munch, J.C., Schloter, M., 2011. Microbial food web dynamics along a soil chronosequence of a glacier forefield. Biogeosciences 8, 3283–3294.

Fierer, N., Strickland, M.S., Liptzin, D., Bradford, M.A., Cleveland, C.C., 2009. Global patterns in belowground communities. Ecol. Lett. 12, 1238–1249.

Finlay, R., Wallander, H., Smits, M., Holmstrom, S., Van Hees, P., Lian, B., Rosling, A., 2009. The role of fungi in biogenic weathering in boreal forest soils. Fungal Biol. Rev. 23, 101–106.

Freney, J.R., Melville, G.E., Williams, C.H., 1975. Soil organic matter fractions as sources of plant available sulfur. Soil Biol. Biochem. 7, 217–221.

Frey, B., Rieder, S.R., Brunner, I., Ploetze, M., Koetzsch, S., Lapanje, A., Brandl, H., Furrer, G., 2010. Weathering-associated bacteria from the Damma glacier forefield: physiological capabilities and impact on granite dissolution. Appl. Environ. Microbiol. 76, 4788–4796.

Frossard, E., Condron, L.M., Oberson, A., Sinaj, S., Fardeau, J.C., 2000. Processes governing phosphorus availability in temperate soils. J. Environ. Qual. 29, 15–23.

Gadd, G.M., Rhee, Y.J., Stephenson, K., Wei, Z., 2012. Geomycology: metals, actinides and biominerals. Environ. Microbiol. Rep. 4, 270–296.

Ghani, A., McLaren, R.G., Swift, R.S., 1992. Sulphur mineralisation and transformations in soils as influenced by additions of carbon, nitrogen and sulphur. Soil Biol. Biochem. 24, 331–341.

Ghani, A., McLaren, R.G., Swift, R.S., 1993a. Mobilization of recently-formed soil organic sulphur. Soil Biol. Biochem. 25, 1739–1744.

Ghani, A., McLaren, R.G., Swift, R.S., 1993b. The incorporation and transformations of sulfur-35 in soil: effects of soil conditioning and glucose or sulphate additions. Soil Biol. Biochem. 25, 327–335.

Gianfreda, L., Ruggiero, P., 2006. Enzyme activities in soil. In: Nannipieri, P., Smalla, K. (Eds.), Nucleic Acids and Proteins in Soil. Springer, Berlin.

Glick, B.R., 2010. Using soil bacteria to facilitate phytoremediation. Biotechnol. Adv. 28, 367–374.

Goeransson, H., Venterink, H.O., Baath, E., 2011. Soil bacterial growth and nutrient limitation along a chronosequence from a glacier forefield. Soil Biol. Biochem. 43, 1333–1340.

Goux, X., Amiaud, B., Piutti, S., Philippot, L., Benizri, E., 2012. Spatial distribution of the abundance and activity of the sulfate ester-hydrolyzing microbial community in a rape field. J. Soils Sediments 12, 1360–1370.

Griffiths, B.S., Philippot, L., 2013. Insights into the resistance and resilience of the soil microbial community. FEMS Microbiol. Rev. 37, 112–129.

Gyaneshwar, P., Kumar, G.N., Parekh, L.J., Poole, P.S., 2002. Role of soil micro-organisms in improving P nutrition of plants. Plant Soil 245, 83–93.

Hallberg, K.B., 2010. New perspectives in acid mine drainage microbiology. Hydrometallurgy 104, 448–453.

Hedrich, S., Schlomann, M., Johnson, D.B., 2011. The iron-oxidizing proteobacteria. Microbiology 157, 1551–1564.

Hildebrandt, U., Regvar, M., Bothe, H., 2007. Arbuscular mycorrhiza and heavy metal tolerance. Phytochemistry 68, 139–146.

Hinsinger, P., Plassard, C., Tang, C.X., Jaillard, B., 2003. Origins of root-mediated pH changes in the rhizosphere and their responses to environmental constraints: a review. Plant Soil 248, 43–59.

Hsieh, Y.J., Wanner, B.L., 2010. Global regulation by the seven-component Pi signaling system. Curr. Opin. Microbiol. 13, 198–203.

Jennings, D.H., 1995. The Physiology of Fungal Nutrition. Cambridge University Press, Cambridge.

Keller, M., Oberson, A., Annaheim, K.E., Tamburini, F., Mäder, P., Mayer, J., Frossard, E., Bünemann, E.K., 2012. Phosphorus forms and enzymatic hydrolysability of organic phosphorus in soils after 30 years of organic and conventional farming. J. Plant Nutr. Soil Sci. 175, 385–393.

Kelly, D.P., Murrell, J.C., 1999. Microbial metabolism of methanesulfonic acid. Arch. Microbiol. 172, 341–348.

Kertesz, M.A., 1999. Riding the sulfur cycle—metabolism of sulfonates and sulfate esters in Gram-negative bacteria. FEMS Microbiol. Rev. 24, 135–175.

Kertesz, M.A., Fellows, E., Schmalenberger, A., 2007. Rhizobacteria and plant sulfur supply. Adv. Appl. Microbiol. 62, 235–268.

Kidd, P., Barcelo, J., Bernal, M.P., Navari-Izzo, F., Poschenrieder, C., Shilev, S., Clemente, R., Monterroso, C., 2009. Trace element behaviour at the root-soil interface: implications in phytoremediation. Environ. Exp. Bot. 67, 243–259.

King, J.E., Quinn, J.P., 1997. The utilization of organosulphonates by soil and freshwater bacteria. Lett. Appl. Microbiol. 24, 474–478.

Kirchman, D.L., 2012. Processes in Microbial Ecology. Oxford University Press, New York.

Kirkby, C.A., Kirkegaard, J.A., Richardson, A.E., Wade, L.J., Blanchard, C., Batten, G., 2011. Stable soil organic matter: a comparison of C:N:P:S ratios in Australian and other world soils. Geoderma 163, 197–208.

Kizewski, F., Liu, Y.T., Morris, A., Hesterberg, D., 2011. Spectroscopic approaches for phosphorus speciation in soils and other environmental systems. J. Environ. Qual. 40, 751–766.

Kraemer, S.M., 2004. Iron oxide dissolution and solubility in the presence of siderophores. Aquat. Sci. 66, 3–18.

Lamers, L.P.M., van Diggelen, J.M., Op den Camp, H.J., Visser, E.J.W., Lucassen, E.C., Vile, M.A., Jetten, M.S.M., Smolders, A.J., Roelofs, J.G., 2012. Microbial transformations of nitrogen, sulfur, and iron dictate vegetation composition in wetlands: a review. Front. Microbiol. 3, 156.

Laslo, E., Gyorgy, E., Mara, G., Tamas, E., Abraham, B., Lanyi, S., 2012. Screening of plant growth promoting rhizobacteria as potential microbial inoculants. Crop. Prot. 40, 43–48.

Lloyd, J.R., Lovley, D.R., Macaskie, L.E., 2003. Biotechnological application of metal-reducing micro-organisms. Adv. Appl. Microbiol. 53, 85–128.

Loper, J.E., Buyer, J.S., 1991. Siderophores in microbial interactions on plant surfaces. Mol. Plant-Microbe Interact. 4, 5–13.

Lovley, D.R., 2012. Electromicrobiology. Annu. Rev. Microbiol. 66, 391–409.

Martin, G., Guggiari, M., Bravo, D., Zopfi, J., Cailleau, G., Aragno, M., Job, D., Verrecchia, E., Junier, P., 2012. Fungi, bacteria and soil pH: the oxalate-carbonate pathway as a model for metabolic interaction. Environ. Microbiol. 14, 2960–2970.

Marzluf, G.A., 1997. Molecular genetics of sulfur assimilation in filamentous fungi and yeast. Annu. Rev. Microbiol. 51, 73–96.

McGill, W.B., Cole, C.V., 1981. Comparative aspects of cycling of organic C, N, S and P through soil organic matter. Geoderma 26, 267–286.

Metcalf, W.W., Griffin, B.M., Cicchillo, R.M., et al., 2012. Synthesis of methylphosphonic acid by marine microbes: a source for methane in the aerobic ocean. Science 337, 1104–1107.

Meyer, B., Kuever, J., 2007. Molecular analysis of the diversity of sulfate-reducing and sulfur-oxidizing prokaryotes in the environment, using *aprA* as functional marker gene. Appl. Environ. Microbiol. 73, 7664–7679.

Miao, Z., Brusseau, M.L., Carroll, K.C., Carreon-Diazconti, C., Johnson, B., 2012. Sulfate reduction in groundwater: characterization and applications for remediation. Environ. Geochem. Health 34, 539–550.

Muyzer, G., Stams, A.J.M., 2008. The ecology and biotechnology of sulphate-reducing bacteria. Nat. Rev. Microbiol. 6, 441–454.

Olsson, P.A., Hammer, E.C., Wallander, H., Pallon, J., 2008. Phosphorus availability influences elemental uptake in the mycorrhizal fungus *Glomus intraradices*, as revealed by particle-induced X-ray emission analysis. Appl. Environ. Microbiol. 74, 4144–4148.

Paris, F., Bonnaud, P., Ranger, J., Lapeyrie, F., 1995. In vitro weathering of phlogopite by ectomycorrhizal fungi. 1. Effect of K^+ and Mg^{2+} deficiency on phyllosilicate evolution. Plant Soil 177, 191–201.

Passman, F.J., Jones, G.E., 1985. Preparation and analysis of *Pseudomonas putida* cells for elemental composition. Geomicrobiol. J. 4, 191–206.

Peltzer, D.A., Wardle, D.A., Allison, V.J., et al., 2010. Understanding ecosystem retrogression. Ecol. Monogr. 80, 509–529.

Pester, M., Schleper, C., Wagner, M., 2011. The Thaumarchaeota: an emerging view of their phylogeny and ecophysiology. Curr. Opin. Microbiol. 14, 300–306.

Philpott, C.C., 2006. Iron uptake in fungi: a system for every source. Biochim. Biophys. Acta 1763, 636–645.

Plassard, C., Dell, B., 2010. Phosphorus nutrition of mycorrhizal trees. Tree Physiol. 30, 1129–1139.

Plassard, C., Louche, J., Ali, M.A., Duchemin, M., Legname, E., Cloutier-Hurteau, B., 2011. Diversity in phosphorus mobilisation and uptake in ectomycorrhizal fungi. Ann. Forest Sci. 68, 33–43.

Plett, J.M., Martin, F., 2011. Blurred boundaries: lifestyle lessons from ectomycorrhizal fungal genomes. Trends Genet. 27, 14–22.

Prietzel, J., Botzaki, A., Tyufekchieva, N., Brettholle, M., Thieme, J., Klysubun, W., 2011. Sulfur speciation in soil by S K-edge XANES spectroscopy: comparison of spectral deconvolution and linear combination fitting. Environ. Sci. Technol. 45, 2878–2886.

Randriamanantsoa, L., Morel, C., Rabeharisoa, L., Douzet, J.M., Jansa, J., Frossard, E., 2013. Can the isotopic exchange kinetic method be used in soils with a very low water extractable phosphate content and a high sorbing capacity for phosphate ions? Geoderma 200, 120–129.

Reid, R.K., Reid, C.P.P., Szaniszlo, P.J., 1985. Effects of synthetic and microbially produced chelates on the diffusion of iron and phosphorus to a simulated root in soil. Biol. Fertil. Soils 1, 45–52.

Richardson, A.E., Hocking, P.J., Simpson, R.J., George, T.S., 2009. Plant mechanisms to optimise access to soil phosphorus. Crop Pasture Sci. 60, 124–143.

Robin, A., Vansuyt, G., Hinsinger, P., Meyer, J.M., Briat, J.F., Lemanceau, P., 2008. Iron dynamics in the rhizosphere: consequences for plant health and nutrition. Adv. Agron. 99, 183–225.

Roden, E.E., 2012. Microbial iron-redox cycling in subsurface environments. Biochem. Soc. Trans. 40, 1249–1256.

Rodriguez, H., Fraga, R., 1999. Phosphate solubilizing bacteria and their role in plant growth promotion. Biotechnol. Adv. 17, 319–339.

Ross, D.J., 1990. Estimation of soil microbial C by a fumigation extraction method—influence of seasons, soils and calibration with the fumigation incubation procedure. Soil Biol. Biochem. 22, 295–300.

Scherer, P., Lippert, H., Wolff, G., 1983. Composition of the major elements and trace elements of 10 methanogenic bacteria determined by inductively coupled plasma emission spectroscopy. Biol. Trace Elem. Res. 5, 149–163.

Schmalenberger, A., Noll, M., 2010. Shifts in desulfonating bacterial communities along a soil chronosequence in the forefield of a receding glacier. FEMS Microbiol. Ecol. 71, 208–217.

Schmalenberger, A., Hodge, S., Bryant, A., Hawkesford, M.J., Singh, B.K., Kertesz, M.A., 2008. The role of *Variovorax* and other *Comamonadaceae* in sulfur transformations by microbial wheat rhizosphere communities exposed to different sulfur fertilization regimes. Environ. Microbiol. 10, 1486–1500.

Schmalenberger, A., Hodge, S., Hawkesford, M.J., Kertesz, M.A., 2009. Sulfonate desulfurization in *Rhodococcus* from wheat rhizosphere communities. FEMS Microbiol. Ecol. 67, 140–150.

Schmalenberger, A., Telford, A., Kertesz, M.A., 2010. Sulfate treatment affects desulfonating bacterial community structures in *Agrostis* rhizospheres as revealed by functional gene analysis based on *asfA*. Eur. J. Soil Biol. 46, 248–254.

Silva, M.C.P., Semenov, A.V., Schmitt, H., Elsas, J.D.V., Salles, J.F., 2013. Microbe-mediated processes as indicators to establish the normal operating range of soil functioning. Soil Biol. Biochem. 57, 995–1002.

Slyemi, D., Bonnefoy, V., 2012. How prokaryotes deal with arsenic. Environ. Microbiol. Rep. 4, 571–586.

Smith, S.E., Smith, F.A., 2011. Roles of arbuscular mycorrhizas in plant nutrition and growth: new paradigms from cellular to ecosystem scales. Annu. Rev. Plant Biol. 62, 227–250.

Smits, M.M., Bonneville, S., Benning, L.G., Banwart, S.A., Leake, J.R., 2012. Plant-driven weathering of apatite—the role of an ectomycorrhizal fungus. Geobiology 10, 445–456.

Smolders, A.J.P., Lucassen, E., Bobbink, R., Roelofs, J.G.M., Lamers, L.P.M., 2010. How nitrate leaching from agricultural lands provokes phosphate eutrophication in groundwater fed wetlands: the sulphur bridge. Biogeochemistry 98, 1–7.

Sorokin, D.Y., 2003. Oxidation of inorganic sulfur compounds by obligately organotrophic bacteria. Microbiology 72, 641–653.

Stolz, J.E., Basu, P., Santini, J.M., Oremland, R.S., 2006. Arsenic and selenium in microbial metabolism. Annu. Rev. Microbiol. 60, 107–130.

Sylvia, D.M., Fuhrmann, J.F., Hartel, P.G., Zuberer, D.A., 1999. Principles and Applications of Soil Microbiology. Prentice-Hall, New Jersey.

Tamburini, F., Pfahler, V., Buenemann, E.K., Guelland, K., Bernasconi, S.M., Frossard, E., 2012. Oxygen isotopes unravel the role of micro-organisms in phosphate cycling in soils. Environ. Sci. Technol. 46, 5956–5962.

Taylor, J.P., Wilson, B., Mills, M.S., Burns, R.G., 2002. Comparison of microbial numbers and enzymatic activities in surface soils and subsoils using various techniques. Soil Biol. Biochem. 34, 387–401.

Tebo, B.M., Johnson, H.A., McCarthy, J.K., Templeton, A.S., 2005. Geomicrobiology of manganese (II) oxidation. Trends Microbiol. 13, 421–428.

Traxler, M.F., Seyedsayamdost, M.R., Clardy, J., Kolter, R., 2012. Interspecies modulation of bacterial development through iron competition and siderophore piracy. Mol. Microbiol. 86, 628–644.

Uroz, S., Calvaruso, C., Turpault, M.P., Frey-Klett, P., 2009. Mineral weathering by bacteria: ecology, actors and mechanisms. Trends Microbiol. 17, 378–387.

Vong, P.C., Dedourge, O., Lasserre-Joulin, F., Guckert, A., 2003. Immobilized-S, microbial biomass-S and soil arylsulfatase activity in the rhizosphere soil of rape and barley as affected by labile substrate C and N additions. Soil Biol. Biochem. 35, 1651–1661.

Waldron, K.J., Robinson, N.J., 2009. How do bacterial cells ensure that metalloproteins get the correct metal? Nat. Rev. Microbiol. 7, 25–35.

Weber, K.A., Achenbach, L.A., Coates, J.D., 2006. Micro-organisms pumping iron: anaerobic microbial iron oxidation and reduction. Nat. Rev. Microbiol. 4, 752–764.

Welc, M., Buenemann, E.K., Fliessbach, A., Frossard, E., Jansa, J., 2012. Soil bacterial and fungal communities along a soil chronosequence assessed by fatty acid profiling. Soil Biol. Biochem. 49, 184–192.

White, A.K., Metcalf, W.W., 2007. Microbial metabolism of reduced phosphorus compounds. Annu. Rev. Microbiol. 61, 379–400.

Williams, P.H., Haynes, R.J., 1993. Forms of sulfur in sheep excreta and their fate after application on to pasture soil. J. Sci. Food Agric. 62, 323–329.

Wolfe-Simon, F., Blum, J.S., Kulp, T.R., et al., 2011. A bacterium that can grow by using arsenic Instead of phosphorus. Science 332, 1163–1166.

Yang, Z.H., Stoven, K., Haneklaus, S., Singh, B.R., Schnug, E., 2010. Elemental sulfur oxidation by *Thiobacillus* spp. and aerobic heterotrophic sulfur-oxidizing bacteria. Pedosphere 20, 71–79.

Zhang, Y., Gladyshev, V.N., 2009. Comparative genomics of trace elements: emerging dynamic view of trace element utilization and function. Chem. Rev. 109, 4828–4861.

Zhang, Q., van der Donk, W.A., 2012. Answers to the carbon-phosphorus lyase conundrum. Chembiochem 13, 627–629.

Zhao, F.J., Wu, J., McGrath, S.P., 1996. Soil organic sulphur and its turnover. In: Piccolo, A. (Ed.), Humic Substances in Terrestrial Ecosystems. Elsevier, Amsterdam, pp. 467–506.

Zhao, F.J., Lehmann, J., Solomon, D., Fox, M.A., McGrath, S.P., 2006. Sulphur speciation and turnover in soils: evidence from sulphur K-edge XANES spectroscopy and isotope dilution studies. Soil Biol. Biochem. 38, 1000–1007.

Zhou, J.Z., He, Q., Hemme, C.L., et al., 2011. How sulphate-reducing micro-organisms cope with stress: lessons from systems biology. Nat. Rev. Microbiol. 9, 452–466.

Zumsteg, A., Bernasconi, S.M., Zeyer, J., Frey, B., 2011. Microbial community and activity shifts after soil transplantation in a glacier forefield. Appl. Geochem. 26, S326–S329.

Chapter 17

Modeling the Dynamics of Soil Organic Matter and Nutrient Cycling

William J. Parton[1], Stephen J. Del Grosso[1,2], Alain F. Plante[3], E. Carol Adair[4] and Susan M. Lutz[1]

[1]*Natural Resource Ecology Laboratory, Colorado State University, Fort Collins, CO, USA*
[2]*USDA Agricultural Research Service and Natural Resource Ecology Laboratory, Colorado State University, Fort Collins, CO, USA*
[3]*Department of Earth and Environmental Science, University of Pennsylvania, Philadelphia, PA, USA*
[4]*Rubenstein School of Environment and Natural Resources, University of Vermont, Burlington, Vermont, USA*

Chapter Contents

I INTRODUCTION

Knowledge of the turnover rates of plant and animal residues, microbial bodies, and soil organic matter (SOM) is a prerequisite for understanding the availability and cycling of nutrients such as C, N, S, and P. This understanding is essential in describing ecosystem dynamics and in calculating crop

Soil Microbiology, Ecology, and Biochemistry. http://dx.doi.org/10.1016/B978-0-12-415955-6.00017-7
505

nutrient needs relative to environmental pollution control. It is also necessary if we are to gain an understanding of how SOM turnover might affect C sequestration and be altered by global change. The significance of microbial decay products involved in SOM decomposition and nutrient cycling can best be determined using mathematical analysis of tracer and nontracer data. This requires knowledge of the various reservoirs or pool sizes in the system under study and the rates at which materials are transformed within and transferred between them.

II   REACTION KINETICS

Understanding the dynamics of nutrient, plant residue, or SOM transformations in the field requires meaningful mathematical expressions for the biological, chemical, and physical processes involved. The reaction rate of decomposition (represented by the change in substrate concentration with time, dS/dt) can be expressed as a function of the concentration of one or more of the substrates being degraded. The order of the reaction is the value of the exponent on the substrate concentration in the equation used to describe the reaction.

A   Zero-Order Reactions

Zero-order reactions are ones in which the rate of transformation of a substrate is unaffected by changes in the substrate concentration (the exponent on S on the right-hand side of the equation is zero, $S^0 = 1$, thus zero-order). Zero-order kinetics can be described using the following equation:

$$\frac{dS}{dt} = -k \tag{17.1}$$

The reaction rate is determined by factors other than the substrate concentration, such as the amount of catalyst. At high substrate concentrations, where substrate levels are not limiting, enzymatic reactions are usually zero order (e.g., nitrification at high NH_4^+ levels and denitrification at high NO_3^- levels). Figure 17.1 shows plots of zero-order reactions compared to plots of other kinetic equations. After integration, the equation can be solved for the substrate concentration as a function of time and becomes

$$S_t = S_0 - kt \tag{17.2}$$

where S_t (concentration) is the amount of substrate remaining at any time, S_0 (concentration) the initial amount of substrate in the system, k (concentration time^{-1}) the rate constant, and t (time) the time since the initiation of the reaction. A useful term to describe the reaction kinetics is the half-life, which is the time required to transform one-half of the initial substrate: $S_t = \frac{S_0}{2}$, then $t_{1/2} = \frac{S_0}{2k}$. The mean residence time (or turnover time) is the time required to transform a quantity of material equal to the starting amount S_0 (i.e., $S_t = S_0$) at steady state: $t_{mrt} = \frac{S_0}{k}$.

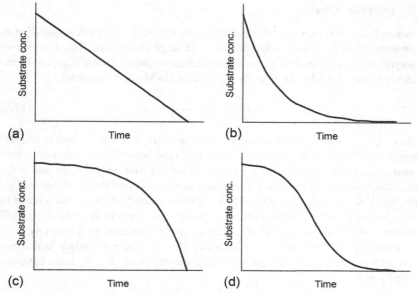

FIG. 17.1 Graphical representation of kinetic equations depicting the shapes of substrate decomposition curves: (a) zero-order, (b) first-order, (c) with exponential growth, and (d) with logistic growth.

B First-Order Reactions

In first-order reactions, the rate of transformation of a substrate is proportional to the substrate concentration. The rate of change of substrate S with time is

$$\frac{dS}{dt} = -kS \tag{17.3}$$

The decrease in the concentration of the substrate with time t is dependent on the rate constant k times the concurrent concentration S of the substrate. After integration, the following equation is obtained:

$$S_t = S_0 e^{-kt} \tag{17.4}$$

where S_t is the concentration of the substrate remaining at any time t. The rate constant k has units of per time (e.g., h^{-1}, day^{-1}, or $year^{-1}$). The half-life, or time required to transform one-half the initial substrate ($S_t = S_0/2$) can be calculated using $t_{1/2} = \frac{\ln 2}{k} \cong \frac{0.693}{k}$. The mean residence time (turnover time of an amount of substrate at steady-state equivalent in size to the starting amount) for first-order reactions is equal to $1/k$. Note that t_{mrt} for zero-order reactions is S_0/k.

C   Enzyme Kinetics

Extracellular enzymes are responsible for the first step in the decomposition of substrate in soil—the depolymerization of large molecules. The kinetics of enzyme reactions can thus be used to model substrate depletion. Enzyme kinetics are represented by the hyperbolic, Michaelis-Menten equation:

$$\frac{\mathrm{d}S}{\mathrm{d}t} = V_m \frac{S}{K_m + S} \qquad (17.5)$$

where V_m is the maximum reaction rate (concentration time^{-1}) and is proportional to the total mass of enzyme in the soil (and hence to the total active biomass), and K_m is the Michaelis-Menten, or half-saturation, constant and is the substrate concentration at which the reaction occurs at half the maximum velocity, $V_m/2$. K_m is inversely related to enzyme-substrate affinity, which tends to have higher values in soil than in aqueous solutions and to decrease when soil slurries are shaken; therefore, K_m is inversely proportional to diffusivity.

Figure 17.2 illustrates how Michealis-Menten kinetics contain both first- and zero-order regions. At high substrate concentrations ($S \gg K_m$), the equation simplifies to:

$$\frac{\mathrm{d}S}{\mathrm{d}t} = V_m \qquad (17.6)$$

which describes a zero-order reaction, and under very low substrate concentrations ($S \ll K_m$), the equation simplifies to:

$$\frac{\mathrm{d}S}{\mathrm{d}t} = \frac{V_m S}{K_m} = k'S \qquad (17.7)$$

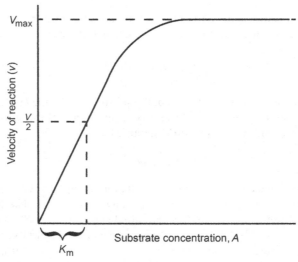

FIG. 17.2   Graphical expression of the Michaelis-Menten kinetic parameters for an enzymatic reaction.

which describes a first-order reaction. Because of the combined linear and exponential forms of S in the equation, the Michaelis-Menten equation cannot be solved for S_t analytically; however, solutions are easily calculated iteratively using spreadsheet or numerical computation software.

D Microbial Growth

The preceding equations have one limitation; none can account for microbial growth. As microorganisms consume a substrate, one portion is used for maintenance energy requirements, and if enough substrate is available, the remainder will be used to support growth. As microorganisms grow and multiply, they will exert an increasing demand on the remaining substrate, thereby changing the kinetics of decomposition. Three equations can be used to describe microbial growth: the exponential, logistic, and Monod equations.

Microbes grow exponentially when rapidly consuming substrate

$$\frac{dN}{dt} = \mu N \tag{17.8}$$

where N is the number (biomass) of cells, and μ is the growth rate. Changes in biomass can be equated to changes in substrate by dividing the biomass by the yield (y), which is the mass of cells generated per mass of substrate consumed:

$$-\frac{dS}{dt} = \frac{dN}{dt}\frac{1}{y} \tag{17.9}$$

Substituting Eq. (17.8) for dN/dt into Eq. (17.9), we obtain the following equation in terms of substrate depletion:

$$-\frac{dS}{dt} = \mu \frac{N_t}{y} \tag{17.10}$$

where N_t is the biomass at any given time. Measuring biomass at all times is unrealistic; therefore, N_t can be described in terms of initial biomass, N_0, such that $N_t = N_0 + N_t - N_0$, and the preceding equation expands to:

$$-\frac{dS}{dt} = \mu \frac{N_0}{y} + \mu \frac{N_t}{y} - \mu \frac{N_0}{y} \tag{17.11}$$

and collecting terms:

$$-\frac{dS}{dt} = \mu \frac{N_0}{y} + \mu \frac{(N_t - N_0)}{y} \tag{17.12}$$

Expressing all terms as substrate, considering that the total amount of substrate consumed since $t = 0$ is $S_0 - S_t = (N_t - N_0)/y$, and letting $X_0 = N_0/y$ where X_0 becomes the amount of substrate needed to produce N_0, then the differential form of the exponential equation for substrate depletion is

$$-\frac{dS}{dt} = \mu(S_0 + X_0 - S) \tag{17.13}$$

Although the exponential equation links substrate depletion with microbial growth, it is well recognized that microbes do not grow exponentially at all times. Rather, microbial populations grow to a limit (K). As the population increases, the growth rate (μ) decreases due to competition among individuals for increasingly scarce substrate resources. Microbial growth to a limit is expressed by the logistic equation:

$$\frac{dN}{dt} = \mu\left(1 - \frac{N}{K}\right)N \tag{17.14}$$

A similar exercise of algebra to express the logistic equation in terms of substrate, considering that K is the maximum biomass (original biomass plus that generated by converting all the original substrate into biomass), yields the differential equation for logistic growth in terms of substrate depletion:

$$-\frac{dS}{dt} = \mu\left(\frac{S}{S_0 + X_0}\right)(S_0 + X_0 - S) \tag{17.15}$$

Initially, $S = S_0$ and the reaction rate is governed primarily by X_0. As the microbial population grows, S decreases such that the first term in the preceding equation decreases while the second term increases. Therefore, two competing trends govern the substrate depletion rate. The rate of substrate depletion is maximized at $-dS/dt_{max} = \mu(S_0 + X_0)/4$. The differential form of the logistic equation can be integrated and solved for S_t to give

$$S_t = S_0 + X_0 - \frac{S_0 + X_0}{1 + \left(\dfrac{S_0}{X_0} - 1\right)e^{-\mu t}} \tag{17.16}$$

Monod kinetics are the most general kinetic expressions because they relate substrate depletion to both changes in population density and changes in substrate concentration. The basic relationship is

$$\mu' = \mu_{max}\frac{S}{K_t + S} \tag{17.17}$$

which appears similar to the Michaelis-Menten equation, but with some subtle differences: μ' is the specific growth rate ($\mu' = \mu/N_t$), μ_{max} is the maximum growth rate when substrate is not limiting, and K_s is the Monod constant, which is similar to the Michaelis-Menten constant. The differential equation describing Monod kinetics with growth in terms of substrate depletion is

$$-\frac{dS}{dt} = \mu_{max}\left(\frac{S}{K_t + S}\right)(S_0 + X_0 - S) \tag{17.18}$$

As with the Michaelis-Menten equation, the integrated form of Monod kinetics cannot be solved for S_t, but can be solved for t. The substrate concentration can then be determined iteratively.

Monod kinetics can be simplified, under certain conditions, to yield the kinetic equations described earlier (Table 17.1). When the initial concentration

TABLE 17.1 The Monod-Kinetic Equation, Its Simplifications, and the Conditions Under Which the Simplifications Can Be Made

Condition	Outcome	Kinetics	Differential Form	Integrated Form	Solve for S?
		Monod	$-\dfrac{dS}{dt} = \mu_{max}\dfrac{S}{K_s+S}(S_0+X_0-S)$	$K_s\ln\left(\dfrac{S_t}{S_0}\right) = (S_0+X_0+K_s)\ln\left(\dfrac{X_t}{X_0}\right) - (S_0+X_0)\mu_{max}t$	N
$S_0 \gg K_s$	$K_s+S\approx S$	Exponential	$-\dfrac{dS}{dt} = \mu_{max}(S_0+X_0-S)$	$S_t \approx S_0+X_0-X_0e^{kt}$	Y
$S_0 \ll K_s$	$K_s+S\approx K_s$	Logistic	$-\dfrac{dS}{dt} = \dfrac{\mu_{max}S}{K_s}(S_0+X_0-S)$	$S_t = S_0+X_0-\dfrac{S_0+X_0}{1+\left(\dfrac{S_0}{X_0}-1\right)e^{-\mu t}}$	Y
$X_0 \gg S_0$	$S_0+X_0-S\approx X_0$	Michaelis-Menten	$-\dfrac{dS}{dt} = \mu_{max}\dfrac{SX_0}{K_s+S} = k'\dfrac{S}{K_s+S}$	$S_0-S_t+K_s\ln\left(\dfrac{S_0}{S_t}\right) = V_m t$	N
$X_0 \gg S_0$ and $S_0 \ll K_s$	$S_0+X_0-S\approx X_0$ $K_s+S\approx K_s$	First-order	$-\dfrac{dS}{dt} = \dfrac{\mu_{max}SX_0}{K_s} = k'S$	$S_t = S_0e^{-kt}$	Y
$X_0 \gg S_0$ and $S_0 \gg K_s$	$S_0+X_0-S\approx X_0$ $K_s+S\approx S$	Zero-order	$-\dfrac{dS}{dt} = \mu_{max}X_0 = k'$	$S_t = S_0 - kt$	Y

of microbial biomass is much greater than the initial substrate concentration (e.g., $X_0 \gg S_0$), then the term $(S_0 + X_0 - S)$ in the Monod equation can be approximated to X_0 and the Monod equation is simplified to the Michaelis-Menten equation used to describe enzyme kinetics.

III MODELING SOIL CARBON AND NUTRIENT DYNAMICS

Simulation models can be used to gain an understanding of the processes and controls involved in C and nutrient cycles to generate data on the size of various SOM pools and the rates at which C and nutrients are transformed and to make predictions when experiments are inappropriate. Although conceptual models may be sufficient for the first task, only quantitative models can achieve the latter tasks. The best, quantitative simulation models of SOM and nutrient dynamics attempt to describe soil biological processes rather than strictly using mathematical expressions and statistical procedures to find best fitting curves. Fitting model equations to C and nutrient mineralization curves provides estimates of the amount of mineralized product released (CO_2) and the rate at which the product (NO_3) is made available to plants or the environment. Models range from single-equation kinetic representations, such as those outlined earlier, to large mechanistic models that account for many components of global ecosystems and require computational software and supercomputers to solve the systems of spatially resolved simultaneous differential equations.

A Analytical Models

The kinetics of plant nutrient transformations have been of interest for a long time, particularly the kinetics of N mineralization. Stanford and Smith (1972) described net N mineralization using a simple, first-order model $N = N_0(1 - e^{-kt})$ where N is the amount of N mineralized at time t, and N_0 is the amount of potential mineralizable N. Several modifications to first-order models, as well as other kinetic models, have been proposed to account for experimental observations of large initial flushes of mineralization or for lags before the initiation of mineralization. In selecting a model for N and S net mineralization, one should generally apply the principle of parsimony: increasing model complexity incrementally to obtain a suitable fit, keeping the number of parameters to a minimum, and knowing that no single model will fit data for all soils under all conditions, whereas some conditions will be adequately described by several models.

Because they play different roles in plants and the environment that lead to differing dynamics, there are fewer short-term or single-season models for C than for N and other plant nutrients. SOM models generally project long-term changes in soil C, but there are a number of ways to describe the short-term decomposition of organic residues during the first few months after introduction to the soil. Field and laboratory experiments have shown that initial decomposition rates of litter are generally independent of the amount of litter biomass added unless it exceeds 1.5% of the dry soil weight. Decomposition of plant residues has been experimentally found to be reasonably well described by first-order rate kinetics. The use of first-order kinetics to describe the decomposition of SOM implies that the microbial inoculum potential of soil is not limiting the decomposition rate (e.g., $X_0 \gg S_0$, from the previous section). This is true, in large part, because soil microbial biomass often has a fast growth rate relative to the length of most decomposition studies.

Several approaches have been used to account for the changing nature of organic matter during decomposition, such as making k a function of time or including additional compartments. Experimental data for the decomposition of added plant residues or manures can be closely fit using the summation of two, first-order equations in the general form:

$$C = Ae^{-k_A t} + Be^{-k_B t} \tag{17.19}$$

where C is the soil-C content at any given time, A and B are the proportions of the two pools, and k_A and k_B are the first-order constants for each of the pools. In the case illustrated in Fig. 17.3, the first pool of the manured treatment represented 5% of the soil C and had a turnover time of 40 days in the laboratory. The second pool represented 45% of the C with a laboratory turnover time of 3 years. The curve representing the fertilized plots showed that the first pool represented 2% of the C with a similar turnover time as the manured (40 days). The second pool representing 48% of the C had a turnover time of 5 years. The remainder of the C was known from ^{14}C dating to have a turnover time of 500-1000 years and

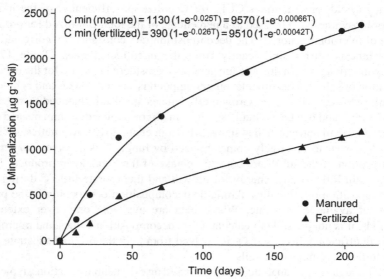

FIG. 17.3   The fit of the sum of two first-order curves describing C mineralization in soil from a manured and fertilized long-term plot during a 220-day incubation.

therefore contributed little CO_2. The example demonstrates how the partitioning of organic matter between labile and more resistant fractions alters the calculated decomposition dynamics.

Differences in the ability of simple models to model short-term versus long-term decomposition dynamics were highlighted by Sleutel et al. (2005), who compared the performance of first-order, sum of first-order, combination of zero-order and first-order, second-order and Monod-kinetic models for extrapolation from short-term data. They concluded that the sum of first-order and Monod models performed best in estimating stable organic C, but did not fit short-term mineralization well.

Only a portion of the actual decomposition is accounted for when determining the decomposition rate (k) by measuring CO_2 output or the amount of C left in the soil. Microorganisms use C compounds for biosynthesis, forming new cellular or extracellular material, and as an energy supply. In the latter process, CO_2, microbial cells and microbial products are produced. Under aerobic conditions, the amount of waste products produced is not usually high, and the amount of biosynthesis, or production of microbial cells, can be calculated from CO_2 data. This requires knowledge of yield or efficiency of substrate conversion to microbial biomass,

$$C = C_i \left(1 + \frac{Y}{100 - Y} \right) \qquad (17.20)$$

where C is the substrate decomposed, C_i the CO_2-C evolved, and Y the efficiency (yield, or sometimes CUE, for C utilization efficiency) of the use of C for biosynthesis, expressed as a percentage of the total C utilized for production of microbial material. The decomposition rate constants (k), corrected for biosynthesis, differ significantly from the uncorrected ones (Table 17.2). Growth efficiencies of 40-60% are generally considered realistic for the decomposition of soluble constituents; other compounds, such as waxes and cellulose, result in lower efficiencies. Growth efficiencies are also influenced by nutrient availability and can be as small as 5-20% in oligotrophic environments and can exceed 80% in nutrient rich systems (Manzoni et al., 2012). Aromatics, such as lignin, appear to be largely cometabolized by fungi. This involves enzymatic degradation of the substrate, but little uptake of the breakdown products. The fungi gain little, if any, energy for growth and incorporate little C during the decomposition of aromatics. Aromatic decomposition occurs only in the presence of available substrate. Where data are available only over extended periods, it is not possible to calculate true decomposition values and microbial growth efficiency because CO_2 is evolved from both the original substrate and the turnover of microbial cells.

Although simple, analytical models, based on exponential equations, typically use fixed values for the decay constant, they can be made more versatile through the use of "modifiers." These modifiers are typically fractional values used to

TABLE 17.2 First-Order Decay Constants With and Without Correction for Microbial Biosynthesis During the Decomposition of Organic Compounds Added to Soil Under Laboratory Conditions

Material	Time of Incubation (Days)	k (Day^{-1})		
		Uncorrected	Corrected for CUE = 20%	Corrected for CUE = 60%
Straw-rye	14	0.02	0.03	0.11
Hemicellulose	14	0.03	0.04	0.11
Lignin	365	0.003	0.006	-
Native grass	30	0.006	0.008	0.02
Fungal cytoplasm	10	0.04	0.05	0.17
Fungal cell wall	10	0.02	0.03	0.07

moderate decomposition under nonideal conditions attributable to litter quality or climatic conditions. As an example, data from the LIDET litter decay study (Parton et al., 2007) was used to develop and test simple litter decay models that account for differences in litter chemistry, as well as climate (Adair et al., 2008). The study measured the amount of C and nutrients remaining in litter bags (annually for 10 years) for root and leaf litter materials decomposing at 28 sites around the world. This data set has been used to test global ecosystem models (Bonan et al., 2013) and suggests that models need to account for the impacts of climate decomposition index (CDI), the initial litter N, lignin, and labile C content to accurately represent litter-decomposition dynamics (Fig. 17.4).

Analyses of the LIDET data showed that CDI was the most correlated variable ($r^2 = 0.44$-0.77) during decomposition because it incorporates the seasonal patterns of temperature and moisture. Figure 17.4b shows the impact of initial N content (higher initial decay rates with high N litter), and the impact of lignin content (lower decay rates with high lignin litter, Fig. 17.4c). Initial N content also has a strong impact on N dynamics (Fig. 17.4d) with low N litter ($< 0.8\%$ N) causing immobilization of N by microbes during the initial phase of decay ($>50\%$ of initial C remaining), while high N litter ($>1.5\%$ N) results in simultaneous release of N and C during litter decay.

Adair et al. (2008) found that a three-pool model best fit the LIDET data.

$$M_t = M_1 e^{-k_1 \text{CDI}jt} + M_2 L_s e^{-k_2 \text{CDI}jt} + M_3 e^{-k_3 \text{CDI}jt} \tag{17.21}$$

where M_t is the fraction of mass remaining at time, t (years), M_p is the initial litter mass of each pool ($p = 1$, 2, or 3), CDI$_j$ = the climate decomposition index

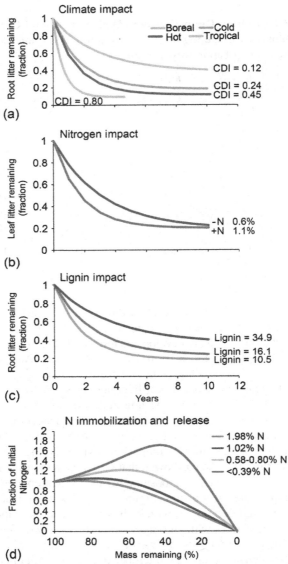

FIG. 17.4 (a) Litter mass remaining vs. time for Big Bluestem fine root decomposing in boreal forest, cold temperate forests, hot temperate forests, and humid tropical forests a); (b) litter mass remaining vs. time for low and high nitrogen surface litter (Chestnut Oak, 1.03% N vs. Western Red Cedar, 0.62% N); (c) mass remaining vs. time for root litters with different lignin content (Big Bluestem, 10.54%, Drypedes, 16.13%, Slash Pine, 34.90%); and (d) the fraction of the initial N in decomposing litter as a function of mass remaining in the litter as a function of the initial N content of the litter for surface litter. The results are all derived from the observed LIDET litter decay study (Parton et al., 2007).

for each site (j), k_p is the decomposition rate of each pool ($k_1 = 3.5$, $k_2 = 0.75$, $k_3 = 0.027$), and L_s is the lignocellulose ratio impact on cellulose decay. This model explained 68% of the variability in the observed global litter-decomposition patterns. An alternative approach to using modifiers to alter decomposition rates modeled using exponential equations is to use simple exponential models in which the decay constant is not constant. Manzoni et al. (2012) reviewed four discrete-compartment models with one or two C pools, two models with a single time-dependent decay rate, and two models based on a continuous distribution of decay rates, and then reported their analytical solutions. They found that differences among some models were marginal, but that all models resulted in apparent decay rates that decreased through time, in contrast to the assumption of constant k adopted in the single-pool exponential decay model.

B Substrate-Enzyme-Microbe Models

Most of the current soil-C models simulate decay of SOM pools using first-order decay rates and assume that microbes decay the material but have little impact on its rate of decay. A number of models (Schimel, 2001) have coupled soil-C decay to microbial physiology and biomass and include the impact of microbial activity on SOM decay rates. Figure 17.5 shows a generalized flow diagram for substrate-enzyme-microbe models (Allison et al., 2010). Key assumptions in these models are that extracellular enzymes control the depolymerization of soil organic matter (SOC) to dissolved organic matter (DOC), and that DOC availability controls decomposition of SOC. Enzyme production is proportional to

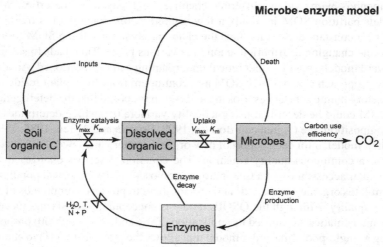

FIG. 17.5 Simplified flow diagram for microbe growth and enzyme kinetic models. *(Redrawn from Allison et al., 2010.)*

microbial biomass. These models typically use Michaelis-Menten functions to represent enzyme reaction rates and microbial uptake of DOC, with the maximum reaction rate, microbial uptake (V_{max}), and half-saturation constant (K_m) being the primary input variables. Other soil environmental variables (soil pH, water, temperature, N, and P) can impact the rate of formation of enzymes and their impact on the decay rate of SOC pools (Sinsabaugh and Shah, 2012). Another important parameter in substrate-enzyme-microbe models is the ratio of microbial growth to C processing costs (where processing costs include any C losses to respiration), frequently called microbial C use efficiency (CUE) or C yield. Conventional simulation models typically represent CUE as the C transfer among pools. Substrate-enzyme-microbe models assume that CUE changes as a function of the soil environmental variables and the SOM pool, whereas conventional SOM models typically use fixed values for CUE. There is great potential to improve conventional SOM models by including the impact of microbial activity, enzyme production, and soil environmental variables on CUE. One recently developed model (Wieder et al., 2013) assumes that SOM content is inversely related to microbial growth efficiency because higher microbial biomass leads to increased decomposition of SOM. Long-term studies showing that SOM is greater in soils with greater OM inputs (e.g., Paustian et al., 1992), and the LIDET (Parton et al., 2007) litter bag experiment showing that more C remains after 10 years with labile (maple leaves) versus structural (wheat straw) litter, are not consistent with this assumption.

C Cohort Models

It is well recognized that SOM is a mixture of a very large number of organic compounds with widely differing chemical and physical properties. Most models partition SOM into only a few discrete pools. Each pool is treated as having a constant decay rate, thus the changing dynamics of the SOM follows from the changing distribution among the various pools. This leads to straightforward models, but their theoretical underpinning can be questioned. An alternative approach is to describe SOM as a continuum from fresh plant residues to refractory humic substances. Bosatta and Ågren proposed that the heterogeneity of SOM could be described using a quality variable (q), which determines the decomposition rate (Ågren and Bosatta, 1998; Bosatta and Ågren, 1985, 1995). In their model, q for each cohort of new organic matter varies over time according to a continuous-quality equation. The equation describes changes in the molecular accessibility of a C atom to decomposers as the microbial population assimilates organic compounds and utilizes them to produce compounds of different quality. Although the QSOIL model is conceptually satisfying, its complex mathematics has limited its application. Yang and Janssen (2000) proposed using a single-pool, first-order model that allows the rate constant (k) to change. This approach is attractive because it is derived only from measured quantities of remaining organic material, and there is no need for other variables that cannot be measured. The key feature of the model is that it takes into account the decrease over time of the average and actual mineralization rates, which the authors refer to as "ageing."

D   Multicompartmental Models

Multicompartmental models seek to simplify the complex nature of SOM by lumping the heterogeneity into a small number of distinct compartments. The distinction between simple, analytical models and multicompartmental models is somewhat arbitrary because the sum of the exponentials model given earlier describes C mineralization from two or three pools or compartments. Generally, compartmental models are needed when a single equation is insufficient to describe the multiple transformation processes that occur simultaneously in soils. A multicompartmental model is depicted graphically as a set of boxes, each of which represents a pool or compartment. Most often, the pools are defined conceptually, but they can also be measurable fractions of SOM. A series of arrows connecting the various pools represent transformations of organic matter or a nutrient element from one form to the other rather than a physical or spatial flux of material. The graphical representation of the model can be described mathematically as a series of simultaneous differential equations. Some simple compartmental models can be solved analytically (e.g., in equation form), such as the sum of exponentials described earlier, but as the models become more complex and include more compartments, it becomes necessary to solve them numerically.

The advent of computers and numerical software has permitted the solution of complex systems models that require iterative solving of multiple equations to address multicompartment dynamics. The modeling of soil biological processes began with ecologists working in natural ecosystems in the 1960s and 1970s. Early modeling in agricultural systems focused on crop production in response to physical parameters, rather than biological processes. The Rothamsted model (ROTHC), one of the first multicompartmental models, was developed using the long-term wheat plots in Rothamsted, England. The initial version used measured C inputs to drive the model. However, later versions have been linked with plant production models (Paul et al., 2003), and the impact of cation exchange capacity on SOM stabilization was included (Jenkinson, 1990). A new emphasis in the 1970s on the environmental impacts of agriculture led to early models of N dynamics, including nitrate leaching and denitrification. Further emphasis on agroecology and SOM management in the 1980s and 1990s led to the development of a large number of models of SOM dynamics.

Most models of SOM dynamics begin by modeling the decay of litter at the soil surface. The models assume that plant material contains a readily decomposable fraction (typically water extractable material—sugars, soluble phenols, and amino acids) and a more resistant fraction containing cellulose and lignin (see Chapter 12). The lignocellulose ratio (lignin/lignin+cellulose) of plant material is positively correlated to the fraction of plant material resistant to decomposition. Most litter decay models include microbes that decompose plant material and form microbial products with a slow decay rate. Many models also assume that the majority of plant lignin (>70%) is directly transformed into organic material with a decay rate similar to microbial products. The division of SOM into various pools is based on stabilization mechanisms, bioavailability, and biochemical and kinetic parameters. Generally, the pools consist of at least one small "active" pool with a rapid turnover rate and one or more larger pools with slower turnover rates ranging from a few decades to thousands of years.

E   Nutrient Dynamics Models

Nutrient cycles in soil are tightly coupled through the nutrient demands of the microbial biomass during decomposition. For this reason, several of the C-centric multicompartmental models of SOM dynamics are also able to model nutrient elements such as N, P, and S. Frissel and Vanveen (1982) classified models of soil N in terms of: (1) their purpose: prediction, management, or scientific understanding; (2) their timespan; (3) whether they were budget-based or dynamic models; and (4) whether the models were dominated by transport processes, SOM dynamics, or soil-plant relations. Models of soil N dynamics generally include descriptions of: (1) physical processes, such as the transport of water, solutes, heat, and gases; (2) biological processes, such as mineralization, immobilization, nitrification, and denitrification; and (3) physicochemical processes, such as volatilization, adsorption, and fixation. Model structures of soil N models, for example, ANIMO as illustrated in Fig. 17.6, generally resemble the conceptual depiction of the soil N cycle.

Similar to N, the demand for P cycling models arose because soil P availability is the major nutrient limiting plant production in many systems and, conversely, because P from nonpoint sources, such as agricultural soils, has a major environmental impact on water quality. Despite this, few models calculating long-term changes in soil P have been developed, partly because of the many complicated solid-phase interactions of P sorption to minerals over and above the biological transformations. Jones et al. (1984) originally developed routines for simulating soil P dynamics, which became incorporated into the Erosion-Productivity Impact Calculator model. These routines have since been incorporated into several other models that describe the transport of soluble and particulate P, adsorption and desorption, mineralization and immobilization between organic and inorganic forms, leaching, plant uptake, and runoff (Fig. 17.7). Lewis and McGechan (2002) compared four P cycling models (ANIMO, GLEAMS, DAYCENT, and MACRO) and concluded that all the existing models have substantial limitations and that a hybrid submodel combining the best features of these models needs to be developed.

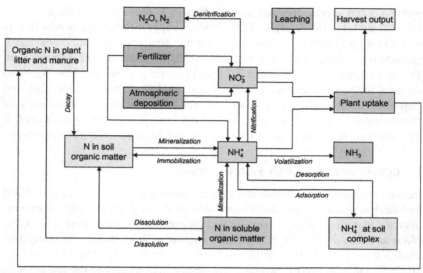

FIG. 17.6 The structure of the ANIMO N submodel for soil N dynamics. *(Redrawn from Wu and McGechan, 1998.)*

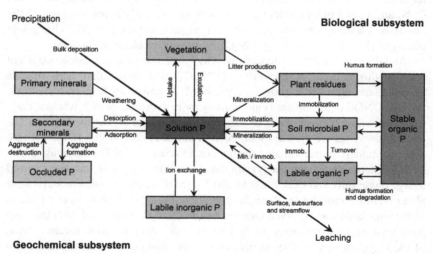

FIG. 17.7 Phosphorus (P) flows between physical, microbial, and plant systems represented in the Century and DayCent models.

Vitousek (2004) published data that show the interaction of plant production, and soil N, P, and C cycling during a soil development chronosequence (see Supplemental Table S17.1 in the online Supplemental Material at http://booksite.elsevier.com/9780124159556). As the soil ages from 3000 to 150,000 years, plant production increases along with soil C, N, and P levels, whereas plant production and soil C, N, and P decrease as soils age from 150,000 to 1.4 million years. These results suggest that plant production is co-limited by N and P availability during the first 150,000 years, whereas P availability constrains plant growth after 150,000 years of soil development. This is confirmed by the fact that N losses (gaseous plus NO_3 and DON losses) increase for soils older than 150,000 years. Modeling predicted the observed changes in plant production and soil C, N, and P levels during the 4 million year chronosequence and suggested that P added from Asian dust balances soil P losses and allows plant production to be sustained at current equilibrium levels.

F   Ecosystem and Earth System Models

Ecosystem models incorporate the various processes and mechanisms of SOM and nutrient dynamics into the broader context of ecosystems and therefore simulate a wider range of processes, such as greenhouse gas emissions (CO_2, CH_4, N_2O) and leaching losses of DOC and nutrients (NO_3). SOM and ecosystem models have recently been linked to Earth system models (ESM; Levis et al., 2012; Bonan et al., 2013), which simulate interactions among the atmosphere, land surface, and soil processes. Bonan et al. (2013) used the LIDET global litter decay data to test simulated C and N dynamics for ESMs. Global scale, soil C, N, and P, and plant production data sets are currently being used to test the ability of ESMs to simulate ecosystem dynamics and the potential changes in trace gas fluxes (CO_2, CH_4, and N_2O) as a result of climatic changes.

Abdalla et al. (2010) compared two widely used ecosystem models (DayCent and DeNitrification-DeComposition [DNDC]) and found that both models needed to be calibrated to properly represent N_2O emissions from Irish pastures, and that DNDC overestimated emissions by more than a factor of 2, whereas DayCent was within 60% of observed values. A comparison of four models simulating a Norway spruce system (Van Oijen et al., 2011) found that BASFOR (the simplest model) underestimated the temporal variance of N_2O emissions. DAYCENT (intermediate complexity) simulated the time series well, but also showed a large phase shift, and COUP and MoBiLE-DNDC (more complex models) were able to remove most biases through calibration. Figure 17.8 shows a comparison of the simulated and observed changes in soil C, N_2O fluxes, and NO_3 leaching from agricultural experiments in the United States using the most recent version of the DayCent model. The results demonstrate that current ecosystem models can often correctly simulate the impact of agricultural management practices on soil-C changes, N_2O fluxes, and NO_3 leaching.

G   Case Studies

In the following paragraphs, we summarize three widely used ecosystem models of varying complexity (Century/DayCent, DNDC, Ecosys) to illustrate some of the differences in modeling approaches.

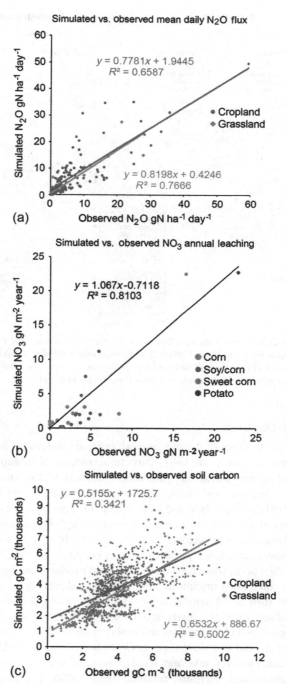

FIG. 17.8 Observed vs. simulated: (a) N_2O emissions, (b) NO_3 leached, and (c) soil carbon levels from different crops and grassland vegetation for the DayCent model.

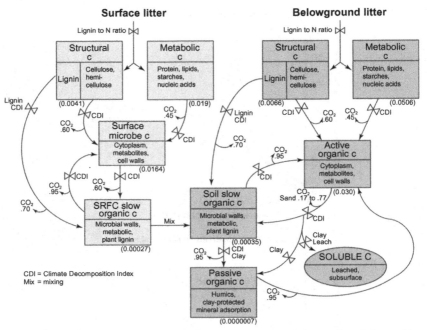

FIG. 17.9   The Century model showing lignin and N controls on the proportion of plant residue structural and metabolic components. Carbon flows entering microbial biomass and three soil components are controlled by decomposition rates (day^{-1}) shown in brackets. Microbial utilization efficiencies and shown on CO_2 loss arrows. *(Redrawn from Parton et al., 1994.)*

1 Century/DayCent

The Century model (Fig. 17.9), divides fresh organic residues on the basis of the lignin-to-N ratio (Parton et al., 1987). Structural components comprising lignin, cellulose, and hemicellulose have maximum decomposition rates of 0.0041-0.0066 day^{-1} (surface and soil), and metabolic components have k's of 0.019 and 0.050 day^{-1} (surface and soil). Lignin content is also used to control the decay rates of the structural components where high lignin results in slower decay rates because lignin is closely associated with the cellulose and hemicellulose components and has a protective effect on them. Surface litter enters a surface microbial pool, whereas surface microbial products and lignin are transferred into surface slow organic matter. Mechanical mixing of surface litter, microbes, and slow SOM matter can occur as a result of cultivation of the soil or diffusion of surface SOM slowly into the soil. Buried litter and roots feed into an active C pool, which comprises microbial biomass and microbial metabolites. The surface and soil slow C pools ($k = 0.00027$ and 0.00035 day^{-1}) have an intermediate turnover rate and receive some C directly from the lignin plant components and microbial products (cell walls). The arrows showing the CO_2 evolved during each transformation is indicative of the microbial growth efficiencies discussed earlier. The first-order maximum decay rates for each of the pools (Fig. 17.9) are a function of soil pH, water, and temperature. The Century model (monthly time step version) uses monthly precipitation and average maximum and minimum air temperature data to calculate a CDI, whereas the DayCent model (daily time step version of Century) uses simulated daily soil water and temperature and pH to control decomposition rates and also includes enhanced decay rates following rainfall events. The other major inputs to the Century/DayCent models include sand, silt, clay, bulk density, and soil depth.

The silt and clay contents play major roles in protecting the slow C pool, while clay content controls formation of passive soil C as a result of interaction of C with mineral surfaces.

DayCent also simulates N gas (N_2O, NO_x, N_2) emissions from nitrification and denitrification (Parton et al., 2001), as well as CH_4 oxidation in upland soils (Del Grosso et al., 2000) and methanogenesis in flooded soils (Cheng et al., 2013). Available NH_4 determines the maximum nitrification rate, which is reduced by multipliers representing water, pH, and temperature limitations on microbial activity. Denitrification is a function of soil NO_3 and labile C availability, water content, and an index of gas diffusivity. Although gas diffusion and redox potential are not represented explicitly in this model, they are included implicitly in that as the index of gas diffusivity decreases, gross denitrification rates and the ratio of N_2/N_2O emissions from denitrification increase.

2 The DeNitrification-DeComposition Model

The DNDC model simulates plant growth and soil processes and was originally developed for agroecosystems, including rice paddies (Li et al., 1992). The model has been extended and can simulate forest and other ecosystems. DNDC has four major submodels: soil climate, plant growth, denitrification, and decomposition, and includes discrete organic matter pools: residue litter, living microbes, humads (i.e., active humus), and passive humus (Li et al., 1992). Each of the pools has two or three subpools with specific C:N ratios and decomposition rates.

The decomposition submodule tracks turnover of litter (leaves, stem, roots) and other organic matter in the soil and provides ammonium and DOC for the nitrification and/or denitrification submodules. The nitrification submodule predicts growth and death of nitrifiers, the nitrification rate, as well as N_2O and NO productions from nitrification regulated by soil temperature, moisture, ammonium, and DOC concentrations. The denitrification submodule simulates denitrification and changes in population size of denitrifiers as a function of soil temperature, moisture, and substrates (DOC, NO_3^-, NO_2^-, NO, and N_2O) concentrations. The denitrification-induced N_2O and NO fluxes are calculated based on the dynamics of soil aeration status, substrate limitation, and gas diffusion. Chemo-denitrification is considered as a source of NO-production in soils. This process is controlled by the availability of nitrite and the soil pH. Because nitrification and denitrification can simultaneously occur in aerobic and anaerobic microsites, a kinetic scheme for anaerobic volumetric fraction (or so-called anaerobic balloon) is used to calculate the anaerobic fraction in a given soil layer depending on O_2-diffusion and the respiratory activity of soil microorganisms and roots (Li et al., 2000). Nitrification occurs in the aerobic fraction of the balloon, whereas denitrification occurs in the anaerobic fraction. The size of the anaerobic balloon is defined by the simulated, oxygen partial pressure, which is calculated based on oxygen diffusion and consumption rates in the soil. As the balloon swells or shrinks, the model dynamically allocates substrates (dissolved organic C, ammonium, nitrate, etc.) into the aerobic and anaerobic fractions. Denitrification rate is driven by Nernst and Michaelis-Menten equations. When the anaerobic balloon swells, several processes will take place, including (1) more substrates (DOC, NO_3^-, NO_2^-, NO, or N_2O) will be allocated within the balloon, (2) rate of the reductive reactions (sequential denitrification reactions) will increase within the constraints imposed by Michaelis-Menten, mediated-microbial growth, and (3) the intermediate product gases (N_2O, NO, etc.) will take longer to diffuse from the anaerobic to the aerobic fraction, increasing the rate at which N gases are further reduced to N_2.

3 Ecosys

Ecosys (Grant et al., 1993; Grant, 2001) is a complete, plant soil system model that includes options to represent the full range of management practices in model simulations, including tillage (defined by dates, depths, degrees of soil mixing), fertilization (dates, application methods and depths, types and amounts of: N, P lime, gypsum, residue, manure), irrigation (dates, times, amounts, chemical composition), planting (species, dates, densities), and harvesting (dates, types). Unlike DayCent and DNDC, all flux equations are solved in three dimensions for each cell of a matrix defined by row, column, and layer position. Soil properties may vary in three dimensions throughout the matrix. In addition to the increased spatial complexity of Ecosys, all solutes (mineral and gaseous) undergo convective-dispersive transport through the simulated soil profile as determined by water transport and soil properties. Algorithms for solution pH, ion speciation and exchange, and precipitation-dissolution reactions (Al, Fe, Ca, Mg, Na, K, Cl, S, P, H, and OH) have been incorporated into Ecosys and coupled through pH and osmotic potential to the biological activity of roots and microbes.

Ecosys simulates activity of several soil microbial populations as a parallel set of substrate-microbe complexes that include the rhizosphere, plant and animal residues, and native organic matter. The activity of each microbial population in the model is driven by the energetics of the oxidation-reduction reactions that it conducts. These energetics depend on the availability of electron acceptors with differing energy yields (O_2, NO_3^-, NO_2^-, N_2O, or reduced C), so that the effect of soil gas exchange on nutrient mineralization can be simulated. Nutrient uptake by plants is driven by soil nutrient concentrations and root O_2 uptake (Grant and Robertson, 1997) and is fully coupled to CO_2 fixation and plant growth. An autotrophic nitrifier population is represented. Simulated microbial activity is coupled to the transport and sequential reduction of O_2, NO_3^-, NO_2^-, and N_2O during C oxidation. By coupling microbial activity to the exchange and transfer of C, O, N, and P in aqueous and gaseous phases, ecosys can reproduce the effects of soil alterations, such as compaction, erosion, or tillage, on the exchange of gaseous C and N with the atmosphere during microbial activity under natural or disturbed soil conditions.

IV MODEL CLASSIFICATION AND COMPARISON

Paustian (1994) classified multicompartmental models of SOM dynamics as either "process-oriented" or "organism-oriented." Organism-oriented models, which are sometimes called "food web models," describe the flow of organic matter and nutrients through different functional or taxonomic groups of soil organisms (Moore and de Ruiter, 2012). Process-oriented models are those that focus on the processes mediating the transformations of organic matter and nutrients, rather than on the activity of specific organisms or groups of organisms. In process-oriented model types, soil organisms, if present, tend to be represented as a generic biomass or as part of a pool of active SOM. This approach precludes the possibility of modeling changes in organic matter dynamics or nutrient cycling that might occur due to changes in the activity or composition of the soil organism community. Schimel (2001) points out that the microbiological underpinnings in process-oriented models are not absent, but are implicit in the equation structure of the model as kinetic constants and response functions.

Several reviews are available that compare many of these models (Frolking et al., 1998; Paustian, 1994; Smith et al., 1997). A review of SOM models was performed by Manzoni and Porporato (2009), who compared and classified ~250 models. They found that most models could be classified based on kinetic and stoichiometric laws, that complexity and number of nonlinearities tended to increase with time, and that complexity is inversely related to spatial and temporal resolution of model operation. They classified models according to the following: (1) the extent to which microbial and enzymatic processes are explicitly represented, (2) the number of state variables used to represent microbial biomass, (3) spatial and temporal resolution of model operation and application, (4) number of partial versus ordinary differential equations, (5) decomposition reaction kinetics, (6) the extent to which growth versus maintenance respiration are represented, and (7) how N mineralization and limitation are represented (see Table A-2 in Manzoni and Porporato, 2009). They concluded that although model complexity has tended to increase with time, most models continue to use first-order kinetics (i.e., decomposition is donor controlled and is not necessary to represent microbial and enzymatic activity explicitly) and discrete SOM pools.

Many models represent vertical spatial variability in the soil profile, yet horizontal variability is almost universally neglected. Even so, some models represent erosion, transport, and deposition of water and nutrients. Very few models describe spatial dynamics of water, nutrients, and organic matter at the microbial scale or within aggregates, and those that do have a large number of equations and compartments and operate at fine temporal resolution.

A large set of models of SOM and nutrient dynamics in the soil-plant system, both standalone models and submodels, have been compared by running simulations on the same data set (de Willigen, 1991; Smith et al., 1997). The conclusions of the N model comparisons were that the models developed in the 1990s adequately predicted aboveground processes, such as plant uptake of N and dry matter production, but the simulation of belowground microbial N transformations was problematic. Wu and McGechan (1998) provided a more detailed examination of four N dynamics models (SOILN, ANIMO, DAISY, and SUNDIAL), where they focused on the equations representing the constituent processes. SOILN was found to have the most detailed treatment of plant uptake. ANIMO had the most complex treatment of animal slurry and also the most mechanistic representation of denitrification. These older models demonstrated individual strengths and weaknesses, but in general, they are more similar than they are different.

V   MODEL PARAMETERIZATION

Although multicompartmental models generally work well at predicting measured changes in total SOM, it is difficult to initialize the different pools using independent measures; therefore, site-specific initialization simulations calibrated to match expected total organic matter levels must be used. Attempts have been made to link conceptual pools of SOM with measurable fractions, leading Elliott et al. (1996) to use the phrase "modeling the measurable or measuring the modelable." Such attempts include development of chemical or physical fractionation procedures to match measurable fractions to model pools or to construct new models that coincide with measurable fractions (see Fig. S17.1 in online Supplemental Material at http://booksite.elsevier. com/978012459556). All the measurable factions contain organic matter with a diversity of turnover times, whereas the models assume pools with uniform decay rates. The biggest problem is to define "slow" or "passive" pools solely in chemical or physical terms (Smith et al., 2002).

Studies of substrate degradation in soil require both short time intervals between early measurements and extending the measurements over time intervals long enough to capture the full range of turnover times. Several tools are available for determining the pool sizes and dynamics of SOM fractions.

The soluble components, and even some of the cellulose of plant residues decompose within hours to days. If measurements are delayed, the degradation of the microbial products, rather than that of the original substrate, is measured. Incubations in the laboratory or in the field, extending to hundreds of days or longer, are needed to evaluate the more persistent components. Generally, incubations that incorporate respiration measurements and curve-fitting analysis are sufficient for determining the short-term dynamics. SOM components also occur in intermediate pools with turnover times of 10-100 years. Where applicable, the turnover rates of these constituents can be measured by using the stable isotope ^{13}C. This isotope accounts for 1.1% of the CO_2-C of the air. Plants with the C_3 photosynthetic pathway discriminate more against this isotope than do plants with the C_4 pathway, resulting in differential ^{13}C enrichment of plant materials. Residue inputs and SOM turnover can be measured using mass spectrometry in soils formed under C_3 plants (cool season grasses, trees) if C_4 plants, such as maize or sorghum, are then grown, or if residues from these plants are added. The reverse sequence of crops will also make measurements possible. For example, Century model simulations of the decrease in soil ^{13}C resulting from the conversion of tropical rain forest to pasture agreed well with observations from several sites in Brazil that were converted at different times and treated as a chronosequence (Fig. 17.10c). The natural abundance of ^{13}C in the plants and soil may provide insufficient differentiation. In this case, artificially labeling plant materials with ^{13}C by growing them in closed chambers with enriched $^{13}CO_2$ or $^{14}CO_2$ may provide a better tracer. Changes in ^{14}C of SOM and respired CO_2 due to elevated atmospheric ^{14}C from atomic bomb testing during the 1950s and 1960s have also been measured and used for model testing. For example, the ForCent (Forest Century) model accurately represented the observed decreases in ^{14}C observed in soil C at ORNL (Fig. 17.10a) and in respired CO_2 at Harvard Forest (Fig. 17.10b). Studies examining the transformations of N, such as plant uptake or mineralization, can be done using the ^{15}N stable isotope tracer.

The slow decomposition rates of the most resistant fractions, making up 50% of the soil C and persisting for hundreds to thousands of years, are not easily measured with normal tracer techniques. For this, we resort to C dating, which utilizes the much longer half-life of naturally occurring ^{14}C. Such studies have found that the average age (mean residence time) of organic matter in the surface of temperate agricultural soils ranged between modern and 1100 years, with an average of 560 years. Deeper in the soil profile (50-100 cm), the average age was 2757 years and ranged from 1500 to 6600 years. It is important to note that studies of nutrient and organic matter turnover are best conducted on well-characterized, long-term plots, where the yield or primary productivity components, soil type, and long-term climate and management controls are known. One such study of different currently and formerly cultivated soils at the KBS LTER site in Michigan used mass balance and incubations to parameterize a three-pool model (Paul et al., 1999). They defined the resistant fraction

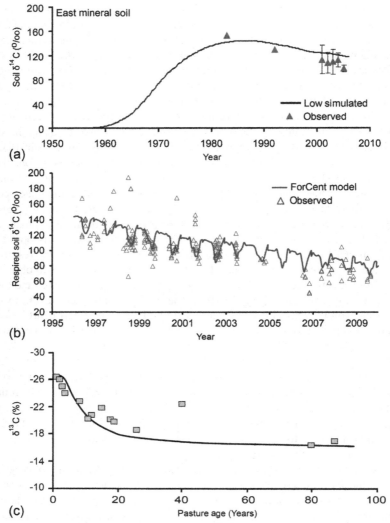

FIG. 17.10   Simulated (ForCent) and observed (a) δ^{14}C in the top (0-30 cm, except 1973 which is 0-15 cm for observed) soil at the low east deciduous forest, Oak Ridge National Laboratory. (*Redrawn from Parton et al., 2010.*) (b) Respired δ^{14}C at Harvard Forest in Massachusetts. (*Redrawn from Savage et al., 2013.*) and (c) δ^{13}C in the top (0-20 cm) soil layer from seven pasture sites in Rondônia, Brazil, treated as a chronosequence (*Redrawn from Cerri et al., 2004.*)

as soil C not solubilized by acid hydrolysis and used ^{14}C dating to determine its turnover rate. Nonlinear regressions of the time series of CO_2 efflux from incubations and mass balance were used to derive pool sizes and turnover rates for the active and slow pools. The resulting model adequately represented field observations of CO_2 emissions (Paul et al., 1999).

VI MODEL SELECTION METHODS

As mechanistic soil-crop-water-atmosphere models become increasingly accepted as tools for analyzing agronomic or environmental issues, users are being faced with an increasing number of models to choose from. Ideally, a simulation model would include all of the processes dictating the dynamics of SOM or nutrient element at a level of detail that represents the current state-of-the-art in understanding, and model selection would be a moot point. Unfortunately, an ideal model for complex and heterogeneous agroecosystems is impractical. Not all system processes are fully understood, and thus, each model includes various assumptions, approximations, and simplifications. In addition, the documentation available for each model often does not refer to the validity, limits, and potential applications of the model, which would provide some guidance in the selection of the most appropriate model. So how does one evaluate the output from models, then select the "best" model to use?

Model output can be evaluated by comparing simulated or predicted values with actual measured or observed values from field experiments not used to originally develop the model. At its simplest, model evaluation can be done visually or graphically. This approach provides a rapid and easy means of evaluating whether or not a model is producing results close to those observed. One problem associated with this approach is that the difference between predicted and observed values cannot easily be quantified. Loague and Green (1991) suggested that model evaluation should include both qualitative graphical and quantitative statistical approaches. Several statistical criteria for evaluating simulated results are outlined in Table 17.3. Maximum error (ME) represents the single largest difference between a pair of predicted and observed values. The root mean squared error (RMSE) represents the total difference between the predicted and observed values, proportioned against the mean observed value. Modeling efficiency (EF) assesses the accuracy of simulations by comparing the variance of predicted from observed values to the variance of observed values from the mean of the observations. The EF is essentially a comparison of the efficiency of the chosen model to the efficiency of a very simple predictive model: the mean of the observations. The coefficient of determination (CD) is a measure of the proportion of the total variance in the observed data that is explained by the predicted data. Last, the coefficient of residual mass (CRM) gives an indication of the consistent errors in the distribution of all simulated values across all measurements. Smith et al. (1997) used many of the preceding statistical criteria to evaluate the effectiveness of various SOM models to simulate SOM dynamics at different sites around the world. This paper has discussed the usefulness of these statistical criteria and demonstrates that six of the nine models considered in the model comparison did an equally good job of representing the observed SOM dynamics.

If several models appear to perform equally well, how might one select the best or most appropriate model among them? Although more complex models

TABLE 17.3 Statistical Criteria for the Evaluation of Model Performance

Criterion	Formula[a]	Optimum
Maximum error (ME)	$\mathrm{Max}|P_i - O_i|_{i=1}^{n}$	0
Root mean square error (RMSE)	$\dfrac{100}{\overline{O}}\sqrt{\sum_{i=1}^{n}(P_i - O_i)^2 \big/ n}$	0
Modeling efficiency (EF)	$\left(\sum_{i=1}^{n}(O_i - \overline{O})^2 - \sum_{i=1}^{n}(P_i - O_i)^2\right)\Big/\sum_{i=1}^{n}(O_i - \overline{O})^2$	1
Coefficient of determination (CDz)	$\sum_{i=1}^{n}(O_i - \overline{O})^2 \Big/ \sum_{i=1}^{n}(P_i - \overline{O})^2$	1
Coefficient of residual mass (CRM)	$\left(\sum_{i=1}^{n}O_i - \sum_{i=1}^{n}P_i\right)\Big/\sum_{i=1}^{n}O_i$	0
Akaike's information criterion (AIC)	$2k - n\log(RSS/n)$	Smallest value

$k = $ # of parameters, RSS = residual sum of squares

From Loague and Green (1991).

[a]P_i, predicted value i; O_i, observed value i; $\overline{O}$, mean of observed values; n, number of data pairs.

are often more accurate than simple models (Del Grosso et al., 2008), this is not always the case, and accuracy may not increase beyond a particular degree of complexity. As models become more complex, it is more difficult to acquire needed input data. Personnel and computation resources required for proper model operation also increase. Model output may be parameterized and compared using observational data and statistical methods such as Akaike's information criterion (AIC), deviance information criterion, or Bayes factors (Csillery et al., 2010; Piou et al., 2009). For example, AIC combines the amount of information lost by using a model to approximate the truth (the Kullback–Leibler, K–L, distance) with maximum likelihood estimation. Maximum likelihood estimates the relative K–L distance between competing models, thereby determining which model is closest to the unknown truth (represented by the data; Burnham and Anderson, 2002; Del Grosso et al., 2005). Thus, AIC ranks a set of *a priori* models based on the support for each found in the data, not only choosing the best model, but also providing information on whether other models in the set are close competitors (Burnham and Anderson, 2002). AIC accounts for goodness of fit as well as overfitting the data because it includes a penalty proportional to the number of optimized parameters in the model.

Such model comparison methodologies allow for the evaluation of the large number of models available to represent various ecosystem processes. Del Grosso et al. (2005) used several CO_2 flux data sets to compare a range of models that are commonly used to characterize the influence of temperature on CO_2 flux rates. They found that variable Q_{10} temperature function best explained CO_2 flux rates. Adair et al. (2008) used AIC and a large, cross-site, long-term data set to compare more than 70, one- to three-pool models of litter decomposition and found that a three-pool model that included a variable Q_{10} temperature function performed best.

Ultimately, a model that performs well (i.e., where modeled output matches observations reasonably well) is not necessarily a "good" model. A good model is conceptually clear and can be easily communicated to others. A bad model may perform well, but because it can only be inspected and modified with great difficulty, there is no way to determine if the validation is pure chance or something meaningful. It is important that the user understand the assumptions and limitations of a selected model at each stage of its use and application.

VII CONCLUSION

Our understanding of the processes and mechanisms leading to SOM and nutrient dynamics in soil continues to improve (Dungait et al. 2012; Schmidt et al. 2011). These papers suggest that the persistence of SOM in the soil is dominated by environmental and biological factors (e.g., the accessibility of the SOM to microbes), not exclusively the chemical recalcitrance of substrate. Although climate (Adair et al., 2008; Follett et al., 2013), soil properties (Jin et al., 2013), and biological factors (Hooper et al., 2012) are important, there is also evidence that litter chemistry impacts decomposition rates (Hobbie, 2005; Newey, 2005). Model formulation and structure must continue to evolve to reflect the most current understanding. Although model output has been shown to effectively simulate measured SOM dynamics, Schmidt et al. (2011) and Dungait et al. (2012) pointed out that we have made little progress during the last 30 years on "reconciling biochemical properties with the kinetically defined pools," and Wutzler and Reichstein (2012) suggest that the "priming effect" (enhanced SOM decay resulting from the addition of fresh SOM) resulting from enhanced root activity or addition of labile plant material needs to be added to SOM models. The current demands being made on models of SOM dynamics include the following improvements:

1. Represent model decay rates and C utilization efficiency as functions of substrate properties, microbial activity, and soil conditions, including temperature, moisture, and pH, rather than fixed values.
2. Include fire-derived C inputs (black C) in SOM models because they are major components of many SOM pools with long mean residence times.

3. Improve model representation of SOM stabilization mechanisms other than biochemical recalcitrance, such as soil aggregate occlusion and mineral association.
4. Improve incorporation of microbial communities and dynamics in SOM models and their impact on stabilization of SOM.
5. Dynamically model the depth distribution of SOM and DOC formation and movement in the soil profile and its role in carbon stabilization
6. Add the potential impact of "soil priming" on the decay rates of SOM pools.

Some of these suggested improvements have been incorporated into the latest versions of ecosystem models such as DayCent (Savage et al. 2013), including (1) representing the major sources of passive C as plant-derived charcoal and adsorption of microbial derived products on clay minerals; (2) representing different microbial growth efficiencies for structural and labile SOM pools; and (3) potential impact of "soil priming" on decay rate of soil SOM pools based on data from Cheng et al. (2014). Several fundamental changes to model structures appear to be imminent in response to the desire to improve the ability of models to predict the responses of SOM to disturbances such as climate change as submodels of SOM dynamics are increasingly incorporated into ESMs.

REFERENCES

Abdalla, M., Jones, M., Yeluripati, J., Smith, P., Burke, J., Williams, M., 2010. Testing DayCent and DNDC model simulations of N_2O fluxes and assessing the impacts of climate change on the gas flux and biomass production from a humid pasture. Atmos. Environ. 44 (25), 2961–2970.

Adair, E.C., Parton, W.J., Del Grosso, S.J., Silver, W.L., Harmon, M.E., Hall, S.A., Burke, I.C., Hart, S.C., 2008. Simple three-pool model accurately describes patterns of long-term litter decomposition in diverse climates. Glob. Chang. Biol. 14, 2636–2660.

Ågren, G.I., Bosatta, E., 1998. Theoretical Ecosystem Ecology. Understanding Element Cycles. Cambridge University Press, Cambridge.

Allison, S.D., Wallenstein, M.D., Bradford, M.A., 2010. Soil-carbon response to warming dependent on microbial physiology. Nat. Geosci. 3, 336–340.

Bonan, G.B., Hartman, M.D., Parton, W.J., Wieder, W.R., 2013. Evaluating litter decomposition in earth system models with long-term litterbag experiments: an example using the Community Land Model version 4 (CLM4). Glob. Chang. Biol. 19, 957–974.

Bosatta, E., Ågren, G.I., 1985. Theoretical analysis of decomposition of heterogeneous substrates. Soil Biol. Biochem. 17, 601–610.

Bosatta, E., Ågren, G.I., 1995. The power and reactive continuum models as particular cases of the Q-theory of organic matter dynamics. Geochim. Cosmochim. Acta 59, 3833–3835.

Burnham, K.P., Anderson, D.R., 2002. Model Selection and Multi-Model Inference: A Practical Information-Theoretic Approach. Springer, Berlin, Germany.

Cerri, C.E.P., Paustian, K., Bernoux, M., Victoria, R.L., Melillo, J.M., Cerri, C.C., 2004. Modeling changes in soil organic matter in Amazon forest to pasture conversion with the Century model. Glob. Change Biol. 10 (5), 815–832.

Cheng, K., Ogle, S.M., Parton, W.J., Pan, G., 2013. Predicting methanogenesis from rice paddies using the DAYCENT ecosystem model. Ecol. Model. 261, 19–31.

Cheng, W., Parton, W.J., Gonzalez-Meler, M.A., Phillips, R., Asao, S., McNickle, G.G., Brzostek, E., Jastrow, J.D., 2014. Synthesis and modeling perspectives of rhizosphere priming. New Phytol. 201, 31–44. http://dx.doi.org/10.1111/nph.12440.

Csillery, K., Blum, M.G., Gaggiotti, O.E., Francois, O., 2010. Approximate Bayesian computation (ABC) in practice. Trends Ecol. Evol. 25, 410–418.

de Willigen, P., 1991. Nitrogen turnover in the soil-crop system; comparison of fourteen simulation models. Fert. Res. 27, 141–149.

Del Grosso, S.J., Parton, W.J., Mosier, A.R., Ojima, D.S., Potter, C.S., Borken, W., Brumme, R., Butterbach-Bahl, K., Crill, P.M., Dobbie, K., Smith, K.A., 2000. General CH_4 oxidation model and comparisons of CH_4 oxidation in natural and managed systems. Glob. Biogeochem. Cycles 14, 999–1019.

Del Grosso, S.J., Parton, W.J., Mosier, A.R., Holland, E.A., Pendall, E., Schimel, D.S., Ojima, D.S., 2005. Modeling soil CO_2 emissions from ecosystems. Biogeochemistry 73, 71–91.

Del Grosso, S.J., Wirth, T., Ogle, S.M., Parton, W.J., 2008. Estimating agricultural nitrous oxide emissions. EOS Trans. Am. Geophys. Union 89, 529–530.

Dungait, J.A.J., Hopkins, D.W., Gregory, A.S., Whitmore, A.P., 2012. Soil organic matter turnover is governed by accessibility and not recalcitrance. Glob. Chang. Biol. 18, 1781–1796.

Elliott, E.T., Paustian, K., Frey, S.D., 1996. Modeling the measurable or measuring the modelable: a hierarchical approach to isolating meaningful soil organic matter fractionations. In: Powlson, D.S. et al., (Ed.), Evaluation of Soil Organic Matter Models, vol. 138. Springer, Berlin, pp. 161–180.

Follett, R.F., Jantalia, C.P., Halvorson, A.D., 2013. Soil carbon dynamics for irrigated corn under two tillage systems. Soil Sci. Soc. Am. J. 77 (3), 951–963.

Frissel, M.J., Vanveen, J.A., 1982. A review of models for investigating the behavior of nitrogen in soil. Philos. Trans. R. Soc. B 296, 341–349.

Frolking, S.E., Mosier, A.R., Ojima, D.S., Li, C., Parton, W.J., Potter, C.S., Priesack, E., Stenger, R., Haberbosch, C., Dörsch, P., Flessa, H., Smith, K.A., 1998. Comparison of N_2O emissions from soils at three temperate agricultural sites: simulations of year-round measurements by four models. Nutr. Cycl. Agroecosyst. 52 (2–3), 77–105.

Grant, R.F., 2001. A review of the Canadian ecosystem model ecosys. In: Shaffer, M.J. (Ed.), Modeling Carbon and Nitrogen Dynamics for Soil Management. CRC Press, Boca Raton, pp. 173–264.

Grant, R.F., Robertson, J.A., 1997. Phosphorus uptake by root systems: mathematical modelling in ecosystems. Plant Soil 188, 279–297.

Grant, R.F., Juma, N.G., McGill, W.B., 1993. Simulation of carbon and nitrogen transformations in soil—mineralization. Soil Biol. Biochem. 25, 1317–1329.

Hobbie, S.E., 2005. Contrasting effects of substrate and fertilizer nitrogen on the early stages of litter decomposition. Ecosystems 8, 644–656.

Hooper, D.U., Adair, E.C., Cardinale, B.J., Byrnes, J.E.K., Hungate, B.A., Matulich, K.L., Gonzalez, A., Duffy, J.E., Gamfeldt, L., O'Connor, M.I., 2012. A global synthesis reveals biodiversity loss as a major driver of ecosystem change. Nature 486, 105–108.

Jenkinson, D.S., 1990. The turnover of organic carbon and nitrogen in soil. Philos. Trans. R. Soc. B 329, 361–368.

Jin, V.L., Haney, R.L., Fay, P.A., Polley, H.W., 2013. Soil type and moisture regime control microbial C and N mineralization in grassland soils more than atmospheric CO_2-induced changes in litter quality. Soil Biol. Biochem. 58, 172–180.

Jones, C.A., Cole, C.V., Sharpley, A.N., Williams, J.R., 1984. A simplified soil and plant phosphorus model. Soil Sci. Soc. Am. J. 48, 800–805.

Levis, S., Bonan, G.B., Kluzek, E., Thornton, P.E., Jones, A., Sacks, W.J., Kucharik, C.J., 2012. Interactive crop management in the Community Earth System Model (CESM1): seasonal influences on land-atmosphere fluxes. J. Clim. 25 (14), 4839–4859.

Lewis, D.R., McGechan, M.B., 2002. A review of field scale phosphorus dynamics models. Bioprocess Biosyst. Eng. 82, 359–380.

Li, C.S., Frolking, S., Frolking, T.A., 1992. A model of nitrous-oxide evolution from soil driven by rainfall events: 1. Model structure and sensitivity. J. Geophys. Res.-Atmos. 97, 9759–9776.

Li, C., Aber, J., Stange, F., Butterbach-Bahl, K., Papen, H., 2000. A process-oriented model of N_2O and NO emissions from forest soils. J. Geophys. Res. 105 (D4), 4369–4384.

Loague, K., Green, R.E., 1991. Statistical and graphical methods for evaluating solute transport models: overview and application. J. Contam. Hydrol. 7, 51–73.

Manzoni, S., Porporato, A., 2009. Soil carbon and nitrogen mineralization: theory and models across scales. Soil Biol. Biochem. 41, 1355–1379.

Manzoni, S., Piñeiro, G., Jackson, R.B., Jobbágy, E.G., Kim, J.H., Porporato, A., 2012. Analytical models of soil and litter decomposition: solutions for mass loss and time-dependent decay rates. Soil Biol. Biochem. 50, 66–76. http://dx.doi.org/10.1016/j.soilbio.2012.02.029.

Moore, J.C., de Ruiter, P.C., 2012. Energetic Food Webs: An Analysis of Real and Model Ecosystems. Oxford University Press, Oxford, UK.

Newey, A., 2005. Decomposition of Plant Litter and Carbon Turnover as a Function of Soil Depth. Australian National University.

Parton, W.J., Schimel, D.S., Cole, C.V., Ojima, D.S., 1987. Analysis of factors controlling soil organic matter levels in Great-Plains grasslands. Soil Sci. Soc. Am. J. 51, 1173–1179.

Parton, W.J., Woomer, P.L., Martin, A., 1994. Modelling soil organic matter dynamics and plant productivity in tropical ecosystems. In: Woomer, P.L., Swift, M.J. (Eds.), The Biological Management of Tropical Soil Fertility. Wiley-Sayce, Chichester, pp. 121–188.

Parton, W.J., Holland, E.A., Del Grosso, S.J., Hartman, M.D., Martin, R.E., Mosier, A.R., Ojima, D.S., Schimel, D.S., 2001. Generalized model for NO_x and N_2O emissions from soils. J. Geophys. Res. 106 (D15), 17403–17420.

Parton, W.J., Silver, W.L., Burke, I.C., Grassens, L., Harmon, M.E., Currie, W.S., King, J.Y., Adair, E.C., Brandt, L.A., Hart, S.C., Fasth, B., 2007. Global-scale similarities in nitrogen release patterns during long-term decomposition. Science 315, 361–364.

Parton, W.J., Hanson, P.J., Swanston, C., Torn, M., Trumbore, S.E., Riley, W., Kelly, R., 2010. For-Cent model development and testing using the enriched background isotope study experiment. J. Geophys. Res. Biogeosci. G04001. http://dx.doi.org/10.1029/2009JG001193.

Paul, E.A., Harris, D., Collins, H.P., Schulthess, U., Robertson, G.P., 1999. Evolution of CO_2 and soil carbon dynamics in biologically managed, row crop agroecosystems. Appl. Soil Ecol. 11, 53–65.

Paul, K.I., Polglase, P.J., Richards, G.P., 2003. Predicted change in soil carbon following afforestation or reforestation, and analysis of controlling factors by linking a C accounting model (CAMFor) to models of forest growth (3PG), litter decomposition (GENDEC) and soil C turnover (RothC). For. Ecol. Manag. 177, 485–501.

Paustian, K., 1994. Modelling soil biology and biochemical processes for sustainable agriculture research. In: Pankhurst, C.E., Doube, D.M., Gupta, V.V.S.R., Grace, P.R. (Eds.), Soil Biota: Management in Sustainable Farming Systems. CSIRO Publishing, Collingwood, pp. 182–193.

Paustian, K., Parton, W.J., Persson, J., 1992. Modeling soil organic matter in organic-amended and nitrogen-fertilized long-term plots. Soil Sci. Soc. Am. J. 56, 476–488.

Piou, C., Berger, U., Grimm, V., 2009. Proposing an information criterion for individual-based models developed in a pattern-oriented modelling framework. Ecol. Model. 220, 1957–1967.

Savage, K.E., Parton, W.J., Davidson, E.A., Trumbore, S.E., Frey, S.D., 2013. Long-term changes in forest carbon under temperature and nitrogen amendments in a temperate northern hardwood forest. Glob. Chang. Biol. 19, 2389–2400.

Schimel, J., 2001. Biogeochemical models: implicit versus explicit microbiology. In: Schulze, E.-D., Harrison, S.P., Heimann, M. et al., (Eds.), Global Biogeochemical Cycles in the Climate System. Academic Press, San Diego, pp. 177–183.

Schmidt, M.W.I., Torn, M.S., Abiven, S., Dittmar, T., Guggenberger, G., Janssens, I.A., Kleber, M., Kögel-Knabner, I., Lehmann, J., Manning, D.A.C., Nannipieri, P., Rasse, D.P., Weiner, S., Trumbore, S.E., 2011. Persistence of soil organic matter as an ecosystem property. Nature 478, 49–56.

Sinsabaugh, R.L., Follstad Shah, J.J., 2012. Ecoenzymatic stoichiometry and ecological theory. Annu. Rev. Ecol. Evol. Syst. 43, 313–343.

Sleutel, S., De Neve, S., Prat Roibas, M.R., Hofman, G., 2005. The influence of model type and incubation time on the estimation of stable organic carbon in organic materials. Eur. J. Soil Sci. 56, 505–514.

Smith, P., Smith, J.U., Powlson, D.S., McGill, W.B., Arah, J.R.M., Chertov, O.G., Coleman, K., Franko, U., Frolking, S., Jenkinson, D.S., Jensen, L.S., Kelly, R.H., Klein-Gunnewiek, H., Komarov, A.S., Li, C., Molina, J.A.E., Mueller, T., Parton, W.J., Thornley, J.H.M., Whitmore, A.P., 1997. A comparison of the performance of nine soil organic matter models using datasets from seven long-term experiments. Geoderma 81, 153–225.

Smith, J.U., Smith, P., Monaghan, R., MacDonald, J., 2002. When is a measured soil organic matter fraction equivalent to a model pool? Eur. J. Soil Sci. 53, 405–416.

Stanford, G., Smith, S.J., 1972. Nitrogen mineralization potentials of soils. Soil Sci. Soc. Am. Proc. 36, 465–472.

Van Oijen, M., Cameron, D.R., Butterbach-Bahl, K., Farahbakhshazad, N., Jansson, P.-E., Kiese, R., Rahn, K.-H., Werner, C., Yeluripati, J.B., 2011. A Bayesian framework for model calibration, comparison and analysis: application to four models for the biogeochemistry of a Norway spruce forest. Agric. For. Meteorol. 151, 1609–1621.

Vitousek, P.M., 2004. Nutrient Cycling and Limitation: Hawai'i as a Model System. Princeton University Press, New Jersey, USA.

Wieder, W.R., Bonan, G.B., Allison, S.D., 2013. Global soil carbon projections are improved by modelling microbial processes. Nat. Clim. Chang. 3 (10), 909–912.

Wu, L., McGechan, M.B., 1998. A review of carbon and nitrogen processes in four soil nitrogen dynamics models. J. Agric. Eng. Res. 69, 279–305.

Wutzler, T., Reichstein, M., 2012. Priming and substrate quality interactions in soil organic matter models. Biogeosci. Discuss. 9, 17176–17201.

Yang, H.S., Janssen, B.H., 2000. A mono-component model of carbon mineralization with a dynamic rate constant. Eur. J. Soil Sci. 51, 517–529.

SUGGESTED READING

Parton, W.J., Neff, J., Vitousek, P.M., 2005. 15 modelling phosphorus, carbon and nitrogen dynamics in terrestrial ecosystems. In: Turner, B.L., Frosserd, E., Baldwin, D.S. (Eds.), Organic Phosphorus in the Environment. CABI Publishing, Oxford, pp. 325–327.

Chapter 18

Management of Soil Biota and Their Processes

Jeffrey L. Smith[1], Harold P. Collins[2], Alex R. Crump[3] and Vanessa L. Bailey[4]

[1]USDA-Agriculture Research Service, Pullman, WA, USA (deceased)
[2]USDA-Agriculture Research Service, Temple, TX, USA
[3]Department of Crop and Soil Sciences, Washington State University, Pullman, WA, USA
[4]Pacific Northwest Laboratory, Richland, WA, USA

Chapter Contents

I INTRODUCTION

Managing microorganisms can mean combating pathogenic or infectious organisms or promoting beneficial organisms or their products. Historically, humans have managed organisms inadvertently, consciously, and sometimes by simple intuition. One of man's first introductions to a microbial process occurred over 100,000 years ago when fruit spoiled and formed wine (Purser, 1977). It was not until the mid-1800s that Louis Pasteur discovered

Soil Microbiology, Ecology, and Biochemistry. http://dx.doi.org/10.1016/B978-0-12-415955-6.00018-9
2015 Published by Elsevier Inc.

the process of fermentation, shedding light on the thousands of years of humans managing organisms (Debré and Forster, 1998). Archaeologists have evidence of cheese consumption dating back to 6000 years BCE and Egyptian tomb murals to 2000 BCE (Smith, 1995). Management of organisms for other human food-stuffs includes yeast (*Saccharomyces cerevisiae*) in bread making and the cultivation of mushrooms. Ancient Chinese and Arab cultures developed a procedure of scraping sores of people infected with mild cases of smallpox and infecting healthy people to ward off a more serious case of the disease. This practice was introduced in Western Europe in the 1800s when Edward Jenner inoculated people with cowpox germs as a prevention against smallpox (Wilson, 1976).

"Chance only favors the prepared mind" describes another manipulation of organisms that led to the discovery and production of penicillin. The accidental contamination of a culture plate of *Staphylococci* by the mold *Penicillium chrysogenum* (formally *Penicillium notatum*) led to the discovery and naming of penicillin by Sir Alexander Fleming in 1928 (Debré and Forster, 1998; Wilson, 1976). This discovery led to the production of penicillin as an antibacterial drug 12 years later by E. Chain and H. W. Florey. This touched off a worldwide search for molds with antibiotic characteristics with the discovery of streptomycin by Dr. Selman Waksman, a soil microbiologist.

Farming, especially in the uplands of Greece, intensified around 800 BCE, giving rise to erosion and decreased fertility (Encyclopedia Britannica, http://www.search.eb.com). The Romans recognized that to maintain fertility, the land needed to be fallowed at some time, the crops rotated, and lime and manure added (Hillel, 1992). They found that growing alfalfa and clover added fertility, as did using green manures such as lupines, but they didn't know why. Parallel with agriculture development was the practice of composting where a soil pit was maintained with human and animal waste, weeds, leaves, and household waste and was watered regularly. The decomposed material was used as fertilizer and mulch that improved the physical, chemical, and biological characteristics of the soil. The management of microorganisms extends to plant pathology. For centuries it was known that there was an association between barberry and stem rust of grain (Walker, 1950). In the 1600s, farmers sought to adopt a barberry-eradication measure without knowing the cause-and-effect relationship.

Many of the agricultural, forestry, and rangeland practices we use today actually cause deleterious effects on microorganisms and their processes. Plowing, clear-cutting forests, and overgrazing rangeland decrease organism populations and promote nutrient loss with an overall decrease of soil quality. Humans, as caretakers of the land, must reverse land degradation to increase soil quality and ecosystem health to provide food and fiber for a growing world population. We can begin this long journey by developing ways in which to manage organisms and their beneficial processes in soil systems.

II CHANGING SOIL ORGANISM POPULATIONS AND PROCESSES

Soil organic matter (SOM), a direct result of microbial activity, plays a role in terrestrial ecosystem development and functioning. The dominant effect that SOM has on ecosystem structure and stability is clear evidence for the need to protect current organic levels and to develop management practices that will enhance soils with declining SOM contents (see Chapter 12 for more detail on organic matter). Organic matter content in soils ranges from less than 0.2% in desert soils to over 80% in peat soils. In temperate regions, SOM ranges between 0.4% and 10.0%, with soils of humid region averaging 3-4% and those in semiarid areas 1-3%. Although it is only a small fraction of the soil, components of SOM are the chief binding agents for soil aggregates that, in turn, control air and water relations for root growth and provide resistance to wind and water erosion. With 95% of soil nitrogen (N), 40% of soil phosphorus (P), and 90% of soil sulfur (S) being associated with the SOM, decomposition and turnover can supply macronutrients needed for plant growth (Smith and Elliot, 1990; Smith et al., 1992). The decomposition, process is controlled by temperature, moisture, soil disturbance, and the quality of SOM as a microbial substrate (Smith, 1994; Smith and Paul, 1990).

The two most significant ecosystem disturbances that directly decrease SOM are tropical and boreal deforestation and the intensive cropping of the world's farmland and forests. The decrease in SOM is paralleled by declines in soil productivity and contributes to increasing global CO_2 concentrations. Recent years have witnessed increased concern about environmental pollution from synthetic chemicals and disposal of wastes. Problems with toxic waste dumps, garbage landfills, and acid rainfall have focused attention on soil, the most widely used depository. Controlling factors and processes may be drastically altered under these circumstances, causing chemical cycling, ecosystem stability, and system resiliency to shift in unpredictable ways.

A Tillage and Erosion

It has long been hypothesized that unsustainable agricultural practices were at least partially responsible for the "collapse" of the Classic Maya civilization during the period 1000 BCE to 90 BCE (Morgan, 2012). Using isotopic methods, Burnett et al. (2012) concluded that there was evidence of deposition from erosional processes from upland to lowland areas of agricultural activity. Tillage affects the amount of SOM buildup or loss in two ways: (1) through the physical disturbance and mixing of soil and the exposure of soil aggregates to disruptive forces, and (2) through incorporation and distribution of plant residues in the soil profile. Decomposition rates of residues are generally slower when left on the soil surface than when buried in soil. The combination of reduced litter decomposition rates and less soil disturbance usually results in greater amounts of SOM in reduced tillage versus conventionally tilled systems (Dalal et al., 1991). Tillage also affects water content, aeration, and the microclimate near the soil surface (Lal et al., 2007). These in turn regulate the soil biota and the biological processes they mediate. Intensive tillage simplifies microbial community structure resulting in lower stability or resiliency in function. Reductions in stability have led to gaps in biochemical functions related to N-transformations during periods of environmental stress or when plants are not present. Reduced tillage and cover crops generate soil-litter conditions; modify microbial biomass, pathogen, nematode, and insect community structure; and regulate microbial mineralization/immobilization rates controlling the loss of C and N from soil (Calderon et al., 2000; Curci et al., 1997).

Between 35% and 50% of the SOM and N were lost during the first 50 years of tillage in the Great Plains of the United States (Bauer and Black, 1981). This loss is most rapid during the first few years of cultivation, and eventually an apparent equilibrium is established, provided constant management practices are employed. The time required to reach this equilibrium will vary depending on climatic conditions, extent of erosion, type and rate of residue return, and soil type. In Canadian soils, over a 60- to 80-year period, C and N losses ranged from 50% to 60% and 40% to 60%, respectively (Campbell et al., 1976; Voroney et al., 1981). A 1986 study (Dalal and Mayer, 1986) found that soils in southern Queensland, Australia, had lost 36% of their C and N over a 20- to 70-year period and that numerous fertility parameters related to SOM levels were declining.

The practice of no tillage was developed to counteract the destructive effects of tillage practices. The data in Supplemental Table 18.1 (Doran, 1980; Staben et al., 1997; see online Supplemental Material at http://booksite.elseiver.com/9780124159556.) are expressed as the ratio of the no-till data to the conventionally tilled or CRP data. Total soil C and N increased in no-till soils, but returning cropland to grassland had little effect in this study. The microbial biomass only increased in the no-till system compared to the conventional system. Returning fields to a less disturbed state significantly increased the numbers of fungi and bacteria as well as the dehydrogenase enzyme activity (a measure of biological activity). Nutrient cycling measured as potentially mineralizable N, increased by 35% in the no-till system, but only by 13% in the converted grass system. The CRP data suggest the effect on organisms is not simply due to increases in soil organic C. In a rice cropping system in China, a compilation of studies showed that soil microbial biomass and activity increased under no tillage (Huang et al., 2013). In reduced tillage systems in Western Europe, D'Haene et al. (2008) found higher stratification of SOC in the soil profile, but not in higher SOC stocks. The higher microbial biomass C and SOC in the top 5 cm of soil resulted in higher C mineralization rates measured in the laboratory.

In a compilation of 106 studies (Wardle, 1995), scientists were able to develop an index of the effects of tillage on detrital food webs and compare

TABLE 18.1 Proposed Soil Physical, Chemical, and Biological Characteristics to be Included as Basic Indicators of Soil Quality

Soil Characteristics	Methodology
Physical	
Soil texture	Hydrometer method
Depth of soil and rooting	Soil coring or excavation
Soil bulk density and infiltration[a]	Field determined using infiltration rings
Water-holding capacity[a]	Field determined after irrigation of rings
Water-retention characteristics	Water content at 33 and 1500 kPa tension
Water content[a]	Gravimetric analysis; wt. loss, 24 h at 105 °C
Soil temperature[a]	Dial thermometer or hand temperature probe
Chemical	
Total organic C and N	Wet or dry combustion, volumetric basis[b]
pH	Field or lab determined, pH meter
Electrical conductivity	Field or lab, pocket conductivity meter
Mineral N (NH_4 and NO_3), P, and K	Field or lab analysis, volumetric basis
Biological	
Microbial biomass C and N	Chloroform fumigation/incubation, volumetric basis
Potentially mineralizable N	Anaerobic incubation, volumetric basis
Soil respiration[a] measured in biomass assay	Field measured using covered infiltration rings, lab
Biomass C/total org. C ratio	Calculated from other measures
Respiration/biomass ratio	Calculated from other measures

Adapted from Doran and Parkin (1996).
[a]*Measurements taken simultaneously in field for varying management conditions, landscape locations, and time of year.*
[b]*Gravimetric results must be adjusted to volumetric basis using field measured soil bulk density for meaningful interpretations.*

conventional tillage to no tillage. In most of the studies, the soil total C and N concentrations were less in the conventionally tilled soil compared to the no-tilled soil as was the soil microbial biomass. However, as in the studies cited earlier, the increase in microbial biomass in the no-till soils was greater than the increase in SOM.

There is a good deal of interest in disturbance effects on trophic levels and on organism diversity within the trophic level. A limited subset of studies tends to show that species diversity of microfaunal groups is unchanged by tillage, whereas the macrofaunal groups can be elevated or reduced in diversity by tillage. This indicates that soil food webs are fairly stable and have a significant amount of resiliency. The bacteria and fungi were generally mildly inhibited by tillage, whereas the microfauna of nematodes and protozoa were mildly to moderately inhibited by tillage. The mesofauna group of collembolan and mites tended to be moderately inhibited, and the macrofauna group of earthworms and beetles were moderately to extremely inhibited by tillage. Thus, the larger organisms were more likely to be reduced than the smaller organisms under tillage systems compared to no-till systems.

Soil compaction has dramatic effects on plant growth, soil biota, and biological processes. It has been shown to affect specific microbial activities (e.g., soil respiration and the denitrification potential through direct effects on soil water content and aeration). Robertson et al. (2000) showed that greenhouse gas emissions were eightfold higher in compacted, conventionally tilled systems than in no-till; however, this observation can be modified by comparing a loam and a clay soil (Rochette et al., 2008). There is increasing evidence for the role of pore size in regulating soil microbial populations and their biochemical processes (Postma and van Veen, 1990). Compaction reduced the predation efficiency of protozoa or nematodes on bacteria due to a decrease in accessible pore space.

Tillage not only reduces SOM through disruption and oxidation, but can also create significant soil erosion. It is estimated that about 20% of the soil C dislocated by erosion will be released into the atmosphere as CO_2, with the rest being deposited over other land areas and into streams and rivers. Because it is mainly surface soil that is lost during erosion events, there is an associated loss of SOM high in biological activity.

Biological activity and biodiversity are important in determining deleterious effects of management on the physical and chemical properties of soil and the resulting effects on the sustainability of agricultural land. More research is needed to identify the relationships between microbial community composition (biodiversity), soil processes, and soil fertility on the health of the soil and its ability to provide a medium for plant growth.

B   Rangeland and Forest Health

Much of the environmental movement of the 1970s and 1980s focused on decreasing the degradation of the environment. In the 1990s, the scientific community turned from monitoring degrading practices to evaluating a holistic attribute—ecosystem health. This began with work on soil quality, much like the air and water quality assessments of the 1980s, and evolved into work such as rangeland and forest ecosystem health. The National Research Council suggested that rangeland health be defined as "the degree to which the integrity of the soil and ecological processes of rangeland ecosystems are sustained." The dictionary definition of healthy is (1) functioning properly or normally in its vital functions; (2) free from malfunction of any kind; and (3) productive of good of any kind. These definitions also apply to forest ecosystems.

At the core of ecosystem health is soil quality, defined as, "The capacity of a soil to function within ecosystem boundaries to sustain biological productivity, maintain environmental quality and promote plant and animal health." The major issues are

1. Productivity: the ability of soil to enhance plant and biological productivity.
2. Environmental quality: the ability of soil to attenuate environmental contaminants, pathogens, and offsite damage.
3. Animal health: the interrelationship between soil quality and plant, animal, and human health.

It is evident that soil biology, specifically microorganisms and their processes, play a major role in determining the health of the ecosystem. The International Standardization Organization has developed standards for the analysis of soils for chemical, biological, and physical properties (Nortcliff, 2002). Attributes that affect the functioning of the soil system are listed in Table 18.1 and, when measured in the context of the proper ecological question, can be used to evaluate ecosystem health. The proper ecological question means, "What is the function of the ecosystem?" or "What is the ecosystem going to be used for?" Thus the soil quality attributes for an urban area may be different than the soil quality attributes for range and pasture land.

Coupling the ecosystem health concept with increasing SOM levels has recently drawn attention due to global warming and greenhouse gas issues. Because the most abundant greenhouse gas is CO_2, the raw material for SOM, increasing SOM will remove CO_2 from the atmosphere. Rangelands and forests have been identified as the ecosystems likely to store significant amounts of additional C in soil and plant material; however, it will be the soil microflora and their processing of plant C that will govern the rate and amount of C that is stored.

The conversion of marginal land and highly erodible land to forest will increase soil C and will also provide an aboveground component of sequestered C to the ecosystem. Increasing intensively managed timberland and canopy cover in urban areas has a large potential effect on soil C and maximum C sequestration. Even with the negative aspects of wood removal and burning, growing short-rotation woody crops for energy has a substantial potential to sequester soil C. The U.S. estimates for these practices on forest and marginal land range from 276 to 529 MMT-C/year; this can be compared to an estimate of 75-208 MMT-C/year from more intensive management of U.S. cropland.

Disturbance of terrestrial ecosystems (forests and grasslands) has been shown to have varying impacts on soil microbial communities depending on the type and severity of the disturbance (Aber and Melillo, 2001). Soil properties drive bacterial community composition, with pH being a dominant factor (Nacke et al., 2011; Wakelin et al., 2008). Wildfires and invasion by exotic plant species can fundamentally change the composition of aboveground plant communities, indirectly influencing belowground microbial dynamics through altering quality and quantity of litter inputs to the system, changing soil moisture and temperature regimens, and altering evapo-transpiration rates (Neary et al., 1999). Forest management strategies can have positive or negative impacts on the quantity and quality of litter residues, rates of C and N mineralization, and microbial community structure and function (Grayston and Renneberg, 2006).

Removal of biomass through timber harvesting and the compaction or physical disturbance of the O horizon reduces populations of G^+ bacteria and fungi while increasing bacterial:fungal ratios (Chatterjee et al., 2008). Fungi are responsible for the breakdown of substrates with high C:N ratios; the result of declining populations would be decreased mineralization and breakdown of high C litter and SOM (Busse et al., 2006). Removal of large logs by harvesting removes ideal habitat for nonsymbiotic, free-living N fixing bacteria (Aber and Melillo, 2001; Wei and Kimmins, 1998). Fire contributes Mg and P to the soil via ash, without which nonsymbiotic fixation can be suppressed. These nutrients tend to be limited in harvested forestry systems.

Fire is a primary cause of disturbance in many ecosystems with the potential to alter both the physical and biological environment. In forest systems, naturally reoccurring fire serves as a regulator between over and understory vegetation, thus altering the substrate available to microorganisms. Alteration of plant communities changes inputs to the soil environment in the form of charcoals, metal oxides, and plant litter (Hart et al., 2005). Fire can have a pronounced impact on soil moisture, temperature, and pH, resulting in changes to the microbial community structure. Higher intensity fires can decrease canopy cover, increasing the amount of solar radiation reaching the soil surface. Increased quantities of photosynthetically active radiation (PAR) reaching the soil surface can lead to a greater abundance of phototrophic bacteria and algae, including *Cyanobacteria*, a free-living N fixer. In addition, fire in a forest ecosystem results in the deposition of ash and charcoal on the soil surface, decreasing albedo. This coupled with the increase of solar radiative input can lead to an increase in soil heating during the day and heat loss at night (Hamman et al., 2007; Hart et al., 2005).

Heating can affect biological properties of soils due to the threshold for living organisms being near 100 °C. High soil temperatures have the greatest effect in the surface layer. This layer can act as an insulating layer protecting microorganisms; however, if this layer combusts, the impact on microbial communities can be dramatic (DeBano et al., 1998). Water content can also play a role in the "O" horizon during fire events. Moist "O" horizons can produce moist heat, effectively imitating "pasteurization," thus leading to higher microbial mortality (Hart et al., 2005). Bacteria are more heat resistant than fungi, thus the fungi: bacteria ratio tends to decrease following fires (Pietikainen and Fritze, 1995).

A number of management practices and land use changes could increase soil C in range and grazing land. Much of the semiarid and arid grazing lands are subject to both wind and water erosion and would benefit by conversion to managed grasslands with greater input intensity. Improving grazing management has been shown to increase soil C compared to nongrazed enclosures. However, the increases in total soil C and N with grazing may be soil texture dependent. The conversion of forest-to-pasture and marginal cropland-to-pasture can also substantially increase soil C as can better management of fertility and plant species composition. Grasslands depend on fire to maintain plant community structure (Docherty et al., 2012; Neary et al., 1999). Differences between grassland and forest fires are severity, intensity, and the quantity of nutrients returned to the system (Dooley and Treseder, 2011). Low-intensity grassland fires alter the physical and chemical soil environment in the top 15 cm of the soil profile, the primary location of most aerobic microbial activity. Belowground communities are resilient to this type of low-intensity fire; however, more intense fires can result in an initial decrease in total microbial biomass (Dooley and Treseder, 2011). Immediately following a fire event, biogeochemical cycling in some systems can accelerate. This is mediated through increased microbial biomass, activity, and changes to the bacterial community structure (Goberna et al., 2012).

Grassland fires can be correlated with significant short-term changes in microbial community structure, biomass C, respiration, and hydrolase activity (Dooley and Treseder, 2011). Combustion of plant residues increases available organic C and NO_3-N, thereby increasing the quantity of substrate available to soil microbial populations. This substrate "pulse" can stimulate both denitrifier and nitrifier activity. Using 16S and 18S regions of rRNA and denaturing gradient gel electrophoresis enabled Goberna et al. (2012) to correlate fire with short-term compositional changes in the bacterial, fungal, and archaeal communities. They showed that species capable of forming heat resistant structures, such as sclerotia (fungi) or endospores (bacteria), recovered rapidly postfire.

III ALTERNATIVE AGRICULTURAL MANAGEMENT

A Organic Agriculture

Organic agriculture strives to integrate human, environmental, and economically sustainable production systems. The term *organic* does not necessarily refer to the types of inputs to the system, but more to the holistic interaction of the plants, soil, animals, and humans in the system. Organic agriculture management promotes maintaining SOM levels for soil fertility, providing plant nutrients through the microbial decomposition of organic materials, and the control of pests, disease, and weeds with crop rotations, natural control agents, and pest-resistant plant varieties (Lampkin et al., 2011; Oelhaf, 1978). It also has beneficial impacts on enhancing soil structure and fertility and increasing water infiltration and storage (Weil and Magdoff, 2004).

Since the 1990s, the land area under organic agriculture production has increased worldwide. Global demand for organic agriculture products has increased, though North America and Europe remain the primary markets (Willer et al., 2009). The United States has seen a sixfold increase in consumer spending on organic products, from $3.6 billion in 1997 to $25.7 billion in 2010 (Willer and Kilcher, 2012). Recent statistics by the International Federation of Organic Agriculture Movements show global land area in certified organic production has increased from 10.6 Mha in 2000 to 37.2 Mha in 2010 (Willer and Kilcher, 2012). Approximately 26% of this growth occurred in Latin America. In 2008, two-thirds of the land under organic management was grassland, being utilized for grazing and the production of animal products (Willer, 2010; Supplemental Fig. 18.1, see online Supplemental Material at http://booksite. elseiver.com/9780124159556.)

During the conversion to organic production, SOM levels and other nutrients can initially decrease (Rees et al., 2001; Wander et al., 1994). However, if manures are incorporated into the transition, increases in these soil properties can be achieved. Additions of animal manures to organic systems have significant impacts on microbial biomass C and enzyme activity (Chaundry et al., 2012). Transition affects a combination of changes in the microbial community

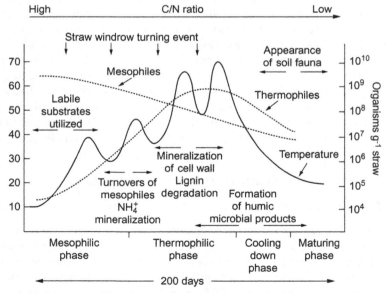

FIG. 18.1 Organisms and processes occurring during composting of straw. The length of time varies with the outside temperature and extent of mixing, but usually involves 200 days. *From W.R. Horwath, personal communication.*

and a physical redistribution of SOM fractions (Clark et al., 1998; Wander and Traina, 1996). In the long term, soils under organic management will tend to increase in SOM. Organically managed soils have a significantly higher microbial biomass, larger fractions of mineralizable C and N, and greater microbial C: mineralizable C ratios than for conventionally farmed soils (Clark et al., 1998; Reganold et al., 1993, 2010).

New technologies, including 454 pyrosequencing and community-level physiological profiling, have been employed to examine the long-term effects of management in organic and conventional production systems, as well as fallow grasslands (Chaundry et al., 2012; Nannipieri et al., 2003). The findings show C source utilization, microbial and functional diversity, and enzyme activities were highest in the organically managed soil (Garcia-Ruiz et al., 2008). Individual management practices favor specific phyla of bacteria, implying that these phyla may play major roles in nutrient cycling and availability, pH, total organic C and N levels, and microbial biomass (Chaundry et al., 2012). Studies using PCR-DGGE (polymerase chain reaction-denaturing gradient gel electrophoresis) show that long-term management has more of an effect on community composition than does short-term application of amendments (Stark et al., 2008).

Because organic systems are often low nutrient input systems, with respect to N, P, and K, the cycling of SOM by microorganisms is important because plants rely solely on nutrients from SOM. Organic management systems have lower pH and electrical conductivity (EC) than conventional and fallow and more available N-P-K, in addition to greater levels of organic C, microbial biomass, and mineralizable C (Chaundry et al., 2012). Organic systems have higher catabolic activity and nutrient cycling stability as microbial diversity increases. This increase in diversity and its relationship to soil function is poorly understood, leading others to conclude that microbial biodiversity is not the primary regulator of SOM dynamics.

Organic systems also exhibit higher enzyme activity levels (Reganold et al., 2010; Stark et al., 2008). When comparing long-term management strategies, organic systems are strongly correlated with higher levels of total C and N, greater microbial biomass, enzyme activity, and greater functional gene abundance and diversity (Reganold et al., 2010; Stark et al., 2007). The use of green manures increases microbial biomass and enzyme activities that are correlated with higher total C and N, as well as gross N mineralization rates (Smith, 1994).

Numerous studies and review articles have been published during the last decade that pertain to the nutritional benefits of organically grown foods, showing mixed results. Higher levels of vitamin C; flavonoids; nutrient elements of Ca, Mg, and Fe; and superior taste are commonly cited as benefits. Comparatively, it has been shown that excess N in conventionally fertilized soils can decrease the nutritional quality and taste of crops and can also increase plants' susceptibility to diseases and insects. Excess N is also responsible for decreasing the vitamin C content of crops. The most cited benefit of organically grown foods is what may or may not be on or in the foodstuffs, such as herbicide and pesticide residues and high concentrations of nitrates. In addition, no genetic modified crops or ingredients are allowed in organic foods, nor is there is any routine use of antibiotics in organic animal production.

A meta-analysis (Smith-Spangler et al., 2012) examined 17 studies looking at the human health effects of consuming organic food. The analysis examined 223 studies that compared nutrient content and pesticide and bacterial contamination of both conventional and organic agricultural products. The analysis found that, when looking at the studies as a whole, there was no conclusive evidence to support the claim that organic food products have a significant health benefit when compared to conventionally produced food. Studies examining individual agricultural products often contradict this meta-analysis. Although the perception that organic food is more nutritious may influence a consumer's choice between organic and conventional agriculture products, the dominant factor consumers take into account when making the decision to purchase organic foods isn't what is in them, but rather what isn't.

B Biodynamic Agriculture and Charcoal

Another example of organic-based food production is biodynamic agriculture, or farming that is rooted in the 1924 lectures of Dr. Rudolf Steiner. The lectures described a holistic view of a farm as an organism, with the plant and animal community of a natural habitat striving for a certain balance where the number of species and individuals are constant. Farm management, including clearing, plowing, cutting, and grazing, along with using monoculture crops, were said to create imbalance in the system, thus creating an unsustainable agricultural system. In similar fashion, the use of pesticides kills both pest and beneficial organisms, creating an imbalanced system, which destroys productivity. Biodynamic farming is a system of organic farming that includes crop diversification and the use of green manures, compost, and manures improved by biodynamic preparations (Sattler and Wistinghauser, 1992). The biodynamic preparations consist of selected plant and animal substances that undergo fermentation for a year and are then used to enhance compost and manure used in the farming operation. The use of biodynamic preparations is the main difference between biodynamic farming and traditional organic agriculture. Studies have shown that biodynamically managed fields maintain higher soil C levels, microbial respiration, mineralizable N, earthworm populations, microbial biomass C and N, and greater enzyme activities. Biodynamic farms have better soil quality, mostly due to enhanced microbial decomposition and stabilization of organic matter. The question of whether organic or biodynamic farmers are in general more dedicated, better managers has not yet been adequately answered.

Another amendment that has its roots in biodynamic agriculture, as well as in ancient "Preta" techniques, is recalcitrant soil C. Soils contain natural recalcitrant materials, including charcoal, especially in areas with a history of burning. This material has been termed char or black C. Biochar is made from any organic material combusted under low oxygen conditions (Johnson et al., 2007). It is hypothesized that it increases soil quality through C sequestration and reduces CO_2 enrichment of the atmosphere (Novak et al., 2009; Smith et al., 2010).

Biochar can improve water-holding capacity, soil aggregation, and soil fertility (Fowles, 2007). It can also reduce potential pollutants (Lehmann et al., 2006) and N_2O production (Singh et al., 2010). Biochar has also been shown to affect microbial communities in soil systems (Bailey et al., 2011; see Burns and Ritz, 2012). Biochar seems to be beneficial to soils; however, production, availability, distribution, and cost are unknown factors. Materials to make biochar are scarce in developing countries and are normally used for heating or cooking. Distribution from areas of high biomass production for biochar to agricultural lands may be cost prohibitive. If the market for soil C sequestration ("C credits") ever materializes, the production of biochar may become cost favorable and environmentally productive.

C Crop Rotations and Green Manures

Cropping rotations have been practiced over the long history of agriculture. Studies dating from the 1840s have shown that N supplied to grain crops was the major reason for using crop rotations containing legumes (Triplett and Mannering, 1978). With the advent of inexpensive N fertilizers, crop rotations containing legumes declined. Only recently has the value of crop rotations, specifically those including legumes, been recognized as critical in maintaining SOM and soil productivity. Studies of the nutrient dynamics in a Canadian Luvisol after 50 years of cropping to a 2-year rotation (wheat-fallow), or a 5-year rotation (wheat-oats-barley-forage-forage; McGill et al., 1986) showed that the soil cropped to the 5-year rotation contained greater amounts of organic C and N. The 5-year rotation doubled the input of C into the soil over the 2-year system and had a greater percentage of the organic C and N in biological form.

In a 10-year study, a low-input diverse crop system with manure and a low-input cash grain system with legumes showed significant increases in SOM compared to a conventional corn/soybean rotation (Wander et al., 1994). The microbial biomass was greater and its activity higher than the conventional rotation of corn/soybeans with chemical inputs. The low-input systems also mineralized significantly more N, and the microbial biomass contained 33 kg N ha^{-1} more than the conventional system. These results suggest that longer cropping system rotations that include forage or legumes will conserve SOM, maintain a greater biological nutrient pool, and put more nutrients into the soil than intensive rotations.

Plant pathogens are an important part of the soil microbial community. As growers reduce tillage and incorporate a greater variety of crops in rotation, they face an increasing number of plant diseases that can cause significant stand and yield reductions. These potential losses, however, may be offset in systems incorporating green manures by: (1) promoting disease-suppressing properties that reduce plant pathogens, either by increasing the levels of SOM that create conditions supporting a greater microbial biomass, competition for resources, antibiosis, or antagonism; or (2) through direct inhibition by production of anti-bacterial/fungal compounds as in the case of *Brassica* cover crops that produce isothiocyanates. Cover crops are known to control disease-causing organisms through competition for resources and space, control of soil micronutrient status, and alteration of root growth.

IV   THE POTENTIAL FOR MANAGING MICROORGANISMS AND THEIR PROCESSES

A   Management of Native and Introduced Microorganisms

There are many cases in which soil organisms have been managed to improve plant growth and serve as biological control agents to suppress plant disease, inhibit weeds, control insects, or detoxify environmental pollutants (Supplemental Table 18.2, see online Supplemental Material at http://booksite. elseiver.com/9780124159556). Beneficial microorganisms promote plant growth through the creation of symbiotic associations with plant roots, release of phytohormones, induction of systemic resistance, suppression of pathogens, production of antibiotics, and reduction of heavy metals toxicity (Bowden and Rovira, 1999). Many of these organisms are naturally present in soil, although under some circumstances, it may be necessary to increase their populations either by modifying the soil environment or through inoculation to enhance their abundance and activity. Significant challenges must be overcome when introducing desired microbial species to natural soils. The most successful introductions have been those associated with rhizobial infection of legumes in which microorganisms are applied to seeds prior to planting. The physiological state and population density of the introduced microorganisms, the native soil conditions (texture, pH, moisture, structure), and the timing of the inoculation must be considered for each unique effect desired (van Veen et al., 1997).

The positive role of symbiotic associations, such as mycorrhizal fungi, with host plants in plant production is well known (see Chapter 10 and Hart and Trevors, 2005), with many cases documenting growth and yield enhancement of infected plants. Benefits of the association are increased nutrient uptake, protection against pathogens, improved tolerance to pollutants, and greater resistance to water stress, high temperatures, and adverse pH. Management strategies that enhance mycorrhizal populations and activity in agricultural fields include reduced tillage, crop rotation, and lower N and P applications.

The other major plant symbionts, rhizobia associated with legumes, provide the N required for plant growth, and in return obtain photosynthetic products from the plant for their own growth. These rhizobia can be inoculated during

TABLE 18.2 Mechanisms of Biocontrol by Soil Bacterial and Fungal Agents

Method of Control	Organism	Mechanism
Competition	Many bacteria and fungal spp.	Niche exclusion, high reproduction rates, nutrient uptake, iron chelating compounds
Antibiosis		
Bacteria	Pseudomonas, Bacillus, Streptomyces, Agrobacterium	Inhibition by production of antimicrobial compounds, e.g., HCN, oomycin A, 2,4-diacetylphloroglucinol, HCN, oomycin A, zwittermycin, phenazine-1-carboxylic acid, pyrrolnitrin, agrocin 84
Fungi	Trichoderma, Talaromyces, Penicillium	Inhibition by production of antimicrobial compounds, e.g., glioviron, gliotoxin, hydrogen peroxide, penicillin
Induced resistance	Bacillus, Pseudomonas, Fusarium, Trichoderma, Pythium	Biochemical changes in plant cell walls that are induced by physical or biochemical agents, e.g., siderophores, chitinases, peroxidase, lipopolysaccharides, phytoalexins, ethylene, salicylic acid
Mycoparasitism	Verticillium, Trichoderma, Coniothyrium (now Leptosphaeria)	The process of mycoparasites is to sense host, directed growth, contact, recognition, attachment, penetration, and exit; production of cell-wall degrading lytic enzymes, chitinases, β-1,3 glucanases, cellulases, proteases

seeding; however, populations may exist in native soils that can outcompete the added organisms. Free-living N_2-fixing bacteria (diazotrophs), such as *Azotobacter* and *Azospirillum*, have in some cases been observed to enhance local N levels in soils (Das and Saha, 2003; Dobbelaere et al., 2003). The most recent studies suggest other mechanisms by which diazotrophs enhance plant growth, such as: (1) synthesis of phytohormones and vitamins (Hernandez-Rodriguez et al., 2010); (2) inhibition of plant ethylene synthesis; and (3) improved nutrient uptake. These organisms may also modify the rhizosphere to stimulate increased nodulation of legumes by rhizobia. The effects of these plant-growth-promoting bacteria may be more profound in nutrient deficient soils compared with richer soils in which microbial activities are not needed to supply the needed nutrients (Egamberdiyeva, 2007).

Many microbial species have been used as plant-growth-promoting organisms, predominantly pseudomonads (e.g. *Pseudomonas fluorescens*, *P. putida*, *P. gladioli*, now *Burkholderia*), bacilli (e.g., *Bacillus subtilis*, *B. cereus*, *B. circulans*), and others (e.g., *Azospirillum*, *Serratia*, *Flavobacterium*, *Alcaligenes*, *Klebsiella*, Enterobacter). Many of these bacteria produce the growth regulators indole acetic acid, auxins, cytokinins, gibberellins, ethylene, or abscisic acid, which stimulate plant growth (Arshad and Frankenberger, 1998; Jiang et al., 2012). *Azospirillum* inoculation increases the density and length of root hairs as well as the elongation of lateral roots, which increases root surface area (Kaymak et al., 2008). Three different strains of fungi (*Aspergillus clavatus*, *Penicillium commune*, and *Thamnidium elegans*) were recently, shown to significantly enhance pea and cotton production through increased ethylene production from L-methionine applications (Mahmood et al., 2008).

Similar results have been observed in *Eucalyptus globulus* with a number of different Bacillus species, *Brevibacillus brevis*, *Paenibacillus lautus* (previously *Bacillus lautus*), and *Stenotrophomona maltophilia* (Diaz et al., 2009). In addition, the secretion of root exudates, low-molecular-weight organic compounds such as oxalate and citrate, stimulates microbial growth in the rhizosphere (Atemkeng et al., 2011), further supporting enriched microbial populations and activities.

Genetic engineering of microorganisms is underway for more targeted biofertilization and biocontrol effects in agriculture, forestry, bioremediation, and environmental management. Public concerns about control of genetically modified organisms need to be considered at all stages of these studies, even if the likely mobility of root-association microorganisms seems slight. Insect movement may lead to transport of bacteria, as has been seen in studies of a genetically engineered strain of *Pseudomonas chlororaphis* by the grasshopper (*Melanoplus femurrubrum*; Snyder et al., 1999). Environmental scientists must keep up with new regulations and new science to mitigate environmental risks posed by GEMs. These should include more extensive studies of GMO risk-benefit balance tailored appropriately for different scenarios, postrelease monitoring, and interdisciplinary regulation (Snow et al., 2005).

B Managing Microbial Populations as Agents of Biological Control

One of the fastest-growing areas in pesticide development is the use of microorganisms for the biological control of pests (biopesticides). The fundamental concept of biological control is the selection and introduction of specific organisms to control a particular pest. The organism(s) selected might be a predator, parasite, competitor, or pathogen of the targeted pest. Biological control occurs widely in nature, masking the fact that most pests never become serious problems because of the controls exerted by other members of the community. Biological control methods are commonly a part of an overall integrated pest management program to reduce the legal, environmental, and public safety hazards of chemicals. Unlike most pesticides, biological control agents are often specific for a particular pest. Successful use of biological control requires a greater understanding of the biology of both the biocontrol agent and the targeted pest. Often, the outcome of using biological control is not as dramatic or quick as the use of pesticides. Most biocontrol agents attack only specific types of plant pathogens, nematodes, or insects. There are three strategies in which organisms are managed to control pest populations.

(1) *Classical biological control* involves collecting organisms pathogenic to the pest from locations where a pest originated and then releasing them in an infected area to control the pest. The natural increase in the biocontrol organisms is relied on to control the pest without future intervention.

(2) *Augmentation* is a method of increasing the population of a natural enemy that attacks a pest. This can be done by mass-producing inoculum in the laboratory and releasing it into the field at the proper time. Control results from and requires the increase of the disease through many disease cycles to reach threshold levels that cause death or control of the target pest. The augmentation method relies on continual human management.

(3) *Conservation* or *enhancement* of indigenous populations is important in any biological control effort. Conservation of natural enemies involves either reducing factors that interfere with their populations or providing needed resources to maintain them, whereas enhancement involves identifying any factors that limit the effectiveness of a particular population and then facilitating those factors to sustain the beneficial species.

Many bacterial and fungal genera have been used extensively for the control of plant pathogens. Various nonpathogenic fungal strains of *Rhizoctonia*, *Phialophora*, *Fusarium*, and *Trichoderma*, as well as mycorrhizal fungi, have been used to reduce damage caused by related strains or other pathogenic fungi. As biocontrol agents, *Trichoderma* species dominate, most likely because of their ease of culture and wide host range. The most commonly targeted pathogens are *Pythium*, *Fusarium*, and *Rhizoctonia* species, reflecting their worldwide importance and relative ease of control (Whipps, 2001). The most commercially successful bacterial-based biocontrol system is the use of the nonpathogenic *Agrobacterium* strains to control crown gall.

The primary modes of action for pathogen control or for management of both bacteria and fungi are competition for C, N, and Fe, or increased colonization of the rhizosphere by the nonpathogenic strains (Table 18.2). For more information on competition and food web interactions, see Chapters 5 and 8. Other mechanisms include antibiosis, induced resistance, and mycoparasitism. Competition for commonly used substrates and for their excretion sites on roots plays an important role in the root zone for controlling pathogens. Organisms capable of rapid growth rates with wide distribution and population numbers throughout the soil have a distinct advantage of colonization and substrate utilization that outcompete many pathogens. Ectomycorrhizal fungi may occupy infection sites thus excluding pathogens due to their physical sheathing morphology. In contrast, arbuscular mycorrhizae provide control of pathogens through induced resistance and improved plant growth rather than niche competition.

The production of antimicrobial metabolites is one of the most studied aspects of biocontrol. Antibiotics produced by a wide range of bacteria and fungi include pyrrolnitrin, phycocyanin, 2,4-diacetylphoroglucinol xanthobactin, and zwittermycin A.

C   Control of Insects

Insect infestations can cause a significant decrease in crop productivity. Infestations have been historically difficult to control without the use of highly toxic synthetic chemicals. The most well-known biopesticides for insect control are the *Bacillus thuringiensis* (*Bt*) formulations for the control of lepidopterous pests. This bacterium produces a protein that by itself is harmless to most insects, but is converted to a potent toxin in the gut of specific target insects following ingestion. *Bt* toxicity depends on recognizing receptors; damage to the gut by the toxin occurs on binding to the receptor. Each insect species possesses different types of receptors that will only match certain toxin proteins. Application of particular Bt toxin proteins has to be carefully matched to the target pest species. Beneficial insects will not be adversely affected by select strains of *Bt*.

The gene coding for this protein has been cloned from various *B. thuringiensis* strains and has been incorporated into the genome of a number of plants (corn, cotton, potato). The insect against which the toxin is active dies soon after feeding on the transformed plant. A drawback to this approach is that resistance of the insect to the toxin may develop. These biological pesticides also degrade rapidly in the environment. Thus, the use of such biological pesticides appears to be more of an environmentally safe alternative to pest control than is the use of synthetic chemical pesticides.

Other examples of biocontrol agents of insects include the fungi *Metarrhizium anisopliae* (formerly Entomophthora anisopliae), which is used to control larvae of grass grubs (*Costelytra*) in pastures, as well as several genera of nematode-trapping or nematophagous fungi (*Arthrobotrys, Dactylella, Verticillium*). Entomopathogenic nematodes of the *Delanus, Neoaplectana, Tetradonema*, and *Heterorhabditis* have been found to control a wide range of insect pests. The success of these nematodes lies in the fact that most plant insect pests spend a part of their life cycle in soil.

D Weed Control

Biological control of weeds studied for a period of over 100 years used fungal plant pathogens to control weeds by a postemergent application of the plant pathogen to weed foliage. The most commonly studied fungi are *Colletotrichum*, *Phytophthora*, *Sclerotinia*, and *Puccinia*. For an in-depth discussion on the use of microbial pesticide control of weeds, see TeBeest (1996), Hajek (2004), and Stubbs and Kennedy (2012).

Classical biological control of weeds uses organisms that have specific pathogenicity to plants in their host range. Examples are the rust fungi and smuts. The first use of rusts as a weed control was in Australia (Hasan, 1972). In an inoculative approach, microorganisms are artificially multiplied and applied to weeds similar to chemical herbicides (Chutia et al., 2007; Hajek, 2004). The goal is to treat the entire population of the target weed species within an area. Supplemental Table 18.3 (see online Supplemental Material at http://booksite.elseiver.com/9780124159556) provides a listing of microorganisms being studied for weed control.

Although many of the organisms discussed have shown excellent efficacy in pure culture or microcosm studies, their adoption for wide use in agriculture is

TABLE 18.3 Nonchemical Approaches for Control of Soil-Borne Diseases

Method	References: see Mazzola et al. (2007) for citations
Host resistance	Fazio et al. (2006), El Mohtar et al. (2007), Herman and Perl-Treves (2007), and Nicol and Rivoal (2008)
Soil solarization	Katan and Devay (1991) and Gamliel et al. (2000)
Crop rotation	Larkin and Honeycutt (2006), Subbarao et al. (2007), and Kirkegaard et al. (2008)
Tillage	Cook et al. (1990) and Roget et al. (1996)
Biological control	Pal and McSpadden-Gardener (2006)
Disease suppressive soils (native) via	Cohen et al. (2005) and Wiggins and Kinkel (2005)
Plant residue amendments	Peters et al. (2003)
Tillage	Shipton et al. (1973)
Cropping sequence	Larkin et al. (1993), Mazzola and Gu (2002), and Mazzola et al. (2004)
Specific plant genotype	

Adapted from Mazzola et al. (2007).

minor. This has been due to strict environmental regulations regarding their release and the technical problems associated with introducing and maintaining populations in the soil. Technical problems include (1) identification of factors that affect survival rates; (2) determination of which strains are best for each crop of interest and field conditions; and (3) field conditions, methods of application, and implementation of management practices that enhance biocontrol.

Suppressive soils. Diseased suppressive soils have been defined as those in which disease development is minimal, even in the presence of a virulent pathogen and a susceptible host (Mazzola et al., 2007; Table 18.3). Specific disease suppression is attributed to the activity of an individual or select group of microorganisms within a soil that are antagonistic toward the target pathogen. A classic example of developing a suppressive soil is from the study of *Gaeumannomyces graminis* var. *Penicillium* (take-all disease of wheat) in continuous monoculture wheat production systems. In monoculture wheat production, incidence of take-all disease commonly increases during the initial few years of production, but at some point during cultivation it declines. The soil remains suppressive to the disease as long as the wheat monoculture is maintained.

Significant disease suppression and development of suppressive soils commonly results from the alteration of soil microbial communities by manipulating the physiochemical and microbiological environment through management practices such as soil amendments, crop rotations, tillage (as discussed earlier), natural or synthetic compounds (soil fumigation), soil solarization, or by the use of genetically resistant plant varieties. Soils that are suppressive to pathogens typically have neutral to alkaline pH and liming of acid; disease prone soils have been shown to reduce the severity of fungal pathogens.

E   Use of Synthetic and Natural Compounds to Modify Soil Communities or Functions

The use of synthetic compounds to control pests began in the 1930s and became more widespread after the end of World War II. "First-generation" pesticides were largely highly toxic compounds, such as arsenic and hydrogen cyanide. The second-generation pesticides largely included synthetic organic compounds. From about 1945 to 1965, organochlorines were used extensively in all aspects of agriculture and forestry for protecting both wooden buildings and humans from a wide variety of insect pests. In recent years, chemical pesticides have become the most important consciously applied form of pest management. However, for some crops in some areas, alternative forms of pest control are still heavily used, such as burning, rotation, or tillage.

Soil fumigation is aimed at controlling pests, or at least for reducing infestations to lower levels, thus enabling successful growth of commercial crops (Ramsay et al., 1992). Soil fumigants are chemicals (solids, liquids, or gases) that, when applied to soil under specified temperature and moisture conditions, generate toxic gases that can kill many kinds of pest organisms as the fumes spread through the soil. Multipurpose soil fumigants used to provide plant disease and nematode controls include chloropicrin, dichloropropene, metam-sodium, metam-potassium, and methyl bromide. Methyl bromide is being phased out because it has been identified as an ozone-depleting substance. Fumigants can reduce populations of nematodes and also many weeds and soil-borne fungi, but can also kill plants at the same rate, so they are applied to the site weeks before planting. Fumigation with sodium N-methyldithiocarbamate (metam-sodium) and the nematicide 1,3-dichloropropene is a commonly used practice in vegetable crop production and has been shown to be an effective method for the control of soil-borne pathogens, weeds, and plant-parasitic nematodes that reduce crop yield and quality (Collins et al., 2006; Hamm et al., 2003; Ingham et al., 2000).

Fumigants have no residual effect, so nematodes or pathogens that either survive or escape treatment (e.g., are too deep to be reached by the fumigant, or are protected inside root tissues and soil aggregates) or are brought in after fumigation, can reinfest the root zone. Synthetic soil fumigants are general biocides that do not target specific genera or pathogens. Therefore, their activity has shown negative effects on both beneficial as well as deleterious populations of soil organisms. Several studies have shown reductions in soil processes involved in the cycling of plant nutrients, symbiosis, and soil tilth with the use of soil fumigants (Bunemann et al., 2006; Ibekwe et al., 2001; Spokas et al., 2007; Stromberger et al., 2005).

Organic pesticides are usually considered as those pesticides that come from natural sources. The fundamental difference between organic and synthetic pesticides is not their toxicity, but rather their origin (e.g., whether they are extracted from natural plants, insects, or mineral ores, or are chemically synthesized. These natural sources are usually plants, as is the case with pyrethrum (pyrethins), rotenone, or ryania (botanical insecticides), or minerals, such as boric acid, cryolite, or diatomaceous earth. Two of the most common organic pesticides, copper and sulfur, are used as fungicides by organic growers. Because they are not as effective as their synthetic counterparts, they are applied at significantly higher rates. Copper fungicides are used to treat foliage, seeds, wood, fabric, and leather as a protectant against blights, downy mildews, and rusts. Copper compounds, especially copper sulfate, are used as fungicides, pesticides, algicides, nutritional supplements in animal feeds, and fertilizers. Copper and sulfur combined accounted for 25% of U.S. pesticide use with over 13 million pounds of copper-based pesticides applied annually to crops in the United States.

The production of secondary plant compounds (e.g., cyanogenic and isothiocyanate) from oil-seed meals of neem (*Azadirachta indica*), castor (*Ricinus communis*), mustard (*Brassica campestris, B. hirta,* or *B. juncea*), and duan (*Eruca sativa*) have been widely studied for efficacy against plant-parasitic nematodes and soil fungi that infest a number of crops (Brown and Morra, 1997; Collins et al., 2006; Mazzola et al., 2007; Sarwar et al., 1998). Populations of plant-parasitic nematodes, *Meloidogyne incognita, Rotylenchulus reniformis, Tylenchorhynchus brassicae, Helicotylenchus indicus,* and the incidence of the pathogenic fungi *Macrophomina phaseolina, Rhizoctonia solani, Phyllosticta phaseolina, Fusarium oxysporum f. ciceri,* and so on, have been reduced by the use of Brassica cover crops or their seed meals.

   Soil solarization is a nonchemical technique used for the control of many soil-borne pathogens and pests (Gamliel et al., 2000; Katan and Devay, 1991). This simple technique captures radiant heat energy from the sun, thereby causing physical, chemical, and biological changes in the soil. Transparent polyethylene plastic placed on moist soil during the hot summer months increases soil temperatures to levels (71 °C) lethal to many soil-borne plant pathogens, weed seeds and seedlings (including parasitic seed plants), nematodes, and some soil-residing mites. Soil solarization also improves plant nutrition by increasing the availability of N and other essential nutrients. Beneficial microbial populations, such as mycorrhizal fungi, *Trichoderma* sp., actinomycetes, and some beneficial bacteria, rapidly recolonize the soil. These in turn may contribute to a biological control of pathogens and pests and/or stimulate plant growth. Soil solarization has also been shown effective at reducing nematode populations, but less dramatically than for fungal pathogens and weeds. Nematodes generally are more tolerant of heat, and control is less effective in soil depths beyond 12 in.

F   Composting

Composting is the biological decomposition of wastes consisting of organic substances of plant or animal origin under controlled conditions to a state sufficiently stable for storage and utilization (Diaz et al., 1993). Composting is perhaps the prime example of management of organisms (Cooperband, 2002). It plays a major role in developing countries, being relied on to provide organic matter and nutrients and to increase soil tilth. It also plays a role in processing human waste. Less than 15% of municipal solid waste is recycled; however, more than 30% of the sewage sludge is beneficially used as composted products (Rynk, 1992; http://compost.css.cornell.edu/OnFarmHandbook).

   Traditionally, yard waste is thought of as "the" compost material; however, manure, meat and dairy waste, wood, sawdust, and crop residue can be composted. Animal carcass composting is receiving significant attention due to the environmental benefits versus burial, which can contribute to groundwater contamination. One important aspect of the material that affects the compost process and product is the C:N ratio of the starting material; ideally it should be 25-30:1. Typical C:N ratios of different materials are shown in Supplemental Table 18.4 (see online Supplemental Material at http://booksite.elseiver.com/9780124159556). Materials can be mixed to adjust the C:N ratio for a consistent product.

   In compost terminology, process strategy refers to the management of the biological and chemical activity of the composting process. Biological processing terminology refers specific stages of composting, such as the active stage

TABLE 18.4 General Compost Properties

% N	>2	Color	Brown black
C:N	<20	Odor	Earthy
% Ash	10-20	% Water-holding capacity	150-200
% Moisture	10-20	CEC (meq 100 g^{-1})	75-100
% P	0.15-1.5	% Reducing sugars	<35

(mesophilic), high rate stage (thermophilic), and the controlled (cooling) and curing stage (maturing; Fig. 18.1). Composting configurations range from windrow or open systems to enclosed systems, with windrow further classified as either static or turned. A static system would be a stationary, undisturbed mound of organic material with air either being forced up through or pulled down through the mound. A turned system uses mixing as the aeration method, which also enhances the uniformity of decomposition and reduction in material particle size (Diaz et al., 1993).

The most prevalent composting technique is aerobic decomposition, carried out by a diverse microbial population that changes composition as the conditions change. It proceeds rapidly and provides a greater reduction in pathogens because higher temperatures are achieved. Physiochemical factors affecting aerobic composting are temperature, moisture, aeration, pH, additives, particle size and the C:N ratio of the composted substrate. Indigenous organism populations are typically used for the composting process; microbial inoculants are only utilized under certain conditions. Figure 18.1 depicts the process of composting straw for 200 days under optimum conditions of temperature and moisture. In the mesophilic stage, metabolism of the labile, C-rich substrates increases rapidly, thus generating heat. At this point, there is a mixture of bacteria, actinomycetes (class Actinobacteria) and fungi contributing to the decomposition process. In the early and transition stages to thermophilic conditions, the windrow is turned, causing a decline in temperature and oxygenation of the inner material and resulting in rapid decomposition and temperature increase. As the temperature reaches 40 °C the system turns from a mesophilic to a thermophilic stage, generally favoring thermophilic bacteria and actinomycetes with *Bacillus* being the dominant genus. Common *Bacillus* species found at this stage, accounting for 10% of the decomposition, are *brevis* (also *Brevibacillus brevis*, some strains now *Aneurinibacillus migulanus*), *circulans*, *coagulans*, *licheniformis*, and *subtilis*. Decomposition will continue in the thermophilic zone until substrates begin to decline; consequently a gradual decrease in temperature will occur. As the temperature declines, the mesophilic organisms reappear, especially fungi that have a preference for the remaining lignin and cellulose substrates. Fungi, responsible for 30-40% of the compost material, include *Absidia* (family *Mucoraceae*), *Mucor* (family *Mucoraceae*), and *Allescheria* (now *Monascus*). The actinomycetes, such as the *Nocardia* spp., *Streptomyces thermofuscus*, and *S. thermoviolaceus*, are important during this phase when humic materials are formed from decomposition and condensation reactions. The actinomycetes are estimated to account for 15-30% of the decomposition of composted material.

The compost has a lower C:N, has higher pH, and can contain considerable NO_3. The end product depends on the original substrate, any added nutrients, degree of maturity, and the composting method; typical properties of composted plant material are listed in Table 18.4. Adding compost to soil increases the SOM, which increases soil structure, water-holding capacity, and infiltration. In addition, compost contains significant amounts of plant nutrients such as N, P, K, and S, and micronutrients, which are slowly released into the soil. As an ancillary benefit, compost contains fairly resistant C compounds (Lynch et al., 2006). Using compost would favor an increase in both the population of fungi and the fungal:bacterial ratio. This can increase C compounds that are agents in binding soil particles into aggregates, resulting in an increase in soil tilth. Recent studies (Bailey et al., 2002) have shown there is increased soil C storage in soils with greater fungal: bacterial ratios.

G Manipulating Soil Populations for Bioremediation Xenobiotics

Bioremediation manages microorganisms to reduce, eliminate, contain, or transform contaminants present in soils, sediments, water, or air. The composting of agricultural residues and sewage treatment is based on the use of microorganisms to catalyze chemical transformations. Composting dates back to 6000 BCE, whereas the modern use of bioremediation began over 100 years ago with the design and operation of the first biological sewage treatment plant in Sussex, England, in 1891. Over the last several decades, *in situ* degradation of biologically foreign chemical compounds (solvents, explosives, polycyclic aromatic hydrocarbons, heavy metals, radionuclides, etc.) has been used as a cost-effective alternative to incineration or burial in landfills (Alexander, 1994). An advantage of bioremediation over other methods is that it transforms contaminants instead of simply moving them from one source to another as in the practice of land filling (Supplemental Table 18.5, see online Supplemental Material at http://booksite.elseiver.com/9780124159556). Another benefit is its relatively low cost compared to other methods of removal. In a bioremediation process, microorganisms break down contaminants to obtain chemical energy. This process involves the manipulation of microorganisms and their metabolic processes (enzymes) to degrade compounds of concern. Supplemental Fig. 18.1 (see online Supplemental Material at http://booksite.elseiver.com/9780124159556) shows the microbial degradation pathway of the pesticide DDT (Bumpus and Aust, 1987).

Phanerochaete chrysosporium and several other species use a peroxidase enzyme system that acts in concert with H_2O_2 (produced by the fungus) to degrade many recalcitrant organics, especially those with structures similar to lignin, which naturally degrade in soil systems. Degradable contaminants include DDT, lindane, chlordane, TNT, and polychlorinated biphenyls (PCBs). Bacteria and fungi have been shown to break down practically all hydrocarbon contamination in the natural environment (Table 18.5).

By studying natural processes, researchers have been able to determine which conditions are necessary for degradation and which organisms are active degraders of specific pollutants and manipulate them. Examples of aerobic bacteria managed for their degradative abilities include *Pseudomonas*, *Alcaligenes*, *Geobacter*, *Sphingomonas*, *Rhodococcus*, and *Mycobacterium*. Many of these organisms are capable of using the contaminant as a sole source of C and energy. In the absence of O_2, anaerobic bacteria have been used in the remediation of PCBs, dechlorination of trichloroethylene (TCE), and chloroform. Ligninolytic fungi, such as *P. chrysosporium*, have the ability to degrade an extremely diverse group of persistent PAHs compounds. Methylotrophs use the methane monooxygenase pathway to degrade a wide range of compounds, including the chlorinated aliphatics, TCE, and 1,2-dichloroethane. *Geobacter sulfurreducens* (Supplemental Fig. 18.2, see online Supplemental Material at http://booksite.elseiver.com/9780124159556) grows with insoluble Mn(IV) oxides as the electron acceptor.

Several factors influence the success of bioremediation and should be considered on a site-by-site basis. Factors include the existence of a microbial population capable of degrading the contaminant, availability of contaminants to the microbial populations, the type of contaminant, and its concentration, and environmental factors, such as soil type, temperature, pH, the presence of oxygen or other electron acceptors, and nutrients. Although microorganisms are present in contaminated soils, they are not necessarily present in the numbers required for intrinsic bioremediation. Their growth and activity are typically stimulated with the addition of nutrients and O_2 (Supplemental Table 18.5, see online Supplemental Material at http://booksite.elseiver.com/9780124159556).

The most common pollutants in the environment are the monoaromatic hydrocarbons produced from gasoline and diesel fuel production (Phelps and Young, 2000). Gasoline contains 20% by volume benzene, toluene, and xylene (BTEX). BTEX compounds enter the environment through accidental surface spills, leakage of underground storage tanks, or are added directly to agricultural soils as the "inactive" ingredients of pesticides. Many pesticide formulations contain 7-14% BTEX by weight as carriers. Benzene in particular is a known carcinogen and is soluble in water at concentrations far greater than the drinking water standard (1 µg/L). For BTEX compounds, the principal concern is their ready migration away from source areas.

Polycyclic aromatic hydrocarbons are the contaminants of concern at manufacturing petroleum plants (Supplemental Table 18.6, see online

TABLE 18.5 Description of Common Bioremediation Approaches

In situ bioremediation of soil	The goal of aerobic *in situ* bioremediation is to supply oxygen and nutrients to the microorganisms in the soil and does not require excavation or removal of contaminated soils. *In situ* techniques can vary in the way they supply oxygen to the organisms that degrade the contaminants. Two such methods are bioventing and injection of hydrogen peroxide. Remediation can take years to reach cleanup goals.
Bioventing	Bioventing systems deliver air from the atmosphere into the soil above the water table through injection wells where contamination is located. Nutrients, nitrogen, and phosphorous may be added to increase the growth rate of the microorganisms.
Injection of hydrogen peroxide	This process delivers oxygen by circulating hydrogen peroxide through contaminated soils to stimulate the activity of indigenous microbial populations.
Ex situ bioremediation of soil	*Ex situ* techniques require excavation and treatment of the contaminated soil. *Ex situ* techniques include slurry- and solid-phase techniques.
Liquid slurry-phase	Contaminated soil is combined with water and other additives in a bioreactor and mixed to keep indigenous microorganisms in contact with the contaminants.
Solid-phase	Solid-phase bioremediation treats soils in aboveground treatment areas equipped with collection systems to prevent contaminants from escaping.
Landfarming	Contaminated soils are excavated and spread on a pad through a system that collects leachates or contaminated liquids that seep out of the contaminated soil. The soil is periodically turned to mix air or provide nutrients.
Soil biopiles	Contaminated soil is piled to several meters over an air distribution system. Moisture and nutrient levels are incorporated to maximize microbial activity.
Composting	Biodegradable contaminates are mixed with straw, hay, or other C-rich compounds to facilitate optimum levels of air and water to microbial populations. The three commonly used designs are static pile composting, mechanically agitated in-vessel composting, and windrow composting

Adapted from US EPA (1996).

Supplemental Material at http://booksite.elseiver.com/9780124159556). Of the PAHs, the high-molecular-weight compounds (those with four or more rings) are of most concern with respect to health risks because a number of these compounds are known carcinogens. PAHs in general are characterized by very low solubility in water.

In addition, these compounds readily form nonaqueous phases in soil and sediment and also bind tightly to soil constituents such as SOM. Despite the limited water solubility of PAH, PAH-contaminated sediment can lead to teratogenicity and toxicity of the water.

Despite the high potential for bioremediation as an effective technology, its use is limited by the lack of understanding of biodegradation processes and inexperience in managing these processes in the field. This includes aspects of cometabolism, inoculation, evolution of biodegradation capabilities, monitoring and process control, measures of effectiveness, and genetic engineering.

V CONCLUDING COMMENTS ON MICROBIAL ECOLOGY

Humans can and have managed microorganisms. However, much of our discussion has centered on "unnatural" systems where we manipulate the organism environment to accomplish the desired outcome. For example, culturing all of the organisms in a gram of soil and using them in a fermentation of grape juice would produce undrinkable wine, but inoculating the fermentation with a specific species of microorganisms produces a quality product. In composting for gardens or commercial production, the operation is usually controlled at least with respect to ingredients, aeration, and moisture.

But what are the options and benefits with organisms and their management in "natural" environments? In natural environments, the mass of organisms and species composition (structure) are governed by the environment, including habitat, moisture, and temperature. Organisms form trophic levels with various modes of interaction and coexistence, from symbiotic to predatory. The community of soil organisms may increase or decrease, and their composition may change principally due to influences of abiotic microclimate and natural input of metabolically available substrates (plant litter, leachates, and exudates). The complexity of the system dictates that introduced organisms must find a niche to survive and be persistent. In general, organisms introduced into a soil system do not usually survive more than a period of days to weeks. The reason for poor survival rates could be due to nutrient competition, susceptibility to predation, and chemical attack due to differences in cell chemistry from culturing or simply nutrient deficiencies.

In natural systems, organisms can be structured compartmentally to be close to some organisms for symbiosis and away from other organisms for protection and in proximity to nutrients and water. An example of organism symbiosis is fungi breaking down the macromolecule cellulose into smaller more "digestible" compounds that can be utilized by other bacteria and fungi that can't utilize cellulose directly. In addition, an organism may need a specific growth factor or vitamin that may be produced by specific bacteria, thus growing in a mixed culture (soil), the first organism receives "nutritional symbiosis." There is, however, a downside to living among myriad different organisms; that is, some are predators, and some will resort to predation to survive. Predation is probably the biggest factor in organism composition changes over time. When carbon inputs from litter reach the soil, bacteria increase in numbers, and then bacterial-feeding protozoa increase and a new short equilibrium will be reached in community composition. This cyclic flux in population structure drives the complex food web system of the soil and dictates nutrient availability to plants. From this perspective, perhaps it is the soil microflora that is doing the "managing" in many ecosystems.

REFERENCES

Aber, J.D., Melillo, J.M., 2001. Terrestrial Ecosystems, second ed. Academic Press, San Diego.

Alexander, M., 1994. Biodegradation and Bioremediation. Academic Press, New York.

Arshad, M., Frankenberger, W.T., 1998. Plant growth-regulating substances in the rhizosphere: microbial production and functions. Adv. Agron. 62, 46–125.

Atemkeng, M.F., Remans, R., Michiels, J., Tagne, A., Ngonkeu, E.L.M., 2011. Inoculation with *Rhizobium etli* enhances organic acid exudation in common bean (*Phaseolus vulgaris* L.) subjected to phosphorus deficiency. Afr. J. Agric. Res. 6, 2235–2242.

Bailey, V.L., Smith, J.L., Bolton Jr., H., 2002. Fungal-to-bacterial ratios in soils investigated for enhanced C sequestration. Soil Biol. Biochem. 34, 997–1007.

Bailey, V.L., Fansler, S.J., Smith, J.L., Bolton Jr., H., 2011. Reconciling apparent variability in effects of biochar amendment on soil enzyme activities by assay optimization. Soil Biol. Biochem. 43, 296–301.

Bauer, A., Black, A.L., 1981. Soil carbon, nitrogen and bulk density comparisons in two cropland tillage systems after 25 years and in virgin grassland. Soil Sci. Soc. Am. J. 45, 1166–1170.

Bowden, G.D., Rovira, A.D., 1999. The rhizosphere and its management to improve plant growth. Adv. Agron. 66, 2–76.

Brown, P.D., Morra, M.J., 1997. Control of soilborne plant pests using glucosinolate-containing plants. Adv. Agron. 61, 167–231.

Bumpus, J.A., Aust, S.D., 1987. Biodegradation of DDT [1,1,1-trichloro-2,2-bis(4-chlorophenyl)ethane] by the white rot fungus *Phanerochaete chrysosporium*. Appl. Environ. Microbiol. 53, 2001–2008.

Bunemann, E.K., Schwenke, G.D., Van Zwieten, L., 2006. Impact of agricultural inputs on soil organisms—a review. Aust. J. Soil Res. 44, 379–406.

Burnett, R.L., Terry, R.E., Sweetwood, R.V., Webster, D., Murtha, T., Silverstein, J., 2012. Upland and lowland soil resources of the ancient Maya at Tikal, Guatemala. Soil Sci. Soc. Am. J. 76, 2083–2096.

Burns, R., Ritz, C., 2012. Biochar. Virtual Special Issue, Soil Biol. Biochem. http://www.journals. elsevier.com/soil-biology-and-biochemistry/virtual-special-issues/virtual-special-issue-on-biochar/.

Busse, M.D., Beattie, S.E., Powers, R.F., Sanchez, F.G., Tiarks, A.E., 2006. Microbial community responses in forest mineral soil to compaction, organic matter removal, and vegetation unmanaged. Can. J. For. Res. 36, 577–588.

Calderon, F.J., Jackson, L.E., Scow, K.M., Rolston, D.E., 2000. Microbial responses to simulated tillage in cultivated and uncultivated soils. Soil Biol. Biochem. 32, 1547–1559.

Campbell, C.A., Paul, E.A., McGill, W.B., 1976. Effect of cultivation and cropping on the amounts and forms of soil N. In: Western Canadian Nitrogen Symposium Proceedings. Calgary, Alberta, pp. 7–101.

Chatterjee, A., Vance, G.F., Pendall, E., Stahl, P.D., 2008. Timber harvesting alters soil carbon mineralization and microbial community structure in coniferous forests. Soil Biol. Biochem. 40, 1901–1907.

Chaundry, V., Rehman, A., Mishra, A., Chaunhan, P.S., Nautiyal, C.S., 2012. Changes in bacterial community structure of agricultural land due to long-term organic and chemical amendments. Microb. Ecol. http://dx.doi.org/10.1007/s00248-012-0025-y.

Chutia, M., Mahanta, J.J., Bhattacharyya, N., Bhuyan, M., Boruah, P., Sarma, T.C., 2007. Microbial herbicides for weed management: prospects, progress and constraints. Plant Pathol. J. 6, 210–218.

Clark, M.S., Horwath, W.R., Shennan, C., Scow, K.M., 1998. Changes in soil chemical properties resulting from organic and low-input farming practices. Soil Sci. Soc. Am. J. 90, 662–671.

Collins, H.P., Alva, A., Boydston, R.A., Cochran, R.L., Hamm, P.B., McGuire, A., Riga, E., 2006. Soil microbial, fungal, and nematode responses to soil fumigation and cover crops under potato production. Biol. Fertil. Soils 42, 247–257.

Cooperband, L., 2002. The Art and Science of Composting. University of Wisconsin-Madison, Madison.

Curci, M., Pizzigallo, M.D.R., Crecchio, C., Mininni, R., 1997. Effects of conventional tillage on biochemical properties of soils. Biol. Fertil. Soils 25, 1–6.

Dalal, R.C., Mayer, R.J., 1986. Long-term trends in fertility of soils under continuous cultivation and cereal cropping in southern Queensland. II. Total organic carbon and its rate of loss from the soil profile. Aust. J. Soil Res. 24, 281–292.

Dalal, R.C., Henderson, P.A., Glasby, J.M., 1991. Organic matter and microbial biomass after 20 yr of zero-tillage. Soil Biol. Biochem. 5, 435–441.

Das, A.C., Saha, D., 2003. Influence of diazotrophic inoculations on nitrogen nutrition of rice. Aust. J. Soil Res. 41, 1543–1554.

DeBano, L.F., Neary, D.G., Folliott, P.F., 1998. Fire's Effects on Ecosystems. John Wiley and Sons Inc., New York, USA

Debré, P., Forster, E., 1998. Louis Pasteur. Johns Hopkins University Press, Baltimore, MD. ISBN: 0-8018-5808-9.

D'Haene, K., Sleutel, S., De Neve, S., Gabriels, D., Hofman, G., 2008. The effect of reduced tillage agriculture on carbon dynamics in silt loam soils. Nutr. Cycl. Agroecosyst. http://dx.doi.org/10.1007/s10705-008-9240-9.

Diaz, L.F., Savage, G.M., Eggerth, L.L., Golueke, C.G., 1993. Composting and Recycling Municipal Solid Waste. Lewis Publishers, Boca Raton, FL.

Diaz, K., Valiente, C., Martinez, M., Castillo, M., Sanfuentes, E., 2009. Root-promoting rhizobacteria in *Eucalyptus globulus* cuttings. World J. Microbiol. Biotechnol. 25, 867–873.

Dobbelaere, S., Vanderleyden, J., Okon, Y., 2003. Plant growth-promoting effects of diazotrophs in the rhizosphere. Crit. Rev. Plant Sci. 22, 107–149.

Docherty, K.M., Balser, T.C., Bohannan, B.J.M., Gutknecht, L.M., 2012. Soil microbial responses to fire and interacting global change factors in a California annual grassland. Biogeochemistry 109, 63–83.

Dooley, S.R., Treseder, K.K., 2011. The effect of fire on microbial biomass: a meta-analysis of field studies. Biogeochemistry. http://dx.doi.org/10.1007/s10533-011-9633-8.

Doran, J.W., 1980. Soil microbial and biochemical changes associated with reduced tillage. Soil Sci. Soc. Am. J. 44, 765–771.

Doran, J.W., Parkin, T.B., 1996. Quantitative indicators of soil quality: a minimum data set. In: Doran, J.W., Jones, A.J. (Eds.), Methods for Assessing Soil Quality. Soil Science Society of America, Madison, WI, pp. 25–37.

Egamberdiyeva, D., 2007. The effect of plant growth promoting bacteria on growth and nutrient uptake of maize in two different soils. Appl. Soil Ecol. 36, 184–189.

Fowles, M., 2007. Black carbon sequestration as an alternative to bioenergy. Biomass Bioenergy 31, 426–432.

Gamliel, A., Grinsten, A., Zilberg, V., Beniches, M., Ucko, O., Katan, J., 2000. Control of soilborne diseases by combining soil solarization and fumigants. Acta Horticult. 71, 157–164.

Garcia-Ruiz, R., Ochoa, V., Hinojosa, M.B., Carreira, J.A., 2008. Suitability of enzyme activities for the monitoring of soil quality improvement in organic agriculture systems. Soil Biol. Biochem. 40, 2137–2145.

Goberna, M., García, C., Insam, H., Hernández, M.T., Verdu', M., 2012. Burning fire-prone Mediterranean shrublands: immediate changes in soil microbial community structure and ecosystem functions. Microb. Ecol. 64, 242–255.

Grayston, S.J., Renneberg, H., 2006. Assessing effects of forest management on microbial community structure in a central European beech forest. Can. J. For. Res. 36, 2595–2604.

Hajek, A.E., 2004. Natural Enemies—An Introduction to Biological Control. Cambridge University Press, New York.

Hamm, P.B., Ingham, R.E., Jaeger, J.E., Swanson, W.H., Volker, K.C., 2003. Soil fumigant effects on three genera of soilborne fungi and their effect on potato yield in the Columbia Basin of Oregon. Plant Dis. 87, 1449–1456.

Hamman, S.T., Burke, I.C., Stromberger, M.E., 2007. Relationships between microbial community structure and soil environmental conditions in a recently burned system. Soil Biol. Biochem. 39, 1703–1711.

Hart, M.M., Trevors, J.T., 2005. Microbe management: application of mycorrhyzal fungi in sustainable agriculture. Front. Ecol. Environ. 3, 533–539.

Hart, S.C., DeLuca, T.H., Newman, G.S., MacKenzie, M.D., Boyle, S.I., 2005. Post-fire vegetation dynamics as drivers of microbial community structure and function in forest soils. For. Ecol. Manag. 220, 166–184.

Hasan, S., 1972. Specificity and host specialization of *Puccinia chondrillina*. Ann. Appl. Biol. 72, 257–263.

Hernandez-Rodriguez, A., Heydrich-Perez, M., Diallo, B., El Jaziri, M., Vandeputte, O.M., 2010. Cell-free culture medium of *Burkholderia cepacia* improves seed germination and seedling growth in maize (*Zea mays*) and rice (*Oryza sativa*). Plant Growth Regul. 60, 191–197.

Hillel, D.J., 1992. Out of the Earth: Civilization and the Life of the Soil. University of California Press, Berkeley.

Huang, M., Jiang, L., Zou, Y., Xu, S., Deng, G., 2013. Changes in soil microbial properties with no-tillage in Chinese cropping systems. Biol. Fertil. Soils 49, 373–377.

Ibekwe, M.A., Papiernik, S.K., Gan, J., Yates, S.R., Yang, C.-H., Crowely, D.E., 2001. Impact of fumigants on soil microbial communities. Appl. Environ. Microbiol. 67, 3245–3257.

Ingham, R.E., Hamm, P.B., Williams, R.E., Swanson, W.H., 2000. Control of *Meloidogyne chitwoodi* in potato with fumigant and nonfumigant nematicides. J. Nematol. 32, 556–565.

Jiang, Y., Wu, Y., Xu, W., Cheng, Y., Chen, J., Xu, L., Hu, F., Li, H., 2012. IAA-producing bacteria and bacterial-feeding nematodes promote *Arabidopsis thaliana* root growth in natural soil. Eur. J. Soil Biol. 52, 20–26.

Johnson, J., Franzlubbers, A.J., Weyers, S., Reicosky, D.C., 2007. Agricultural opportunities to mitigate greenhouse gas emissions. Environ. Pollut. 150, 107–124.

Katan, J., DeVay, J.E., 1991. Soil Solarization. CRC Press, Boston.

Kaymak, H.C., Yarali, F., Guvenc, I., Donmez, M.F., 2008. The effect of inoculation with plant growth rhizobacteria (PGPR) on root formation of mint (*Mentha piperita* L.) cuttings. Afr. J. Biotechnol. 7, 4479–4483.

Lal, R., Reicosky, D.C., Hanson, J.D., 2007. Evolution of the plow over 10,000 years and the rationale for no-till farming. Soil Till. Res. 93, 1–12.

Lampkin, N.H., Measures, M., Padel, S. (Eds.), 2011. Organic Farm Management Handbook, ninth ed. Organic Research Centre, Newbury.

Lehmann, J., Gaunt, J., Rondon, M., 2006. Bio-char sequestration in terrestrial ecosystems—a review. Mitig. Adapt. Strategies Glob. Chang. 11, 403–427.

Lynch, D.H., Voroney, R.P., Warman, P.R., 2006. Use of 13C and 15N natural abundance techniques to characterize carbon and nitrogen dynamics in composting and in compost-amended soils. Soil Biol. Biochem. 38, 103–114.

Mahmood, M.H., Khalid, A., Khalid, M., Arshad, M., 2008. Response of etiolated pea seedlings and cotton to ethylene produced from L-methionine by soil microorganisms. Pak. J. Bot. 40, 859–866.

Mazzola, M., Brown, J., Izzo, A.J., Cohen, M.F., 2007. Mechanism of action and efficacy of seed meal-induced pathogen suppression differ in a Brassicaceae species and time-dependent manner. Phytopathology 97, 454–460.

McGill, W.B., Cannon, K.R., Robertson, J.A., Cook, F.D., 1986. Dynamics of soil microbial biomass and water soluble organic C in Breton L after 50 years of cropping to 2 rotations. Can. J. Soil Sci. 66, 1–19.

Morgan, J., 2012. Invisible artifacts: uncovering secrets of ancient Maya agriculture with modern soil science. Soil Horizon. http://dx.doi.org/10.2136/sh2012-53-6-if.

Nacke, H., Thurmer, A., Wollherr, A., Will, C., Hodac, L., Herold, N., Schoning, I., Schrumpf, M., Daniel, R., 2011. Pyrosequencing-based assessment of bacterial community structure along different management types in German forest and grassland soils. PLoS ONE 6, e17000. www.plosone.org.

Nannipieri, P., Ascher, J., Ceccherini, M.T., Landi, L., Pietramellara, G., Renella, G., 2003. Microbial diversity and soil functions. Eur. J. Soil Sci. 54, 655–670.

Neary, D.G., Klopatek, C.C., DeBano, L.F., Fgolliott, P.F., 1999. Fire effects on belowground sustainability: a review and synthesis. For. Ecol. Manag. 122, 51–71.

Nortcliff, S., 2002. Standardization of soil quality attributes. Agric. Ecosyst. Environ. 88, 161–168.

Novak, J.M., Busscher, W.J., Laird, D.A., Ahmedna, M., Watts, D.W., Niandou, M., 2009. Impact of biochar amendment on fertility of a southeastern coastal plain soil. Soil Sci. 174, 105–112.

Oelhaf, R.C., 1978. Organic Agriculture. John Wiley and Sons, New York.

Phelps, C.D., Young, L.Y., 2000. Biodegradation of BTEX under anerobic conditions: a review. Adv. Agron. 70, 330–359.

Pietikainen, J., Fritze, H., 1995. Clear-cutting and prescribed burning in coniferous forest: comparison of effects soil fungal and total microbial biomass, respiration activity and nitrification. Soil Biol. Biochem. 27, 101–109.

Postma, J., Van Veen, J.A., 1990. Habitable pore space and survival of *Rhizobium leguminosarum biovar trifolii* introduced into soil. Microb. Ecol. 19, 149–161.

Purser, J.E., 1977. The Winemakers of the Pacific Northwest. Harbor House, Vashon Island.

Ramsay, C., Haglund, B., Santo, G., 1992. Bulletin 21: Soil Fumigation. Washington State University Press, Pullman.

Rees, R.M., Ball, B.C., Campbell, C.D., Watson, C.A., 2001. Sustainable Management of Soil Organic Matter. CAB International, Wallingford.

Reganold, J.P., Palmer, A.S., Lockhart, J.C., Macgregor, A.N., 1993. Soil quality and financial performance of biodynamic and conventional farms in New Zealand. Science 260, 344–349.

Reganold, J.P., Andrews, P.K., Reeve, J.R., Carpenter-Boggs, L., Schadt, C.W., Alldredge, R., Ross, C.F., Davies, N.M., Zhou, J., 2010. Fruit and soil quality of organic and conventional strawberry agroecosystems. PLoS ONE 5 (9), e12346. http://dx.doi.org/10.1371/journal.pone.0012346.

Robertson, G.P., Paul, E.A., Harwood, R.R., 2000. Greenhouse gases in intensive agriculture: contributions of individual gases to the radiative forcing of the atmosphere. Science 289, 1922–1925.

Rochette, P., Angers, D.A., Chantigny, M.H., Bertrand, N., 2008. Nitrous oxide emissions respond differently to no-till in a loam and heavy clay soil. Soil Sci. Soc. Am. J. 72, 1363–1369.

Rynk, R., 1992. On-Farm Composting Handbook. NRAES Cooperative Extension, Ithaca.

Sarwar, M., Kirkegaard, J.A., Wong, P.T.W., Desmarchelier, J.M., 1998. Biofumigation potential of brassicas: III. In vitro toxicity of isothiocyanates to soilborne fungal pathogens. Plant Soil 201, 103–112.

Sattler, F., Wistinghauser, E., 1992. Bio-Dynamic Farming Practice. Bio-Dynamic Agriculture Association, West Midlands.

Singh, B.P., Hatton, B.J., Singh, B., Cowie, A.L., Kathuria, A., 2010. Influence of bio-chars on nitrous oxide emission and nitrogen leaching from two contrasting soils. J. Environ. Qual. 39, 1224–1235.

Smith, J.L., 1994. Cycling of nitrogen through microbial activity. In: Hatfield, J.L., Stewart, B.A. (Eds.), Soil Biology: Effects on Soil Quality. CRC Press, Boca Raton, pp. 91–120.

Smith, J.H., 1995. Cheesemaking in Scotland—A History. The Scottish Dairy Association, Glasgow.

Smith, J.L., Elliot, L.F., 1990. Tillage and residue management effects on soil organic matter dynamics in semi-arid regions. In: Stewart, B.A. (Ed.), Dryland Agriculture: Strategies for Sustainability, vol. 13. Springer Verlag, New York, pp. 69–88.

Smith, J.L., Paul, E.A., 1990. The significance of soil microbial biomass estimations. In: Bollag, G.M., Stotzkey, G. (Eds.), Soil Biochemistry, vol. 6. Marcel Dekker, New York, pp. 357–396.

Smith, J.L., Papendick, R.I., Bezdicek, D.F., Lynch, J.M., 1992. Soil organic matter dynamics and crop residue management. In: Metting, B. (Ed.), Soil Microbial Ecology. Marcel Dekker, New York, pp. 65–94.

Smith, J.L., Collins, H.P., Bailey, V.L., 2010. The effect of young biochar on soil respiration. Soil Biol. Biochem. 42, 2345–2347.

Smith-Spangler, C., Brandeau, M.L., Hunter, G.E., Bavinger, J.C., Pearson, M., Eschbach, P.J., Sundaram, V., Liu, H., Schirmer, P., Stave, C., Olkin, I., Btavata, D.M., 2012. Are organic foods safer or healthier than conventional alternatives? A systematic review. Ann. Intern. Med. 157, 348–366.

Snow, A.A., Andow, D.A., Gepts, P., Hallerman, E.M., Power, A., Tiedje, J.M., Wolfenbarger, L.L., 2005. Genetically engineered organisms and the environment: current status and recommendations. Ecol. Appl. 15, 377–404.

Snyder, W.E., Tonkyn, D.W., Kluepfel, D.A., 1999. Transmission of a genetically engineered rhizobacterium by grasshoppers in the laboratory and field. Ecol. Appl. 9, 245–253.

Spokas, K., King, J., Wang, D., Papiernik, S., 2007. Effects of soil fumigants on methanotrophic activity. Atmos. Environ. 41, 8150–8162.

Staben, M.L., Bezdicek, D.F., Smith, J.L., Fauci, M.F., 1997. Microbiological assessment of soil quality in conservation reserve program and wheat-fallow soil. Soil Sci. Soc. Am. J. 61, 124–130.

Stark, C., Condron, L.M., Stewart, A., Di, H.J., O'Callaghan, M., 2007. Influence of organic and mineral amendments on microbial properties and processes. Appl. Soil Ecol. 35, 79–93.

Stark, C.H., Condron, L.M., O'Callaghan, M., Stewart, A., Di, H.J., 2008. Differences in soil enzyme activities, microbial community structure and short term nitrogen mineralization resulting from farm management history and organic matter amendments. Soil Biol. Biochem. 40, 1352–1363.

Stromberger, M.E., Klose, S., Ajwa, H., Trout, T., Fennimore, S., 2005. Microbial populations and enzyme activities in soils fumigated with methyl bromide alternatives. Soil Sci. Soc. Am. J. 69, 1987–1999.

Stubbs, T.L., Kennedy, A.C., 2012. Microbial weed control and microbial herbicides. In: Alvarez-Fernandez, R. (Ed.), Herbicides—Environmental Impact Studies and Management Approaches, 135–166. Tech. http://dx.doi.org/10.5772/32705, http://www.intechopen.com/books/herbicides-environmental-impact-studies-and-management-approaches/microbial-weed-control-and-microbial-herbicides.

TeBeest, D.O., 1996. Biological control of weeds with plant pathogens and microbial pesticides. Adv. Agron. 56, 115–135.

Triplett Jr., G.B., Mannering, J.V., 1978. Crop residue management in crop rotation and multiple cropping systems. In: Oschwald, W.R. (Ed.), Crop Residue Management Systems. American Society of Agronomy, Madison, WI, pp. 198–206, Special Publication 31.

US EPA, 1996. A citizen's guide to bioremediation. Solid Waste and Emergency Response. Technology Innovation Office Technology Fact Sheet. EPA 542-F-96-007.

van Veen, J.A., van Overbeek, L.S., van Elsas, J.D., 1997. Fate and activity of microorganisms introduced into soil. Microbiol. Mol. Biol. Rev. 61, 121–127.

Voroney, R.P., van Veen, J.A., Paul, E.A., 1981. Organic C dynamics in grassland soils. II. Model validation and simulation of the long-term effects of cultivation and rainfall erosion. Can. J. Soil Sci. 61, 211–224.

Wakelin, S.A., Macdonald, L.M., Rogers, S.L., Gregg, A.L., Bolger, T.P., Baldock, J.A., 2008. Habitat selective factors influencing the structural composition and functional capacity of microbial communities in agricultural soils. Soil Biol. Biochem. 40, 803–813.

Walker, J.C., 1950. Plant Pathology. McGraw Hill, New York.

Wander, M.M., Traina, S.J., 1996. Organic matter fractions from organic and conventional managed soils: I. Carbon and nitrogen distribution. Soil Sci. Soc. Am. J. 60, 1081–1087.

Wander, M.M., Traina, S.J., Stinner, B.R., Peters, S.E., 1994. Organic and conventional management effects on biologically active soil organic matter pools. Soil Sci. Soc. Am. J. 58, 1130–1139.

Wardle, D.A., 1995. Impacts of disturbance on detritus food webs in agro-ecosystems of contrasting tillage and weed management practices. Adv. Ecol. Res. 26, 105–182.

Wei, X., Kimmins, J.P., 1998. Asymbiotic nitrogen fixation in harvested and wildfire-killed lodgepole pine forests in the central interior of British Columbia. For. Ecol. Manag. 109, 343–353.

Weil, R.R., Magdoff, F., 2004. Significance of soil organic matter to soil quality and health. In: Magdoff, F., Weil, R.R. (Eds.), Soil Organic Matter in Sustainable Agriculture. CRC Press, Boca Raton, pp. 1–43.

Whipps, J.M., 2001. Microbial interactions and biocontrol in the rhizosphere. J. Exp. Bot. 52, 487–511.

Willer, H., 2010. The World of Organic Agriculture 2010: a summary. In: Willer, H., Kilcher, L. (Eds.), The World of Organic Agriculture. Statistics and Emerging Trends 2010. International Federation of Organic Agriculture Movements, Bonn, Germany.

Willer, H., Kilcher, L. (Eds.), 2012. The World of Organic Agriculture: Statistics and Emerging Trends 2012. International Federation of Organic Agriculture Movements, Bonn, Germany.

Willer, H., Rohwedder, M., Wynen, E., 2009. Current statistics. In: Willer, H., Kilcher, L. (Eds.), The World of Organic Agriculture—Statistics and Emerging Trends 2009. International Federation of Organic Agriculture Movements, Bonn, Germany, pp. 25–58.

Wilson, D., 1976. In Search of Penicillin. Random House, New York.

SUGGESTED READING

Bailey, K., 2004. Microbial weed control: an off-beat application of plant pathology. Can. J. Plant Pathol. 26, 239–244.

Evidente, A., Cimmino, A., Andolfi, A., Vurro, M., Zonno, M., Motta, A., 2008. Phyllostoxin and phyllostin, bioactive metabolites produced by *Phyllosticta cirsii*, a potential mycoherbicide for *Cirsium arvense* biocontrol. J. Agric. Food Chem. 56, 884–888.

Flores-Vargas, R., O'Hara, G., 2006. Isolation and characterization of rhizosphere bacteria with potential for biological control of weeds in vineyards. J. Appl. Microbiol. 100, 946–954.

Ghorbani, R., Scheepens, P., Zweerde, W., Leifert, C., McDonald, A., Seel, W., 2002. Effects of nitrogen availability and spore concentration on the biocontrol activity of *Ascochyta caulina* in common lambsquarters (*Chenopodium album*). Weed Sci. 50, 628–633.

Guske, S., Schulz, B., Boyle, C., 2004. Biocontrol options for *Cirsium arvense* with indigenous fungal pathogens. Weed Res. 44, 107–116.

Hasan, S., 1988. Biocontrol of Weeds with Microbes, Biocontrol of Plant Diseases. CRC Press, Boca Raton, FL pp. 129–151.

Imaizumi, S., Honda, M., Fujimori, T., 1999. Effect of temperature on the control of annual bluegrass (*Poa annua* L.) with *Xanthomonas campestris* pv. poae (JT-P482). Biol. Control 16, 13–17.

Kennedy, A., Stubbs, T., 2007. Management effects on the incidence of jointed goatgrass inhibitory rhizobacteria. Biol. Control 40, 213–221.

Koepf, H.H., Pettersson, B.D., Schaumann, W., 1976. Bio-Dynamic Agriculture: An Introduction. Anthroposophic Press, Spring Valley, NY.

Lawrie, J., Greaves, M., Down, V., Morales-Aza, B., Lewis, J., 2002. Outdoor studies of the efficacy of *Alternaria alternata* in controlling *Amaranthus retroflexus*. Biocontrol Sci. Tech. 12, 83–94.

Leth, V., Netland, J., Andreasen, C., 2008. *Phomopsis cirsii*: a potential biocontrol agent of *Cirsium arvense*. Weed Res. 48, 533–541.

Mejri, D., Gamalero, E., Tombolini, R., Musso, C., Massa, N., Berta, G., Souissi, T., 2010. Biological control of great brome (*Bromus diandrus*) in durum wheat (*Triticum durum*): specificity, physiological traits and impact on plant growth and root architecture of the fluorescent pseudomonad strain X33d. BioControl 55, 561–572.

Tichich, R., Doll, J., 2006. Field-based evaluation of a novel approach for infecting Canada thistle (*Cirsium arvense*) with *Pseudomonas syringae* pv. *tagetis*. Weed Sci. 54, 166–171.

Trujillo, E.E., 2005. History and success of plant pathogens for biological control of introduced weeds in Hawaii. Biol. Control 33, 113–122.

Vidali, M., 2001. Bioremediation. An overview. Pure Appl. Chem. 73, 1163–1172.

Yuzikhin, O., Mitina, G., Berestetskiy, A., 2007. Herbicidal potential of stagonolide, a new phytotoxic nonenolide from *Stagonospora cirsii*. J. Agric. Food Chem. 55, 7707–7711.

Zermane, N., Souissi, T., Kroschel, J., Sikora, R., 2007. Biocontrol of broomrape (Orobanche crenata Forsk. and Orobanche foetida Poir.) by *Pseudomonas fluorescens* isolate Bf7-9 from the faba bean rhizosphere. Biocontrol. Sci. Technol. 17, 483–497.

Index

Note: Page numbers followed by "*f*" indicate figures, and "*t*" indicate tables.

Printed in the United States
By Bookmasters